College Algebra

College Algebra

C.L. Johnston

Dellen Publishing Company
San Francisco

Printed in the United States of America

10 9 8 7 6 5 4 3 2 1

Library of Congress Cataloging in Publication Data

Johnston, Carol Lee.
College algebra.

Includes index.
1. Algebra. I. Title.
QA154.2.J6 512.9 77-15783

ISBN 0-89517-002-7

Contents

Preface

Our major objective throughout the writing of this book has been to present the concepts of college algebra and analytic geometry to the student in an easily understood language supplemented with illustrations, solved problems, and other aids. We assume that the students taking this course have had at least an introductory course in algebra, plane geometry and plane trigonometry.

Some of the special features of this book are:

1. The contents are arranged in small, easy-to-read segments using a one-concept-at-a-time approach. Each essential manipulation is followed by several specific examples. And examples are followed by exercises that employ the concept under discussion.
2. Liberal use of visual aids, such as shading, dotted lines, and arrows, helps clarify concepts and manipulations.
3. The text contains approximately 500 complete examples and 1,900 exercises.
4. Each exercise set is divided into two equal parts marked by the letters A and B in the left margin. The answers are given in the back of the book for the exercises of the first group. The answers for the second group of exercises are *not* given in this book. They are included, with complete solutions, in a solutions manual that is available at a nominal cost. The first problem of each of the groups (A and B) is the same kind of problem. The second problem of each group is the same kind of problem, and so on, throughout the two groups of exercises. This makes it easy for a student to find the solution to a problem in the solutions manual similar to the problem in part A with which he or she might be having difficulty.
5. A chapter summary and a set of review exercises are included at the end of each chapter.

6. A diagnostic test is included at the end of each chapter. Complete solutions for the diagnostic test (with section references) are given in the back of the book.
7. An instructor's manual contains two different tests for each chapter and two final examinations. These may be easily removed and duplicated for class use. The tests include adequate space for students to work the problems. Answer keys and solutions for these tests are provided in the manual.
8. The use of calculators is encouraged and exercises to be solved with a calculator are provided. These are marked with a special symbol in the left margin to indicate that a calculator may be used in the solution. If calculators are not available, appropriate tables, which are included in the back of the book, can be used.
9. The instructor has the option of having students use calculators exclusively, or tables exclusively, or both calculator and tables in doing computational work.
10. Important concepts and algorithms are enclosed in boxes for easy identification and quick location.
11. "Words of Caution" help students avoid common mathematical errors. These are marked with a special symbol in the left margin.
12. Much of the analytic geometry is integrated with the algebra.
13. Symbols and basic formulas are listed inside the cover so as to be immediately available when needed.
14. The text can be easily adapted for use in a variety of instructional programs: conventional lecture courses, learning laboratories, or self-study programs.

I wish to thank Don Dellen of the Dellen Publishing Company for his encouragement and for soliciting critical reviews of the manuscript from Dale R. Bedgood, East Texas State University; Charles M. Biles, Humboldt State University; Boyd L. Cardon, Ricks College; John Distad, University of Alaska; Thomas Drouet, East Los Angeles College; Paul Drowne, Mount Wachusett Community College; Clifton Gary, Oscar Rose College; O. T. Gilbert, Johnson County Community College; Donald W. Gladstone, Orange County Community College; Frederic Gooding, Gallaudet College; Joel Greenstein, New York City Community College; Mark Hale, University of Florida; Nancy J. Halford, Rio Hondo College; George L. Holloway, California State University, Northridge; James Kinney, George C. Wallace State Junior College; Rodney Kohler, Central Oregon Community College; Jeanne Lazaris, East Los Angeles College; Doyle McCown, Oklahoma State University Technical Institute; William Mech, Boise State University; William A. Neal, Fresno City College; Kenneth J. Shabell, Riverside City College; Robert Shloming, Essex County College; Rose Marie Smith, Texas Woman's University; Lynn Tooley, Bellevue Community College; Jack Twitchell, Mesa Community College; Irene Verner, Baruch College, City University of New York; Richard G. Vinson, University of South Alabama; and Jack D. Wilson, Murray State University. I also want to thank Linda Thompson and Phyllis Niklas for their meticulous editing of the final manuscript. To all of these I wish to express my sincere appreciation for their valuable help.

Finally, I owe a great deal to my students who always inspire one to produce a book worthy of their time and to my wife for her encouragement and moral support over long periods of writing.

Whittier, California *C. L. Johnston*

College Algebra

1 Basic concepts of real numbers

It is our purpose in this course to prepare you for a successful study of calculus. We assume that you have had the equivalent of two years of high school algebra and a course in plane geometry. However, for various reasons, you may find a brief review of some of your earlier work in mathematics helpful. For that reason much of the work of the early chapters of this book is devoted to a review of your former mathematics.

A mathematical system usually begins with some undefined words or terms, some defined words or terms, a set of elements, one or more relations between the elements, one or more operations defined on that **set**, and some statements we accept as true without proof (**axioms**). Using these tools, we can (usually) prove many other statements (**theorems**), and thus build a mathematical structure.

In this chapter we give a brief review of some information on sets that we shall be using in this course and some of the properties of the real number system. In algebra we work with letters that represent numbers and must therefore deal with the letters according to the laws of numbers.

1 Sets

A set is a collection of particular things. For example, we speak of the set of numbers between 2 and 7, the set of points on a line segment, or within a circle, and so on. In this course we shall be concerned only with sets of numbers. The objects that make up the set are called its **elements** (or **members**). Sets are often represented by capital letters or by listing their elements within braces $\{\quad\}$. For example, $A = \{1, 2, 3\}$. If the braces have nothing within them, the set is called the **empty set** (or **null set**). An example of an empty set would be the set of all people in your class who are 10 feet tall. The empty set is represented by the symbol $\{\quad\}$ or preferably, $\varnothing$.

A WORD OF CAUTION The set $\{0\}$ is not an empty set because it contains the element "0." Also, the set $\{\varnothing\}$ is not an empty set because it contains the symbol $\varnothing$.

ROSTER METHOD OF REPRESENTING A SET

A class roster is a list of the members of the class. The roster of a set of numbers is a list of the numbers in the set. If, for example, the numbers in a set are 1, 3, 5, and 7, the **roster method** of representing these numbers is $\{1, 3, 5, 7\}$.

The expression $B = \{1, 3, 5\}$ is read "B is the set whose elements are 1, 3, and 5." The symbol $\in$ is read "is an element of." It is used in set notation to denote that an element is a member of a particular set. For example, in the set $D = \{5, 8, 9\}$ the symbol $5 \in D$ tells us that "5 is an element of D." In like manner the symbols $8 \in D$ and $9 \in D$ tell us that 8 and 9 are elements of D.

The slash line (/) negates most symbols. The symbol $\notin$ is read "is *not* an element of." For example, if $A = \{8, 9\}$, then $4 \notin A$. This is read "4 is not an element of A."

SET-BUILDER METHOD OF REPRESENTING A SET

A set can be represented by giving a property of its members. For example, the expression

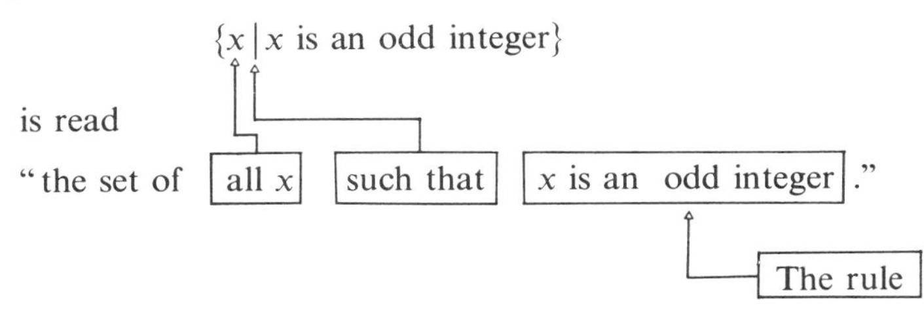

EXAMPLE 1 Write $\{x \mid x$ is an even number between 1 and 9$\}$ in roster notation.

SOLUTION $\{x \mid x$ is an even number between 1 and 9$\} = \{2, 4, 6, 8\}$.

EXAMPLE 2 Write $\{3, 5, 7\}$ in set-builder notation.

SOLUTION $\{3, 5, 7\} = \{x \mid x$ is an odd number between 2 and 8$\}$.

SUBSETS

Set A is a **subset** of set B if every element of set A is contained in set B. Thus, if $A = \{1, 2, 3\}$ and $B = \{1, 2, 3, 4\}$, then A is a subset of B. This is written "$A \subseteq B$." Every set is a subset of itself. The null set, $\varnothing$, is, by definition, a subset of every set.

EXAMPLE 3 The subsets of the set $\{3, 4, 6\}$ are $\{3\}$, $\{4\}$, $\{6\}$, $\{3, 4\}$, $\{3, 6\}$, $\{4, 6\}$, $\{3, 4, 6\}$, and $\varnothing$.

EQUAL SETS

Two sets A and B are said to be equal, written $A = B$, if and only if $A \subseteq B$ and $B \subseteq A$; that is, if they contain the same elements. The order in which the elements are written and the number of times a particular element is repeated within a given set does not matter.

EXAMPLE 4 Examples of equal sets.

a. $\{1, 2, 3\} = \{2, 1, 3\} = \{3, 1, 2\} = \{1, 2, 2, 3, 2\}$
b. $\{a, c, e\} = \{a, e, c, e, a\}$

UNION AND INTERSECTION OF SETS

Sets are often represented by closed geometric figures, where it is understood that all points within each respective figure represent the elements of a set. For example, in Figure 1 we show that sets A and B are subsets of U.

In Figure 2, the interiors of circles A and B represent different sets.

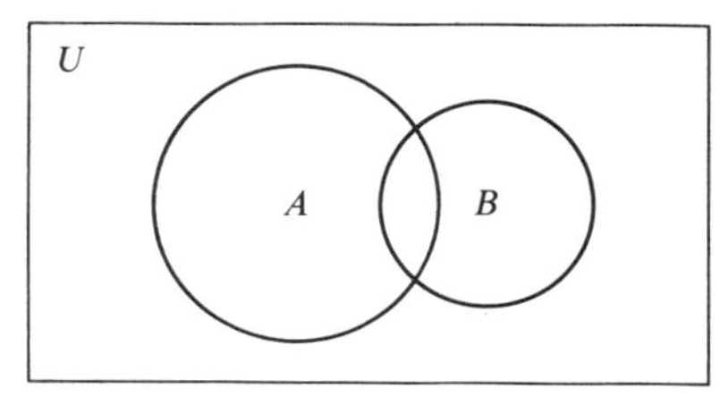

FIGURE 1
$A \subseteq U$ **and** $B \subseteq U$

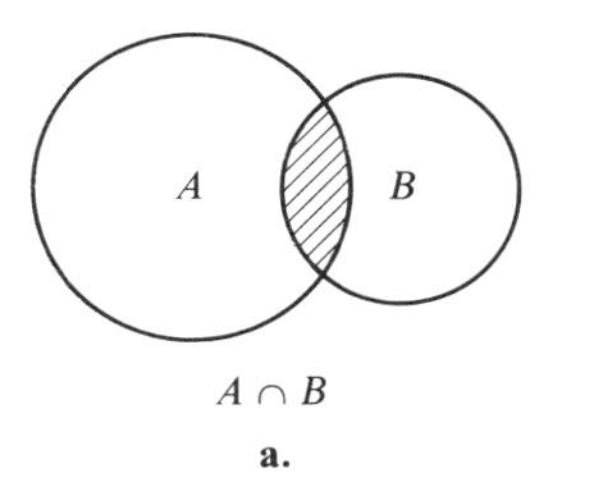

$A \cap B$
a.

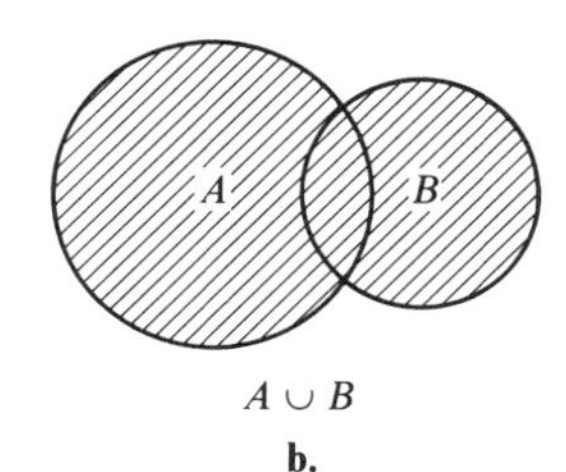

$A \cup B$
b.

FIGURE 2

The shaded area in Figure 2a represents the **intersection** of sets A and B, written $A \cap B$. This is read "A intersect B."

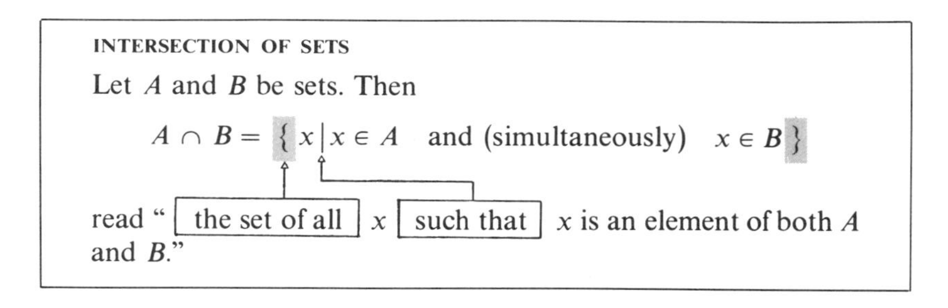

INTERSECTION OF SETS

Let A and B be sets. Then

$$A \cap B = \{x \mid x \in A \quad \text{and (simultaneously)} \quad x \in B\}$$

read "the set of all x such that x is an element of both A and B."

EXAMPLE 5 Example of the intersection of sets: If $A = \{1, 2, 3, 4, 5\}$ and $B = \{4, 5, 6, 7\}$, then $A \cap B = \{4, 5\}$.

EXAMPLE 6 If $B = \{4, 5, 6, 7\}$ and $C = \{8, 9, 10\}$, then $B \cap C = \varnothing$.

In Figure 2b, the interiors of circles A and B each represent a different set of points. When all the points of A are combined with all the points of B, as shown by the entire shaded region, we have the **union** of A and B. This is written $A \cup B$ and read "A union B."

UNION OF SETS

Let A and B be different sets. Then

$$A \cup B = \{x \mid x \in A \quad \text{or} \quad x \in B \text{ (or both)}\}.$$

EXAMPLE 7 Examples of the union of sets.

a. If $A = \{1, 2, 3, 4, 5\}$ and $B = \{4, 5, 6, 7\}$, then

$$A \cup B = \{1, 2, 3, 4, 5, 6, 7\}.$$

b. If $B = \{4, 5, 6, 7\}$ and $C = \{8, 9, 10\}$, then

$$B \cup C = \{4, 5, 6, 7, 8, 9, 10\}.$$

DISJOINT SETS

Disjoint sets are sets that have no elements in common. Their intersection is the empty set.

EXAMPLE 8 Examples of disjoint sets.

a. $\{3, 5, 7\}$ and $\{4, 6, 8\}$ are disjoint sets.
b. $\{1, a, b\}$ and $\{2, x, y\}$ are disjoint sets.

Exercises 1

A

1. Is the collection consisting of "Δ, $\sqrt{3}$, A, and B" a set?
2. What are the elements of the set $\{1, b, 3, \square\}$?
3. If $A = \{3, 4, 5\}$ and $B = \{3, 7, 8\}$,
   a. write $A \cup B$ in roster notation.
   b. write $A \cap B$ in roster notation.
4. Write all the subsets of the set $\{4, e\}$.
5. Are the sets $\{5, 4, 5\}$ and $\{4, 5\}$ equal sets?
6. Which of the sets $E = \{a, c\}$, $F = \{b, f\}$, and $G = \{c, d\}$ are disjoint sets?
7. Is the null set $\varnothing$ a subset of $\{0, 1, 2\}$? Explain.

B

1. Is the collection consisting of "$\sqrt{2}$, 7, Δ, and k" a set?
2. What are the elements of the set $\{0, 3, a\}$?
3. If $A = \{a, b, c\}$ and $D = \{1, 2, 3\}$,
   a. write $A \cup D$ in roster notation.
   b. write $A \cap D$ in roster notation.
4. Write all the subsets of the set $\{a, 5, c\}$.
5. Are the sets $\{4, 9\}$ and $\{9, 4\}$ equal sets?
6. Which of the sets $A = \{5, 7\}$, $B = \{6, 7\}$, and $C = \{4, 8\}$ are disjoint sets?
7. Is the set $\{0\}$ equal to the set $\{\ \ \}$? Explain.

2 Real numbers

NUMBER LINE

A **number line** (or **line graph**) is a graphic representation of numbers by points on a line (see Figure 3). We construct a number scale on a horizontal line by first selecting a zero point and then marking equal

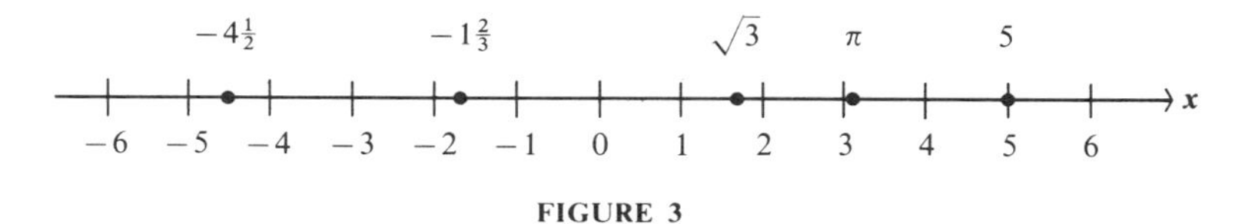

FIGURE 3

Number line

distances to the right and left of 0. The numbers to the right are positive and those to the left are negative. We will assume that you are familiar with symbols such as 5, π, $\sqrt{3}$, $-\frac{2}{3}$, $-4\frac{1}{2}$.

The number corresponding to a particular point on a number line is called the **coordinate** of that point, and the point is called the **graph** of the number. The number line has an infinite number of points and it continues with no bounds in both the positive and negative directions. For every point on the number line, there is a corresponding number. We use the arrowhead at the right end of the number line to indicate that the numbers get larger as we count from left to right.

IMPORTANT SETS

Digits are the ten basic symbols used to write numbers. The set of digits (D) is

$$D = \{0, 1, 2, 3, 4, 5, 6, 7, 8, 9\}.$$

The set of **natural numbers** (N, or **counting numbers**) is

$$N = \{1, 2, 3, 4, 5, 6, 7, 8, 9, 10, 11, 12, \ldots\}.$$

Read "and so on" [referring to $\ldots$]

When zero (0) is included with the natural numbers, we have the infinite set of **whole numbers** (W). This set is indicated by

$$W = \{0, 1, 2, 3, 4, 5, 6, \ldots\}.$$

The set of **integers** (J) is,

$$J = \{\ldots, -3, -2, -1, 0, 1, 2, 3, \ldots\}.$$

The set of **rational numbers** (Q) is

$$Q = \left\{x \,\middle|\, x = \frac{a}{b},\quad a, b \in J,\quad b \neq 0\right\}.$$

EXAMPLE 1 Examples of rational numbers: $\frac{2}{3}$, 0.1 because $0.1 = \frac{1}{10}$, $4.1 = \frac{41}{10}$, $\frac{0}{5} = 0$, 0.3333 ... because $0.3333\ldots = \frac{1}{3}$, 4 because $4 = \frac{40}{10} = \frac{12}{3}$, etc.

Irrational numbers are those numbers that are not rational numbers. They are represented by the letter H.

EXAMPLE 2 Examples of irrational numbers.

$$\sqrt{3}, \quad \sqrt[4]{5}, \quad \pi, \quad \sqrt[3]{7}, \quad -\sqrt{2}$$

It is possible to locate points on a number line corresponding to irrational numbers by construction. In Figure 4, we show how to locate the points corresponding to $\sqrt{2}$ and $\sqrt{3}$. We first construct right

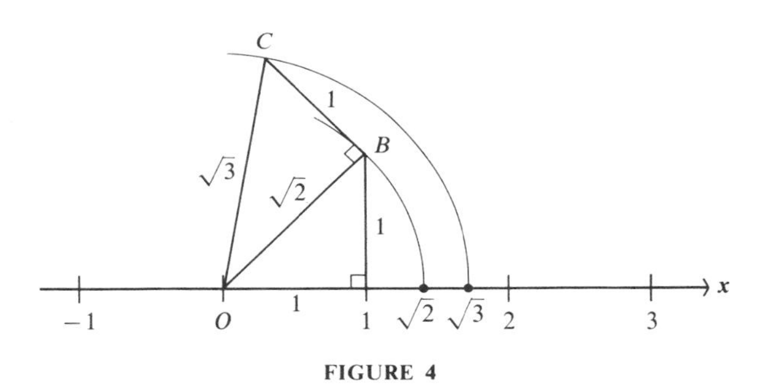

FIGURE 4

triangles, as shown in Figure 4. Using the Pythagorean Theorem, which states that the length of the hypotenuse of a right triangle is equal to the square root of the sum of the squares of the lengths of its legs, we find:

$$OB = \sqrt{1^2 + 1^2} = \sqrt{2} \qquad \text{and} \qquad OC = \sqrt{(\sqrt{2})^2 + 1^2} = \sqrt{3}.$$

With O as center and OB and OC as radii, we strike arcs to locate the points on the number line corresponding to $\sqrt{2}$ and $\sqrt{3}$.

Pi (π), which is the ratio of the circumference of any circle to its diameter, is another irrational number. Every point on the number line corresponds to either a rational or irrational number.

The **real number** system (R) is the set

$$R = Q \cup H.$$

That is, the real numbers include all rational numbers and all irrational numbers.

The sets of rational numbers and irrational numbers are **disjoint** (have no members in common). The sets of digits, natural numbers, whole numbers, integers, rational numbers, and irrational numbers are subsets of the real number system. This is shown in Figure 5.

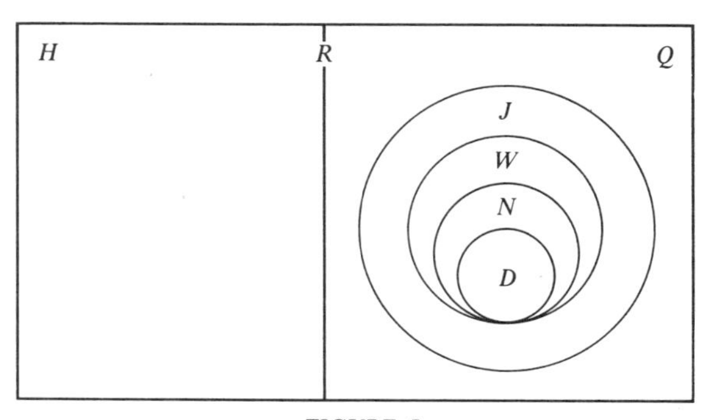

FIGURE 5

The real number system and its subsets

FINITE SET OF NUMBERS

A set is called a **finite set** if, in counting the elements of a set, the counting will eventually come to an end.

EXAMPLE 3 Examples of finite sets.

a. $D =$ the set of digits

b. $A = \{3, 6, 9\}$

c. $B =$ the set of whole numbers less than 1,000

INFINITE SET OF NUMBERS

A set is called an **infinite set** if, in counting the elements of a set, the counting will never come to an end. The integers, whole numbers, rational numbers, and irrational numbers are all examples of infinite sets of numbers.

3 Inequality symbols

The symbols $<$ ("is less than") and $>$ ("is greater than") are called **inequality** symbols. A statement of inequality tells us that one quantity is not equal to another quantity. If a is less than b, we write $a < b$; if a is greater than b, we write $a > b$. If b is between a and c and a is less than c, we write $a < b < c$. Also, $a < b$ and $b < c$.

Numbers on the number line get larger as we move from left to right. Therefore, for any two numbers on the number line, the one on the left is the smaller of the two. See Figure 6.

The symbol $a \leq b$, which is read "a is less than or equal to b,"

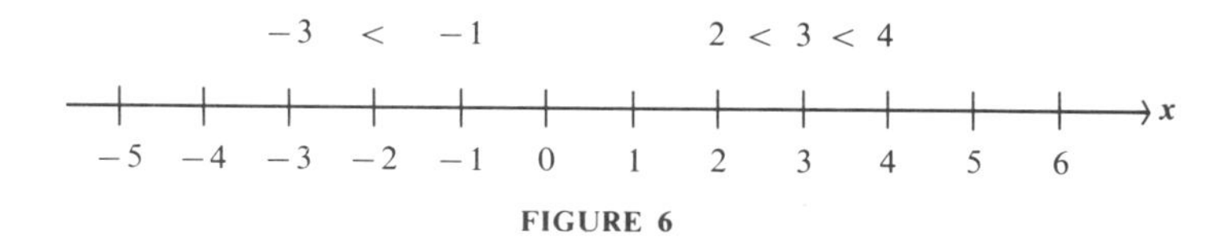

FIGURE 6

means that a either equals b or is to the left of b on the number line. The symbol $a \geq b$, which is read "a is greater than or equal to b," means that a either equals b or is to the right of b on the number line.

The symbol $b > 0$ is often used to indicate that b is a positive number. Similarly, the symbol $b < 0$ indicates that b is a negative number. Note that zero is neither positive nor negative.

4 Absolute value of a number

The symbol $|x|$ is read "the absolute value of x."

ABSOLUTE VALUE OF A NUMBER

For $a \in R$, $$|a| = \begin{cases} a & \text{if} \quad a \geq 0 \\ -a & \text{if} \quad a < 0 \end{cases}$$

This means that the absolute value of a real number is never negative; it is always zero or positive.

EXAMPLE 1 Find $|5|$.

SOLUTION Since $5 > 0$, we use the rule that when $a \geq 0$, $|a| = a$. Therefore, $|5| = 5$.

EXAMPLE 2 Find $|-5|$.

SOLUTION Since $-5 < 0$, we use the rule $|a| = -a$ when $a < 0$. Therefore, $|-5| = -(-5) = 5$.

EXAMPLE 3 Find $|0|$.

SOLUTION Since $0 \geq 0$, we use the rule $|a| = a$ when $a \geq 0$. Therefore, $|0| = 0$.

EXAMPLE 4 Find $|\sqrt{3} - 1|$. (*Note:* $\sqrt{3} \doteq 1.7$. See Table 1. We use the symbol "$\doteq$" as an approximation symbol.)

SOLUTION Since $\sqrt{3} \doteq 1.7$, $\sqrt{3} - 1 > 0$. We use the rule $|a| = a$ when $a \geq 0$. Therefore, $|\sqrt{3} - 1| = \sqrt{3} - 1$.

EXAMPLE 5 Find $|1 - \sqrt{3}|$.

SOLUTION Since $\sqrt{3} \doteq 1.7$, $1 - \sqrt{3} < 0$. We use the rule $|a| = -a$ when $a < 0$. Therefore, $|1 - \sqrt{3}| = -(1 - \sqrt{3}) = \sqrt{3} - 1$.

Exercises 4

A

1. Of the numbers $\frac{3}{5}$, 2, 0, $-\sqrt{3}$, $-\pi$, which are
   a. whole numbers? **b.** rational numbers? **c.** integers?
   d. irrational numbers? **e.** real numbers? **f.** digits?

2. Insert either $>$ or $<$ in each blank to make the statement correct.
   a. $8 \_ 4$ b. $-2 \_ -3$ c. $-5 \_ 0$
   d. $3 \_ |-6|$
3. Write each of the following without the absolute value symbol. (Use Table 1 when needed.)
   a. $|7|$ b. $|-3|$ c. $|5-\sqrt{5}|$ d. $|2-\sqrt{7}|$
4. Write $C = \{x \mid x > 6, \quad x \in D\}$ in roster notation.
5. Write $A = \{x \mid -2 < x < 3, \quad x \in W\}$ in roster notation.

B

1. Of the numbers $2, \pi, 0, \sqrt{5}, -\frac{3}{4}, -6$, which are
   a. digits? b. whole numbers? c. integers?
   d. rational numbers? e. irrational numbers?
   f. real numbers?
2. Insert either $>$ or $<$ in each blank to make the statement correct.
   a. $5 \_ 7$ b. $-5 \_ -3$ c. $0 \_ -1$
   d. $|-4| \_ 2$
3. Write each of the following without the absolute value symbol. (Use Table 1 when needed.)
   a. $|6|$ b. $|-4|$ c. $|2-\sqrt{2}|$ d. $|1-\sqrt{5}|$
4. Write $A = \{x \mid x < 4, \quad x \in D\}$ in roster notation.
5. Write $C = \{x \mid 0 < x < 5, \quad x \in J\}$ in roster notation.

5 Some properties of operation for real numbers

In algebra, we work with letters that represent numbers; we must therefore be able to deal with the letters according to the laws of numbers. For that reason, we list some of the properties of real numbers.

PROPERTIES OF REAL NUMBERS

If $a, b, c \in R$, then

1. $a + b$ is a real number	Closure for addition
2. ab is a real number	Closure for multiplication
3. $a + b = b + a$	Commutative property of addition
4. $ab = ba$	Commutative property of multiplication
5. $(a + b) + c = a + (b + c)$	Associative property of addition
6. $(ab)c = a(bc)$	Associative property of multiplication
7. $a(b + c) = ab + ac$ $(b + c)a = ba + ca$	Distributive laws

There exists a unique number 1 such that:

8. $a \cdot 1 = 1 \cdot a = a$ — Multiplicative identity

9. $a\left(\frac{1}{a}\right) = 1$ and $\left(\frac{1}{a}\right)a = 1$ — Multiplicative inverse axiom
$a \neq 0$

There exists a unique number 0 such that:

10. $a + 0 = 0 + a = a$ — Additive identity

11. $a + (-a) = (-a) + a = 0$ — Negative, or additive inverse axiom

12. $0 \cdot a = a \cdot 0 = 0$ — Multiplication property of zero

If a is any real number except 0, then

13. $\frac{0}{a} = 0$ — Division involving zero

$\frac{a}{0}$ is not defined

$\frac{0}{0}$ cannot be determined

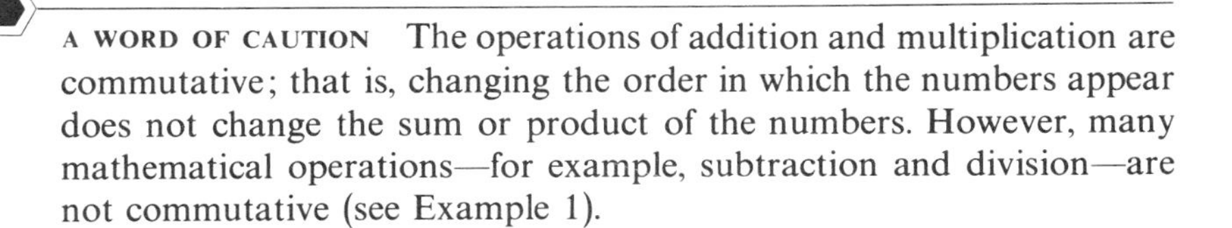

A WORD OF CAUTION The operations of addition and multiplication are commutative; that is, changing the order in which the numbers appear does not change the sum or product of the numbers. However, many mathematical operations—for example, subtraction and division—are not commutative (see Example 1).

EXAMPLE 1

a. $6 - 2 \neq 2 - 6$ **b.** $8 \div 4 \neq 4 \div 8$

Students sometimes confuse the associative property of multiplication with the distributive law. We illustrate this error in Example 2.

EXAMPLE 2

Correct	Incorrect
$4(3 \cdot 2) = 4 \cdot 3 \cdot 2 = 24$	$4(3 \cdot 2) \neq 4(3)4(2) = 96$
$4(3 + 2) = 4(3) + 4(2) = 20$	

The distributive law only applies when the sign of operation is addition or subtraction.

To help avoid confusing the associative and commutative properties, remember that when the commutative property is used, the order of the numbers is changed. When the associative property is used, the grouping of the numbers is changed.

LAWS OF OPERATION FOR SUMS, DIFFERENCES, PRODUCTS AND QUOTIENTS OF SIGNED NUMBERS

1. To add two numbers having like signs, add their absolute values and prefix the common sign (see Example 3).
2. To add two numbers having unlike signs, subtract the smaller absolute value from the larger absolute value and prefix the sign of the number having the larger absolute value (see Example 4).
3. To subtract one number from another, change the subtraction sign to an addition sign and change the sign of the number being subtracted; then proceed as in addition (see Example 5).
4. To multiply (or divide) two numbers first multiply (or divide) their absolute values. Then prefix a negative sign if the two numbers have unlike signs or a positive sign if they have like signs (see Examples 6 and 7).

EXAMPLE 3 Adding two numbers with like signs.

a. $3 + 5 = 8$ **b.** $(-3) + (-5) = -8$

EXAMPLE 4 Adding two numbers with unlike signs.

a. $7 + (-4) = +(7 - 4) = 3$ **b.** $-7 + 4 = -(7 - 4) = -3$
c. $-2 + 8 = +(8 - 2) = 6$ **d.** $2 + (-8) = -(8 - 2) = -6$

EXAMPLE 5 Subtracting two numbers.

a. $6 - (-4) = 6 + (+4) = 6 + 4 = 10$
b. $-6 - (-4) = -6 + (+4) = -6 + 4 = -2$
c. $6 - (+4) = 6 + (-4) = 2$
d. $-6 - (+4) = -6 + (-4) = -10$

EXAMPLE 6 Multiplying two numbers.

a. $8(-4) = -32$ **b.** $(-8)(+4) = -32$
c. $(+8)(+4) = +32$ **d.** $(-8)(-4) = +32$

EXAMPLE 7 Dividing two numbers.

a. $10 \div (2) = +5$ **b.** $(-10) \div (-2) = +5$
c. $10 \div (-2) = -5$ **d.** $(-10) \div (2) = -5$

Exercises 5

Perform the indicated operations or write "not possible" or "undefined."

A

1. $7 + (-3)$	**2.** $-6 + (-2)$	**3.** $-8 + 5$
4. $-11 - 9$	**5.** $10 - (-4)$	**6.** $24(-2)$
7. $12 \div (-4)$	**8.** $(-14)(-3)$	**9.** $(-27)(-9)$
10. $4(5 \cdot 3)$	**11.** $7(8 + 2)$	**12.** $8(5 - b - 3)$
13. $5\|-6\|$	**14.** $0 \div 3$	**15.** $5 \div 0$

B

1. $6 + (-4)$	**2.** $-5 + (-3)$	**3.** $-7 + 9$
4. $-10 - 4$	**5.** $8 - (-2)$	**6.** $12(-3)$
7. $-9 \div (3)$	**8.** $(-14)(-2)$	**9.** $(-32) \div (-4)$
10. $5(3 \cdot 6)$	**11.** $6(3 + 4)$	**12.** $7(x + y - 6)$
13. $8\|-4\|$	**14.** $0 \div 5$	**15.** $6 \div 0$

6 Integral exponents

In this section, we will review some of the laws of exponents.

DEFINITION If a is any real number and n is a positive integer, then

$$a^n = \overbrace{a \cdot a \cdots a}^{n}, \qquad n \text{ factors of } a;$$

n is called the **exponent**, a the **base**, and a^n the ***n*th power** of a.

EXAMPLE 1

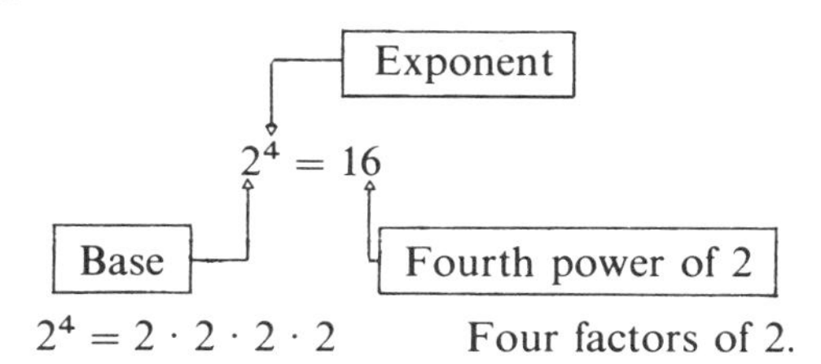

$2^4 = 2 \cdot 2 \cdot 2 \cdot 2$ Four factors of 2.

PROPERTIES OF EXPONENTS

Let a and b be any two real numbers (not zero) and m and n be any two integers.

	Examples
1. $a^m \cdot a^n = a^{m+n}$	$2^4 \cdot 2^5 = 2^{4+5} = 2^9$
2. $(a^m)^n = a^{mn}$	$(3^2)^4 = 3^{2 \cdot 4} = 3^8$

3. **a.**	$\dfrac{a^m}{a^n} = a^{m-n}$, if $m > n$	$\dfrac{6^5}{6^2} = 6^{5-2} = 6^3$
b.	$\dfrac{a^m}{a^n} = \dfrac{1}{a^{n-m}}$, if $m < n$	$\dfrac{2^3}{2^8} = \dfrac{1}{2^{8-3}} = \dfrac{1}{2^5} = \dfrac{1}{32}$
c.	$\dfrac{a^m}{a^n} = 1$, if $m = n$	$\dfrac{3^4}{3^4} = 1$
4.	$(ab)^n = a^n b^n$	$(2x)^4 = 2^4 x^4 = 16x^4$
5.	$\left(\dfrac{a}{b}\right)^n = \dfrac{a^n}{b^n}$	$\left(\dfrac{2}{3}\right)^3 = \dfrac{2^3}{3^3} = \dfrac{8}{27}$
6.	$a^0 = 1$, if $a \neq 0$ (definition)	$5^0 = 1$
7.	$a^{-n} = \dfrac{1}{a^n}$, if $a \neq 0$ (definition)	$4^{-2} = \dfrac{1}{4^2} = \dfrac{1}{16}$
	$a^n = \dfrac{1}{a^{-n}}$, if $a \neq 0$	$4^2 = \dfrac{1}{4^{-2}} = 16$
8.	An odd power of a negative number is negative; an even power of a negative number is positive.	$(-2)^5 = -32$ $(-2)^4 = 16$

We illustrate these properties as follows.

PROPERTY 1 If we wish to multiply a^2 by a^3, we have, from the definition of exponents,

$$a^2 \cdot a^3 = (a \cdot a)(a \cdot a \cdot a) = a^{2+3} = a^5.$$

PROPERTY 2 Using the same reasoning, we write

$$\begin{aligned} (a^m)^n &= a^m \cdot a^m \cdot a^m \cdots a^m && n \text{ factors of } a^m \\ &= \underbrace{a^{m+m+\cdots+m}}_{n \text{ times}} && m \text{ is added } n \text{ times} \\ &= a^{mn}. \end{aligned}$$

PROPERTY 3 If we wish to divide a^6 by a^2, or a^2 by a^6, we have, from the definition of exponents,

$$\frac{a^6}{a^2} = a^{6-2} = a^4,$$

or

$$\frac{a^2}{a^6} = \frac{1}{a^{6-2}} = \frac{1}{a^4}.$$

By Property 7, $1/a^4 = a^{-4}$. Thus, in general,

$$\frac{a^m}{a^n} = a^{m-n}.$$

PROPERTY 6 We must define a^0 to be 1 for consistency in applying Property 3 when $m = n$ and in applying Property 1 when $n = 0$. We use the following examples to illustrate that Property 6 is valid.

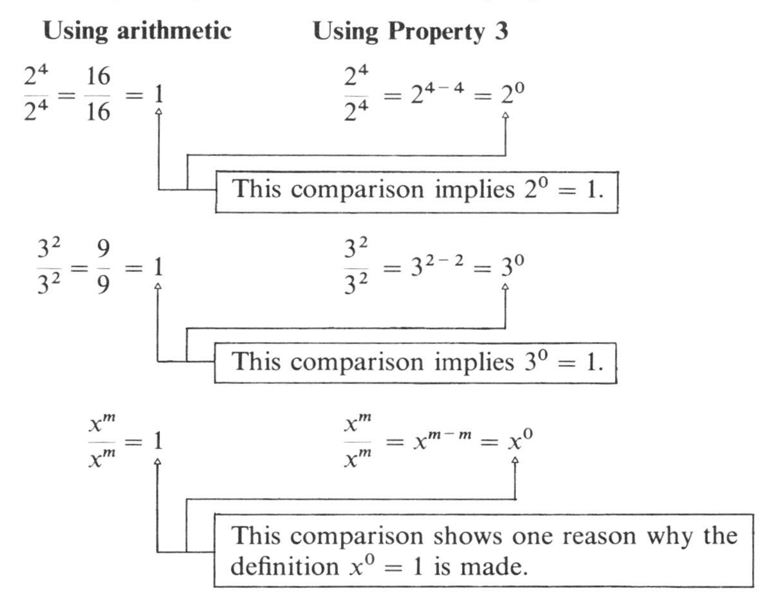

The above comparisons implied that $2^0 = 1$, $3^0 = 1$, and for any nonzero number x, $x^0 = 1$ (0^0 is not defined).

Also if $x^0 = 1$, then when $n = 0$ in Property 1, we have

$$x^m \cdot x^0 = x^{m+0}$$

$$x^m \cdot 1 = x^m \qquad \text{Substituting 1 for } x^0.$$

Properties 6 and 7 are definitions. After accepting these two properties, we no longer need Properties 3b and 3c.

Exercises 6

Find the value of each expression.

A

1. $2^3 \cdot 2^4$
2. $(3^2)^3$
3. $\frac{10^5}{10^2}$
4. $(2x)^3$
5. $\left(\frac{3}{4}\right)^3$
6. 9^0
7. 3^{-2}
8. $(-3)^3$

9. $(-1)^{22}$ **10.** $(-1)^{79}$ **11.** $(3x)^{-2}$ **12.** $\left(\frac{3}{5}\right)^{-1}$

13. $(-3y^2)^3$ **14.** $\frac{1}{3^{-4}}$ **15.** $(2a^2b^3c^4)^3$

16. Express each answer to three significant digits. We assume a^n holds true when $a \in R$ and $n \in Q$.*

a. $(1.189)^{6.715}$ **b.** $(3.1416)^{2.6182}$

c. $(0.7432)^{-1.964}$

B

1. $3^2 \cdot 3^5$ **2.** $(2^2)^3$ **3.** $\frac{10^6}{10^4}$ **4.** $(3x)^2$

5. $\left(\frac{2}{5}\right)^2$ **6.** 8^0 **7.** 2^{-3} **8.** $(-2)^5$

9. $(-1)^{30}$ **10.** $(-1)^{99}$ **11.** $(2x)^{-3}$ **12.** $\left(\frac{2}{3}\right)^{-2}$

13. $(-2a^3)^4$ **14.** $\frac{1}{5^{-2}}$ **15.** $(2ab^2c^3)^4$

16. Express each answer to three significant digits. We assume a^n holds true when $a \in R$ and $n \in Q$.

a. $(2.156)^{4.561}$ **b.** $(3.1416)^{2.5863}$

c. $(0.6845)^{-2.564}$

7 Roots of numbers

The values of x that satisfy the equation $x^2 = a$ are called the **square roots** of a. Thus 2 and -2 are square roots of 4 because $(-2)^2 = 4$ and $2^2 = 4$.

Every positive number has two real square roots. These two square roots have the same absolute value but have opposite signs. The positive root is called the **principal square root** of a and is indicated by the symbol $\sqrt{a}$. The negative square root is indicated by the symbol $-\sqrt{a}$.

EXAMPLE 1

a. $\sqrt{4} = 2$ The principal square root of 4 is 2.

b. $-\sqrt{4} = -2$ The negative square root of 4 is -2.

c. $\sqrt{25} = 5$ The principal square root of 25 is 5.

d. $-\sqrt{25} = 5$ The negative square root of 25 is -5.

* The calculator symbol shown at the left will be used throughout this book to indicate exercises and examples that should be worked using a hand calculator.

A WORD OF CAUTION The roots of the equation $x^2 = 9$ are 3 and -3, but the value of $\sqrt{9}$ is 3, the principal square root.

DEFINITION If x is any real number, then

$$\sqrt{x^2} = |x|.$$

EXAMPLE 2 $\sqrt{(-3)^2} = |-3| = 3$

The symbol $\sqrt[n]{a}$ is read "nth root of a."

If n is an odd integer, the real number a has only one real nth root. This root is called the **principal *n*th root of *a*** and is indicated by the symbol $\sqrt[n]{a}$. This principal nth root of a is positive, negative, or zero according to whether a is positive, negative, or zero.

EXAMPLE 3 Examples of principal nth roots.

a. $\sqrt[3]{64} = 4$ **b.** $\sqrt[5]{-32} = -2$ **c.** $\sqrt[7]{0} = 0$

If n is an even integer and a is positive, then a has two, and only two, real nth roots. These roots have the same absolute value, but one is positive and the other is negative. The **positive root** is the principal nth root of a.

EXAMPLE 4

a. $\sqrt[4]{16} = 2$ The principal fourth root of 16 is 2.
b. $-\sqrt[4]{16} = -2$ The other fourth root of 16 is -2.
c. $\sqrt[6]{64} = 2$ The principal sixth root of 64 is 2.
d. $\sqrt[8]{1} = 1$ The principal eighth root of 1 is 1.
e. $\sqrt{0.09} = 0.3$ The principal square root of 0.09 is 0.3.

DEFINITION For $a \in R$, $a > 0$, $n \in N$, and $b > 0$,

$$\sqrt[n]{a} = b \text{ because } b^n = a.$$

EXAMPLE 5 $\sqrt[3]{8} = 2$ because $2^3 = 8$.

If n is an even integer and a is negative, then there are no real nth roots of a. In this case all nth roots of a are imaginary roots. (Imaginary roots will be discussed in Chapter 4.)

Exercises 7

A In Exercises 1–10, find the principal root.

1. $\sqrt{64}$ **2.** $-\sqrt{36}$ **3.** $\sqrt[3]{-8}$ **4.** $-\sqrt[4]{16}$

5. $\sqrt{(-7)^2}$ 6. $\sqrt[9]{-1}$ 7. $\sqrt{0.09}$ 8. $\sqrt[4]{0.0016}$
9. $\sqrt[5]{-0.00032}$ 10. $\sqrt{(-8)^2}$
11. Find two square roots of 64. Which is the principal root?
12. Find the principal cube root of -27.
13. Find the principal square root of 81.

14. Find the following roots. Assume the given numbers to be exact numbers and express the roots to two decimal places.

 a. $\sqrt{4.86}$ b. $\sqrt[4]{28.6}$ c. $\sqrt[5]{-96.4}$

B In Exercises 1–10, find the principal roots.

1. $\sqrt{49}$ 2. $-\sqrt{9}$ 3. $\sqrt[3]{-27}$ 4. $-\sqrt[4]{81}$
5. $\sqrt{(-5)^2}$ 6. $\sqrt[7]{-1}$ 7. $\sqrt{0.04}$ 8. $\sqrt[3]{0.125}$
9. $\sqrt[5]{-0.00001}$ 10. $\sqrt{(-6)^2}$
11. Find two square roots of 49. Which is the principal square root?
12. Find the principal cube root of -8.
13. Find the principal square root of 100.

14. Find the following roots. Assume the given numbers to be exact numbers and express the roots to two decimal places.

 a. $\sqrt{27.5}$ b. $\sqrt[4]{107.8}$ c. $\sqrt[5]{-74.8}$

8 Combining like terms and removing grouping symbols

The following symbols are commonly used to indicate grouping of numbers, letters, and terms.

GROUPING SYMBOLS			
Parentheses	( )	Braces	{ }
Brackets	[ ]	Bar (used with fractions and radicals)	——

Before showing the use of these grouping symbols, we will briefly review the meaning of coefficients, variables, factors, and terms.

The word **term** has many meanings, depending on how it is used. For example, **a term of a fraction** is the numerator or denominator of the fraction. **An algebraic term** is a term containing only algebraic symbols and numbers. For example, $7x$, x^2, $(3xy + z)$, $(x + y)(x - y)$, and $\sqrt{5x^2 + y}$ are algebraic terms. In the algebraic expression $4x^2y - 3xy + 2y^2$, the **terms** are $4x^2y$, $-3xy$, and $2y^2$. The plus and minus signs in an algebraic expression separate the expression into pieces called *terms*. There is one exception: an expression enclosed within grouping symbols is considered as a single term. Each plus and minus sign is a part of the term that follows it. The plus and minus signs are algebraic symbols. Terms having identical literal parts are called **like terms**.

In this chapter, we will consider the **coefficient** to be the number part of the term, usually written before the literal part. The coefficients in $2x$ and $3(x + y)$ are 2 and 3, respectively. A **variable** is a symbol that can be replaced by one or more values in a given expression.

To combine like terms, add the numerical coefficients of terms with identical literal parts, and multiply each sum by its corresponding literal part.

EXAMPLE 1 Combine the like terms of $5xy^2 + 3xy - 2xy^2 + 7xy$.

SOLUTION First identify like terms. (It might help to draw lines under them.) Then combine the like terms.

$$\begin{aligned} &\underline{\underline{5xy^2}} + \underline{3xy} - \underline{\underline{2xy^2}} + \underline{7xy} \\ &\quad = \underline{\underline{5xy^2}} - \underline{\underline{2xy^2}} + \underline{3xy} + \underline{7xy} \\ &\quad = (5 - 2)xy^2 + (3 + 7)xy \qquad \text{The distributive property.} \\ &\quad = 3xy^2 + 10xy \end{aligned}$$

The quantities that are multiplied together to give a product are called the **factors** of that product. A term often has both literal and numerical factors.

EXAMPLE 2 Terms and their factors.

a. $7x$

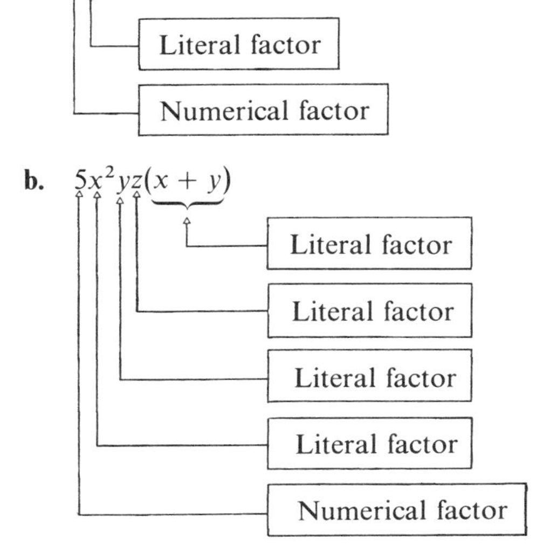

To multiply two terms, find the product of their numerical coefficients, and then multiply that product by the product of their literal factors.

EXAMPLE 3 Products of single terms.

a. $(2x^2y)(5xy^3) = (2 \cdot 5)(x^2 \cdot x)(y \cdot y^3)$

$= 10x^3y^4$

b. $(3ab^2)(-4bx) = [3 \cdot (-4)](a)(b^2 \cdot b)(x)$

$= -12ab^3x$

TO REMOVE A PAIR OF GROUPING SYMBOLS

A. Preceded by a plus sign (or no sign, which is interpreted as a +).

1. If the first enclosed term does not have a written sign, write in a plus sign for it.
2. Leave the signs of all enclosed terms unchanged.
3. Remove the grouping symbols and the signs that precede them.
4. If there are any like terms, combine them.

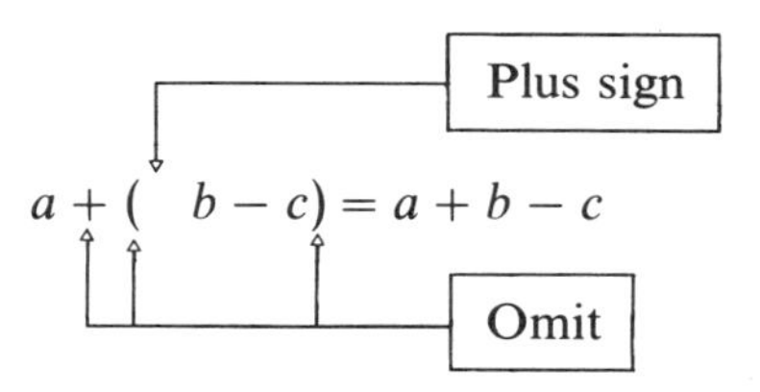

B. Preceded by a minus sign.

1. If the first enclosed term does not have a written sign, write in a plus sign for it.
2. Change the sign of each enclosed term (including any plus sign written in for the first term).
3. Remove the grouping symbols and the signs that precede them.
4. If there are any like terms, combine them.

Plus sign

$$a - (\ b - c) = a - (+b - c) = a + (-b + c) = a - b + c$$

Omit

C. Preceded by a factor: Use the distributive property.

D. When pairs of grouping symbols are enclosed within other pairs of grouping symbols, first remove the innermost pair, then the next innermost pair, and so on until all pairs of grouping symbols have been removed.

EXAMPLE 4 Remove grouping symbols and combine like terms.

a. $5 + (a - 3) = 5 + a - 3$
$= 2 + a$

b. $6x - (2x - 3) = 6x + (-2x + 3)$
$= 6x - 2x + 3$
$= (6 - 2)x + 3$
$= 4x + 3$

c. $18 - \{10 - [8 + (3 - 5) - 4] + 7\}$
$= 18 - \{10 - [8 + (-2) - 4] + 7\}$
$= 18 - \{10 - [8 - 2 - 4] + 7\}$
$= 18 - \{10 - [2] + 7\}$
$= 18 - \{10 - 2 + 7\}$
$= 18 - \{15\}$
$= 18 - 15$
$= 3$

d. $5 - \{7x - [4x + 5y - 2(3x - 4y)] - 6\}$
$= 5 - \{7x - [4x + 5y - 6x + 8y] - 6\}$
$= 5 - \{7x - 4x - 5y + 6x - 8y - 6\}$
$= 5 - 7x + 4x + 5y - 6x + 8y + 6$
$= 11 - 9x + 13y$

e. $2x^2y - [4xy^2 - 2x(3y^2 - xy - 4) - 8x]$
$= 2x^2y - [4xy^2 - 6xy^2 + 2x^2y + \cancel{8x} - \cancel{8x}]$ $\quad (8x - 8x = 0)$
$= 2x^2y - 4xy^2 + 6xy^2 - 2x^2y$
$= 2xy^2$

Exercises 8

Remove the grouping symbols and simplify.

A

1. $20 - \{10 - [7 + (2 - 6) - 3] + 5\}$
2. $8 - 6b - 3[a - 2(3a + b - 1)]$
3. $2m(3m - n) - 3m[2m - n - (4 - n)]$
4. $3b - \{-2[-3(a + b) + 5] - (2a + 3b)\}$
5. $6ab^2(2a^2 - b) - 2ab(6a^2b - 3b^2 - 5)$
6. $(\frac{5}{7}x^2y^3z^4)(\frac{14}{15}x^3y^2z^5)(3xy^2z)$
7. $6 - 2\{3 - 2[5 - 3(x - 2) - 4] - 2\} - 1$
8. $xy^3 - x[x - y(3x^2 - y^2) + 3x^2y] + y^2$
9. $4a^2 - a\{a - b[b - a(2a - b) + 1] - b(2a^2 - 1)\} - ab^2(1 + a)$
10. $5x^2 - y\{1 - x[y - 2x(x^2 + y^2) + 2x^3]\} - xy^2(1 - 2xy)$

B

1. $10 - \{8 - [6 + (3 - 4) - 5] + 2\}$
2. $5 - 6b - 2[a - 3(2a + b)]$
3. $2x(x - 3y) - 3y[y - 2x - (3 + y)]$
4. $B - \{B - [4 - 2(A - 3B) - 7B] - 2A\}$
5. $3x^2y(2y - x) - 3xy(2xy - x^2 - 1)$
6. $(\frac{5}{8}a^3c^2)(\frac{16}{15}b^3c^4)(3a^2b^4)$
7. $5 - \{2 - 3[4 - 2(x - 1) - 3] - 4\} - 3$
8. $mn^2 - m[m - n(2m^2 + n^2) + n^3 - m] - 2m^3n$
9. $3x^2y^2 - \{x - y[-x(x^2 - y) + x^3] + xy^2\} + x$
10. $5a^2b^2 - a\{-a[-2a(a^3 - b^3) + 2a^4] + a^2b^3\} - a^3b^3$

9 Order of operations

Addition, subtraction, multiplication, division, powers, and roots are performed in the following order.

> **ORDER OF OPERATIONS WITH REAL NUMBERS**
>
> First: Simplify expressions within grouping symbols by using the following rules.
>
> Second: Powers and roots are evaluated in any order.
>
> Third: Multiplication and division are performed in order from left to right.
>
> Last: Additions and subtractions are performed in order from left to right.

EXAMPLE 1 $20 \div 2 \cdot 5$

$= 10 \cdot 5 = 50$

Division is performed before multiplication because it is written first.

EXAMPLE 2 $5\sqrt{36}$

$$= 5(6) = 30$$

Square root is performed before multiplication.

EXAMPLE 3 $4 + 2(3)$

$$= 4 + 6 = 10$$

Multiplication is performed before addition.

EXAMPLE 4

$8 \cdot 3^2 \div 4 + 3\sqrt{25}$ — Powers and roots are performed first.

$= 8 \cdot 9 \div 4 + 3(5)$ — Multiplication is performed next because it is on the left.

$= 72 \div 4 + 3(5)$ — Division and multiplication are performed before addition.

$= 18 + 15 = 33$

EXAMPLE 5

$$7 \cdot 4 - [6 - 2(12 - \tfrac{10}{2}) - 4] \div 2$$

First remove grouping symbols, starting with the inner pair, then the next inner pair, and so on until all have been removed.

$$= 7 \cdot 4 - [6 - 2(12 - 5) - 4] \div 2$$
$$= 7 \cdot 4 - [6 - 2(7) - 4] \div 2$$
$$= 7 \cdot 4 - (6 - 14 - 4) \div 2$$
$$= 7 \cdot 4 - (-12) \div 2$$
$$= 7 \cdot 4 + 12 \div 2$$
$$= 28 + 6 = 34$$

The bar ——— in expressions such as

$$\frac{(16)(12)}{(4)(3)}$$

is equivalent to a division sign and is a grouping symbol. Otherwise, we would have $16 \cdot 12 \div 4 \cdot 3 = 144$. If only multiplication and division are involved in problems written this way, any order of operation can be used.

EXAMPLE 6

$$\frac{(\overset{4}{\cancel{16}})(\overset{4}{\cancel{12}})}{(\cancel{4})\ (\cancel{3})} = 4 \cdot 4 = 16$$

or

$$\frac{(16)(12)}{(4)\ (3)} = \frac{(16)(\cancel{12})}{\cancel{12}} = 16$$

Exercises 9

A

1. Which is correct? Explain.
   a. $18 \div 2 \cdot 3 = 9 \cdot 3 = 27$ **b.** $18 \div 2 \cdot 3 = 18 \div 6 = 3$
2. Which is correct? Explain.
   a. $16 - 2 \cdot 4 = 14 \cdot 4 = 56$ **b.** $16 - 2 \cdot 4 = 16 - 8 = 8$
3. Which is correct? Explain.
   a. $3 \cdot 5^2 = 15^2 = 225$ **b.** $3 \cdot 5^2 = 3 \cdot 25 = 75$
4. Which is correct? Explain.
   a. $-3^2 = 9$ **b.** $-3^2 = -9$

Find the value of each expression.

5. $8^2 \div 4 - 3\sqrt{25}$
6. $(5 + 3)8$
7. $(-20) \div (-2) - 5^2$
8. $16 - [3(-2)^3 - \sqrt[3]{-27}]$
9. $\dfrac{(52)(44)}{(22)(13)} - (-6)^2$
10. $\sqrt{(25)^2 + (7)^2 - 2(25)(7)(1)}$

11. Evaluate $\sqrt{(2.514)^2 + (3.015)^2 - 2(2.514)(3.015)(0.8028)}$. Express your answer to four significant digits.

B

1. Which is correct? Explain.
   a. $12 \div 2 \cdot 3 = 6 \cdot 3 = 18$ **b.** $12 \div 2 \cdot 3 = 12 \div 6 = 2$
2. Which is correct? Explain.
   a. $7 + 2 \cdot 3 = 9 \cdot 3 = 27$ **b.** $7 + 2 \cdot 3 = 7 + 6 = 13$
3. Which is correct? Explain.
   a. $2 \cdot 3^2 = 2 \cdot 9 = 18$ **b.** $2 \cdot 3^2 = 6^2 = 36$
4. Which is correct? Explain.
   a. $-2^2 = -4$ **b.** $-2^2 = 4$

Find the value of each expression.

5. $6^2 \div 3 - 4\sqrt{100}$
6. $(6 + 2)7$
7. $(-16) \div (-4) - 2^2$
8. $12 - [2(-3)^2 - \sqrt[3]{-8}]$
9. $\dfrac{(100)(64)}{(16)(50)} - (-3)^2$
10. $\sqrt{4^2 + 5^2 - 2(4)(5)(1)}$

11. Evaluate $\sqrt{(3.466)^2 + (2.304)^2 - 2(3.466)(2.304)(0.5807)}$. Express your answer to four significant digits.

10 Scientific notation

Because most of you will be using calculators in this course, many of which display very large or very small numbers in scientific notation, we shall discuss **scientific notation** at this time. Scientific notation is commonly used in calculations and in writing extremely large or small numbers.

We use a **caret** ($\wedge$) as an aid in explaining the conversion of a number to scientific notation. The caret shows the position of the decimal point in the transformed number.

The student is reminded that if the decimal point is not written in a number, it is understood to follow the number.

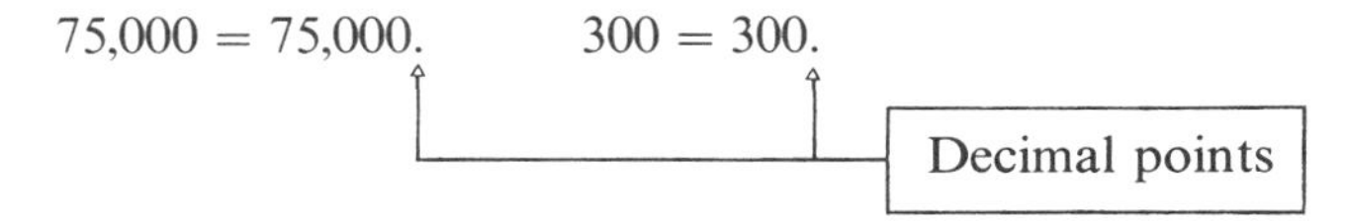

CONVERTING A NUMBER TO SCIENTIFIC NOTATION

A number is said to be written in scientific notation when it is in the form

$$a \times 10^n \quad \text{where} \quad 1 \leq a < 10, \text{ and } n \text{ is an integer.}$$

To convert a number into the $a \times 10^n$ form:

1. Place a caret to the right of the first left nonzero digit of the number being converted (see Example 1).
2. Determine the value of n by counting the number of places to the right or left from the caret to the decimal point. If the counting is to the right, n will be positive. If the counting is to the left, n will be negative (see Examples 1 and 2).
3. Rewrite the number with a decimal point in place of the caret and multiply this number by 10^n. The value of n is the number of positive or negative places counted in Step 2 (see Example 1).

EXAMPLE 1 Converting numbers to scientific notation.

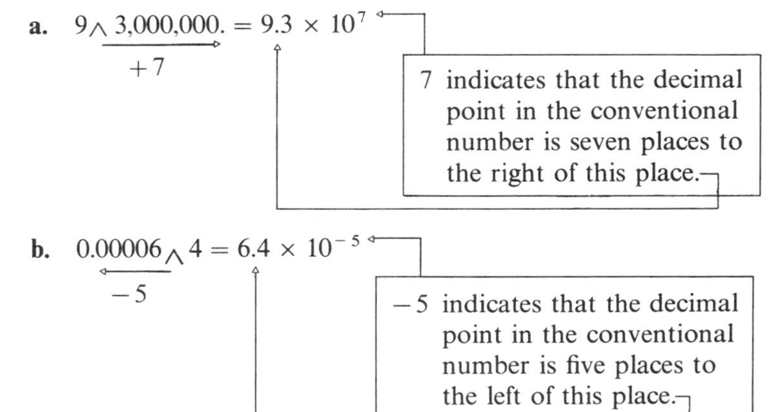

EXAMPLE 2 Converting numbers from scientific notation to conventional form. (The exponent of 10 tells how many places to move the decimal point. Move the decimal point to the right if n is positive and to the left if n is negative.)

a. $1.75 \times 10^8 = 1_\wedge\underset{+8}{\underrightarrow{75000000.}} = 175{,}000{,}000$

b. $2.96 \times 10^{-7} = \underset{-7}{\underleftarrow{0.0000002}}_\wedge 96 = 0.000000296$

APPROXIMATIONS

It is often desirable to express a number with fewer digits than were originally given. This process is called **rounding the number**. We shall use the following rules to round a number.

> RULES FOR ROUNDING NUMBERS
>
> **1.** If the first digit of the part to be discarded amounts to less than 5 units in the first discarded place, keep the last retained digit unchanged (see Example 3a).
>
> **2.** If the first digit of the part to be discarded amounts to more than 5 units in the first discarded place, increase the last retained digit by one unit (see Examples 3b and 3e).
>
> **3.** If the first digit of the part to be discarded is 5 or is 5 followed by zeros, round the preceding digit to the nearest even number (see Examples 3c, 3d, and 3f).
>
> **4.** If the first digit to be dropped is 5 followed by nonzero numbers, increase the retained digit by 1 (see Example 3i).
>
> **5.** When the decimal point of the number being rounded is to the right of the place to which you are rounding, replace the nonzero digits to the right of that place with zeros (see Examples 3e and 3g).

EXAMPLE 3 Examples of rounding numbers.

	Rounded to:
a. $3.1416 \doteq 3.14$	3 significant digits or hundredths
b. $3.1416 \doteq 3.142$	4 significant digits or thousandths
c. $47.45 \doteq 47.4$	3 significant digits or tenths
d. $47.35 \doteq 47.4$	3 significant digits or tenths
e. $92{,}860{,}000 \doteq 93{,}000{,}000$	Millions
f. $0.185 \doteq 0.18$	Hundredths
g. $186{,}284 \doteq 186{,}000$	Thousands

h. $1.96 \doteq 2.0$ Tenths
i. $2.65001 \doteq 2.7$ Tenths

SIGNIFICANT DIGITS

Applied problems are usually stated in terms of data obtained by measurement or observation. The accuracy of such data is subject to the mechanical limitations of the instruments used. It is common practice to say that a number is expressed accurate to, for example, three decimal places or to four significant digits.

To find the **significant digits** of a number written in decimal form, we begin at the left with the first nonzero digit and end with the last digit obtained by counting or by measurement. In the number 27.43, the significant digits are 2, 7, 4, and 3. In the number 0.00412, the significant digits are 4, 1, and 2. In the number 0.315, the significant digits are 3, 1, and 5. In the number 3.510, the significant digits are 3, 5, 1, and 0; the final zero is significant since it indicates that the degree of precision is thousandths. Zeros used to fix the position of the decimal point as in 0.00412, are not significant.

In a measurement such as 180,000, it is impossible to tell which of the zeros are significant until further information is given concerning the accuracy of the measurement. We can use scientific notation to indicate which, if any, of the zeros in the number 180,000 are significant. For example, if we want to show that 180,000 is accurate to four significant digits, we write 1.800×10^5. In the absence of other information, the last significant digit of 1,800 is 8, that of 1,492 is 2, and that of 0.00326 is 6.

In summary, zeros that are doing nothing more than playing the role of place holders are not significant.

EXAMPLE 4

0.004 One significant digit

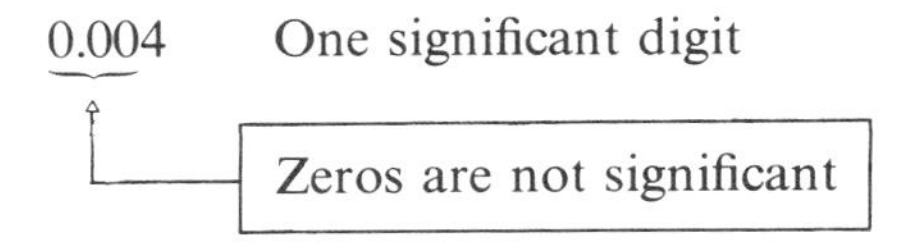

Also, zeros that follow nonzero digits or zero digits are not significant unless they are preceded by a decimal point or they are between significant digits.

EXAMPLE 5

a. 93,000,000 Two significant digits

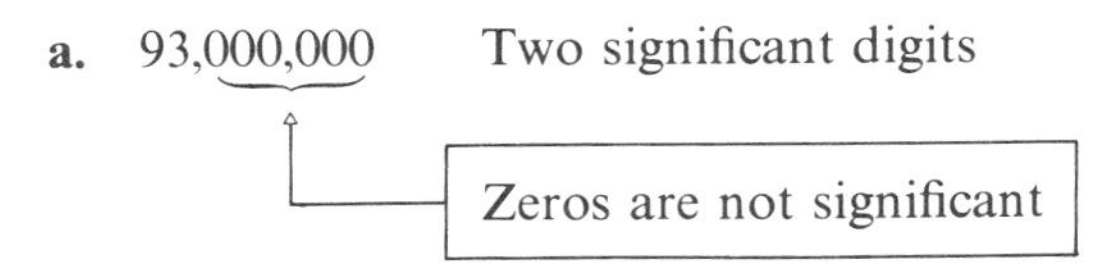

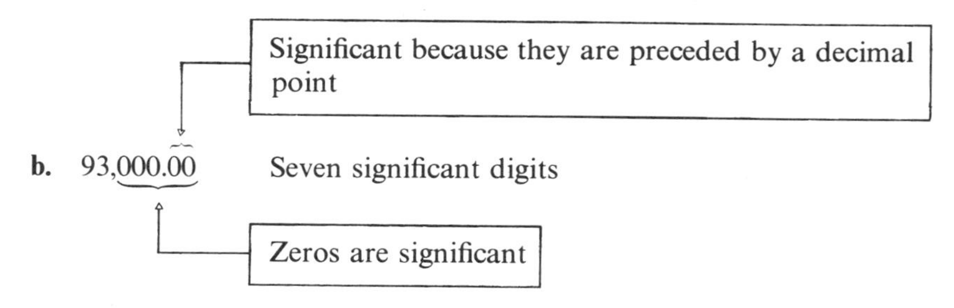

EXAMPLE 6

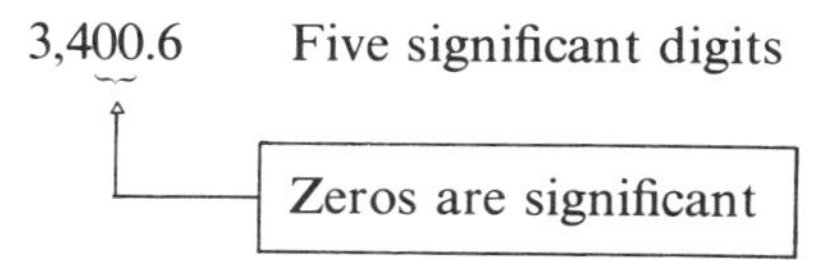

EXAMPLE 7 Sirius, the brightest star in our winter night sky, is 50,561,000,000,000 miles from the earth. Express this number in scientific notation rounded to three significant digits.

SOLUTION

$$5_{\wedge}\underrightarrow{0561000000000}. = 5.06 \times 10^{13}$$

13 places

EXAMPLE 8 Evaluate the following expression.

$$\frac{(865{,}194)(3{,}495)}{(8{,}966)(57.40)}$$

SOLUTION 5,876 (accurate to four significant digits)

Exercises 10

A Complete the following table.

	Common notation (decimal form)	Scientific notation
1.	57,000	________
2.	0.0041	________
3.	________	2.8×10^{-4}
4.	________	9.7×10^{5}
5.	40,700,000	________
6.	________	1.4×10^{-2}

In Exercises 7–10, round each number as indicated.

7. 0.00678 2 significant digits
8. 12.1462 Thousandths
9. 0.555 Hundredths
10. 0.469450 4 significant digits
11. The total area of the earth is 196,940,000 square miles. Express this number in scientific notation, rounded to three significant digits.

12. The sun is our closest star. The next closest star is Alpha Centauri, which is 4.3 light-years from us. Find this distance, in miles, by multiplying $4.3 \times 186{,}284 \times 60 \times 60 \times 24 \times 365.25$. Assume these numbers to be exact and write your answer in scientific notation, rounded to three significant digits.

B Complete the following table.

	Common notation (decimal form)	Scientific notation
1.	410,000	______
2.	0.00036	______
3.	______	5.3×10^3
4.	______	1.8×10^{-4}
5.	3,050,000	______
6.	______	7.14×10^{-5}

In Exercises 7–10, round each number as indicated.

7. 0.07568 3 significant digits
8. 5.50716 Thousandths
9. 0.56257 Hundredths
10. 4,250,000 2 significant digits
11. The mass of a hydrogen atom is approximately 1.7×10^{-24} grams. Express this number in decimal form.

12. Find the distance, in miles, that light travels in one year by multiplying $186{,}284 \times 60 \times 60 \times 24 \times 365.25$. Assume these numbers to be exact and write your answer in scientific notation, rounded to three significant digits.

11 Evaluating formulas

A formula is a general rule stated in mathematical language. Formulas play an important role in the physical, economic, and social world in which we live. All the formulas used in this section are formulas taken from some field of endeavor. You no doubt will recognize some of the formulas that we use.

EXAMPLE 1 Given the formula

$$C = \frac{5}{9}(F - 32),$$

find C when $F = -13$. (This formula converts Fahrenheit temperatures to temperatures in degrees Celsius.)

SOLUTION

$$\begin{aligned} C &= \frac{5}{9}(F - 32) \\ &= \frac{5}{9}(-13 - 32) \\ &= \frac{5}{9}(-45) = -25^\circ C \end{aligned}$$

EXAMPLE 2 Given the formula $A = P(1 + rt)$, find A when $P = \$1{,}000$, $r = 0.08$, and $t = 1.5$. (This is a simple interest formula.)

SOLUTION

$$\begin{aligned} A &= P(1 + rt) \\ &= 1{,}000[1 + (0.08)(1.5)] \\ &= 1{,}000[1 + 0.12] \\ &= 1{,}000[1.12] = \$1{,}120 \end{aligned}$$

Notice [ ] were used in place of ( ) to clarify the grouping.

EXAMPLE 3 Given the formula

$$S = \frac{a(1 - r^n)}{1 - r},$$

find S when $a = -4$, $r = \frac{1}{2}$, and $n = 3$. (This is a formula for the sum of n terms of a geometric series.)

SOLUTION

$$S = \frac{a(1 - r^n)}{1 - r}$$

$$= \frac{(-4)\left[1 - \left(\frac{1}{2}\right)^3\right]}{1 - \frac{1}{2}} \qquad \left(\frac{1}{2}\right)^3 = \left(\frac{1}{2}\right)\left(\frac{1}{2}\right)\left(\frac{1}{2}\right) = \frac{1}{8}$$

$$= \frac{(-4)\left[1 - \frac{1}{8}\right]}{1 - \frac{1}{2}} = \frac{(-4)\left[\frac{7}{8}\right]}{\frac{1}{2}} = -7$$

Exercises 11

A In Exercises 1–8, evaluate each formula using the values given with the formula. Round answers to three significant digits.

1. $A = \frac{1}{2}bh$ $\qquad b = 13,\ h = 14.6$
2. $C = \frac{5}{9}(F - 32)$ $\qquad F = -10$
3. $HP = \frac{Fs}{550t}$ $\qquad F = 186,\ s = 1{,}320,\ t = 606$
4. $t = \pi\sqrt{\frac{L}{g}}$ $\qquad L = 128.64,\ g = 32.16,\ \pi = 3.14$
5. $CF = \frac{wv^2}{gr}$ $\qquad w = 4{,}400,\ v = 88,\ g = 32,\ r = 880$
6. $V = \sqrt{2gS}$ $\qquad S = 169,\ g = 32$
7. $V = \frac{4}{3}\pi r^3$ $\qquad r = 12,\ \pi = 3.14$
8. $S = R\left[\frac{(1 + i)^n - 1}{i}\right]$ $\qquad R = 100,\ n = 2,\ i = 0.01$

9. The formula

$$R = \frac{Ai(1 + i)^n}{(1 + i)^n - 1}$$

is used to calculate the monthly payment on a home-owner's mortgage.

$R =$ Monthly payment
$i =$ Interest rate per month expressed as a decimal
$n =$ Total number of *months* during which payments are made
$A =$ The amount of the mortgage

Find the monthly payment on a $9\frac{1}{2}\%$ *per year*, 25-year, \$35,000 loan. Round your answer to the nearest cent. (*Hint:* $i = 0.095/12 = 0.0079$, the interest rate per month.)

B In Exercises 1–8, evaluate each formula using the values given with the formula. Round answers to three significant digits.

1. $I = Prt$ $\quad P = 1{,}500,\ r = 0.095,\ t = 2.5$
2. $S = P(1 + i)^n$ $\quad P = 200,\ i = 0.01,\ n = 3$
3. $HP = \dfrac{Fs}{550t}$ $\quad F = 185,\ s = 2{,}450,\ t = 33$
4. $t = \pi\sqrt{\dfrac{L}{g}}$ $\quad L = 96,\ g = 32,\ \pi = 3.14$
5. $CF = \dfrac{wv^2}{gr}$ $\quad w = 3{,}850,\ v = 44,\ g = 32.16,\ r = 275$
6. $V = \sqrt{2gS}$ $\quad S = 100,\ g = 32$
7. $V = \dfrac{4}{3}\pi r^3$ $\quad r = 10,\ \pi = 3.14$
8. $S = \dfrac{a(1 - r^n)}{1 - r}$ $\quad a = -9,\ r = \dfrac{1}{3},\ n = 3$

---

9. The formula

$$R = \frac{Ai(1 + i)^n}{(1 + i)^n - 1}$$

is used to calculate the monthly payment on a home-owner's mortgage.

$R =$ Monthly payment
$i =$ Interest rate per month expressed as a decimal
$n =$ Total number of *months* during which payments are made
$A =$ The amount of the mortgage

Find the monthly payment on a 9% *per year*, 30-year, \$24,000 loan. Round your answer to the nearest cent.

Summary

A **set** is a collection of things. The objects that make up the collection are called **elements** or **members** of the set. Sets are usually represented by capital letters, by listing their elements within braces { }, or both.

The following symbols were used in this chapter.

$\{\ \}$, $\varnothing$	The null set or empty set	$\cap$	Intersection of
$\in$	Is an element of	$<$	Is less than
$\notin$	Is not an element of	$>$	Is greater than
$=$	Is equal to	$\leq$	Is less than or equal to
$\neq$	Is not equal to	$\geq$	Is greater than or equal to
$\doteq$	Is approximately equal to	$\lvert a \rvert$	Absolute value of a
$\subseteq$	Is a subset of	$\sqrt{a}$	Square root of a
$\cup$	Union of	$\sqrt[n]{a}$	nth root of a

The following sets are subsets of the real numbers, R. $R = Q \cup H$.

Digits	$D = \{0, 1, 2, 3, 4, 5, 6, 7, 8, 9\}$
Natural numbers	$N = \{1, 2, 3, \ldots\}$
Whole numbers	$W = \{0, 1, 2, 3, \ldots\}$
Integers	$J = \{\ldots, -3, -2, -1, 0, 1, 2, 3, \ldots\}$
Rational numbers	$Q = \left\{x \,\middle\vert\, x = \frac{a}{b},\quad a, b \in J,\quad b \neq 0\right\}$
Irrational numbers	$H = \{x \mid x \in R,\quad x \notin Q\}$

A **finite set** of elements is a set where the counting of the elements will eventually come to an end. An **infinite set** of elements is a set where the counting of the elements will never come to an end.

The **absolute value** of a real number is

$$|a| = \begin{cases} a & \text{if } a \geq 0 \\ -a & \text{if } a < 0 \end{cases}$$

The operations of addition and multiplication are commutative and associative. There exists a unique number 1, called the **identity element of multiplication**, such that

$$a \cdot 1 = 1 \cdot a = a$$

There exists a unique number 0, called the **identity element of addition**, such that

$$a + 0 = 0 + a = a$$

and

$$a \cdot 0 = 0 \cdot a = 0.$$

If a is any real number except 0, then

$$\frac{0}{a} = 0 \quad \text{and} \quad \frac{a}{0} \text{ is not defined.}$$

For the properties of the real number system, see Section 5.

The symbol $\sqrt[n]{a}$ is read "nth root of a." If n is an odd integer, the real number a has only one real nth root. This root is called the **principal *n*th root of *a***. If n is an even integer and if a is positive, then a has two, and only two, real nth roots. These two roots have the same absolute value, but one is positive and the other is negative. The positive root is called the principal nth root of a.

For the laws of exponents, see Section 6.

For rules about the removal of grouping symbols, see Section 8.

The order of operations used in evaluating an expression is as follows.

1. Expressions within grouping symbols are simplified first.
2. Powers and roots are done in any order.
3. Multiplication and division are done in order from left to right.
4. Finally, additions and subtractions are performed in order from left to right.

For discussion of **significant digits**, see Section 10.

A number is said to be written in **scientific notation** when it is in the form

$$a \times 10^n \quad \text{where } |a| < 10,\ a \neq 0, \text{ and } n \text{ is an integer.}$$

For discussion and examples showing conversion from decimal notation to scientific notation, see Section 10.

Review exercises

A

1. If $A = \{3, 5, 7\}$, $B = \{2, 4, 6\}$, and $C = \{7, 8, 9\}$, find each of the following.
   a. $A \cup C$ **b.** $B \cap C$
2. Which of the numbers $-2, 0, \frac{3}{5}, \sqrt{2}$, are
   a. rational numbers? **b.** digits? **c.** real numbers?
3. Evaluate.
   a. $|-5|$ **b.** $|\sqrt{7} - 1|$ **c.** $-|1 - \sqrt{7}|$
4. Write $A = \{x \mid -3 < x < 1, \quad x \in J\}$ in roster notation.
5. Write 0.00078 in scientific notation.
6. Perform the indicated operations.
   a. $15 \div 5(-3) - 2\sqrt{16}$
   b. $4(-1)^3 - 5^2 + 4 + \sqrt{(-3)^2}$
   c. $7|-4| + (10^2)(10^{-2})$
7. Remove the grouping symbols and combine like terms.
   a. $20 - \{5x - 3[4 - (2x + 3)] - 10x\} + x$
   b. $(\frac{5}{6}xy^3)(-2yz^2)(\frac{3}{5}x^2z)$

8. Round each of the following numbers as indicated.
   a. 0.04825 3 significant digits
   b. 928,756 Thousands
   c. 0.00614 Thousandths
9. Convert each of the following numbers to scientific notation.
   a. 34,500,000 **b.** 0.00052 **c.** 50,006
10. Write 74,000,000 in scientific notation, showing that the number is accurate to three significant digits.

---

11. Solve the formula

$$K = \frac{l' - l}{l(t' - t)}$$

for K when $l' = 12.1$, $l = 11.8$, $t = 25$, and $t' = 137$. Express your answer to three significant digits.

B

1. If $A = \{1, 3, 4\}$, $B = \{1, 4, 5\}$, and $C = \{2, 3\}$, find each of the following.
   a. $A \cup B$ **b.** $B \cap C$
2. Which of the numbers $\frac{2}{3}$, 4, $\sqrt{3}$, $-\pi$, are
   a. rational numbers? **b.** real numbers?
   c. irrational numbers?
3. Evaluate.
   a. $|-2|$ **b.** $|1 - \sqrt{2}|$ **c.** $|-\sqrt{3} - 1|$
4. Write $B = \{x \mid 5 < x < 10,\ x \in J\}$ in roster notation.
5. Write 370,000,000 in scientific notation.
6. Perform the indicated operations.
   a. $7|-2| + 3\sqrt{16} - \sqrt[3]{-8}$
   b. $20 \div 2(-5) - 3^2 + \sqrt{(-4)^2}$
   c. $5(-2^3)^2 + (-1)^{17}$
7. Remove the grouping symbols and combine like terms.
   a. $10 - \{4x - 2[3 - (x - 4)] + 2\} - 5x$
   b. $(\frac{2}{3}x^4yz^2)(9xy^2z)(-\frac{1}{2}xy^2)$
8. Round each of the following numbers as indicated.
   a. 0.0585 2 significant digits
   b. 186,356 Hundreds
   c. 4.075 Hundredths
9. Convert each of the following numbers to scientific notation.
   a. 785,000,000,000 **b.** 0.00614 **c.** 7,006
10. Write 1,500,000 in scientific notation, showing that the number is accurate to three significant digits.

11. Solve the formula

$$\frac{1}{R} = \frac{1}{r_1} + \frac{1}{r_2}$$

for R when $r_1 = 23$ and $r_2 = 47$. Express your answer to two significant digits.

Diagnostic test

The purpose of this test is to see how well you understand the work in this chapter. Allow yourself approximately 50 minutes to do the test. Solutions to the problems, together with section references, are given in the Answers section at the end of the book. We suggest that you study the sections referred to for the problems you do incorrectly.

1. If $A = \{1, 3, 5\}$, $B = \{2, 4, 6\}$, and $C = \{5, 6, 7\}$, find each of the following.

a. $B \cup C$ **b.** $A \cap C$ **c.** $A \cap B$

2. State which of the following sets are finite and which are infinite.

a. The set of whole numbers.
b. $\{x \mid -3 < x < 10, \quad x \in J\}$
c. $\{x \mid 2 < x < 6, \quad x \in R\}$

3. Perform the indicated operations.

a. $5|-2| - 5^2$
b. $30 \div (-2)(5) - 4\sqrt{36}$
c. $0(15)^2 + (-1)^{25} + \frac{0}{6}$

4. a. Find two square roots of 16.
b. What is the principal square root of 16?

5. What is the principal 5th root of -32?

6. Find the value of each of the following expressions. If no value is defined, state that fact.

a. 3^0 **b.** $\dfrac{2^{-4}}{2^{-6}}$ **c.** $10^6 \cdot 10^{-4}$
d. $(2^{-3})^{-1}$ **e.** 0^5 **f.** $\frac{4}{0}$
g. $\sqrt[3]{-64}$ **h.** $\sqrt{49}$ **i.** $\sqrt{(-6)^2}$

7. Given the numbers -5, 0, 2.4, $\frac{2}{7}$, $\sqrt{3}$, and π,

a. which are irrational numbers?
b. which are integers?
c. which are whole numbers?
d. which are real numbers?

8. State which of the following statements are true and which are false.

a. $\{x \mid -2 < x < 2, \quad x \in J\} = \{-1, 0, 1\}$

b. The integers are a subset of the digits.

c. The irrational numbers are a subset of the real numbers.

9. Round each of the following numbers as indicated.

a. 0.15645 4 significant digits

b. 186,384 Tens

c. 7.249 Tenths

10. Remove the grouping symbols and combine like terms.

a. $12 - \{2x - 3[x - 2(4 - x)] - 24\} - 4x$

b. $3x(4x^2y - 5y^2) - xy(12x^2 - 5y)$

11. Write 2,300,000 in scientific notation, showing that the number is accurate to four significant digits.

---

12. Find the value of S in the formula

$$S = \frac{a(1 - r^n)}{1 - r}$$

when $a = -16$, $r = \frac{1}{2}$, and $n = 5$.

2 Polynomials

Polynomials have a similar importance in algebra that whole numbers have in arithmetic. Because much of the work in algebra involves operations with polynomials, it is important that you be able to work efficiently with them.

1 Definitions

An **algebraic expression** is any expression of variables and real number constants used in a number of additions, subtractions, multiplications, divisions, raising to integral or fractional powers, or taking of roots. The polynomial in x is one of the simplest algebraic expressions.

EXAMPLE 1 Examples of algebraic expressions.

a. $\dfrac{3xy^2 - 2x}{x + y}$ **b.** $x^2 - 5x + \dfrac{1}{\sqrt{2x}}$

c. $\dfrac{2^{1/2}xz^2 + \left(\dfrac{-x}{x+z}\right)^3}{\sqrt[5]{x^2 + y}}$ **d.** $3x^2 + x + 4$

> **POLYNOMIALS IN X**
>
> A polynomial in x is an algebraic expression of the form
>
> $$a_0 x^n + a_1 x^{n-1} + a_2 x^{n-2} + \cdots + a_{n-1} x^1 + a_n, \quad a_0 \neq 0,$$
>
> in which n is a whole number and the coefficients $a_0, a_1, \ldots, a_n$ are real numbers. The exponents of x cannot be negative.

The equation formed by equating a polynomial to zero is called a **polynomial equation**.

EXAMPLE 2 The equation $6x^3 + 7x^2 - 11x - 12 = 0$ is an example of a polynomial equation in x.

The **degree of a term** is the **sum** of the exponents of its variables.

EXAMPLE 3

a. $4xy^2z$ Fourth-degree term

b. $7x^3y^2z^4$ Ninth-degree term

c. $5x$ First-degree term

d. 3 Zero-degree term (because 3 can be written $3x^0$)

e. 3^7xy^2z Fourth-degree term (the exponent 7 does not affect the degree)

The **degree of a polynomial** in simplified form is the same as that of its highest-degree term. Polynomials may contain more than one variable, such as x and y, or x, y, and z. In such cases, they are called polynomials in x and y, or in x, y, and z.

A **monomial** is a polynomial consisting of a single term. The **zero monomial** is defined to be the number zero and has no degree. Any real number can be thought of as a monomial because it can be written in the form ax^0, where a is any real number. A **polynomial is any sum or difference of monomials**. If every coefficient in a polynomial is an integer, it is said to be a polynomial over the integers. In like manner, if the coefficients are rational numbers, it is a polynomial over the rationals.

A **binomial** is a polynomial of two terms, such as $2x + 7y$ or $3 - 2(x + y)$. A **trinomial** is a polynomial of three terms, such as $3x^2 - 5x + 2$.

EXAMPLE 4 Examples of polynomials.

a.	$4x^2y$	Third-degree polynomial (monomial) in two variables
b.	$5xy + z$	Second-degree polynomial (binomial) in three variables
c.	$4x^2y - 3xy^3 + 7$	Fourth-degree polynomial (trinomial) in two variables
d.	$x + 2y + 3z + w$	First-degree polynomial in four variables

A WORD OF CAUTION Not all algebraic expressions are polynomials. The variables included in the terms of the polynomials must have only nonnegative integers as exponents.

EXAMPLE 5 Algebraic expressions which are **not** polynomials.

a.	$6x^2y^{-1}$	Not a monomial because y has a negative exponent.
b.	$4y + (1/z^2)$	Not a binomial because $1/z^2 = z^{-2}$. The exponent is negative.
c.	$5x + 3x^{1/2} + 4$	Not a polynomial because x has a fractional exponent.
d.	$5\sqrt[3]{z} + 6x$	Not a binomial since z has an exponent of $\frac{1}{3}$; $\sqrt[3]{z} = z^{1/3}$.

Symbols such as $P(x)$, read "P of x," $Q(x)$, read "Q of x," and $f(x)$, read "f of x," are often used to represent a polynomial containing x as the only variable. Similarly, $P(x, y)$ can be used to represent a polynomial containing the variables x and y.

EXAMPLE 6 If $P(x) = 5x^2 - 7x + 3$, find $P(2)$, $P(1)$, and $P(0)$.

SOLUTION Since $P(x)$ represents $5x^2 - 7x + 3$, $P(2)$ represents the value of the expression $5x^2 - 7x + 3$ when $x = 2$. Therefore, to find $P(2)$, we substitute 2 in place of x throughout the expression, as shown below.

$$\begin{aligned} P(x) &= 5x^2 - 7x + 3 \\ P(2) &= 5(2)^2 - 7(2) + 3 \\ &= 5 \cdot 4 - 7 \cdot 2 + 3 \\ &= 20 - 14 + 3 = 9 \\ P(1) &= 5(1)^2 - 7(1) + 3 \\ &= 5 - 7 + 3 = 1 \\ P(0) &= 5(0)^2 - 7(0) + 3 = 3 \end{aligned}$$

EXAMPLE 7

$$\begin{aligned} P(x, y) &= x^3 + 2x^2y - y^3 \\ P(2, 3) &= 2^3 + 2(2)^2(3) - 3^3 && \text{Replace } x \text{ by 2 and } y \text{ by 3.} \\ &= 8 + 24 - 27 = 5 \end{aligned}$$

EXAMPLE 8 Given $P(x) = 3x^2 - 5x$, find $P(-2)$ and $P(4a)$.

SOLUTION

$$\begin{aligned} P(x) &= 3x^2 - 5x \\ P(-2) &= 3(-2)^2 - 5(-2) && \text{Replace } x \text{ by } -2. \\ &= 3 \cdot 4 - 5(-2) \\ &= 12 + 10 = 22 \\ P(4a) &= 3(4a)^2 - 5(4a) && \text{Replace } x \text{ by } 4a. \\ &= 3(16a^2) - 5(4a) \\ &= 48a^2 - 20a \end{aligned}$$

Exercises 1

A In Exercises 1–3, identify each algebraic expression as a monomial, binomial, or trinomial, and give the degree of the polynomial.

1. $3x^4y^2z - 4xy^2$ **2.** $4x^2 - 12x + 9$ **3.** $5xy^2z^3$

4. Is the expression

$$\frac{\sqrt{x + y}}{3} + \frac{x^2}{4}$$

a polynomial? Give a reason for your answer.

5. If $P(x) = 3x^2 - 4x - 5$, find each of the following.
   a. $P(-3)$ **b.** $P(0)$ **c.** $P(2a^2)$
6. If $P(x) = x^{16} - 1$, find each of the following.
   a. $P(1)$ **b.** $P(0)$ **c.** $P(-1)$
7. If $Q(x, y) = x^3 + 3x^2y + 3xy^2 + y^3$, find each of the following.
   a. $Q(2, -1)$ **b.** $Q(0, -3)$ **c.** $Q(-1, -2)$

B In Exercises 1–3, identify each algebraic expression as a monomial, binomial, or trinomial, and give the degree of the polynomial.

1. $5x^3 + 2x$ **2.** $x^2 + 2xy + y^2$ **3.** $7x^2yz^3$
4. Is the expression

$$7x^2y + \frac{\sqrt{x}}{3y^2}$$

   a polynomial? Give a reason for your answer.
5. If $P(x) = 5x^3 - 2x + 4$, find each of the following.
   a. $P(-2)$ **b.** $P(0)$ **c.** $P(3a^2)$
6. If $P(x) = x^{15} - 1$, find each of the following.
   a. $P(1)$ **b.** $P(-1)$ **c.** $P(0)$
7. If $Q(x, y) = x^3 - 3x^2y + 3xy^2 - y^3$, find each of the following.
   a. $Q(1, 2)$ **b.** $Q(-1, 0)$ **c.** $Q(-2, -1)$

---

2 Addition, subtraction, and multiplication of polynomials

Polynomials are usually written in descending powers of one of the variables.

EXAMPLE 1 Arrange the polynomial $7x^2 - 2x^4 + 3x^3 + 4 - 5x$ in descending powers of x.

SOLUTION

$$-2x^4 + 3x^3 + 7x^2 - 5x + 4$$

EXAMPLE 2 Arrange the polynomial $3xy - 2x^2y^3 + xy^4 + x^2y + 5$ in descending powers of y.

SOLUTION

$$xy^4 - 2x^2y^3 + x^2y + 3xy + 5$$

Since y is the same power in both terms, the higher-degree term is written first.

To find the sum of any number of polynomials, simply combine like terms.

EXAMPLE 3 Find the sum of the polynomials $5x^2 + 7x^3 - 3x - 4$ and $3x^3 + 2x - 5$.

SOLUTION Remove parentheses, combine like terms, and write in descending powers of x.

$$\begin{aligned}(5x^2 + 7x^3 - 3x - 4) &+ (3x^3 + 2x - 5)\\ &= 7x^3 + 5x^2 - 3x - 4 + 3x^3 + 2x - 5\\ &= (7 + 3)x^3 + 5x^2 + (-3 + 2)x + (-4 - 5)\\ &= 10x^3 + 5x^2 - x - 9\end{aligned}$$

To subtract one polynomial from another, change the signs of the terms in the polynomial being subtracted, and then proceed as in addition.

EXAMPLE 4 Subtract $7x^4 - x^2 + x - 3$ from $2x^3 - 4x^2 + x - 5$.

SOLUTION

$$\begin{aligned}(2x^3 - 4x^2 + x - 5) &- (7x^4 - x^2 + x - 3)\\ &= 2x^3 - 4x^2 + x - 5 - 7x^4 + x^2 - x + 3\\ &= -7x^4 + 2x^3 - 3x^2 - 2\end{aligned}$$

EXAMPLE 5 Find the product of $2x^2 + 5$ and $3x - 4$.

SOLUTION

$$(2x^2 + 5)(3x - 4) = (2x^2 + 5)3x + (2x^2 + 5)(-4)$$

Use the distributive law.

$$= 6x^3 + 15x - 8x^2 - 20$$

$$= 6x^3 - 8x^2 + 15x - 20$$

Arrange terms in descending powers of x.

The work may also be arranged as follows

$$\begin{array}{llll} 2x^2 + 5 & & & \\ 3x \quad - 4 & & & \\ \hline 6x^3 \qquad + 15x & & = & 3x(2x^2 + 5) \\ \qquad - 8x^2 \qquad - 20 & & = & -4(2x^2 + 5) \\ \hline 6x^3 - 8x^2 + 15x - 20 & & & \end{array}$$

EXAMPLE 6 Find the product of $5 + 3x^2 - 2x$ and $2 - 4x + x^3$.

SOLUTION First, we arrange each polynomial in descending powers of x. This gives $3x^2 - 2x + 5$ and $x^3 - 4x + 2$.

$$\begin{array}{r}
3x^2 - 2x \quad + 5 \\
x^3 - 4x \quad + 2 \\
\hline
3x^5 - 2x^4 + 5x^3 \qquad\qquad\qquad\qquad\quad \\
- 12x^3 + 8x^2 - 20x \qquad\quad \\
+ 6x^2 - 4x + 10 \\
\hline
3x^5 - 2x^4 - 7x^3 + 14x^2 - 24x + 10
\end{array}$$

The following products occur so often in algebraic calculations that they should be memorized.

SPECIAL PRODUCT RULES

	Factors	**Products**
1.	$a(b + c)$	$= ab + ac$
2.	$(x + a)(x + b)$	$= x^2 + (a + b)x + ab$
3.	$(a + b)(a - b)$	$= a^2 - b^2$
4.	$(a + b)^2$	$= a^2 + 2ab + b^2$
5.	$(a - b)^2$	$= a^2 - 2ab + b^2$
6.	$(ax + b)(cx + d)$	$= acx^2 + (ad + bc)x + bd$
7.	$(a + b)^3$	$= a^3 + 3a^2b + 3ab^2 + b^3$
8.	$(a - b)^3$	$= a^3 - 3a^2b + 3ab^2 - b^3$

EXAMPLE 7 Examples of special products.

a. $(x + 5)(x - 3) = x^2 + (5 - 3)x + (5)(-3)$
$= x^2 + 2x - 15$ Rule 2

b. $(4x^3 + 3y^2)(4x^3 - 3y^2) = (4x^3)^2 - (3y^2)^2$
$= 16x^6 - 9y^4$ Rule 3

c. $(3x^4 + 5)^2 = (3x^4)^2 + 2(3x^4)(5) + 5^2$
$= 9x^8 + 30x^4 + 25$ Rule 4

d. $\left(\frac{2x^2}{3} + \frac{3}{5}\right)\left(\frac{2x^2}{3} - \frac{3}{5}\right) = \left(\frac{2x^2}{3}\right)^2 - \left(\frac{3}{5}\right)^2$
$= \frac{4x^4}{9} - \frac{9}{25}$ Rule 3

e. $(3x + 2)(5x - 4) = (3)(5)x^2 + [3(-4) + 2(5)]x + (2)(-4)$
$= 15x^2 - 2x - 8$ Rule 6

f. $(3x^2 - 2y^3)^3$
$= (3x^2)^3 - 3(3x^2)^2(2y^3) + 3(3x^2)(2y^3)^2 - (2y^3)^3$
$= 27x^6 - 54x^4y^3 + 36x^2y^6 - 8y^9$ Rule 8

A WORD OF CAUTION The student is reminded that $(a + b)^n \neq a^n + b^n$ unless $n = 1$.

EXAMPLE 8 $(a + b)^2 = a^2 + 2ab + b^2$
$(a + b)^2 \neq a^2 + b^2$

Exercises 2

A In Exercises 1–10, find the indicated products.

1. $(a + 9)(a - 3)$
2. $4x^2y(5x^3 - 3y^2)$
3. $(x - 9)(x + 3)$
4. $\left(\frac{x}{2} + \frac{2y^2}{3}\right)\left(\frac{x}{2} - \frac{2y^2}{3}\right)$
5. $(2a + 3b)(3a - 5b)$
6. $(5x + 3y)^2$
7. $\left(\frac{2x}{3} - 3\right)^2$
8. $(4x - 1)^3$
9. $\left(m + \frac{r}{3}\right)^2$
10. $\left(5b - \frac{3c}{2}\right)^2$

In Exercises 11–16, perform the indicated operations.

11. $(3x^2 + 2)(x - 5)$
12. $(3x - 1)(9x^2 + 3x + 1)$
13. $(a + b)(a^2 - ab + b^2)$
14. $(4x - 1)^2 - (4x + 1)^2$
15. $(2a + b)^3 - (2a - b)^3$
16. $(2 + x^3 - x^2)(x^2 - 2 + x^3)$

B In Exercises 1–10, find the indicated products.

1. $(x + 7)(x - 2)$
2. $3a^2b^5(4a^3 + 2b)$
3. $(x + 8)(x - 3)$
4. $\left(\frac{2a}{5} - \frac{b^2}{3}\right)\left(\frac{2a}{5} + \frac{b^2}{3}\right)$
5. $(2x + 3)(2x - 4)$
6. $(4x - 5)^2$
7. $\left(\frac{5z}{2} + 4\right)^2$
8. $(2x + 3)^3$
9. $\left(w + \frac{v}{5}\right)^2$
10. $\left(6a - \frac{b}{4}\right)^2$

In Exercises 11–16, perform the indicated operations.

11. $(2x^3 + 1)(x + 4)$

12. $(2x + 1)(4x^2 - 2x + 1)$

13. $(x - y)(x^2 + xy + y^2)$

14. $(3x + 1)^2 - (3x - 1)^2$

15. $(x - 2y)^3 - (x + 2y)^3$

16. $(4 - 2x^3 + 5x)(5x - 2 + x^3)$

3 Quotients of polynomials

The names of the parts of a division problem are shown below.

$$\text{Divisor} \overline{\big)\,\text{Dividend}}^{\text{Quotient}}$$

Remainder

If r is some constant, and if $Q(x)$ is the quotient obtained by dividing a polynomial $P(x)$ by $x - r$, and if R is the constant remainder after the division, then we have the following definition.

STANDARD DIVISION FORMULA

We can divide a polynomial by a binomial of degree one, as follows.

$$P(x) = (x - r) \cdot Q(x) + R$$

Stated in words,

Dividend = Divisor × Quotient + Remainder.

The above definition is true for all values of any variable. If $P(x)$ is a polynomial of degree n, then $Q(x)$ is a polynomial of degree $n - 1$ (see Example 1). If $R = 0$, then $x - r$ is an exact divisor of $P(x)$, which means that both $x - r$ and $Q(x)$ are factors of $P(x)$.

The **check** for a division may be obtained from the following relation.

Dividend = Divisor × Quotient + Remainder

We will use a simple example to illustrate the above relation.

EXAMPLE 1 Find $7 \div 3$.

SOLUTION

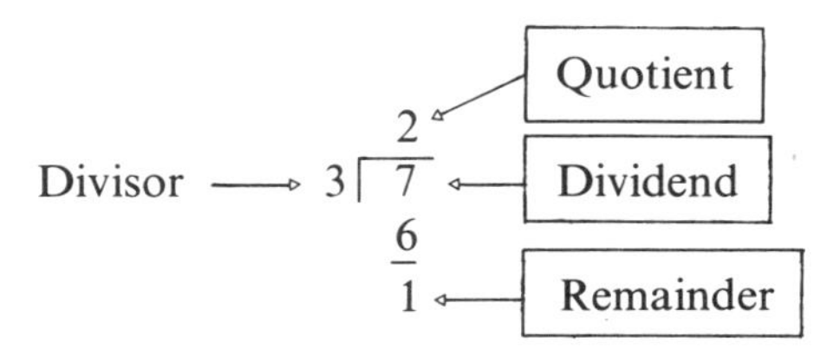

CHECK

$$\begin{aligned}\text{Dividend} &= \text{Divisor} \times \text{Quotient} + \text{Remainder}\\ 7 &= 3 \times 2 + 1\\ 7 &= 7\end{aligned}$$

To divide a polynomial by a monomial, divide each term of the polynomial by the monomial and add the results.

EXAMPLE 2 Divide $10x^3y^2 - 6x^2y^2 + 3xy^3$ by $2xy^2$.

SOLUTION

$$\frac{10x^3y^2 - 6x^2y^2 + 3xy^3}{2xy^2} = \frac{10x^3y^2}{2xy^2} - \frac{6x^2y^2}{2xy^2} + \frac{3xy^3}{2xy^2}$$

$$= 5x^2 - 3x + \frac{3y}{2}$$

To divide one polynomial by another polynomial, arrange the terms of each polynomial in descending powers of the same variable. If they have more than one variable, arrange the terms in descending powers of one of the variables, leaving a space for any missing terms (terms with coefficient zero). Continue dividing until the remainder is of lower degree than that of the divisor or is zero.

EXAMPLE 3 Divide $3x + 6x^3 + 2 - 5x^2$ by $2x - 1$.

SOLUTION First we arrange both polynomials in decreasing powers of x.

$$(6x^3 - 5x^2 + 3x + 2) \div (2x - 1)$$

Next, divide the first term of the divisor ($2x$) into the first term of the dividend ($6x^3$), obtaining $(6x^3/2x) = 3x^2$. Multiply the divisor $(2x - 1)$ by $3x^2$, obtaining $6x^3 - 3x^2$. This is subtracted from the dividend, as shown. Next divide $-2x^2$ by the first term in the divisor. This gives $(-2x^2/2x) = -x$, which is the next term in the quotient. Continue in this way until the degree of the remainder (3) is less than the degree of the divisor.

$$
\begin{array}{r|l}
 & 3x^2 - x + 1 \\ \cline{2-2}
2x - 1 & 6x^3 - 5x^2 + 3x + 2 \\
 & 6x^3 - 3x^2 \qquad \leftarrow (3x^2)(2x - 1) \\ \cline{2-2}
 & - 2x^2 + 3x \\
 & - 2x^2 + x \qquad \leftarrow (-x)(2x - 1) \\ \cline{2-2}
 & 2x + 2 \\
 & 2x - 1 \qquad \leftarrow 1(2x - 1) \\ \cline{2-2}
 & 3
\end{array}
$$

Therefore,

$$(6x^3 - 5x^2 + 3x + 2) \div (2x - 1) = 3x^2 - x + 1 + \frac{3}{2x - 1}.$$

CHECK

$$
\begin{aligned}
\text{Dividend} \quad &= \text{Divisor} \times \text{Quotient} + \text{Remainder} \\
6x^3 - 5x^2 + 3x + 2 &= (2x - 1) \times (3x^2 - x + 1) + 3 \\
&= 6x^3 - 2x^2 + 2x - 3x^2 + x - 1 + 3 \\
&= 6x^3 - 5x^2 + 3x + 2
\end{aligned}
$$

EXAMPLE 4 Divide $3x^3 - 7x^2 + 6x - 3$ by $x - 2$.

SOLUTION

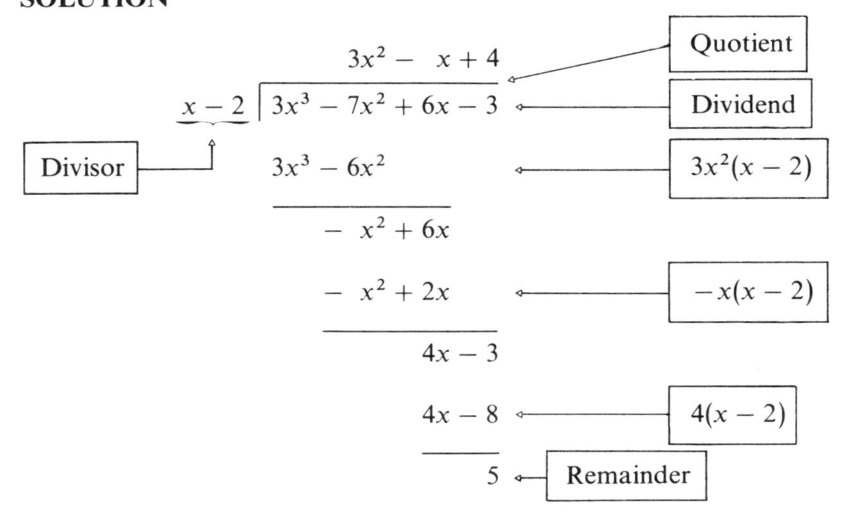

Therefore,

$$(3x^3 - 7x^2 + 6x - 3) \div (x - 2) = 3x^2 - x + 4 + \frac{5}{x - 2}.$$

EXAMPLE 5 Divide $(x^2y^3 - x^2y + x^3y^2 + 2y)$ by $(xy - 1)$.

SOLUTION

a. We can arrange the polynomials in decreasing powers of x.

$$\begin{array}{r|l}
 & x^2y + xy^2 + y \\
\hline
xy - 1 & x^3y^2 + x^2y^3 - x^2y + 2y \\
 & x^3y^2 \qquad\quad - x^2y \\
\hline
 & \qquad x^2y^3 \qquad\quad + 2y \\
 & \qquad x^2y^3 - xy^2 \\
\hline
 & \qquad\qquad xy^2 + 2y \\
 & \qquad\qquad xy^2 - y \\
\hline
 & \qquad\qquad\qquad 3y
\end{array}$$

b. We can also arrange the polynomials in descending powers of y.

$$\begin{array}{r|l}
 & xy^2 + x^2y + y \\
\hline
xy - 1 & x^2y^3 + x^3y^2 - x^2y + 2y \\
 & x^2y^3 - xy^2 \\
\hline
 & \qquad x^3y^2 + xy^2 - x^2y + 2y \\
 & \qquad x^3y^2 \qquad\quad - x^2y \\
\hline
 & \qquad\qquad xy^2 \qquad\quad + 2y \\
 & \qquad\qquad xy^2 \qquad\quad - y \\
\hline
 & \qquad\qquad\qquad\qquad 3y
\end{array}$$

Therefore,

$$(x^2y^3 + x^3y^2 - x^2y + 2y) \div (xy - 1) = xy^2 + x^2y + y + \frac{3y}{xy - 1}.$$

Note that in both cases, we have the same quotient and remainder.

SYNTHETIC DIVISION

Synthetic division is a simplified method of dividing a polynomial in one variable, say x, by x minus a constant.

Figures 1, 2, and 3 show how synthetic division is related to conventional division and how it shortens the work in a division problem of this kind.

Write the dividend in descending powers of x, using zero coefficients for any missing power.

FIGURE 1 We first divide in the usual way.

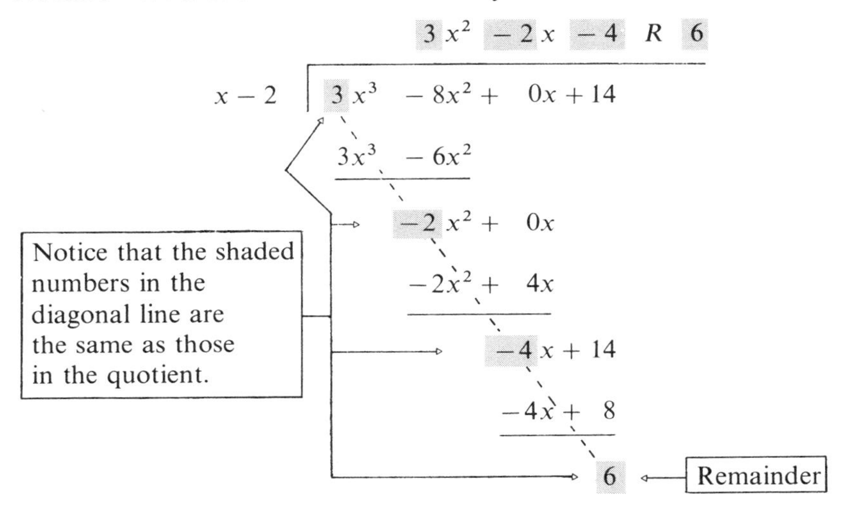

In Figure 2, we rewrite the above division omitting the first term of the divisor, since it is always x, and omitting all powers of x since each is indicated by the position of its coefficient in the computation. Notice that the zero coefficient is included to hold the position for the missing x term of the dividend. We also omit the plus signs and the quotient, since the quotient can be determined from the shaded numbers shown on the dotted diagonal lines of Figures 1 and 2.

FIGURE 2

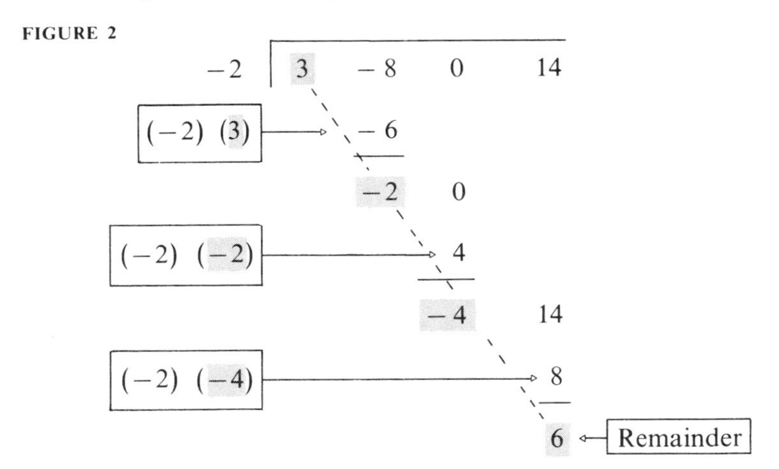

To make the work more compact, the numbers of Figure 2 are moved as shown in Figure 3. We also change -2 in the divisor to $+2$, which changes the products -6, 4, and 8 to $+6$, -4, and -8, respectively. This is done so that we can *add* rather than *subtract* the numbers. These latter products are then added to the numbers above them rather than being subtracted, as was done in Figure 2.

FIGURE 3

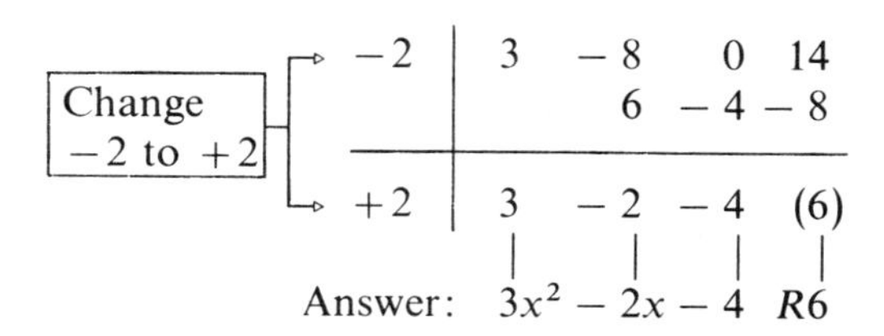

Therefore,

$$\frac{3x^3 - 8x^2 + 14}{x - 2} = 3x^2 - 2x - 4 + \frac{6}{x - 2}.$$

EXAMPLE 6 We use synthetic division to divide $2x^3 - 9x^2 + 14x - 11$ by $x - 3$.

STEP 1

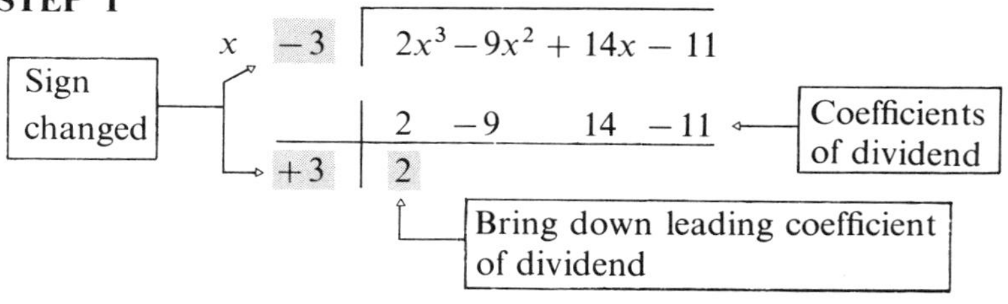

STEP 2

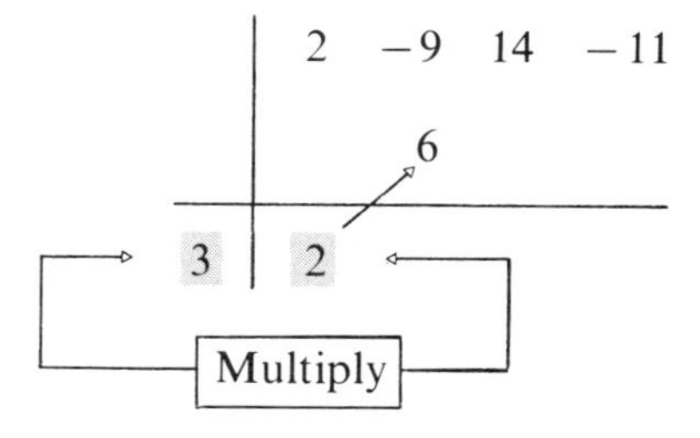

STEP 3

Add

	2	−9	14	−11
		6		
3	2	−3		

STEP 4

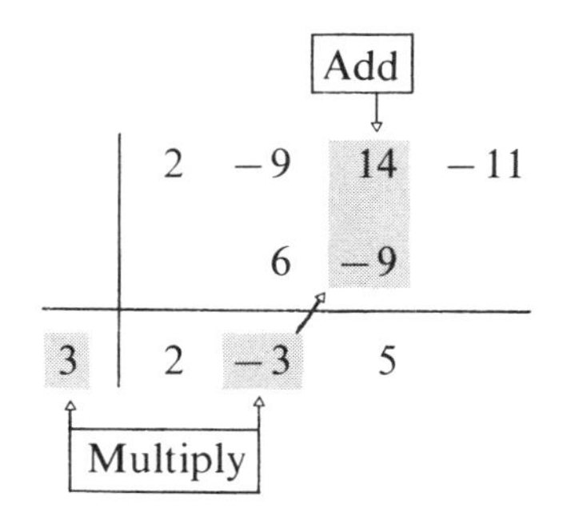

STEP 5

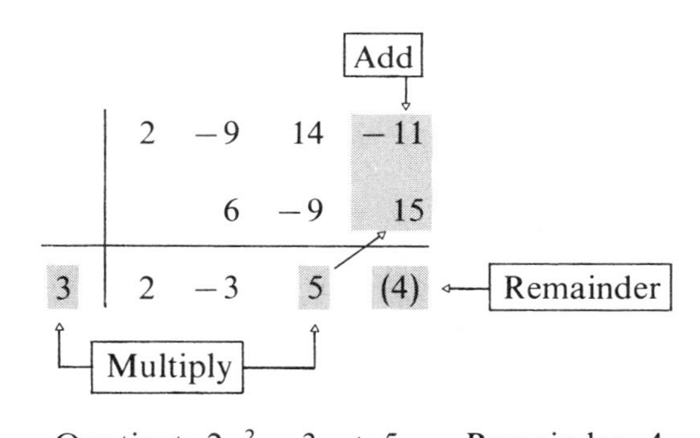

Quotient: $2x^2 - 3x + 5$ Remainder: 4

Therefore,

$$\frac{2x^3 - 9x^2 + 14x - 11}{x - 3} = 2x^2 - 3x + 5 + \frac{4}{x - 3}.$$

Notice that the degree of the quotient is one degree less than that of the dividend.

EXAMPLE 7 Divide $(4x^2 - 3x + 1)$ by $(x + 1)$.

SOLUTION

$$x + 1 \;\overline{\big)\; 4x^2 - 3x + 1}$$

	4	−3	1
		−4	7
−1	4	−7	(8) ← Remainder

Quotient: $4x - 7$ Remainder: 8

Therefore,

$$\frac{4x^2 - 3x + 1}{x + 1} = 4x - 7 + \frac{8}{x + 1}.$$

EXAMPLE 8 Divide $(2x^6 + 4x^5 - 3x^2 - 7x + 2)$ by $(x + 2)$.

SOLUTION

$$x + 2 \overline{\big) \; 2x^6 + 4x^5 + 0x^4 + 0x^3 - 3x^2 - 7x + 2}$$

	2	4	0	0	−3	−7	2
		−4	0	0	0	6	2
−2	2	0	0	0	−3	−1	(4)

Quotient: $2x^5 - 3x - 1$ Remainder: 4

Therefore,

$$\frac{2x^6 + 4x^5 - 3x^2 - 7x + 2}{x + 2} = 2x^5 - 3x - 1 + \frac{4}{x + 2}.$$

EXAMPLE 9 Divide $(x^3 + 3x^2 - 7x + 15)$ by $(x + 5)$.

SOLUTION

	1	3	−7	15
		−5	10	−15
−5	1	−2	3	(0)

← When the remainder is zero the division is exact.

Quotient: $x^2 - 2x + 3$ Remainder: 0

Therefore,

$$\frac{x^3 + 3x^2 - 7x + 15}{x + 5} = x^2 - 2x + 3.$$

Thus $x^3 + 3x^2 - 7x + 15 = (x + 5)(x^2 - 2x + 3)$. That is, $x + 5$ and $x^2 - 2x + 3$ are factors of $x^3 + 3x^2 - 7x + 15$. This application of synthetic division to factoring will be discussed in Section 6 of this chapter.

Exercises 3

A In Exercises 1–6, perform each of the divisions and check by using the definition, Dividend = Divisor × Quotient + Remainder.

1. $\dfrac{8x^2 + 4}{4}$

2. $\dfrac{6a^5b^2 - 9a^3b^4 + 3a^2b}{3a^2b}$

3. $\dfrac{10x^2 + 13x - 3}{2x + 3}$ 4. $\dfrac{10x^2 + x - 2}{5x - 2}$

5. $\dfrac{3x^3 - 10x^2 - 7x + 20}{3x - 4}$ 6. $\dfrac{27x^3 + 1}{3x + 1}$

In Exercises 7 and 8, perform the divisions.

7. $(x^4 - 9x^2 + 2) \div (3x - 1 + x^2)$
8. $(x^2 + 7x - 4x^3 + x^5 + x^4 - 4) \div (3 + x^3 - 2x)$

In Exercises 9–15, use synthetic division to perform the divisions.

9. $\dfrac{2x^2 - 7x - 15}{x - 5}$ 10. $(3x^2 + 10x - 8) \div (x + 4)$
11. $(5x^2 + 3x - 14) \div (x + 2)$
12. $(3x^2 + x + 2x^4 - 3x^3 + 4) \div (x - 1)$
13. $\dfrac{27x^4 - 6x^2 - \frac{2}{3}}{x - \frac{1}{3}}$ 14. $\dfrac{x^6 + 1}{x + 1}$ 15. $\dfrac{x^5 - 32}{x - 2}$

B In Exercises 1–6, perform each of the divisions and check by using the definition, Dividend = Divisor × Quotient + Remainder.

1. $\dfrac{6x^2 + 3}{3}$ 2. $\dfrac{2xy^5 - 10x^2y^4 + 4xy^2}{2xy^2}$

3. $\dfrac{6x^2 + x - 12}{3x - 4}$ 4. $\dfrac{12x^2 + x - 35}{4x + 7}$

5. $\dfrac{3x^3 + 4x^2 - 13x + 6}{3x - 2}$ 6. $\dfrac{8x^3 - 27}{2x - 3}$

In Exercises 7 and 8, perform the divisions.

7. $(4x - 4x^2 + x^4 + 5) \div (x^2 + 1 - 2x)$
8. $(4x^2 + 12 + 2x^5 - x^3 - 15x) \div (2 + x^3 - 3x)$

In Exercises 9–15, use synthetic division to perform the divisions.

9. $\dfrac{3x^2 - x - 10}{x - 2}$ 10. $(5x^2 + 12x - 9) \div (x + 3)$
11. $(4x^2 + 13x - 35) \div (x + 5)$
12. $(4x + 5x^3 - 4x^2 + 7 + 4x^4) \div (x + 2)$
13. $\dfrac{4x^3 - x^2 + \dfrac{3x}{2} + \dfrac{5}{2}}{x - \dfrac{1}{2}}$ 14. $\dfrac{x^5 + 1}{x + 1}$ 15. $\dfrac{x^7 - 1}{x - 1}$

4 Basic types of factoring

A prime number is a natural number greater than 1 which has no natural number factor other than itself and 1. The numbers 2, 3, and 5 are the prime factors of 30.

A composite number is a natural number greater than 1 that has at least one natural number factor other than itself and 1. The numbers 4, 6, 8, 9, and 10 are examples of composite numbers.

The numbers 2 and 3 are factors of 6 since $2 \times 3 = 6$. Also, 6 and 1 are factors of 6 since $6 \times 1 = 6$. However, 2 and 3 are called the **prime factors** of 6.

Note that the set of prime numbers and the set of composite numbers are disjoint sets.

The process of expressing a polynomial as a product is called **factoring the polynomial**. The product of $x(x+1)(x-1)$ is $x^3 - x$. Therefore, x, $(x+1)$, and $(x-1)$ are factors of $x^3 - x$. A **prime polynomial** is a polynomial which has no polynomial factors except itself and constants. The factors x, $(x+1)$, and $(x-1)$ are the **prime factors** of $x^3 - x$ because they cannot be factored into other polynomials. The product $(x+1)(x-1)$ yields $x^2 - 1$, which is a factor of $x^3 - x$; however $x^2 - 1$ is a **composite factor** of $x^3 - x$ because $x^2 - 1$ factors into $(x+1)(x-1)$.

Because factoring is the reverse of finding products, factoring will be easier for you if you have memorized the special products of the last section. We shall, however, cover other types of factoring not listed with the special products.

BASIC TYPES OF FACTORING

1. Common monomial factor	$ab + ac = a(b + c)$
2. Difference of two squares	$a^2 - b^2 = (a + b)(a - b)$
3. Perfect square trinominals	$a^2 + 2ab + b^2 = (a + b)^2$ $a^2 - 2ab + b^2 = (a - b)^2$
4. Trinomials	$acx^2 + (ad + bc)x + bd$ $= (ax + b)(cx + d)$
5. Sum or difference of two cubes	$a^3 + b^3 = (a + b)(a^2 - ab + b^2)$ $a^3 - b^3 = (a - b)(a^2 + ab + b^2)$
6. Factoring by grouping	$ac + bc + ad + bd$ $= c(a + b) + d(a + b)$ $= (a + b)(c + d)$

TYPE 1 Common monomial factor To factor this kind of expression, you must be able to find the greatest common divisor (GCD) of the terms. **The greatest common divisor** is the greatest factor that is a divisor

of each term. In many cases, the GCD can be determined by inspection. For more complicated problems, factor each term into prime factors, and then from these prime factors find the largest factor that is common to each term.

EXAMPLE 1 Factor $2a^2b^2 - 6ab^3 + 8ab^2c$.

SOLUTION Factor each term into prime factors.

$$2a^2b^2 - 6ab^3 + 8ab^2c = 2\ a\ a\ b^2 - 2 \cdot 3\, a\ \ b^2\, b + 2 \cdot 2^2\ a\ \ b^2\ c$$

$$\text{GCD} = 2ab^2$$

Next divide each term by the GCD to find the remaining factor.

$$\frac{2a^2b^2}{2ab^2} - \frac{6ab^3}{2ab^2} + \frac{8ab^2c}{2ab^2} = a - 3b + 4c$$

Therefore $2a^2b^2 - 6ab^3 + 8ab^2c = (2ab^2)(a - 3b + 4c)$.

EXAMPLE 2 Factor $3x^5 + 6x^3 + 9x^2$.

SOLUTION By inspection, we observe that the factor of highest degree that divides each term is $3x^2$.

$$3x^5 + 6x^3 + 9x^2 = 3x^2\left(\frac{3x^5}{3x^2} + \frac{6x^3}{3x^2} + \frac{9x^2}{3x^2}\right) = 3x^2(x^3 + 2x + 3)$$

TYPE 2 Difference of two squares The formula is $a^2 - b^2 = (a + b)(a - b)$. Both first and last terms are squares, separated by a minus sign.

EXAMPLE 3 Factor $9x^2 - 25y^2$.

SOLUTION $9x^2 - 25y^2 = [\sqrt{9x^2} + \sqrt{25y^2}][\sqrt{9x^2} - \sqrt{25y^2}]$

$$= (3x + 5y)(3x - 5y)$$

TYPE 3 Perfect square trinomials The formulas we use are

$$a^2 + 2ab + b^2 = (a + b)^2, \text{ or } (a + b)(a + b)$$

$$a^2 - 2ab + b^2 = (a - b)^2, \text{ or } (a - b)(a - b).$$

EXAMPLE 4 Factor $9x^2 + 12xy + 4y^2$.

SOLUTION $a = \sqrt{9x^2} = 3x$ and $b = \sqrt{4y^2} = 2y$. The middle term is $2ab = 2(3x)(2y) = 12xy$. Therefore, $9x^2 + 12xy + 4y^2 = (3x + 2y)^2$.

EXAMPLE 5 $9x^2 - 12xy + 4y^2 = (3x - 2y)^2$

TYPE 4 Trinomials If the leading coefficient of the trinomial is 1, we use the formula $x^2 + (a + b)x + ab = (x + a)(x + b)$.

EXAMPLE 6 Factor $x^2 + 7x + 12$.

SOLUTION We want to find a and b so that $x^2 + 7x + 12 = (x + a) \times (x + b) = x^2 + (a + b)x + ab$. The signs in the trinomial are both plus signs, so the signs in both factors must be plus signs. The product of a and b must be 12 and their sum must be 7.

$$\text{Factors of 12 are} \begin{cases} 2 \cdot 6 & \text{Sum} = 8 \\ 1 \cdot 12 & \text{Sum} = 13 \\ 3 \cdot 4 & \text{Sum} = 7 \end{cases}$$

Therefore, a is 3 and b is 4, and $x^2 + 7x + 12 = (x + 3)(x + 4)$.

EXAMPLE 7 Factor $x^2 - 7x + 12$.

SOLUTION Since the last sign is a plus, both signs in the binomial factors will be the same, and since the middle sign is minus, both signs of the binomial factors must be negative. We see that a and b must be factors of 12, and their sum must be -7. Therefore, a is -3 and b is -4, and $x^2 - 7x + 12 = (x - 3)(x - 4)$.

EXAMPLE 8 Factor $x^2 - x - 12$.

SOLUTION Since the last sign is a minus, the signs in the binomial factors must be opposite. We see that a and b must be factors of 12, and they must have a sum of -1. Therefore a is 3 and b is -4, and $x^2 - x - 12 = (x + 3)(x - 4)$.

If the leading coefficient is greater than one, we use the formula $acx^2 + (ad + bc)x + bd = (ax + b)(cx + d)$.

EXAMPLE 9 Factor $6x^2 + 19x + 15$.

SOLUTION The signs in the trinomial are both plus signs, so the signs in both binomial factors must be plus.

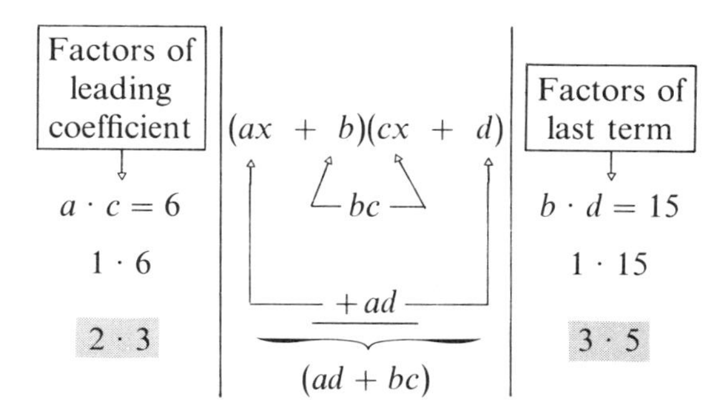

The sum of the inner and outer products must equal the middle term of the trinomial. By factoring the first coefficient and the last term of the trinomial we find all possible values of a, b, c, and d. Then by trial and error, we find the correct combination.

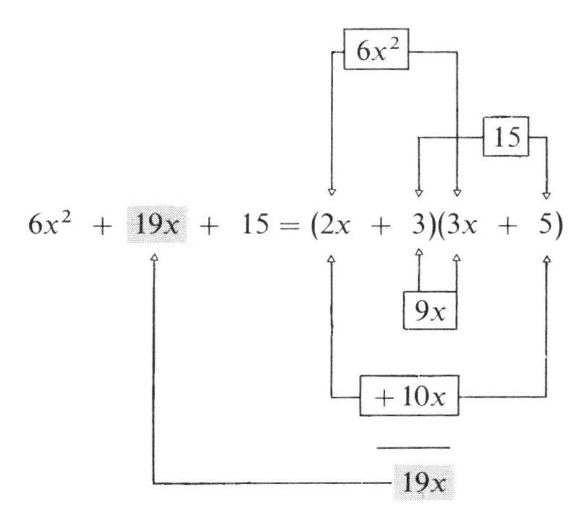

If one or more of the signs of the trinomial are negative, we use the same procedure, finding the correct combination by trial.

EXAMPLE 10 Factor $6x^2 - x - 15$.

SOLUTION Because of the negative sign before 15, b and d will have opposite signs. Make a blank outline; then fill in the numbers, by trial and error, that make the middle term $-x$. Then, by inspection, fill in the signs of the numbers in the factors.

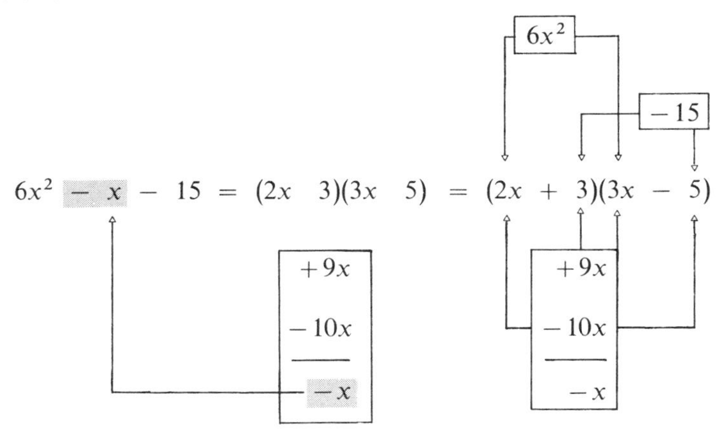

Therefore, $6x^2 - x - 15 = (2x + 3)(3x - 5)$.

EXAMPLE 11 Factor $2(x + y)^2 - 5(x + y) - 12$.

SOLUTION We first substitute another letter, u, for $(x + y)$ in each place where $(x + y)$ appears. We then factor the resulting expression and, finally, replace u by $(x + y)$.

$$\begin{aligned} &2(x+y)^2 - 5(x+y) - 12 \\ &\qquad = 2u^2 - 5u - 12 && \text{Substitute } u \text{ for } (x+y). \\ &\qquad = (2u+3)(u-4) && \text{Factor.} \\ &\qquad = [2(x+y)+3][(x+y)-4] && \text{Substitute } (x+y) \text{ for } u. \\ &\qquad = [2x+2y+3][x+y-4] \end{aligned}$$

Therefore, $2(x + y)^2 - 5(x + y) - 12 = (2x + 2y + 3)(x + y - 4)$.

TYPE 5 Sum or difference of two cubes We use the formulas

$$a^3 + b^3 = (a + b)(a^2 - ab + b^2)$$
$$a^3 - b^3 = (a - b)(a^2 + ab + b^2).$$

EXAMPLE 12 Factor $x^3 + 27$.

SOLUTION

$$\begin{aligned} x^3 + 27 &= (x)^3 + (3)^3 \\ &= [(x) + (3)][(x^2 - x(3) + (3)^2] \\ &= (x + 3)(x^2 - 3x + 9) \end{aligned}$$

EXAMPLE 13 Factor $8u^3 - 125v^3$.

SOLUTION

$$\begin{aligned} 8u^3 - 125v^3 &= (2u)^3 - (5v)^3 \\ &= [(2u) - (5v)][(2u)^2 + (2u)(5v) + (5v)^2] \\ &= (2u - 5v)(4u^2 + 10uv + 25v^2) \end{aligned}$$

TYPE 6 Factoring by grouping (usually with four or more terms) Many expressions that do not look like any of the types we have given can be reduced to one of these types by a suitable grouping of terms.

EXAMPLE 14 Factor $3ac - 2bd + 6bc - ad$.

SOLUTION We factor the terms as follows, using common monomial factors.

$$\begin{aligned} (6bc - 2bd) + (3ac - ad) &= 2b(3c - d) + a(3c - d) \\ &= (3c - d)(2b + a) \end{aligned}$$

Therefore, $3ac - 2bd + 6bc - ad = (3c - d)(2b + a)$.

EXAMPLE 15 Factor $x^2 + y^2 - 4 + 2xy$.

SOLUTION We rearrange the terms as follows. The first three terms form a perfect trinomial square. We then have the difference of two squares.

$$\begin{aligned}(x^2 + 2xy + y^2) - 4 \\ &= (x + y)^2 - 4 \\ &= [(x + y) + 2][(x + y) - 2] = (x + y + 2)(x + y - 2)\end{aligned}$$

Therefore, $x^2 + y^2 - 4 + 2xy = (x + y + 2)(x + y - 2)$.

Exercises 4

Factor each polynomial into prime factors.

A

1. $12y - 4$
2. $8m^2n^3 - 4m^3n^2 + 2m^2n^2$
3. $x^2 + 2x - 24$
4. $12a^2 - 2a - 30$
5. $16x^2y^2 - 9$
6. $25r^2 - 10r + 1$
7. $16 - 2n^3$
8. $6ac - 9bc + 12bd - 8ad$
9. $xy - x - y + 1$
10. $16a^2 - 25b^2c^2$
11. $9x^2 + 30xy + 25y^2$
12. $6(x + y)^2 + (x + y) - 12$
13. $9u^2 - 1$
14. $9a^2 + b^2 - 1 - 6ab$
15. $(4 - x)^3 - (4 + x)^3$

B

1. $10x + 5$
2. $3a^2b^3 + 6a^3b - 9ab$
3. $x^2 - x - 12$
4. $10u^2 + uv - 3v^2$
5. $64y^2 - 1$
6. $49z^2 - 14z + 1$
7. $32 + 4m^3$
8. $8ac - 3bd - 2ad + 12bc$
9. $xy + x - y - 1$
10. $9x^2y^2 - z^2$
11. $16x^2 + 8xy + y^2$
12. $48x^2 - 4x - 2$
13. $3(x - y)^2 + 7(x - y) + 2$
14. $4x^2 + 4x - 9z^2 + 1$
15. $(x + 2)^3 + (x - 2)^3$

5 Factoring by completing the square

When using the method of **completing the square**, we add a term to the given binomial or trinomial to make it a perfect square. We then subtract the same term. This leaves the value of the original expression unchanged. For this method to work, the first and last terms must be squared numbers of at least fourth degree. We will use examples to show the procedure.

EXAMPLE 1 Factor $4x^4 + 3x^2 + 1$.

SOLUTION By taking the square root of the first and last terms of the trinomial ($\sqrt{4x^4} = 2x^2$ and $\sqrt{1} = 1$), we have the binomial $2x^2 + 1$.

Squared, $(2x^2 + 1)^2 = 4x^4 + 4x^2 + 1$. This shows that x^2 must be added to the middle term of the original expression to make it a perfect square.

$$
\begin{array}{ll}
4x^4 + 3x^2 + 1 & \text{Add and subtract } x^2 \\
\quad\;\; + x^2 \qquad - x^2 & \text{(the value of } 4x^4 + 3x^2 + 1 \\
\hline
\text{Trinomial square} \rightarrow (4x^4 + 4x^2 + 1) - x^2 & \text{is unchanged).} \\
= (2x^2 + 1)^2 - x^2 & \text{Difference of two squares.} \\
= [(2x^2 + 1) + x][(2x^2 + 1) - x] & \text{Factor the difference of two squares.} \\
= (2x^2 + x + 1)(2x^2 - x + 1) & \text{Rearrange terms.}
\end{array}
$$

Therefore, $4x^4 + 3x^2 + 1 = (2x^2 + x + 1)(2x^2 - x + 1)$.

EXAMPLE 2 Factor $x^4 + 4$.

SOLUTION $\sqrt{x^4} = x^2$ and $\sqrt{4} = 2$; $(x^2 + 2)^2 = x^4 + 4x^2 + 4$. This shows that $4x^2$ must be added to and then subtracted from the original expression.

$$
\begin{array}{ll}
x^4 \qquad + 4 & \text{Add and subtract } 4x^2. \\
\quad + 4x^2 \qquad - 4x^2 & \\
\hline
\text{Trinomial square} \rightarrow (x^4 + 4x^2 + 4) - 4x^2 & \\
= (x^2 + 2)^2 - 4x^2 & \text{Difference of two squares.} \\
= [(x^2 + 2) + 2x][(x^2 + 2) - 2x] & \\
= (x^2 + 2x + 2)(x^2 - 2x + 2) & \text{Rearrange terms.}
\end{array}
$$

Therefore, $x^4 + 4 = (x^2 + 2x + 2)(x^2 - 2x + 2)$.

EXAMPLE 3 Factor $2x^8 - 6x^4 + 18$.

SOLUTION First factor out 2.

$$2x^8 - 6x^4 + 18 = 2(x^8 - 3x^4 + 9)$$

Next we complete the square to factor $x^8 - 3x^4 + 9$.

$$\sqrt{x^8} = x^4 \text{ and } \sqrt{9} = 3;\ (x^4 + 3)^2 = x^8 + 6x^4 + 9$$

When we add $9x^4$ to $-3x^4$, we get the required $6x^4$ needed in the perfect trinomial square.

$$x^8 - 3x^4 + 9 \qquad \text{Add and subtract } 9x^4.$$

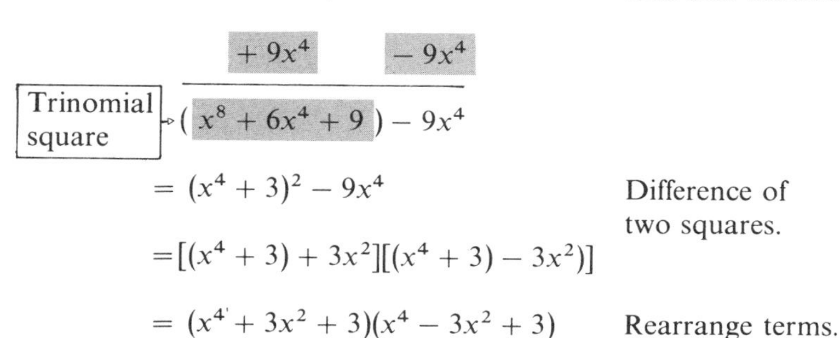

$$+ 9x^4 \qquad - 9x^4$$

Trinomial square → $(x^8 + 6x^4 + 9) - 9x^4$

$$= (x^4 + 3)^2 - 9x^4 \qquad \text{Difference of two squares.}$$

$$= [(x^4 + 3) + 3x^2][(x^4 + 3) - 3x^2)]$$

$$= (x^4 + 3x^2 + 3)(x^4 - 3x^2 + 3) \qquad \text{Rearrange terms.}$$

Therefore, $2x^8 - 6x^4 + 18 = 2(x^4 + 3x^2 + 3)(x^4 - 3x^2 + 3)$.

Exercises 5

Factor each polynomial into prime factors by completing the square.

A

1. $x^4 + 9x^2 + 25$ **2.** $9x^4 + 5x^2 + 1$
3. $m^4 + m^2 + 25$ **4.** $9r^4 - 3r^2 + 1$
5. $y^4 + 64z^4$ **6.** $16w^4 - 17w^2z^2 + z^4$
7. $8m^4n + 2n^5$ **8.** $50x^4y + 32x^2y^3 + 8y^5$
9. $4x^4 - 4x^2y^2 + 64y^4$

B

1. $x^4 + 4x^2 + 16$ **2.** $4x^4 + 11x^2 + 9$
3. $25a^4 + a^2 + 1$ **4.** $16b^4 - 17b^2 + 1$
5. $4m^4 + n^4$ **6.** $x^4 - 3x^2y^2 + 9y^4$
7. $5xy^4 + 20x$ **8.** $4x^5y + 11x^3y^3 + 9xy^5$
9. $2u^4 - 6u^2v^2 + 18v^4$

6 Factoring by using synthetic division

THEOREM When the remainder in any division problem is zero, the divisor and quotient are factors of the dividend.

PROOF

Dividend = (Divisor) · (Quotient) + Remainder

Dividend = (Divisor) · (Quotient) + 0

Dividend = (Divisor) · (Quotient)

Factors of dividend

EXAMPLE 1 Consider the product $(x-2)(3x+4) = 3x^2 - 2x - 8$. The factors of $3x^2 - 2x - 8$ are $x - 2$ and $3x + 4$. When $3x^2 - 2x - 8$ is divided by either factor, the other factor is the quotient. We use synthetic division to make this division.

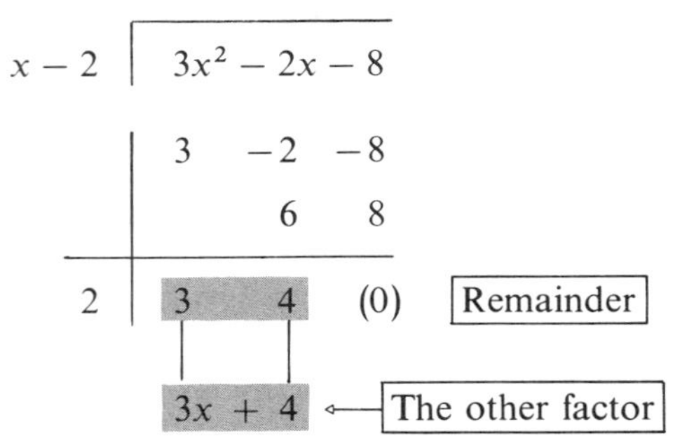

Notice that the remainder is zero.

EXAMPLE 2 Consider the following product.

$$(x-1)(x+2)(x+5) = x^3 + 6x^2 + 3x - 10$$

The factors of $x^3 + 6x^2 + 3x - 10$ are $x - 1$, $x + 2$, and $x + 5$. We use synthetic division as a check to show that $x + 2$ and $x - 1$ are factors of the product $x^3 + 6x^2 + 3x - 10$.

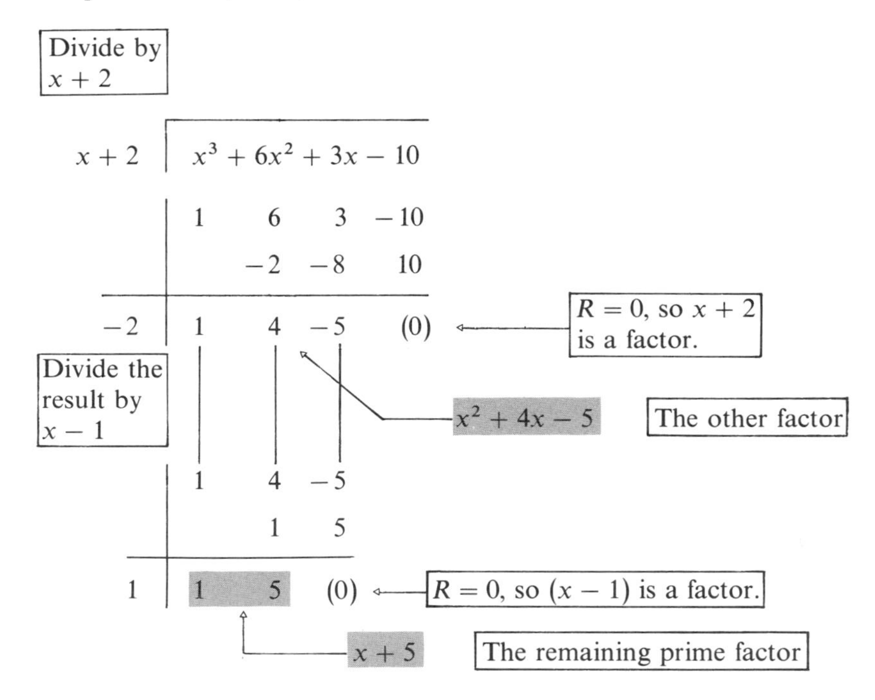

Therefore, $x^3 + 6x^2 + 3x - 10 = (x + 2)(x - 1)(x + 5)$.

In Example 2, we knew the factors and demonstrated how synthetic division could be used to check the factoring. Notice that in multiplying

$$x - 1,\ x + 2, \text{ and } x + 5 \text{ to get } x^3 + 6x^2 + 3x - 10$$

we multiplied

$$-1,\ 2, \text{ and } 5 \text{ to get } -10.$$

The numbers -1, 2, and 5, which appeared in the binomial factors, are factors of 10.

> When the leading coefficient of the polynomial to be factored is 1, the constants in the binomial factors of the polynomial are factors of the constant term of the polynomial.

In a more general way, the product $(x - r_1)(x - r_2)(x - r_3)$ gives a last term of $-r_1r_2r_3$. This means r_1, r_2, and r_3 are factors of the last term of the polynomial.

When an nth-degree polynomial $P(x)$ is divided by $x - r$ and a remainder of zero is obtained, the resulting quotient is a polynomial of $(n - 1)$ degree. This quotient is often called the **depressed polynomial**.

EXAMPLE 3 Factor $x^4 - 4x^3 - 7x^2 + 34x - 24$.

SOLUTION The trial factors are found from the factors of 24. The factors of 24 are: ± 1, ± 2, ± 3, ± 4, ± 6, ± 8, ± 12, ± 24.

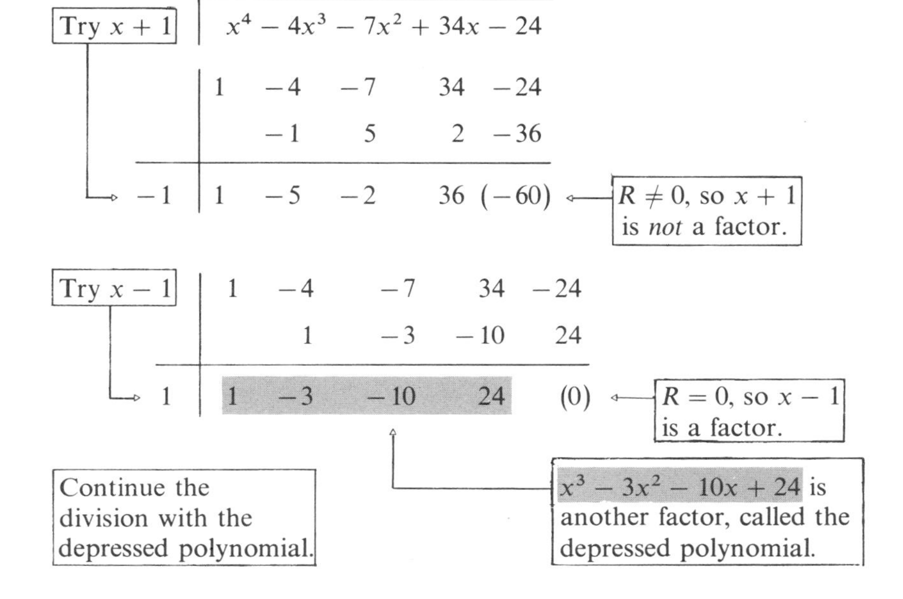

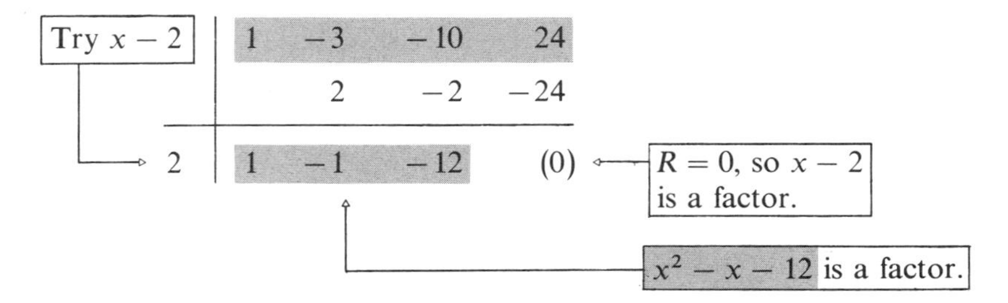

We factor this last trinomial by inspection.

$$x^2 - x - 12 = (x - 4)(x + 3)$$

Therefore, $x^4 - 4x^3 - 7x^2 + 34x - 24 = (x - 1)(x - 2)(x - 4)(x + 3)$.

The student should keep in mind that the depressed polynomial may be factored by another method, as was done in Example 3.

The above work can be shortened as follows. In this form, we do the multiplication and additions in our head.

$$x^4 - 4x^3 - 7x^2 + 34x - 24$$

	1	−4	−7	34	−24
1	1	−3	−10	24	(0)
2	1	−1	−12	(0)	
4	1	3	(0)		

$(x - 1)\ (x - 2)\ (x - 4)\ (x + 3)$ — Factors

At this time we will use synthetic division only as an aid in factoring polynomials having a leading coefficient of one.

Exercises 6

Use synthetic division as an aid in factoring the following expressions.

A

1. $x^3 - 2x^2 - 5x + 6$
2. $x^3 - 7x^2 + 14x - 8$
3. $x^3 - 7x - 6$
4. $x^3 + x^2 - 4x - 4$
5. $x^4 - 2x^3 - 9x^2 + 2x + 8$
6. $x^4 + 4x^3 - 2x^2 - 12x + 9$
7. $3x^4 - 12x^3 + 3x^2 + 18x$
8. $x^5 + 3x^4 - 5x^3 - 15x^2 + 4x + 12$

B

1. $x^3 - 4x^2 + x + 6$
2. $x^3 - 6x^2 + 11x - 6$
3. $x^3 - 7x + 6$
4. $x^3 + 4x^2 - 9$
5. $x^4 + 4x^3 + 3x^2 - 4x - 4$
6. $x^4 - 2x^3 - 3x^2 + 4x + 4$
7. $2x^4 - 4x^3 - 10x^2 + 12x$
8. $x^5 + 5x^4 - 5x^3 - 25x^2 + 4x + 20$

Summary

An **algebraic expression** is any expression of variables and real number constants used in a number of additions, subtractions, multiplications, divisions, raising to integral or fractional powers, or taking of roots.

A **monomial** is an algebraic expression consisting of a single term that is the product of real numbers and nonnegative integral powers of one or more variables. The **degree** of a nonzero monomial is the sum of the exponents of its variables.

A **polynomial** is any sum of monomials. The **degree of a polynomial** in simplified form is the same as that of its highest-degree term. A polynomial with two terms is called a **binomial**. A polynomial with three terms is called a **trinomial**.

The standard **division formula** is

$$\text{Dividend} = \text{Divisor} \times \text{Quotient} + \text{Remainder}.$$

Synthetic division is a shortened method of dividing a polynomial in one variable, such as x, by $x - r$, where r is a positive or negative constant. See Section 3.

Methods of Factoring

1. Common monomial	$ab + ac = a(b + c)$
2. Difference of two squares	$a^2 - b^2 = (a + b)(a - b)$
3. Sum of two cubes	$a^3 + b^3 = (a + b)(a^2 - ab + b^2)$
4. Difference of two cubes	$a^3 - b^3 = (a - b)(a^2 + ab + b^2)$
5. Trinomial square	$\begin{cases} a^2 + 2ab + b^2 = (a + b)^2 \\ a^2 - 2ab + b^2 = (a - b)^2 \end{cases}$
6. General trinomial	$acx^2 + (ad + bc)x + bd = (ax + b)(cx + d)$
7. Completing the square	See Section 5.
8. Grouping	See Section 4.
9. Synthetic division	See Section 6.

Review exercises

A

1. Given $f(x) = x^3 - 3x^2 - 5$, find each of the following.
 a. $f(-1)$ **b.** $f(2)$ **c.** $f(0)$ **d.** $f(a - 1)$

2. Given $f(x, y) = x^2 - 2xy + 3$, find each of the following.
 a. $f(-1, 2)$ **b.** $f(0, 0)$

In Exercises 3–8, perform the indicated operations and simplify.

3. $(4x + 7)(2x - 5)$

4. $(4x^2 - 3y^2)(4x^2 + 3y^2)$

5. $(12a^3b - 8a^2b^2 - 4a^2b) \div 4a^2b$

6. $(3a - 4b)(9a^2 + 12ab + 16b^2)$
7. $(14x^2 + 8x^3 + 25 - 5x) \div (4x^2 - 3x + 5)$
8. Subtract $(b - 8ab^2 - 3a + 7)$ from the sum of $(4a^2b - ab^2 + 10)$ and $(7a - 4 + 5ab^2 + b)$.
9. Use synthetic division to find the quotient and remainder for $(x^3 + 6x^2 + 7x + 10) \div (x + 4)$.

In Exercises 10–15, factor each polynomial into prime factors.

10. $x^4 - 16$
11. $1 - 27a^3$
12. $6x^2 + 7x - 5$
13. $4m^2 + 4mn + n^2 - 2m - n$
14. $9a^4 + 2a^2b^2 + b^4$
15. $x^4 - x^3 - 7x^2 + x + 6$

B

1. Given $P(x) = 2x^3 - x^2 + 4$, find each of the following.
   a. $P(2)$ **b.** $P(-1)$ **c.** $P(0)$ **d.** $P(a + 1)$
2. Given $f(x, y) = x^2 - y^2$, find each of the following.
   a. $f(1, -2)$ **b.** $f(0, -1)$

In Exercises 3–8, perform the indicated operations and simplify.

3. $(3x - 1)(5x + 2)$
4. $(3x^2 + 1)(3x^2 - 1)$
5. $(2x + 3)(4x^2 - 6x + 9)$
6. $(6x^2y^3 - 3xy^2 - 12x^3y^4) \div 3xy$
7. $(x^2 + 6x^3 - 2 - 5x) \div (3x^2 - x - 2)$
8. Subtract $(4x^3 - 2x^2y + 3x - 4)$ from the sum of $(5x^2y + 7x - 3x^3)$ and $(8 - 6x^3 + 4x^2y - 5x)$.
9. Use synthetic division to find the quotient and remainder for $(x^3 + 4x^2 + 6x - 7) \div (x + 3)$.

In Exercises 10–15, factor each polynomial into prime factors.

10. $18x^2 - 50$
11. $8a^3 + 1$
12. $10x^2 + x - 21$
13. $4a^2 - b^2 - 4a + 1$
14. $4x^4 + 3x^2y^2 + 9y^4$
15. $x^5 - x^4 - 9x^3 + x^2 + 20x + 12$

Diagnostic test

The purpose of this test is to see how well you understand the work in this chapter. Allow yourself approximately 50 minutes to do the test. Solutions to the problems, together with section references, are given in the Answers section at the end of the book. We suggest that you study the sections referred to for the problems you do incorrectly.

1. Which of the following expressions are polynomials?

$$3x^2 + 2x, \quad 5xy^{-1}, \quad x + \frac{1}{x}, \quad 5x^3y - 2xy^2 + y^3$$

2. Given $f(x) = x^2 - x + 1$, find each of the following.
a. $f(-2)$ **b.** $f(x + h)$

3. Is the expression

$$\frac{x^2 + y^2}{x + y}$$

an algebraic expression? Give a reason for your answer.

Perform the indicated operations and simplify.

4. $(3x^2 - \frac{1}{2})(3x^2 + \frac{1}{2})$

5. $(7x - 2y)(5x + 3y)$

6. $(3a - 4b)(9a^2 + 12ab + 16b^2)$

7. $\dfrac{4 - 12x}{4}$

8. $(12x^3y^2 + 8x^2y^3 - 4xy) \div 4x^2y^2$

9. Subtract $(3u^2v - 5uv^2 + 7)$ from $(10uv^2 + 8u^2v - 3)$.

10. $(10x^4 - 2x^3 - 25x^2 - 15x + 4) \div (5x + 4 - 2x^3)$

Factor each expression into prime factors.

11. $3x^3 - 12x$

12. $15x^2 - 4x - 4$

13. $2ax - 2bx + ay - by$

14. $x^4 + 5x^3 - 20x - 16$

15. $8a^3 + 27b^3$

16. $4x^4 + 11x^2y^2 + 9y^4$

3 Fractions

In this chapter, we review arithmetic fractions, define a **rational fraction** (often called an **algebraic fraction**), and discuss the operations on fractions.

1 Definitions

A **fraction** is an indicated quotient of two quantities. The dividend is called the **numerator** and the divisor is called the **denominator**. A **simple fraction** such as $\frac{3}{4}$ is a fraction whose numerator (3) and denominator (4) are both integers. In contrast, a **complex fraction** such as

$$\frac{\frac{5}{6}}{7}, \quad \frac{3}{\frac{1}{2}}, \quad \frac{\frac{3}{5}}{\frac{2}{7}}$$

has a fractional numerator, denominator, or both. If the numerator of a simple fraction is 1, it is a **unit fraction**. Two simple fractions are **similar** if they have the same denominators. A fraction whose numerator and denominator are rational numbers is a **rational fraction**. A **proper fraction** is a fraction in which the numerator is less in absolute value than the denominator. An **improper fraction** is a fraction in which the numerator is greater than (or equal to) the denominator in absolute value. Thus $\frac{2}{3}$ is a proper fraction, and $\frac{5}{4}$ and $\frac{6}{6}$ are improper fractions.

> FORM OF A RATIONAL FRACTION
>
> $$\frac{P(x)}{D(x)}, \quad D(x) \neq 0,$$
>
> where $P(x)$ and $D(x)$ are polynomials.

Any polynomial can be considered to be a rational fraction, because it is the quotient of itself and 1 (see Example 1).

In this chapter, we will assume that any values of the variables that make the denominators zero are excluded.

EXAMPLE 1 Examples of rational fractions.

a. $\dfrac{x}{x+y}$ **b.** $\dfrac{x^2+xy+y^2}{x-y}$ **c.** $\dfrac{1}{x+z}$

d. $xz+5$, which equals $\dfrac{xz+5}{1}$ **e.** 3, which equals $\dfrac{3x^0}{1}$

f. $\dfrac{x^2+xy}{x^3-x^2+1}$

A rational fraction is a **proper fraction** if the numerator is of lower degree than the denominator and an **improper fraction** if the degree of the numerator is greater than or equal to the degree of the denominator.

EXAMPLE 2 Examples of proper rational fractions.

a. $\frac{x}{x^2 + 1}$ **b.** $\frac{1}{x}$ **c.** $\frac{1{,}000x^2 + x}{x^3 + x^2 + 100}$ **d.** $\frac{x^2}{x^3 y}$

EXAMPLE 3 Examples of improper rational fractions.

a. $\frac{x^2 + 1}{x}$ **b.** $\frac{x - 1}{x + 1}$ **c.** $\frac{7}{7}$ **d.** $xy + 3$

Equivalent fractions are fractions having the same quotient and remainder. For example, $\frac{2}{3}$ is equivalent to $\frac{4}{6}$ and to $\frac{6}{9}$.

In arithmetic, a number that is a product of integers is a **multiple** of each of its factors. Thus 12 is a multiple of 1, 2, 3, 4, 6, and 12. In general, an arithmetic or algebraic product is said to be a **multiple of each of its factors**.

A relation of multiples is shown in Example 4.

EXAMPLE 4

a. 2, 4, 6, 8, 10, 12, 14, ... are multiples of 2.

b. 3, 6, 9, 12, 15, ... are multiples of 3.

Notice that both 6 and 12 are common multiples of both 2 and 3.

c. $x + 1$, $x^2 - 1$, $(x + 1)(x^2 - 1)$, ... are multiples of $x + 1$.

d. $x - 1$, $x^2 - 1$, $(x - 1)(x^2 - 1)$, ... are multiples of $x - 1$.

Notice that $x^2 - 1$ is a common multiple of both $x + 1$ and $x - 1$.

The **least common multiple (LCM)** of two or more different expressions is the expression with the fewest factors that is exactly divisible by each of the given expressions. For example, the LCM of 2, 3, 4, and 6 is 12; 12 is the smallest number exactly divisible by 2, 3, 4 and 6. **The LCM of a set of algebraic expressions is the product of the distinct prime factors, each taken the greatest number of times it occurs in any one of the expressions.**

EXAMPLE 5 Find the LCM of $x^2 - 1$, $x^2 + 2x + 1$, and $6x + 6$.

STEP 1 Factor each expression into its prime factors.

$$x^2 - 1 = (x + 1)(x - 1)$$
$$x^2 + 2x + 1 = (x + 1)(x + 1)$$
$$6x + 6 = 2 \cdot 3(x + 1)$$

STEP 2 The LCM is the product of the distinct factors, each taken the greatest number of times it occurs in any one expression.

$$\text{LCM} = 2 \cdot 3(x - 1)(x + 1)(x + 1)$$
$$= 6(x - 1)(x + 1)^2$$

SIMPLIFYING FRACTIONS

The value of a fraction is not changed if both numerator and denominator are multiplied or divided by the same nonzero expression.

A fraction is **simplified**, or **reduced to lowest terms**, if all of the factors common to both numerator and denominator have been removed by dividing both numerator and denominator by those common factors.

EXAMPLE 6 Examples of fractions reduced to lowest terms. Each numerator and denominator is written as a product of prime factors, then divided by the common factors.

a. $\dfrac{105}{120} = \dfrac{\cancel{3} \cdot \cancel{5} \cdot 7}{\cancel{3} \cdot \cancel{5} \cdot 2^3} = \dfrac{7}{8}$

b. $\dfrac{x^2 - x - 6}{x^2 - 4} = \dfrac{\cancel{(x+2)}(x - 3)}{\cancel{(x+2)}(x - 2)} = \dfrac{x - 3}{x - 2}$

THE THREE SIGNS OF A FRACTION

Every fraction has three signs associated with it, the sign of the numerator, the sign of the denominator, and the sign of the fraction. If the signs of any two of these parts, or of any even number of factors, are changed in the fraction, the sign of the fraction remains unchanged.

EXAMPLE 7 Examples of changing signs in fractions.

a. $-\dfrac{4}{-5} = -\dfrac{-4}{5} = \dfrac{4}{5}$

b. $-\dfrac{x - 1}{1 - x} = -\dfrac{\cancel{(x - 1)}}{(-1)\cancel{(x - 1)}} = -\dfrac{1}{-1} = 1$

(*Note*: $1 - x = (1)(1 - x) = (-1)(x - 1)$.
That is, signs of both factors were changed.)

c. $\dfrac{(x+2)(3-x)}{x-3} = \dfrac{(x+2)(-1)(x-3)}{x-3} = -x-2$

d. $-\dfrac{(3)(-4)}{(-2)(-6)} = \dfrac{(-3)(-4)}{(-2)(-6)} = \dfrac{12}{12} = 1$

Exercises 1

A

1. State which of the following fractions are proper rational fractions and which are improper rational fractions.

a. $\dfrac{10}{10}$ **b.** $\dfrac{7}{5}$ **c.** $\dfrac{5}{5}$ **d.** $\dfrac{x^2}{x^2+1}$

e. $\dfrac{x^3-1}{x^4+1}$ **f.** $x+1$ **g.** $\dfrac{x^4}{x^5-1}$

h. $\dfrac{3x^2-9}{x^3-7x^2+8x+10}$

In Exercises 2–4, find the LCM for each group of numbers or expressions.

2. 4, 6, 10, 12 **3.** $3x^2, x^2-9, x^2-4x+3$

4. $2x-2, x^2-2x+1, x^2-1$

In Exercises 5–10, reduce each fraction to lowest terms.

5. $-\dfrac{-6}{18}$ **6.** $\dfrac{5-x}{x-5}$

7. $\dfrac{x^2-4}{x^3-8}$ **8.** $\dfrac{6x^2+13x-5}{4x^2+16x+15}$

9. $\dfrac{2x^3+5x^2+x-2}{2x^3-5x^2-4x+3}$ **10.** $\dfrac{x^3-6x^2+11x-6}{x^3-x^2-4x+4}$

B

1. State which of the following fractions are proper rational fractions and which are improper rational fractions.

a. $\dfrac{3}{4}$ **b.** $\dfrac{6}{5}$ **c.** $\dfrac{3}{3}$ **d.** $\dfrac{10}{x}$

e. $\dfrac{x^2+1}{x}$ **f.** $x-10$ **g.** $\dfrac{1{,}000x}{x-1}$

h. $\dfrac{x^3+y^3}{10x^2+8x+3}$

In Exercises 2–4, find the LCM for each group of numbers or expressions.

2. 6, 9, 12, 15 **3.** $2x^3, x^2-1, x^2-2x+1$

4. $3x+6, x^2+4x+4, x^2-4$

In Exercises 5–10, reduce each fraction to lowest terms.

5. $-\dfrac{-8}{12}$ **6.** $\dfrac{x-3}{3-x}$

7. $\dfrac{x^2-1}{x^3-1}$ **8.** $\dfrac{6x^2-5x-4}{9x^2-6x-8}$

9. $\dfrac{2x^3+x^2-10x}{x^3-7x^2+10x}$ **10.** $\dfrac{x^3+x^2-x-1}{x^3-3x^2-x+3}$

2 Multiplication and division of fractions

> **MULTIPLYING FRACTIONS**
>
> 1. Factor the numerators and denominators of the fractions.
> 2. Divide out factors that are common to both numerators and denominators.
> 3. The answer is the product of the factors remaining in the numerator divided by the product of the factors remaining in the denominator.

Multiplication of fractions is often made easier if you first factor the numerators and denominators of the fractions, if possible, then reduce in the multiplication process. We illustrate with examples.

EXAMPLE 1 Multiply the fractions.

a. $5 \cdot \dfrac{3}{10} = \dfrac{\cancel{5}}{1} \cdot \dfrac{3}{2 \cdot \cancel{5}} = \dfrac{3}{2}$

b.
$$\frac{x}{x^2-1} \cdot (x-1) = \frac{x}{(x+1)(x-1)} \cdot \frac{x-1}{1}$$
$$= \frac{x\cancel{(x-1)}}{(x+1)\cancel{(x-1)}}$$
$$= \frac{x}{x+1}$$

c. $\dfrac{6ab}{c} \cdot \dfrac{c^3}{3a^2} = \dfrac{6abc^3}{3a^2c} = \dfrac{2 \cdot \cancel{3}\cancel{a}b\cancel{c}c^2}{\cancel{3}\cancel{a}a\cancel{c}} = \dfrac{2bc^2}{a}$

d.
$$\frac{x^3+y^3}{x^2-y^2} \cdot \frac{x^2-2xy+y^2}{x^2-xy+y^2}$$
$$= \frac{\cancel{(x+y)}\cancel{(x^2-xy+y^2)}}{\cancel{(x+y)}\cancel{(x-y)}} \cdot \frac{\cancel{(x-y)}(x-y)}{\cancel{x^2-xy+y^2}}$$
$$= x-y$$

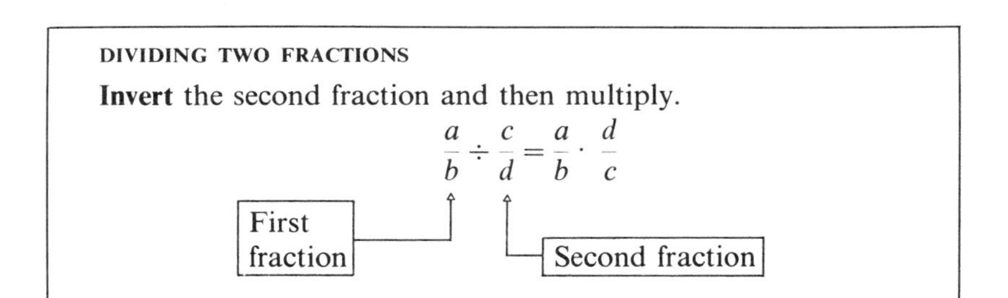

We demonstrate why, in the division of fractions, we **invert** the second fraction.

PROOF

The value of this fraction is 1.

$$\frac{a}{b} \div \frac{c}{d} = \frac{\frac{a}{b}}{\frac{c}{d}} = \frac{\frac{a}{b}}{\frac{c}{d}} \cdot \frac{\frac{d}{c}}{\frac{d}{c}} = \frac{\frac{a}{b} \cdot \frac{d}{c}}{\frac{\cancel{c}}{\cancel{d}} \cdot \frac{\cancel{d}}{\cancel{c}}} = \frac{\frac{a}{b} \cdot \frac{d}{c}}{1} = \frac{a}{b} \cdot \frac{d}{c}$$

Thus $\boxed{\frac{a}{b} \div \frac{c}{d}} = \boxed{\frac{a}{b} \cdot \frac{d}{c}}$.

EXAMPLE 2 Quotients of fractions.

a. $\frac{3}{4} \div 6 = \frac{3}{4} \div \frac{6}{1} = \frac{\overset{1}{\cancel{3}}}{4} \cdot \frac{1}{\underset{2}{\cancel{6}}} = \frac{1}{8}$

b. $(2x - 1) \div \frac{4x - 2}{x} = \frac{\cancel{(2x - 1)}x}{(1)(2)\cancel{(2x - 1)}} = \frac{x}{2}$

c. $\frac{2x^2 + x - 6}{2x^2 + 7x + 5} \div \frac{x^2 - 2x - 8}{2x^2 - 3x - 20}$

$$= \frac{2x^2 + x - 6}{2x^2 + 7x + 5} \cdot \frac{2x^2 - 3x - 20}{x^2 - 2x - 8}$$

$$= \frac{(2x - 3)\cancel{(x + 2)}\cancel{(x - 4)}\cancel{(2x + 5)}}{\cancel{(2x + 5)}(x + 1)\cancel{(x + 2)}\cancel{(x - 4)}}$$

$$= \frac{2x - 3}{x + 1}$$

Exercises 2

Perform the indicated operations.

A

1. $10 \div \frac{5}{3}$
2. $(4x - 2)\left(\frac{x}{2}\right)$
3. $\frac{8a^2b}{3c} \cdot \frac{9c^2}{4ab^2}$
4. $\frac{4x + 2}{2x - 10}(3x - 15)$
5. $\frac{3x - 6y}{x + 5y} \div \frac{4x^2 - 16y^2}{3x^2 - 75y^2}$
6. $\frac{3x^2 - 3xy}{x - 5y} \div \frac{2x^2 - 2xy}{x^2 - 6xy + 5y^2}$
7. $\frac{14x - 4}{25x - 5} \cdot \frac{5x^2 + 9x - 2}{7x^2 + 12x - 4}$
8. $\frac{2x^2 - 2x - 4}{x^2 + xy + y^2} \cdot \frac{x^4 + x^2y^2 + y^4}{x^2 - 4x + 4} \div \frac{2x^2 - 2xy + 2y^2}{x^2 + 3x - 10}$
9. $\frac{7a^4b^2}{8a^3b^3} \div \frac{7ab}{16a^3b^5} \div \frac{2a^5b^2}{3a^2b}$
10. $\frac{6x^2 + 13x - 5}{x^3 - 8} \div \frac{3x^2 + 5x - 2}{x^2 + 2x + 4} \cdot \frac{x^2 - 4}{2x^2 + 7x + 5}$

B

1. $\frac{3}{8} \div 4$
2. $\frac{x}{3}(6x + 3)$
3. $\frac{10uv^2}{3xy^2} \cdot \frac{6x^2y^2}{5u^2v^2}$
4. $\frac{6x - 14}{4x + 8}(x^2 + 5x + 6)$
5. $\frac{4x + 8y}{x + 4y} \div \frac{3x^2 - 12y^2}{2x^2 - 32y^2}$
6. $\frac{2x + y}{x^2 + 2xy} \div \frac{2x^2 - xy + y^2}{x^2 + xy - 2y^2}$
7. $\frac{4x^2 + x - 3}{5x + 5} \cdot \frac{x^2 - x - 6}{4x^2 + 5x - 6}$
8. $\frac{12x^3 - 12xy^2}{x - y + 1} \cdot \frac{x^2 - y^2 + x + y}{x^2 - y^2} \div \frac{4x}{9y^3}$
9. $\frac{9x^5y^2}{5x^2y} \div \frac{3x^3y}{(-20xy^2)} \div \frac{3xy}{2}$
10. $\frac{x^3 - 1}{x - 1} \div \frac{x^2 + x + 1}{x^2 - 1} \cdot \frac{2x^2 + 3x + 1}{2x^2 - x - 1}$

3 Adding and subtracting fractions

> **The sum or difference** of two or more fractions having the same denominator is a fraction with the same denominator as the given fractions, and whose numerator is the sum or difference of the numerators of the given fractions.

After performing an operation, the answer often can, and must, be reduced.

EXAMPLE 1 Find the sum of

$$\frac{2x+y}{x+y} - \frac{x-2y}{x+y} + \frac{x}{x+y}.$$

SOLUTION

$$\frac{2x+y}{x+y} - \frac{x-2y}{x+y} + \frac{x}{x+y} = \frac{(2x+y)-(x-2y)+(x)}{x+y}$$

$$= \frac{2x+y-x+2y+x}{x+y} = \frac{2x+3y}{x+y}$$

Cannot be reduced.

Before we can add or subtract fractions that have different denominators, we must replace them by equivalent fractions with equal denominators. For example, to add $\frac{2}{3}$ and $\frac{1}{2}$, we change each fraction to sixths. Six is called the **lowest common denominator (LCD)** of the fractions $\frac{2}{3}$ and $\frac{1}{2}$.

> **The lowest common denominator (LCD)** is the expression with the fewest factors that is exactly divisible by each of the denominators of a given set of fractions. It is the **least common multiple (LCM)** of the denominators of a set of fractions. (For a discussion of LCM, see Section 1.)

We find the LCD in the same way we found the LCM in Section 1 of this chapter.

> TO ADD UNLIKE FRACTIONS
>
> **1.** Find the LCD.
> **2.** Convert all fractions to equivalent fractions having the LCD as denominator.
> **3.** Add the similar fractions.
> **4.** Reduce the resulting fraction to lowest terms.

EXAMPLE 2 Add

$$\frac{1}{24} + \frac{4}{9}.$$

SOLUTION

$$\left.\begin{aligned} 24 &= 2^3 \cdot 3 \\ 9 &= \phantom{2^3 \cdot{}} 3^2 \end{aligned}\right\} \text{LCD} = 2^3 \cdot 3^2.$$

Next we change each fraction to an equivalent fraction having the LCD $2^3 \cdot 3^2$.

$$\frac{1}{24} = \frac{1}{2^3 \cdot 3} = \frac{1}{2^3 \cdot 3} \cdot \frac{3}{3} = \frac{3}{2^3 \cdot 3^2} = \frac{3}{72}$$

$$\frac{4}{9} = \frac{4}{3^2} = \frac{4}{3^2} \cdot \frac{2^3}{2^3} = \frac{32}{2^3 \cdot 3^2} = \frac{32}{72}$$

Therefore,

$$\frac{1}{24} + \frac{4}{9} = \frac{3}{72} + \frac{32}{72} = \frac{35}{72}.$$

EXAMPLE 3 Perform the indicated operation.

$$\frac{y}{x-y} - \frac{x}{y-x}$$

SOLUTION

$$\frac{y}{x-y} - \frac{x}{y-x} = \frac{y}{x-y} + \frac{x}{x-y}$$ Change sign of second fraction and its denominator.

$$= \frac{y+x}{x-y} \quad \text{or} \quad \frac{x+y}{x-y}$$

EXAMPLE 4 Simplify

$$\frac{1}{x^2-1} + \frac{2}{3x+3} - \frac{3}{x^2+2x+1}.$$

STEP 1 Factor the denominators and use the factors to determine the LCD.

$$x^2 - 1 = (x+1)(x-1)$$
$$3x + 3 = 3(x+1)$$
$$x^2 + 2x + 1 = (x+1)^2$$

STEP 2 The LCD is the product of all distinct prime factors, each taken the greatest number of times it occurs in any one denominator.

$$\text{LCD} = 3(x-1)(x+1)^2$$

STEP 3 Convert each fraction to an equivalent fraction with the LCD.

$$\frac{1}{(x^2-1)}\cdot\frac{3(x+1)}{3(x+1)}+\frac{2}{3(x+1)}\cdot\frac{(x+1)(x-1)}{(x+1)(x-1)}-\frac{3}{(x+1)^2}\cdot\frac{(3)(x-1)}{(3)(x-1)}$$

Multiply fractions by appropriate factors to give each the same denominator.

STEP 4 Write the sum of the numerators over the common denominator.

$$\frac{(1)(3)(x+1)+2(x+1)(x-1)-3(3)(x-1)}{3(x-1)(x+1)^2}$$
$$=\frac{3x+3+2x^2-2-9x+9}{3(x-1)(x+1)^2}$$
$$=\frac{2x^2-6x+10}{3(x-1)(x+1)^2}$$

Always try to factor to see if the fraction can be reduced.

Therefore,

$$\frac{1}{x^2-1}+\frac{2}{3x+3}-\frac{3}{x^2+2x+1}=\frac{2x^2-6x+10}{3(x-1)(x+1)^2}$$
$$=\frac{2(x^2-3x+5)}{3(x-1)(x+1)^2}$$

A WORD OF CAUTION The sum in Example 4 is *not* $2x^2 - 6x + 10$. The denominator must *not* be discarded. Students often confuse a sum of this type with a fractional equation. In solving fractional equations, we multiply each term by the LCD of the terms, which does eliminate the denominators. The denominators are *not* eliminated in the addition of fractions.

A **mixed expression**, such as $a + b/c$, may be added by first writing the quantity a as $a/1$ and adding the fractions according to the given rule.

EXAMPLE 5 Perform the indicated operations:

$$x+y+\frac{x^2-xy}{x-y}.$$

SOLUTION In this problem the last fraction reduces:

$$\frac{x^2 - xy}{x - y} = \frac{x\cancel{(x - y)}}{\cancel{x - y}} = x.$$

Therefore,

$$x + y + \frac{x^2 - xy}{x - y} = x + y + x = 2x + y.$$

In case we did not reduce the last fraction and added in the usual way, the same answer would be obtained. We demonstrate as follows:

STEP 1 $x + y + \dfrac{x^2 - xy}{x - y} = \dfrac{x + y}{1} + \dfrac{x^2 - xy}{x - y}$

STEP 2 $\text{LCD} = x - y$

STEP 3 $\dfrac{(x + y)}{1} \boxed{\dfrac{(x - y)}{(x - y)}} + \dfrac{x^2 - xy}{x - y}$

$$= \frac{(x + y)(x - y) + (x^2 - xy)}{x - y}$$

$$= \frac{x^2 - y^2 + x^2 - xy}{x - y}$$

$$= \frac{2x^2 - xy - y^2}{x - y} = \frac{\cancel{(x - y)}(2x + y)}{\cancel{x - y}} = 2x + y$$

ALGEBRAIC SUM OF TWO FRACTIONS

The sum of *only two* fractions occurs so often that rules for finding the sum are helpful.

RULES

1. $\dfrac{a}{b} + \dfrac{c}{d} = \dfrac{ad + bc}{bd}$

2. $\dfrac{a}{b} - \dfrac{c}{d} = \dfrac{ad - bc}{bd}$

These rules are especially helpful when b and d have no common factors. When b and d have common factors, the sum found this way can always be reduced. The verification of the first of these two rules is as follows.

$$\frac{a}{b} = \frac{a \cdot d}{b \cdot d} = \frac{ad}{bd}$$

$$\frac{c}{d} = \frac{c \cdot b}{d \cdot b} = \frac{bc}{bd}$$

Therefore,

$$\frac{a}{b} + \frac{c}{d} = \frac{ad}{bd} + \frac{bc}{bd} = \frac{ad + bc}{bd}.$$

The verification of the second of these two rules is left for the student.

EXAMPLE 6 Applications of the rules.

a. $\frac{2}{3} + \frac{1}{4} = \frac{8 + 3}{3(4)} = \frac{11}{12}$

b. $\frac{2}{3} - \frac{1}{4} = \frac{8 - 3}{3(4)} = \frac{5}{12}$

c. $\frac{x}{y} - \frac{y}{x} = \frac{x^2 - y^2}{xy}$

d. $x + y - \frac{y}{x - y} = \frac{x + y}{1} - \frac{y}{x - y}$

$$= \frac{(x + y)(x - y) - y}{(1)(x - y)} = \frac{x^2 - y^2 - y}{x - y}$$

e. $\frac{1}{4} + \frac{5}{6} = \frac{6 + 20}{(4)(6)} = \frac{26}{24} = \frac{13}{12}$

f. $2\frac{3}{4} = 2 + \frac{3}{4} = \frac{2}{1} + \frac{3}{4} = \frac{8 + 3}{(1)(4)} = \frac{11}{4}$

Exercises 3

Perform the indicated operations and simplify.

A

1. $\frac{4}{7} + \frac{2}{5}$

2. $\frac{5}{4} - \frac{3}{5}$

3. $\frac{2}{5} + \frac{3}{10}$

4. $\frac{m - 5}{5} + \frac{5}{m + 5}$

5. $\frac{a + b}{4} - \frac{a - b}{3}$

6. $\frac{a - b}{a + b} - \frac{a + b}{a - b}$

7. $\frac{5x - 2}{15} + \frac{3 - x}{12}$

8. $\frac{3}{z^2} - \frac{4}{z}$

9. $\frac{3}{z + 4} + \frac{z}{z - 1}$

10. $\frac{H}{H + 2} - 4$

11. $\dfrac{x+1}{x^2-x-6}-\dfrac{x}{x^2-2x-3}$ **12.** $\dfrac{x}{x+4}-\dfrac{x}{x-4}+\dfrac{32}{x^2-16}$

13. $\dfrac{x-1}{x^2+2x+1}-\dfrac{x+1}{x-1}+\dfrac{x-1}{x+1}$

14. $\dfrac{x+2}{x^3+8}-\dfrac{1}{x^2-2x+4}+\dfrac{4}{x+2}$

15. $\dfrac{1}{(a-b)(b-c)}-\dfrac{1}{(b-a)(c-b)}-\dfrac{1}{(b-c)(b-a)}$

16. $\dfrac{x-2}{x^2}-\dfrac{x-1}{3x}-\dfrac{4x-1}{3x^2-3x}$

B

1. $\dfrac{3}{5}+\dfrac{2}{3}$ **2.** $\dfrac{6}{7}-\dfrac{2}{3}$

3. $\dfrac{2}{3}+\dfrac{5}{6}$ **4.** $\dfrac{x-2}{4}+\dfrac{1}{x+2}$

5. $\dfrac{a-b}{3}-\dfrac{a+b}{4}$ **6.** $\dfrac{x+y}{x-y}-\dfrac{x-y}{x+y}$

7. $\dfrac{z+5}{6}+\dfrac{3z-1}{4}$ **8.** $\dfrac{3}{x^2}+\dfrac{2}{x}$

9. $\dfrac{x}{x+1}+\dfrac{2}{x-1}$ **10.** $5+\dfrac{x}{x-2}$

11. $\dfrac{x+1}{x^2+5x+6}-\dfrac{x}{x^2+4x+3}$ **12.** $\dfrac{2x}{x-3}-\dfrac{2x}{x+3}+\dfrac{36}{x^2-9}$

13. $\dfrac{x+2}{x^2-4x+4}+\dfrac{x+2}{x-2}-\dfrac{x-2}{x+2}$

14. $\dfrac{x+1}{x^3+1}-\dfrac{1}{x^2-x+1}+\dfrac{5}{x+1}$

15. $\dfrac{1}{(x-y)(y+z)}-\dfrac{1}{(x-z)(y-x)}-\dfrac{1}{(z-x)(y+z)}$

16. $\dfrac{x-3}{x^2}+\dfrac{x-1}{4x}-\dfrac{2x+3}{4x^2+12x}$

4 Complex fractions

Recall that a complex fraction is a fraction which has a fraction as its numerator, denominator, or both (see Section 1). We will give two methods for simplifying complex fractions.

SIMPLIFYING COMPLEX FRACTIONS

Method I Write the numerator as a simple fraction and the denominator as a simple fraction. Then divide the numerator by the denominator (see Example 1).

Method II Multiply both numerator and denominator of the complex fraction by the LCM of the denominators of all the fractions that appear in either the numerator or the denominator of the complex fraction (see Example 2).

The secondary denominators of a complex fraction are the denominators of the simple fractions in the complex fraction. The secondary denominators in the following example are a, b, and $(a + b)$.

EXAMPLE 1 Simplify

$$\frac{\dfrac{a+b}{b} - \dfrac{b}{a+b}}{\dfrac{1}{b} + \dfrac{2}{a}}.$$

SOLUTION Convert the numerator to a single fraction.

$$\frac{a+b}{b} - \frac{b}{a+b} = \frac{(a+b)^2 - b^2}{b(a+b)} = \frac{a^2 + 2ab}{b(a+b)}$$

Convert the denominator to a single fraction.

$$\frac{1}{b} + \frac{2}{a} = \frac{a+2b}{ab}$$

Then divide.

$$\frac{\dfrac{a+b}{b} - \dfrac{b}{a+b}}{\dfrac{1}{b} + \dfrac{2}{a}} = \frac{\dfrac{a^2+2ab}{b(a+b)}}{\dfrac{a+2b}{ab}}$$

$$= \frac{a^2+2ab}{b(a+b)} \div \frac{a+2b}{ab}$$

$$= \frac{a\cancel{(a+2b)}}{\cancel{b}(a+b)} \cdot \frac{a\cancel{b}}{\cancel{a+2b}} = \frac{a^2}{a+b}$$

EXAMPLE 2 Simplify

$$\frac{\dfrac{2}{x} - \dfrac{1}{y} + \dfrac{3}{z}}{\dfrac{1}{xy} + \dfrac{2}{xz} - \dfrac{1}{yz}}.$$

SOLUTION The LCM of the secondary denominators is xyz. We multiply each term of the numerator and denominator by $xyz/1$.

This is equivalent to multiplying by 1.

$$\frac{\frac{xyz}{1}}{\frac{xyz}{1}} \cdot \frac{\left(\frac{2}{x} - \frac{1}{y} + \frac{3}{z}\right)}{\left(\frac{1}{xy} + \frac{2}{xz} - \frac{1}{yz}\right)} = \frac{\frac{xyz}{1}\left(\frac{2}{x}\right) + \frac{xyz}{1}\left(-\frac{1}{y}\right) + \frac{xyz}{1}\left(\frac{3}{z}\right)}{\frac{xyz}{1}\left(\frac{1}{xy}\right) + \frac{xyz}{1}\left(\frac{2}{xz}\right) + \frac{xyz}{1}\left(-\frac{1}{yz}\right)}$$

$$= \frac{2yz - xz + 3xy}{z + 2y - x}$$

Always check the final fraction to see if it is reduced to lowest terms.

Method II for simplifying a complex fraction is usually shorter when the denominators of the complex fraction have several like factors. We will illustrate this in Example 3.

EXAMPLE 3 Simplify

$$\frac{\frac{1}{2} + \frac{x}{4(x+y)}}{\frac{3}{4} - \frac{x}{x+y}}.$$

The LCD is $4(x + y)$. We multiply numerator and denominator by $4(x + y)/1$.

Multiply by LCD of denominators in the complex fraction.

$$\frac{\frac{1}{2} + \frac{x}{4(x+y)}}{\frac{3}{4} - \frac{x}{x+y}} \cdot \frac{\frac{4(x+y)}{1}}{\frac{4(x+y)}{1}} = \frac{\frac{\overset{2}{\cancel{4}}(x+y)}{1} \cdot \frac{1}{\cancel{2}} + \frac{\cancel{4(x+y)}}{1} \cdot \frac{x}{\cancel{4(x+y)}}}{\frac{\cancel{4}(x+y)}{1} \cdot \frac{3}{\cancel{4}} + \frac{4\cancel{(x+y)}}{1} \cdot \left(\frac{-x}{\cancel{x+y}}\right)}$$

$$= \frac{2x + 2y + x}{3x + 3y - 4x} = \frac{3x + 2y}{3y - x}$$

When negative exponents (Section 6, Chapter 1) are used in expressions such as

$$\frac{x + 1}{1 - x^{-2}},$$

the fraction is considered to be a complex fraction.

EXAMPLE 4 Simplify

$$\frac{x+1}{1-x^{-2}}.$$

SOLUTION

$$\frac{x+1}{1-x^{-2}} = \frac{x+1}{1-\dfrac{1}{x^2}}$$

$$= \frac{x^2(x+1)}{x^2\left(1-\dfrac{1}{x^2}\right)}$$ Multiply the numerator and denominator by the LCD x^2.

$$= \frac{x^2(x+1)}{x^2-1}$$

$$= \frac{x^2\cancel{(x+1)}}{\cancel{(x+1)}(x-1)}$$

$$= \frac{x^2}{x-1}$$

EXAMPLE 5 Write

$$\frac{x^{-2}-y^{-2}}{x^{-1}+y^{-1}}$$

in simplified form with positive exponents.

SOLUTION We will show three ways of working this problem.
First method

$$\frac{x^{-2}-y^{-2}}{x^{-1}+y^{-1}} = \frac{\dfrac{1}{x^2}-\dfrac{1}{y^2}}{\dfrac{1}{x}+\dfrac{1}{y}}$$

$$= \frac{\boxed{\dfrac{x^2y^2}{1}}\left(\dfrac{1}{x^2}-\dfrac{1}{y^2}\right)}{\boxed{\dfrac{x^2y^2}{1}}\left(\dfrac{1}{x}+\dfrac{1}{y}\right)}$$

$$= \frac{y^2-x^2}{xy(y+x)}$$

$$= \frac{\cancel{(y+x)}(y-x)}{xy\cancel{(y+x)}} = \frac{y-x}{xy}$$

Second method

$$\frac{x^{-2} - y^{-2}}{x^{-1} + y^{-1}} = \frac{\frac{1}{x^2} - \frac{1}{y^2}}{\frac{1}{x} + \frac{1}{y}}$$

$$= \frac{\left(\frac{1}{x}\right)^2 - \left(\frac{1}{y}\right)^2}{\frac{1}{x} + \frac{1}{y}}$$

$$= \frac{\cancel{\left(\frac{1}{x} + \frac{1}{y}\right)}\left(\frac{1}{x} - \frac{1}{y}\right)}{\cancel{\frac{1}{x} + \frac{1}{y}}}$$

$$= \frac{1}{x} - \frac{1}{y}$$

$$= \frac{y - x}{xy}$$

Third method

$$\frac{x^{-2} - y^{-2}}{x^{-1} + y^{-1}} = \frac{(x^{-1})^2 - (y^{-1})^2}{x^{-1} + y^{-1}}$$

$$= \frac{\cancel{(x^{-1} + y^{-1})}(x^{-1} - y^{-1})}{\cancel{x^{-1} + y^{-1}}}$$

$$= (x^{-1} - y^{-1})$$

$$= \frac{1}{x} - \frac{1}{y}$$

$$= \frac{y - x}{xy}$$

EXAMPLE 6 Simplify

$$3 - \cfrac{1}{6 - \cfrac{7}{2 - \frac{1}{4}}}.$$

This is called a **continued** fraction.

SOLUTION Start at the bottom and work up.

$$2-\frac{1}{4}=\frac{7}{4}$$

Next,

$$\frac{7}{\frac{7}{4}}=\frac{7}{1}\cdot\frac{4}{7}=4.$$

Then

$$\frac{1}{6-4}=\frac{1}{2},$$

and finally,

$$3-\frac{1}{2}=\frac{5}{2}$$

Exercises 4

Simplify each of the following complex fractions. Write the final answers in a form using positive exponents.

A

1. $\dfrac{\frac{3}{4}-\frac{1}{5}}{\frac{1}{4}+\frac{3}{5}}$ **2.** $\dfrac{a-\frac{2}{a}}{a+\frac{1}{a}}$

3. $\dfrac{c}{\frac{3}{c}-2}$ **4.** $\dfrac{\frac{n}{m}-\frac{m}{n}}{\frac{2}{m}+\frac{2}{n}}$

5. $\dfrac{\frac{c}{3}-\frac{3}{c}}{\frac{1}{3}+\frac{1}{c}}$ **6.** $\dfrac{a+5+\frac{4a+23}{a-3}}{a+1-\frac{2a+9}{a-3}}$

7. $\dfrac{\frac{x}{y}-\frac{y}{x}}{\frac{1}{x}+\frac{1}{y}}$ **8.** $4-\dfrac{3}{2-\dfrac{1}{3-\frac{2}{5}}}$

9. $\dfrac{\frac{2-5x}{3-5x}-\frac{4-x}{1-x}}{\frac{16x-10}{5x^2-8x+3}}$ **10.** $\dfrac{16x^{-2}-9y^{-2}}{4x^{-1}+3y^{-1}}$

B

1. $\dfrac{\frac{2}{3}-\frac{1}{4}}{\frac{3}{4}+\frac{1}{2}}$ **2.** $\dfrac{x+\frac{1}{x}}{x-\frac{1}{x}}$

3. $\dfrac{a}{3 + \dfrac{2}{a}}$

4. $\dfrac{\dfrac{3}{c}}{\dfrac{6}{c}}$

5. $\dfrac{\dfrac{x}{y} - \dfrac{y}{x}}{\dfrac{1}{x} + \dfrac{1}{y}}$

6. $\dfrac{x + 3 + \dfrac{x+1}{x+2}}{x - 2 + \dfrac{2x+5}{x+2}}$

7. $\dfrac{x - \dfrac{4}{x^3}}{\dfrac{1}{2} - \dfrac{1}{x^2}}$

8. $2 - \dfrac{1}{3 - \dfrac{2}{2 - \frac{1}{3}}}$

9. $\dfrac{\dfrac{3-4x}{5-4x} - \dfrac{7-x}{2-x}}{\dfrac{22x - 29}{4x^2 - 13x + 10}}$

10. $\dfrac{4x^{-2} - y^{-2}}{2x^{-1} + y^{-1}}$

Summary

A **fraction** is an indicated quotient of two quantities. The **dividend is called the numerator** and the **divisor is called the denominator**. A fraction has no meaning when its denominator is zero.

A **simple fraction** is a fraction where the numerator and denominator are both integers, with the denominator not zero. A **proper fraction** is a fraction in which the numerator is less in absolute value than the denominator. An **improper fraction** is a fraction in which the numerator is greater than or equal to the denominator in absolute value.

A **complex fraction** is a fraction having a fraction in its numerator, denominator, or both.

A **rational algebraic fraction** is of the form $P(x)/D(x)$ where $P(x)$ and $D(x)$ are polynomials and $D(x) \neq 0$.

In general, an arithmetic or algebraic product is a **multiple** of any of its factors.

The **least common multiple** (LCM) of a set of quantities is the product of all their different prime factors, each taken the greatest number of times it occurs in any one of the quantities. The **lowest common denominator** (LCD) is the least common multiple (LCM) of the denominators of a set of fractions.

The **value of a fraction is not changed** when its numerator and denominator are both multiplied or both divided by the same nonzero quantity. A fraction is said to be **simplified**, reduced to lowest terms, when all of the factors common to both its numerator and denominator have been removed by dividing both by their common factors.

OPERATIONS ON FRACTIONS

To multiply fractions, multiply the numerators and the denominators to obtain the numerator and denominator, respectively, of the product.

To divide one fraction by another, invert the divisor and then proceed as in multiplication.

To find the sum of two or more unlike fractions:

1. Find the LCD.
2. Convert all fractions to equivalent fractions having the LCD as denominators.
3. Write the sum of the converted numerators over the LCD.
4. Simplify the results if possible.

Review exercises

A

1. Of the fractions

$$\frac{x^3-1}{x^2}, \frac{3}{5}, \frac{x+6}{x}, \frac{12}{12}, \frac{x+1}{x^2-1}, \frac{x+100}{x^2},$$

a. which are proper fractions?

b. which are improper fractions?

2. Find the LCM of the expressions $5x^2$, $(2x+4)$, and (x^2+4x+4).

In Exercises 3–5, reduce each fraction to lowest terms.

3. $\dfrac{m-4}{4-m}$ **4.** $\dfrac{x^3+8}{x^2-2x+4}$ **5.** $\dfrac{3x^2+x-10}{9x^2-12x-5}$

In Exercises 6–10, perform the indicated operations and simplify. Write all answers with positive exponents.

6. $1-x^{-3}$ **7.** $\dfrac{1}{x^2-4x-5}-\dfrac{4}{x^2-3x-10}$

8. $\dfrac{\dfrac{x}{y}+\dfrac{x+y}{x-y}}{\dfrac{x-y}{x+y}-\dfrac{x}{y}}$ **9.** $\dfrac{2x^{-1}+3y^{-1}}{4x^{-2}-9y^{-2}}$

10. $\left[\left(\dfrac{a}{b}-\dfrac{b}{a}\right)-\left(\dfrac{1}{a}+\dfrac{1}{b}\right)\right]\div\dfrac{a-b-1}{ab^2+a^2b}$

B

1. Of the fractions

$$\frac{2}{3}, \frac{1}{x}, \frac{x^2+1}{x}, \frac{5}{5}, \frac{x^3-1}{x^3+1}, \frac{x}{x^2-10},$$

   a. which are proper fractions?

   b. which are improper fractions?

2. Find the LCM of the expressions 2, $x^2 - 1$, and $3x^2 - 6x + 3$.

In Exercises 3–5, reduce each fraction to lowest terms.

3. $\dfrac{a-b}{b-a}$ **4.** $\dfrac{x^3-1}{x^2-1}$ **5.** $\dfrac{3ac-3ad+4bc-4bd}{4ac-4ad+bc-bd}$

In Exercises 6–10, perform the indicated operations and simplify. Write all answers with positive exponents.

6. $x + x^{-1}$

7. $\dfrac{3}{x^2-2x-8} - \dfrac{2}{x^2-6x+8}$

8. $\dfrac{\dfrac{x}{x+1} + \dfrac{x-1}{x}}{\dfrac{x}{x+1} - \dfrac{x-1}{x}}$

9. $\dfrac{xy^{-1} - yx^{-1}}{x^{-2} - y^{-2}}$

10. $\left[\left(\dfrac{y}{x} - \dfrac{x}{y}\right) + \left(\dfrac{1}{x} - \dfrac{1}{y}\right)\right] \div \dfrac{2x-2y}{xy^2 - x^2y}$

Diagnostic test

The purpose of this test is to see how well you understand the work in this chapter. Allow yourself approximately 55 minutes to do the test. Solutions to the problems, together with section references, are given in the Answers section at the end of the book. We suggest that you study the sections referred to for the problems you do incorrectly.

1. Of the fractions

$$\frac{x+10}{x}, \frac{x}{x^2+1}, \frac{7}{4}, \frac{x^2}{x^3+8},$$

   a. which are proper fractions?

   b. which are improper fractions?

2. Find the LCM of the expressions $x^2 - 4$, $x^2 - x - 6$, and $x^2 - 5x + 6$.

3. Which are rational fractions?

$$\frac{2x-1}{x^2+1}, \frac{1+x^{-1}}{\sqrt{x+2}}, x^2 - 4$$

In Exercises 4–12, perform the indicated operations and simplify. Write all answers with positive exponents.

4. $\dfrac{a}{a+b} - \dfrac{a-b}{a}$

5. $(x^{-2} + y^{-2})(x^2y^2)$

6. $\left(x - \dfrac{4}{x}\right)\left(\dfrac{x}{4 - x^2}\right)$

7. $x + 1 - \dfrac{x^3}{x^2 - x + 1}$

8. $\left(\dfrac{x}{y} + \dfrac{x+y}{x}\right)\left(\dfrac{x}{x^3 - y^3}\right)$

9. $\dfrac{x + 2 - \dfrac{x}{x-1}}{x - 3 + \dfrac{x}{x-1}}$

10. $\dfrac{\dfrac{x-1}{x+1} + \dfrac{x+1}{x-1}}{\dfrac{x+1}{x-1} - \dfrac{x-1}{x+1}}$

11. $\dfrac{x-1}{x+2} - \dfrac{3x^2}{3x^2 + 5x - 2} + \dfrac{2}{3x - 1}$

12. $x^{-1}y(x + y)^{-1}(x^2 - y^2)(y^2 - xy)^{-1}$

4 Rational exponents, radicals, and complex numbers

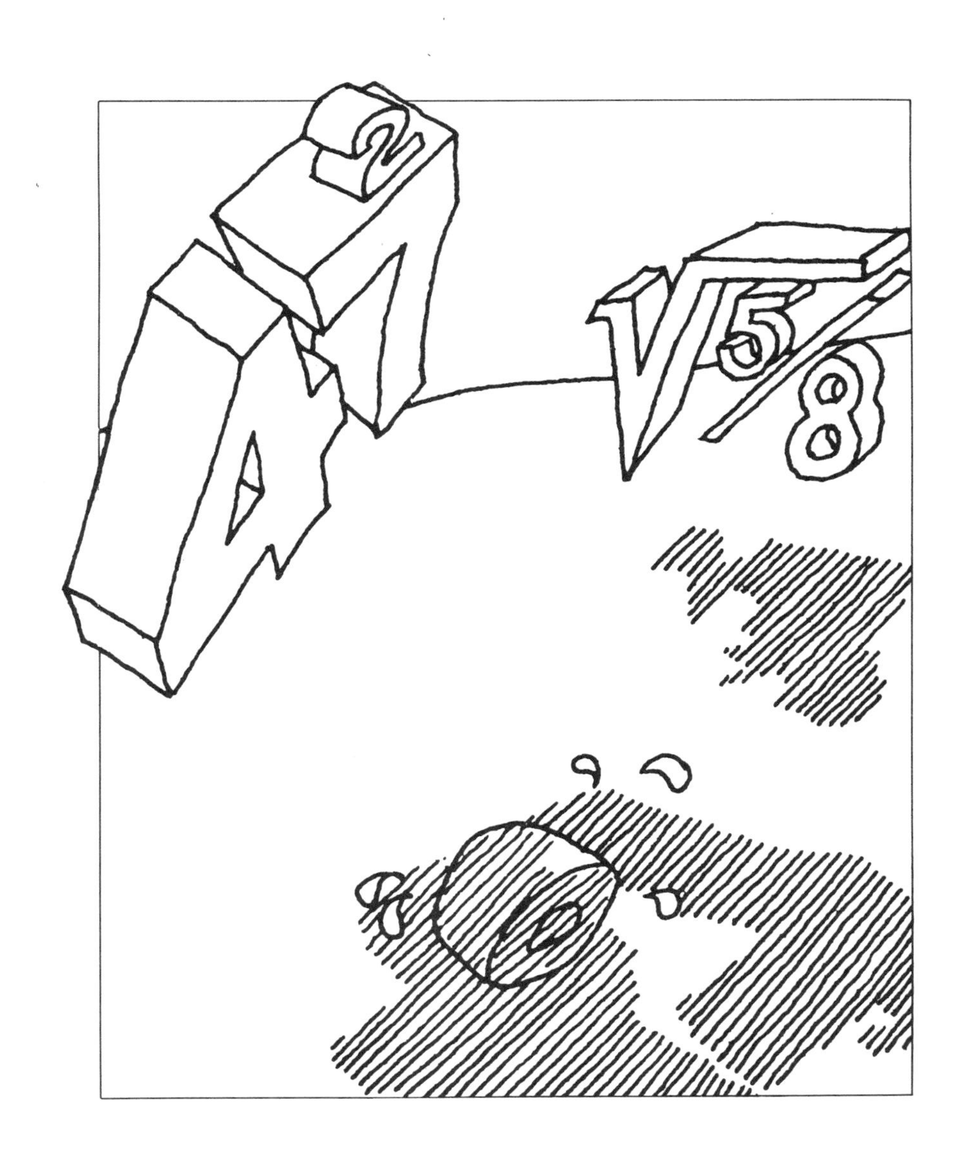

In Chapter 1, we defined the real number system and included some of the laws of integral exponents and radicals. In this chapter, we will discuss rational exponents, relate radical and exponential expressions, and extend our number system to include the complex numbers.

1 Rational exponents

In Section 6 of Chapter 1, we included eight laws of exponents for integral exponents. In this section, we will extend those laws to include fractional, or rational, exponents.

We use the law $(a^m)^n = a^{mn}$ and substitute m/n for m.

$$(a^{m/n})^n = a^{(m/n)(n/1)} = a^m \qquad n \neq 0, 1$$

From the definition of the nth root of a number, it follows that $a^{m/n}$ must be the nth root of a^m. From this definition, we define $a^{m/n}$ as the principal root of a^m.

DEFINITIONS

If $m, n \in J$, where $n \neq 0, 1$ and $a \in R$:

1. $$a^{m/n} = \sqrt[n]{a^m} = (\sqrt[n]{a})^m$$
   When n is odd, a is unrestricted. When n is even, $a \geq 0$.
2. When $m = 1$,
   $$a^{1/n} = \sqrt[n]{a}.$$
   When n is odd, a is unrestricted. When n is even, $a \geq 0$.
3. When n is even and $a < 0$,
   $$\sqrt[n]{a^n} = |a|$$
   $$\sqrt{a^2} = |a|.$$

The laws of exponents given in Section 6 of Chapter 1 can be used when the exponents are rational numbers. This is proved in more advanced courses of mathematics.

In this section, we will exclude exercises in which a is negative when n is even. Exercises of this kind will be discussed in Section 3 of this chapter.

EXAMPLE 1

a. $2^{2/5} = \sqrt[5]{2^2} = \sqrt[5]{4}$

b. $(-27)^{1/3} = \sqrt[3]{-27} = -3$

c. $16^{1/4} = \sqrt[4]{16} = 2$

d. $8^{2/3} = \sqrt[3]{8^2} = \sqrt[3]{64} = 4$, or $8^{2/3} = (\sqrt[3]{8})^2 = (2)^2 = 4$

e. $(9m^2)^{1/2} = \sqrt{9m^2} = 3m$

f. $(-8)^{2/3} = \sqrt[3]{(-8)^2} = \sqrt[3]{64} = 4,$
or $(-8)^{2/3} = (\sqrt[3]{-8})^2 = (-2)^2 = 4$

g. $2^{3/5} = \sqrt[5]{2^3} = \sqrt[5]{8}$

EXAMPLE 2

a. $(3x^{1/2})^2 = 3^2(x^{1/2})^2 = 9x$

b. $\dfrac{x^{5/6}}{x^{1/3}} = x^{(5/6)-(1/3)} = x^{(5/6)-(2/6)} = x^{3/6} = x^{1/2}$, or $\sqrt{x}$

c. $(x^{1/2} + y^{1/2})^2 = (x^{1/2})^2 + 2x^{1/2}y^{1/2} + (y^{1/2})^2$
Perfect square trinomial

$$= x^{(1/2)\cdot(2/1)} + 2x^{1/2}y^{1/2} + y^{(1/2)\cdot(2/1)}$$
$$= x + 2x^{1/2}y^{1/2} + y, \text{ or } x + 2\sqrt{xy} + y$$

d. $(x^{1/2} + y^{1/2})(x^{1/2} - y^{1/2}) = (x^{1/2})^2 - (y^{1/2})^2 = x - y$
Difference of two squares

e. $a^{1-x} \cdot a^x = a^{1-x+x} = a^1 = a$

f. $\dfrac{(-8)^{1/3}(m^{2/a})^{3a/1} \cdot (m^{x/3})^{6/x}}{(4)^{1/2}(m^{1/2})^4}$

$$= \frac{\sqrt[3]{-8}(m^{(2/a)\cdot(3a/1)})(m^{(x/3)\cdot(6/x)})}{\sqrt{4m^2}}$$
$$= \frac{(-2)m^6 \cdot m^2 \cdot m^{-2}}{2}$$
$$= (-1)m^{6+2-2}$$
$$= -m^6$$

THEOREM If $-n$ is the exponent of a rational number, then $+n$ is the exponent of the reciprocal of that rational number.

$$\left(\frac{a}{b}\right)^{-n} = \left(\frac{b}{a}\right)^n \qquad n \in Q, \quad a \neq 0, \quad b \neq 0$$

DEMONSTRATION Let a, b, and n be rational numbers, where a and b are not zero. Then

$$\left(\frac{a}{b}\right)^{-n} = \frac{a^{-n}}{b^{-n}} = \frac{\frac{1}{a^n}}{\frac{1}{b^n}} = \frac{1}{a^n} \div \frac{1}{b^n} = \frac{1}{a^n} \cdot \frac{b^n}{1} = \frac{b^n}{a^n} = \left(\frac{b}{a}\right)^n$$

EXAMPLE 3

a. $\left(\frac{2}{3}\right)^{-3} = \left(\frac{3}{2}\right)^{3} = \frac{3^3}{2^3} = \frac{27}{8}$

b. $\left(\frac{9}{25}\right)^{-1/2} = \left(\frac{25}{9}\right)^{1/2} = \frac{\sqrt{25}}{\sqrt{9}} = \frac{5}{3}$

c. $\left(\frac{1}{27x^6}\right)^{-1/3} = \left(\frac{27x^6}{1}\right)^{1/3} = \sqrt[3]{27x^6} = 3x^2$

EXAMPLE 4 Rewrite using only positive exponents.

a. $x^{-3}y^2z^{-4} = \frac{1}{x^3} \cdot y^2 \cdot \frac{1}{z^4} = \frac{y^2}{x^3z^4}$

b. $3a^{-1} + b^0 = 3 \cdot \frac{1}{a} + 1 = \frac{3}{a} + 1$

c.
$$\begin{aligned}\left(\frac{x^2y^3z^{-6}}{9x^{-2}y^{-3}z^2}\right)^{-1/2} &= \left(\frac{x^{2-(-2)}y^{3-(-3)}z^{-6-(2)}}{9}\right)^{-1/2} \\ &= \left(\frac{x^4y^6z^{-8}}{9}\right)^{-1/2} \\ &= \left(\frac{x^4y^6}{9z^8}\right)^{-1/2} \\ &= \left(\frac{9z^8}{x^4y^6}\right)^{1/2} \\ &= \frac{\sqrt{9z^8}}{\sqrt{x^4y^6}} \\ &= \frac{3z^4}{x^2y^3}\end{aligned}$$

This problem can also be done by raising each factor to the $\frac{1}{2}$ power and then simplifying.

EXAMPLE 5 Writing expressions with a denominator of 1, using negative exponents when necessary.

a. $\frac{8}{x^2y^{-3}} = 8x^{-2}y^3$

b.
$$\begin{aligned}\frac{10a}{\sqrt[3]{8a^{3/2}}} &= \frac{10a}{8^{1/3}(a^{3/2})^{1/3}} \\ &= \frac{10a}{\sqrt[3]{8} \cdot a^{(3/2)(1/3)}} \\ &= \frac{10a}{2a^{1/2}} \\ &= 5a^{1-(1/2)} \\ &= 5a^{1/2}, \text{ or } 5\sqrt{a}\end{aligned}$$

Exercises 1

A In Exercises 1–6, find the value of each expression.

1. $36^{1/2}$ **2.** $9^{3/2}$

3. $\dfrac{1}{16^{3/4}}$ **4.** $\left(\dfrac{16}{25}\right)^{-1/2}$

5. $32^{2/5} + 4^0 - \left(\dfrac{4}{9}\right)^{1/2}$ **6.** $\left(\dfrac{7^{-1} - 2^{-3}}{7^{-1} + 2^{-3}}\right)^{-1}$

In Exercises 7–12, write each expression with positive exponents and then express the results as a single fraction.

7. x^3y^{-5} **8.** $a^{-3}b^{-2}c^4$

9. $2x^{-1} + 3y^{-3}$ **10.** $3m^{-1} - (3m)^{-1}$

11. $2e^x - \dfrac{1}{2e^{-x}}$ **12.** $\dfrac{u^0 - v^{-1}}{u^{-1} + v^0}$

In Exercises 13–16, write each expression with a denominator of 1, using negative exponents when necessary. Simplify when possible.

13. $\dfrac{5}{u^3v}$ **14.** $\dfrac{3xy^2}{w^5z^3}$

15. $\dfrac{5x}{\sqrt{25y}}$ **16.** $\dfrac{10a}{5\sqrt[3]{a^{3/7}}}$

In Exercises 17–22, perform the indicated operations.

17. $x^{3/5}(x^{2/5} + x^{3/10})$ **18.** $(a^{1/2} - 3b^{1/2})^2$

19. $(27a^6)^{2/3}$ **20.** $(m^{6x-12})^{1/6}$

21. $(x^{-2} + y^{-2})(x^{-2} - y^{-2})$

22. $(x^{1/3} + y^{1/3})(x^{2/3} - x^{1/3}y^{1/3} + y^{2/3})$

B In Exercises 1–6, find the value of each expression.

1. $81^{1/2}$ **2.** $27^{2/3}$

3. $\dfrac{1}{25^{1/2}}$ **4.** $\left(\dfrac{49}{36}\right)^{-1/2}$

5. $5^0 + 8^{1/3} - \left(\dfrac{25}{4}\right)^{1/2}$ **6.** $\left(\dfrac{2^{-3} - 3^{-2}}{2^{-3} + 3^{-2}}\right)^{-1}$

In Exercises 7–12, write each expression with positive exponents and then express the result as a single fraction.

7. a^2b^{-3} **8.** $x^{-2}yz^{-1}$

9. $2m^{-1} + 3n^{-2}$ **10.** $2a^{-1} - (2a)^{-1}$

11. $5x^n - \dfrac{1}{5x^{-n}}$ **12.** $\dfrac{x^0 - y^{-1}}{x^{-1} + y^0}$

In Exercises 13–16, write each expression with a denominator of 1, using negative exponents when necessary. Simplify when possible.

13. $\dfrac{6}{x^2y^3}$ **14.** $\dfrac{2mn}{r^3t^2}$

15. $\dfrac{4x}{\sqrt{16y}}$ **16.** $\dfrac{8a}{2\sqrt[3]{a^{3/5}}}$

In Exercises 17–22, perform the indicated operations.

17. $a^{2/3}(a^{1/3} + a^{1/6})$ **18.** $(5x^{1/2} - y^{1/2})^2$

19. $(16x^4)^{3/4}$ **20.** $(x^{8n-4})^{1/4}$

21. $(x^{-1} + y^{-1})(x^{-1} - y^{-1})$

22. $(x^{1/3} - y^{1/3})(x^{2/3} + x^{1/3}y^{1/3} + y^{2/3})$

2 Radicals

A **radical** is an indicated root of a quantity. Examples of radicals are $\sqrt{2}$, $\sqrt[3]{8}$, and $\sqrt[5]{6x^4y}$. The symbol for the indicated root is called a **radical sign** and is written $\sqrt{\ }$. To indicate a particular root, a number (**the index**) is written over the sign. Examples of these are $\sqrt[2]{\ }$, $\sqrt[3]{\ }$, and $\sqrt[n]{\ }$, which indicate square root, cube root, and nth root. In the case of square root, the index is usually omitted. The quantity under a radical sign, such as the 2 in $\sqrt{2}$, or the $(a + b)$ in $\sqrt{a + b}$ is called the **radicand**. The bar above the radicand is a grouping symbol and is frequently included as part of the radical.

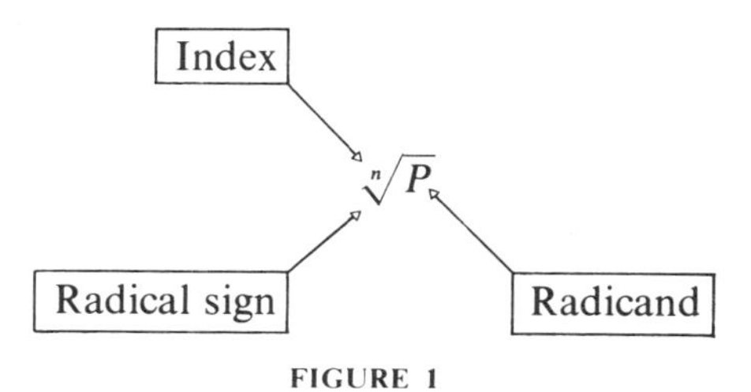

FIGURE 1

LAWS OF RADICALS

In working with radicals, it is often necessary to change the form in which the radical is written. The rules we follow in making these changes can be obtained from the laws of exponents by using the following relation.

$$\sqrt[n]{a^m} = a^{m/n}$$

Radical form ($\sqrt[n]{a^m}$) Exponential form ($a^{m/n}$)

In this section, we assume that the letters a, b, and c used in the four basic laws of radicals represent real numbers and that they are positive when n is even. We also assume n is a nonzero positive integer.

FOUR BASIC LAWS OF RADICALS

1. $\sqrt[n]{a^n} = a$, when $a \geq 0$ for all $n > 1$
   $\sqrt[n]{a^n} = |a|$, when $a < 0$ and n is even
   $(\sqrt[n]{a})^n = a$
2. $\sqrt[n]{ab} = \sqrt[n]{a}\sqrt[n]{b}$
3. $\sqrt[n]{\dfrac{a}{b}} = \dfrac{\sqrt[n]{a}}{\sqrt[n]{b}}$ $\quad b \neq 0$
4. $\sqrt[cn]{a^{cm}} = \sqrt[n]{a^m}$

The first of the above rules is equivalent to the principal nth root of a^n, since a is positive and n is even. We can show this rule by using equivalent exponential expressions for the radical expressions.

$$\sqrt[n]{a^n} = (a^n)^{1/n} = a^{n/n} = a^1 = a$$

$$(\sqrt[n]{a})^n = (a^{1/n})^n = a^{(1/n)(n/1)} = a^1 = a$$

The remaining rules can be obtained from the rules of exponents as follows.

2. $\sqrt[n]{ab} = (ab)^{1/n} = a^{1/n} \cdot b^{1/n} = \sqrt[n]{a}\sqrt[n]{b}$
3. $\sqrt[n]{\dfrac{a}{b}} = \left(\dfrac{a}{b}\right)^{1/n} = \dfrac{a^{1/n}}{b^{1/n}} = \dfrac{\sqrt[n]{a}}{\sqrt[n]{b}}$
4. $\sqrt[cn]{a^{cm}} = a^{\not{c}m/\not{c}n} = a^{m/n} = \sqrt[n]{a^m}$

A RADICAL IS SIMPLIFIED IF:

1. The radicand is positive and has no prime factors with exponents as great as, or greater than, the index of the radical.
2. The radicand contains no fractions.
3. The denominator contains no radicals.
4. The exponents of the prime factors in the radicand and the index of the radical have no common factors.

EXAMPLE 1 Simplifying radical expressions.

a. $\sqrt[4]{x^4} = x$

b. $(\sqrt[3]{y})^3 = y$

c. $\sqrt{32} = \sqrt{16 \cdot 2} = \sqrt{16}\sqrt{2} = 4\sqrt{2}$

d. $\sqrt[3]{8x^5} = \sqrt[3]{8x^3x^2}$

$= \sqrt[3]{2^3}\sqrt[3]{x^3}\sqrt[3]{x^2} = 2x\sqrt[3]{x^2}$

e. $\sqrt[3]{-16x^4y^7} = \sqrt[3]{2(-8)\cdot x^3\cdot x\cdot y^6\cdot y}$

$= \sqrt[3]{(-2)^3}\sqrt[3]{x^3}\sqrt[3]{y^6}\sqrt[3]{2xy} = -2xy^2\sqrt[3]{2xy}$

f. $\sqrt{a^2+2ab+b^2} = \sqrt{(a+b)^2} = a+b$

g. $\sqrt[6]{a^4} = \sqrt[\cancel{2}\cdot 3]{a^{\cancel{2}\cdot 2}} = \sqrt[3]{a^2}$

EXAMPLE 2 Finding products and simplifying.

a. $\sqrt{2}\sqrt{8} = \sqrt{16} = 4$

b. $\sqrt[3]{3}\sqrt[3]{9} = \sqrt[3]{3\cdot 9} = \sqrt[3]{27} = \sqrt[3]{3^3} = 3$

c. $\sqrt{\dfrac{12}{5}}\sqrt{\dfrac{15}{2}} = \sqrt{\dfrac{12}{5}\cdot\dfrac{15}{2}} = \sqrt{18} = \sqrt{9\cdot 2} = 3\sqrt{2}$

d. $\sqrt[4]{x}\sqrt[4]{x^5} = \sqrt[4]{x^6} = \sqrt[4]{x^4\cdot x^2} = x\sqrt[4]{x^2} = x\sqrt[\cancel{2}\cdot 2]{x^{\cancel{2}}} = x\sqrt{x}$

e. $3\sqrt{2x}(5\sqrt{8x} - 7\sqrt{2x}) = 3\sqrt{2x}(5\sqrt{8x}) + 3\sqrt{2x}(-7\sqrt{2x})$

$= 3\cdot 5\sqrt{2x\cdot 8x} + 3(-7)\sqrt{2x\cdot 2x}$

$= 15\sqrt{16x^2} - 21\sqrt{4x^2}$

$= 15(4x) - 21(2x)$

$= 60x - 42x = 18x$

EXAMPLE 3 Using exponents in simplifying radicals.

a. $\sqrt{2}\sqrt[3]{4} = 2^{1/2}\cdot 4^{1/3} = 2^{3/6}\cdot 4^{2/6} = \sqrt[6]{2^3\cdot 4^2}$

$= \sqrt[6]{2^3(2^2)^2} = \sqrt[6]{2^7} = \sqrt[6]{2^6\cdot 2^1} = 2\sqrt[6]{2}$

b. $\dfrac{\sqrt[3]{x^2}}{\sqrt[4]{x}} = \dfrac{x^{2/3}}{x^{1/4}} = x^{(2/3)-(1/4)} = x^{5/12} = \sqrt[12]{x^5}$

Sometimes we need to rationalize the denominator of a fraction containing radicals in the denominator and not in the numerator, or of a fraction with radicals in both its numerator and denominator. To find products or quotients of radicals, the indices must be the same. We use Law 4 of radicals to make them the same.

EXAMPLE 4 Finding products and quotients of radical expressions.

a. $\dfrac{\sqrt[n]{x^{n+1}}}{\sqrt[2n]{x^{2n-1}}} = \dfrac{\sqrt[2n]{(x^{n+1})^2}}{\sqrt[2n]{x^{2n-1}}}$

$= \sqrt[2n]{\dfrac{(x^{n+1})^2}{x^{2n-1}}}$

$= \sqrt[2n]{\dfrac{x^{2n+2}}{x^{2n-1}}}$

$= \sqrt[2n]{x^{(2n+2)-(2n-1)}} = \sqrt[2n]{x^3}$

b. $$\frac{\sqrt{xy^2}\sqrt[3]{x^4y}}{\sqrt[6]{x^2y^5}} = \frac{\sqrt[6]{(xy^2)^3}\sqrt[6]{(x^4y)^2}}{\sqrt[6]{x^2y^5}}$$
$$= \sqrt[6]{\frac{x^3y^6x^8y^2}{x^2y^5}}$$
$$= \sqrt[6]{x^9y^3}$$
$$= \sqrt[6]{x^6x^3y^3}$$
$$= x\sqrt[6]{x^3y^3}$$
$$= x\sqrt{xy}$$

RATIONALIZING THE DENOMINATOR

When a radical has a fractional radicand, it is often convenient to change its form so that the denominator is rational. This procedure is known as **rationalizing the denominator**. This can be done by multiplying both the numerator and the denominator of the fraction by an expression that will make the denominator a perfect *n*th power. This makes it possible to remove the radical from the denominator.

EXAMPLE 5 Examples of rationalizing the denominator and simplifying.

a. $$\sqrt[3]{\frac{5}{4}} = \sqrt[3]{\frac{5}{4}\cdot\frac{2}{2}} = \sqrt[3]{\frac{10}{8}} = \frac{\sqrt[3]{10}}{\sqrt[3]{8}} = \frac{\sqrt[3]{10}}{2} = \frac{1}{2}\sqrt[3]{10}$$

b. $$\frac{6}{\sqrt{2}} = \frac{6}{\sqrt{2}}\cdot\frac{\sqrt{2}}{\sqrt{2}} = \frac{6\sqrt{2}}{\sqrt{4}} = \frac{\overset{3}{\cancel{6}}\sqrt{2}}{\underset{1}{\cancel{2}}} = 3\sqrt{2}$$

c. $$\sqrt{\frac{a}{a+b}} = \sqrt{\frac{a}{a+b}\cdot\frac{a+b}{a+b}} = \sqrt{\frac{a(a+b)}{(a+b)^2}} = \frac{\sqrt{a(a+b)}}{a+b}$$

d. $$\sqrt[4]{\frac{32x^7}{125y^5}} = \sqrt[4]{\frac{(16)(2)x^4x^3}{125y^4y}\cdot\frac{5y^3}{5y^3}}$$
$$= \sqrt[4]{\frac{2^4 10x^4x^3y^3}{(5^4)y^4y^4}}$$
$$= \frac{2x\sqrt[4]{10x^3y^3}}{5y^2}$$

e. $$\frac{x+y}{\sqrt[3]{x-y}} = \frac{x+y}{\sqrt[3]{x-y}}\cdot\frac{\sqrt[3]{(x-y)^2}}{\sqrt[3]{(x-y)^2}}$$
$$= \frac{(x+y)\sqrt[3]{(x-y)^2}}{\sqrt[3]{(x-y)^3}}$$
$$= \frac{(x+y)\sqrt[3]{(x-y)^2}}{x-y}$$

f. We can use a calculator to evaluate $8/\sqrt[6]{7}$ by using the y^x key.

Key operators:

6 [1/x] [STO] [1] 7 [y^x] [RCL] [1] [=] [1/x] [x] 8 [=]

See this: 5.784160211

EXAMPLE 6 Rationalizing a denominator containing a binomial. (This often gives a much simpler expression.)

a.

$$\frac{6}{\sqrt{5}-\sqrt{3}} = \frac{6}{\sqrt{5}-\sqrt{3}} \cdot \frac{\sqrt{5}+\sqrt{3}}{\sqrt{5}+\sqrt{3}}$$

$$= \frac{6(\sqrt{5}+\sqrt{3})}{(\sqrt{5})^2-(\sqrt{3})^2}$$

$$= \frac{6(\sqrt{5}+\sqrt{3})}{5-3}$$

$$= \frac{\overset{3}{\cancel{6}}(\sqrt{5}+\sqrt{3})}{\underset{1}{\cancel{2}}}$$

$$= 3(\sqrt{5}+\sqrt{3})$$

b.

$$\frac{2\sqrt{5}+\sqrt{2}}{2\sqrt{5}-\sqrt{2}} = \left(\frac{2\sqrt{5}+\sqrt{2}}{2\sqrt{5}-\sqrt{2}}\right)\frac{2\sqrt{5}+\sqrt{2}}{2\sqrt{5}+\sqrt{2}}$$

$$= \frac{20+4\sqrt{10}+2}{(2\sqrt{5})^2-(\sqrt{2})^2}$$

$$= \frac{22+4\sqrt{10}}{20-4}$$

$$= \frac{\overset{1}{\cancel{2}}(11+2\sqrt{10})}{\underset{8}{\cancel{16}}}$$

$$= \frac{11+2\sqrt{10}}{8}$$

Exercises 2

A In Exercises 1–8, simplify each expression.

1. $\sqrt[3]{8y^3}$
2. $(\sqrt{b})^2$
3. $\sqrt[4]{48}$
4. $\sqrt{18a^5}$
5. $\sqrt[3]{-27x^4}$
6. $\sqrt{9x^2+12x+4}$
7. $\sqrt[6]{m^9}$
8. $\sqrt[n]{u^{3n}v^{2n}}$

In Exercises 9–15, perform the indicated operations and simplify.

9. $\sqrt[4]{2}\sqrt[4]{8}$

10. $\sqrt[3]{x}\sqrt[4]{x^3}$

11. $\sqrt{\dfrac{3}{8}}\sqrt{\dfrac{32}{27}}$

12. $(\sqrt{20} - \sqrt{5})^2$

13. $4\sqrt{3x}(5\sqrt{4x} + \sqrt{3x})$

14. $\sqrt[5]{xy^2}\sqrt[4]{x^2y^2}$

15. $\dfrac{\sqrt[4]{ab^2}\sqrt{a^3b}}{\sqrt[8]{a^5b^2}}$

In Exercises 16–20, rationalize each denominator and simplify the results.

16. $\dfrac{9}{\sqrt{3}}$

17. $\sqrt{\dfrac{6}{3m}}$

18. $\dfrac{b}{\sqrt{b+1}}$

19. $\dfrac{-14}{\sqrt{5} - 2\sqrt{3}}$

20. $\dfrac{\sqrt{a+b} + \sqrt{a}}{\sqrt{a+b} - \sqrt{a}}$

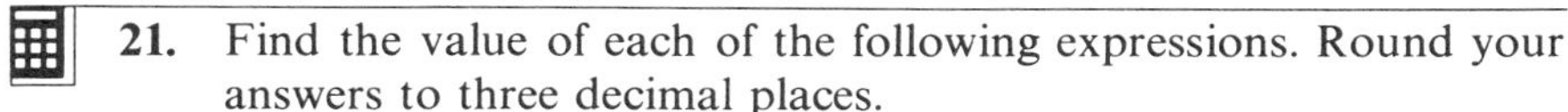

21. Find the value of each of the following expressions. Round your answers to three decimal places.

a. $\sqrt[6]{205}$ b. $\sqrt[5]{84}$ c. $\sqrt[3.4]{0.6874}$ d. $\dfrac{2}{\sqrt{5}}$

B In Exercises 1–8, simplify each expression.

1. $\sqrt[4]{16x^4}$

2. $(\sqrt[5]{a})^5$

3. $\sqrt[3]{54}$

4. $\sqrt{8x^3}$

5. $\sqrt[5]{-32u^6v^7}$

6. $\sqrt{m^2 + 4m + 4}$

7. $\sqrt[4]{p^6}$

8. $\sqrt[n]{u^{2n}v^n}$

In Exercises 9–15, perform the indicated operations and simplify.

9. $\sqrt[3]{2}\sqrt[3]{4}$

10. $\sqrt[4]{x^3}\sqrt{x}$

11. $\sqrt[3]{\dfrac{3}{5}}\sqrt[3]{\dfrac{40}{81}}$

12. $(\sqrt{2} + \sqrt{8})^2$

13. $5\sqrt{2x}(3\sqrt{8x} - \sqrt{2x})$

14. $\sqrt[3]{x^2y}\sqrt[4]{x^6y^2}$

15. $\dfrac{\sqrt[3]{ab}\sqrt[3]{a^2b^3}}{\sqrt[9]{a^3b^5}}$

In Exercises 16–20, rationalize each denominator and simplify the results.

16. $\dfrac{8}{\sqrt{2}}$

17. $\sqrt{\dfrac{3}{2m}}$

18. $\dfrac{a}{\sqrt{a-1}}$

19. $\dfrac{-12}{2\sqrt{3}+3\sqrt{2}}$

20. $\dfrac{\sqrt{x+1}-\sqrt{x}}{\sqrt{x+1}+\sqrt{x}}$

21. Find the value of each of the following expressions. Round your answers to three decimal places.

a. $\sqrt[5]{100}$ **b.** $\sqrt[6]{75}$ **c.** $\sqrt[1.5]{0.4965}$ **d.** $\dfrac{1}{\sqrt{3}}$

3 Complex numbers

So far, we have worked only with real numbers. In this section, we will discuss a number system known as the **complex number system**. The real number system is a subset of the complex number system.

In the real number system, we are unable to find the value of such numbers as $\sqrt{-b}$, where b is positive. To operate with such numbers, a new symbol $i = \sqrt{-1}$, called the **imaginary unit**, is defined.

DEFINITION

$$\sqrt{-1} = i \qquad i^2 = (\sqrt{-1})^2 = -1$$

Using this definition, we have solutions for equations such as

$$x^2 + 1 = 0$$
$$x^2 = -1$$
$$x = \pm\sqrt{-1} = \pm i$$

CHECK $x^2 + 1 = 0$

$$(\pm i)^2 + 1 = 0$$
$$-1 + 1 = 0$$
$$0 = 0$$

Whole powers of i have only four different values, namely, i, -1, $-i$, and 1 (see Example 1).

EXAMPLE 1 Powers of i.

$i^1 = i$ $\qquad$ $i^5 = i^4 i = (1)i = i$

$i^2 = -1$ $\qquad$ $i^6 = (i^2)^3 = (-1)^3 = -1$

$i^3 = i^2 i = (-1)i = -i$ $\qquad$ $i^7 = i^6 i = (-1)i = -i$

$i^4 = (i^2)^2 = (-1)^2 = 1$ $\qquad$ $i^8 = (i^4)^2 = (1)^2 = 1$

If a and b are real numbers, the expression $a + bi$ is called a **complex number**. The **real part** of the complex number is a and the **imaginary part** is bi. Complex numbers include all real numbers and all imaginary numbers. That is, if $b = 0$, the complex number $a + bi = a + 0i = a$ is a real number. If $b \neq 0$, the complex number is said to be imaginary. In particular, if $a = 0$, and $b \neq 0$, the complex number is called a **pure imaginary** number. The letter C will be used to designate the set of complex numbers.

THE SET OF COMPLEX NUMBERS

$$C = \{a + bi \mid a, b \in R, \quad i = \sqrt{-1}, \quad i^2 = -1\}$$

EXAMPLE 2 Examples of complex numbers.

a. $7 = 7 + 0i$

b. $a = a + 0i$ Real numbers

c. $6i = 0 + 6i$

d. $bi = 0 + bi$ Pure imaginary number

e. $6 - 3i$

f. $0 = 0 + 0i$

FUNDAMENTAL PRINCIPLE OF OPERATIONS WITH COMPLEX NUMBERS

Algebraic operations involving integral powers of complex numbers can be performed by writing i in place of $\sqrt{-1}$, operating with i as if it were a real number, and, finally, replacing i^2 by -1.

In multiplying numbers like $\sqrt{-a} \cdot \sqrt{-b}$, where $a > 0$ and $b > 0$, first rewrite each factor in the form

$$\sqrt{-a} = \sqrt{(-1)(a)} = \sqrt{-1} \cdot \sqrt{a} = i\sqrt{a},$$
$$\sqrt{-b} = \sqrt{(-1)(b)} = \sqrt{-1} \cdot \sqrt{b} = i\sqrt{b},$$

and then multiply the numbers as follows.

$$(i\sqrt{a})(i\sqrt{b}) = i^2\sqrt{ab} = -\sqrt{ab}$$

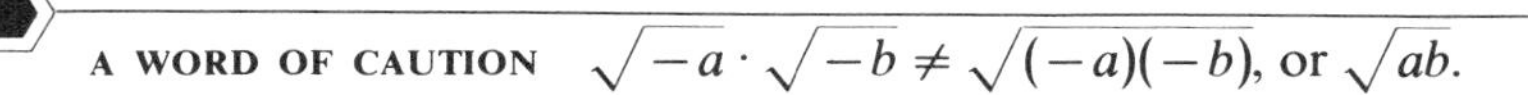

A WORD OF CAUTION $\sqrt{-a} \cdot \sqrt{-b} \neq \sqrt{(-a)(-b)}$, or $\sqrt{ab}$.

EXAMPLE 3 Simplifying imaginary numbers.

a.
$$\begin{aligned}\sqrt{-4} &= \sqrt{(4)(-1)} \\ &= \sqrt{4} \cdot \sqrt{-1} \\ &= 2i\end{aligned}$$

b. $\sqrt{-2}\sqrt{-8} = \sqrt{(2)(-1)}\sqrt{8(-1)}$
$= \sqrt{2}\sqrt{-1}\sqrt{8}\sqrt{-1}$
$= \sqrt{2} \cdot i \cdot \sqrt{8} \cdot i$
$= \sqrt{2 \cdot 8} \cdot i^2$
$= \sqrt{16} \cdot i^2$
$= 4(-1) = -4$

c. $3\sqrt{-5a} = 3\sqrt{5a(-1)}$
$= 3\sqrt{5a}\sqrt{-1}$
$= 3\sqrt{5a} \cdot i$, or $3i\sqrt{5a}$, $\quad a > 0$

d. $\sqrt{-54} = \sqrt{(9)(6)(-1)}$
$= \sqrt{9}\sqrt{6}\sqrt{-1}$
$= 3i\sqrt{6}$

e. $(\sqrt{-2})^5 = (\sqrt{2(-1)})^5$
$= (\sqrt{2}\sqrt{-1})^5$
$= (2^{1/2})^5 i^5$
$= 2^{5/2} i^5$
$= 2^2 \cdot 2^{1/2} \cdot i$
$= 4i\sqrt{2}$

DEFINITION OF EQUALITY

Two complex numbers are said to be equal if and only if their real parts are equal and the coefficients of the imaginary unit i are equal.

$$x + yi = a + bi, \text{ if and only if } x = a \text{ and } y = b$$

EXAMPLE 4 Solve $y + 2i = (2 + xi)(3 - 2i)$ for the real numbers x and y.

SOLUTION

$y + 2i = (2 + xi)(3 - 2i)$
$= 6 + 3xi - 4i - 2xi^2$ — Multiply
$= 6 + (3x - 4)i - 2x(-1)$ — $i^2 = -1$

$y + 2i = (6 + 2x) + (3x - 4)i$ — Collect terms

By the definition of equality: $y = 6 + 2x$ and $2 = 3x - 4$ — Real parts equal and imaginary parts equal

$6 = 3x$

$x = 2$

$y = 6 + 2x = 6 + 2(2) = 10$

CONJUGATE OF A COMPLEX NUMBER

The conjugate of a complex number $a + bi$ is the complex number $a - bi$, which is obtained by changing the sign of the coefficient of the imaginary part.

> The conjugate of $a + bi$ is $a - bi$.

EXAMPLE 5 Conjugates of complex numbers.

	Complex number	**Conjugate**
a.	$2 + 3i$	$2 - 3i$
b.	$-5i$	$5i$
c.	4	4

Since a real number can be written as $a + 0i$, its conjugate is $a - 0i = a$. Thus any real number is its own conjugate.

4 Algebraic operations on complex numbers

ADDITION

To add two complex numbers, add the real parts and add the imaginary parts.

> ADDITION
>
> $$(a + bi) + (c + di) = (a + c) + (b + d)i$$

EXAMPLE 1 $(3 + 2i) + (4 - 5i) = (3 + 4) + (2 - 5)i = 7 - 3i$

SUBTRACTION

To subtract one complex number from another, subtract its real and imaginary parts from the corresponding parts of the other number.

> SUBTRACTION
>
> $$(a + bi) - (c + di) = (a - c) + (b - d)i$$

EXAMPLE 2 $(3 + 2i) - (4 - 5i) = (3 - 4) + [2 - (-5)]i = -1 + 7i$

MULTIPLICATION

To multiply two complex numbers, multiply as you would in multiplying two binomials, and replace i^2 by -1.

MULTIPLICATION

$$(a + bi)(c + di) = ac + (ad + bc)i + bdi^2$$
$$= (ac - bd) + (ad + bc)i$$

EXAMPLE 3 $(3 + 2i)(4 - 5i) = 12 - 7i - 10i^2$

$$= 12 - 7i - 10(-1)$$
$$= 22 - 7i$$

DIVISION

To divide one complex number by another, multiply both dividend and divisor by the conjugate of the divisor.

DIVISION

$$\frac{a + bi}{c + di} = \frac{(a + bi)(c - di)}{(c + di)(c - di)} = \frac{(ac + bd) + (bc - ad)i}{c^2 + d^2}$$

EXAMPLE 4 $\dfrac{3 + 2i}{4 - 3i} = \dfrac{(3 + 2i)}{(4 - 3i)}\dfrac{(4 + 3i)}{(4 + 3i)}$

$$= \frac{6 + 17i}{16 + 9}$$
$$= \frac{6}{25} + \frac{17}{25}i$$

Exercises 4

A

1. Write the conjugate of each of the following complex numbers.
   a. $7 + 2i$ **b.** -3 **c.** $-5i$ **d.** $8 - i$

In Exercises 2–6, simplify each of the imaginary numbers.

2. $\sqrt{-3}\sqrt{-3}$
3. $\sqrt{-8}\sqrt{-6}$
4. $(\sqrt{-3})^5$
5. $\sqrt{-50}$
6. $\sqrt{-5}\sqrt{10}\sqrt{-2}$

In Exercises 7–18, perform the indicated operations. Simplify and write the results in the form $a + bi$.

7. $(2 + 3i) + (7 - i)$
8. $(8 + 3i) - (-2 - 5i)$
9. $(4 - 3i)(4 + 3i)$
10. $3i(5 - 2i)$

11. $(4 - i)^2$
12. $\sqrt{-3}(4 + 3i)$
13. $(3 - i)^3$
14. i^{25}
15. i^{64}
16. $\dfrac{34}{4 + i}$
17. $\dfrac{1 + i}{1 - i}$
18. $\dfrac{5 - 2i}{5 + 2i}$

B

1. Write the conjugate of each of the following complex numbers.
   a. $5 - i$ **b.** $3i$ **c.** 7 **d.** $-6 + i$

In Exercises 2–6, simplify each of the imaginary numbers.

2. $\sqrt{-5}\sqrt{-5}$
3. $\sqrt{-3}\sqrt{-6}$
4. $(\sqrt{-2})^3$
5. $\sqrt{-27}$
6. $\sqrt{2}\sqrt{-32}\sqrt{-2}$

In Exercises 7–18, perform the indicated operations. Simplify and write the results in the form $a + bi$.

7. $(3 + 5i) + (4 - 2i)$
8. $(6 - i) - (3 + 4i)$
9. $(5 + 2i)(5 - 2i)$
10. $2i(3 - 7i)$
11. $(3 - 2i)^2$
12. $\sqrt{-5}(3 - 2i)$
13. $(2 + i)^3$
14. i^{15}
15. i^{48}
16. $\dfrac{26}{5 + i}$
17. $\dfrac{1 - i}{1 + i}$
18. $\dfrac{3 + 2i}{3 - 2i}$

Summary

The parts of a radical expression are illustrated below.

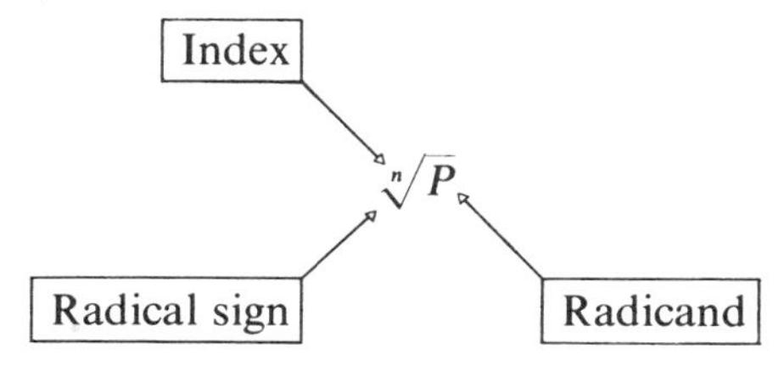

In the following relations, we assume that a and b are real numbers and m and n are integers, with restrictions as indicated.

1. $a^{1/n} = \sqrt[n]{a}$, $n \neq 0, 1$; when n is odd, a is unrestricted; when n is even, $a \geq 0$
2. $a^{m/n} = \sqrt[n]{a^m} = (\sqrt[n]{a})^m$, $n \neq 0, 1$; when n is odd, a is unrestricted; when n is even, $a \geq 0$

3. $a^{m/n} = (a^m)^{1/n} = (a^{1/n})^m, \quad n \neq 0$
4. $\left(\frac{a}{b}\right)^{-n} = \left(\frac{b}{a}\right)^n, \quad a \neq 0, b \neq 0$

The **four basic laws of radicals** are the following.

1. $\sqrt[n]{a^n} = a, \quad$ when $a \geq 0$ for all $n > 1$
   $\sqrt[n]{a^n} = |a|, \quad$ when n is even and $a < 0$
   $(\sqrt[n]{a})^n = a$
2. $\sqrt[n]{ab} = \sqrt[n]{a}\sqrt[n]{b}$
3. $\sqrt[n]{\frac{a}{b}} = \frac{\sqrt[n]{a}}{\sqrt[n]{b}}, \quad b \neq 0$
4. $\sqrt[cn]{a^{cm}} = \sqrt[n]{a^m}$

A radical is simplified if the following conditions hold.

1. The radicand is positive and has no prime factors with exponents as great as, or greater than, the index of the radical.
2. The radicand contains no fractions.
3. The denominator contains no radicals.
4. The exponents of the prime factors in the radicand and the index of the radical have no common factors.

The set of **complex numbers**, C:

$$C = \{a + bi \mid a, b \in R, \quad i = \sqrt{-1}, \quad i^2 = -1\}.$$

When $b = 0$, $a + bi$ is a real number. When $b \neq 0$, $a + bi$ is called **imaginary**. If $a = 0$ and $b \neq 0$, then $a + bi$ is called a **pure imaginary number**. Two complex numbers are defined to be equal if and only if their real parts are equal and the coefficients of the imaginary unit i are equal. For example, $x + yi = a + bi$ means $x = a$ and $y = b$.

The **conjugate** of a complex number $a + bi$ is the complex number $a - bi$, which is obtained by changing the sign of the coefficient of the imaginary part.

To add, subtract, multiply, and divide complex numbers, see Section 4.

The **complex number system** is shown in the following chart:

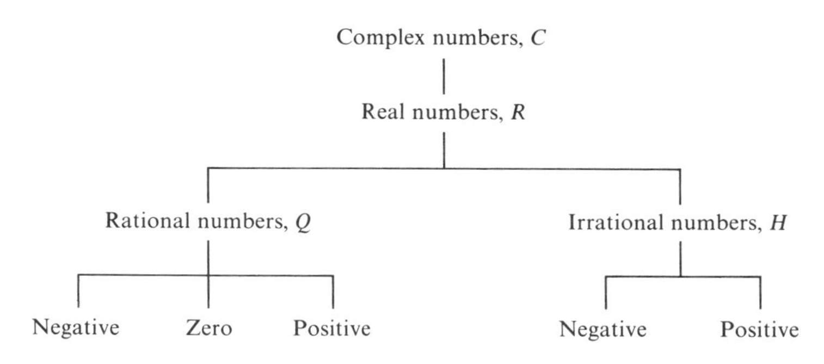

Review exercises

A In Exercises 1–8, find the value of each expression.

1. $49^{1/2}$

2. $\left(\frac{4}{9}\right)^{3/2}$

3. $\left(\frac{27}{8}\right)^{-1/3}$

4. $\sqrt[3]{-125}$

5. i^{12}

6. $(2i)^5$

7. $(-i)^3$

8. $\left(\frac{5^{-1}-3^{-2}}{5^{-1}+3^{-2}}\right)^{-1}$

In Exercises 9–20 simplify each expression. If the result is not exact, write the answer in radical form.

9. $\sqrt{18}$

10. $\sqrt{-25}$

11. $\sqrt[3]{-54}$

12. $\sqrt[4]{32}$

13. $\sqrt[5]{\frac{x^4y^6}{8}}\sqrt[5]{\frac{x^6}{4y^8}}$

14. $\sqrt{a^{4n}b^{3n}}$

15. $\sqrt{(-6)^2}$

16. $(9x^4)^{1/2}$

17. $\sqrt{-64x^6}$

18. $\frac{\sqrt{x}}{\sqrt[4]{x}}$

19. $\sqrt[4]{(x+y)^5}$

20. $\frac{\sqrt{xy^2}\sqrt[10]{xy}}{\sqrt[5]{x^2y}}$

In Exercises 21–26, perform the indicated operations and simplify. Where appropriate, write the answer in the form $a + bi$.

21. $(5\sqrt{2}-2\sqrt{5})(5\sqrt{2}+2\sqrt{5})$

22. $\sqrt{-3}\sqrt{-8}$

23. $(4-3i)^2$

24. $3i(2i-\sqrt{-3})$

25. $(4x^{1/3}+y^{1/3})(16x^{2/3}-4x^{1/3}y^{1/3}+y^{2/3})$

26. $\frac{a^{-3}+b^{-3}}{a^{-1}+b^{-1}}$

In Exercises 27–35, rationalize the denominator and simplify.

27. $\frac{9}{\sqrt{3}}$

28. $\frac{5}{i}$

29. $\frac{14}{2+i\sqrt{3}}$

30. $\sqrt[4]{\frac{a^8}{b^4}}$

31. $\frac{\sqrt{x}-\sqrt{y}}{\sqrt{x}+\sqrt{y}}$

32. $\sqrt[4]{5x^2+\frac{3}{16x^4y^4}}$

33. $5\sqrt{\frac{3}{2}}-3\sqrt{\frac{1}{6}}$

34. $6\sqrt[3]{\frac{a^4}{54}}+6a\sqrt[3]{\frac{a}{2}}$

35. $\frac{26}{3-2i}$

B In Exercises 1–8, find the value of each expression.

1. $100^{1/2}$
2. $\left(\frac{8}{27}\right)^{2/3}$
3. $\left(\frac{81}{16}\right)^{-1/4}$
4. $\sqrt[3]{-64}$
5. i^{20}
6. $(2i)^6$
7. $(-i)^2$
8. $\left(\frac{4^{-1} - 2^{-3}}{4^{-1} + 2^{-3}}\right)^{-1}$

In Exercises 9–20, simplify each expression. If the result is not exact, write the answer in radical form.

9. $\sqrt{20}$
10. $\sqrt{-9}$
11. $\sqrt[3]{-16}$
12. $\sqrt[5]{64}$
13. $\sqrt[4]{\frac{2x}{2y^3}}\sqrt[4]{\frac{8x^3}{3y}}$
14. $\sqrt[n]{x^{3n}y^n}$
15. $\sqrt{(-5)^2}$
16. $(4x^2)^{1/2}$
17. $\sqrt{-125x^4}$
18. $\frac{\sqrt{x}}{\sqrt[3]{x}}$
19. $\sqrt[3]{(a-b)^4}$
20. $\frac{\sqrt{ab}\sqrt[4]{a^3b^3}}{\sqrt[3]{a^2b^2}}$

In Exercises 21–26, perform the indicated operations and simplify. Where appropriate, write the answer in the form $a + bi$.

21. $(2\sqrt{5} + 3\sqrt{2})(3\sqrt{5} - 2\sqrt{2})$
22. $\sqrt{-6}\sqrt{-10}$
23. $(4 + 3i)^2$
24. $2i(3 - \sqrt{-2})$
25. $(2x^{1/3} - y^{1/3})(4x^{2/3} + 2x^{1/3}y^{1/3} + y^{2/3})$
26. $\frac{x^{-2} - y^{-2}}{x^{-1} + y^{-1}}$

In Exercises 27–35, rationalize the denominator and simplify.

27. $\frac{6}{\sqrt{2}}$
28. $\frac{3}{i}$
29. $\frac{36}{4 + i\sqrt{2}}$
30. $\sqrt[3]{\frac{x^6}{y^3}}$
31. $\frac{\sqrt{x} + \sqrt{y}}{\sqrt{x} - \sqrt{y}}$
32. $\sqrt[3]{2x + \frac{3}{8x^3y^3}}$
33. $5\sqrt{\frac{1}{10}} - 3\sqrt{\frac{5}{2}}$
34. $\sqrt[3]{a^4} + 2a\sqrt[3]{8a}$
35. $\frac{10}{1 + 3i}$

Diagnostic test

The purpose of this test is to see how well you understand the work in this chapter. Allow yourself approximately 50 minutes to do the test. Solutions to the problems, together with section references, are given in the Answers section at the end of the book. We suggest that you study the sections referred to for the problems you do incorrectly.

1. Find the value of each expression.

   a. $(81)^{-1/2}$ **b.** $\left(\frac{27}{8}\right)^{2/3}$ **c.** $\sqrt{(-3)^2}$

   d. i^6 **e.** $\left(\frac{6^{-1} - 2^{-3}}{6^{-1} + 2^{-3}}\right)^{-1}$

2. Write each expression in simplified radical form.

   a. $\sqrt{98}$ **b.** $\sqrt[3]{-32x^5y^7}$

   c. $\frac{4}{\sqrt[3]{2}}$ **d.** $\sqrt[n]{a^{3n}b^n}$

3. Perform the indicated operations and express the results in simplified radical form.

   a. $(2 + \sqrt{3})(4 - 3\sqrt{3})$ **b.** $\frac{6}{2 - \sqrt{5}}$

   c. $\frac{3 + \sqrt{2}}{3 + \sqrt{8}}$ **d.** $2\sqrt[4]{\frac{y}{81x^2}} - \frac{1}{3}\sqrt[4]{\frac{y^2}{x^2y}}$

4. Find the value of $\sqrt{25 - x^2}$ when $x = 3$.

5. Simplify $\frac{(x^n)^{n+1}}{x^{n^2}}$.

6. Simplify $\frac{\sqrt{x + y}}{\sqrt{x + y} - \sqrt{y}}$.

In Exercises 7–10, perform the indicated operations, simplify, and write the results in the form $a + bi$.

7. $(2 - 3i)^2$
8. $(6 - 5i) - (5 + 6i)$
9. $\frac{10i}{3 + \sqrt{-1}}$
10. $i(2 - \sqrt{-50})$

5 Equations and inequalities

Word statements from many fields of knowledge can often be written in symbolic language as equations or inequalities. One of the main concerns of algebra is to be able to solve such equations and inequalities. In this chapter, we briefly review basic equations and inequalities and then extend the discussion to more complicated forms.

An equation is a statement of equality between two expressions. *An equation has three parts.*

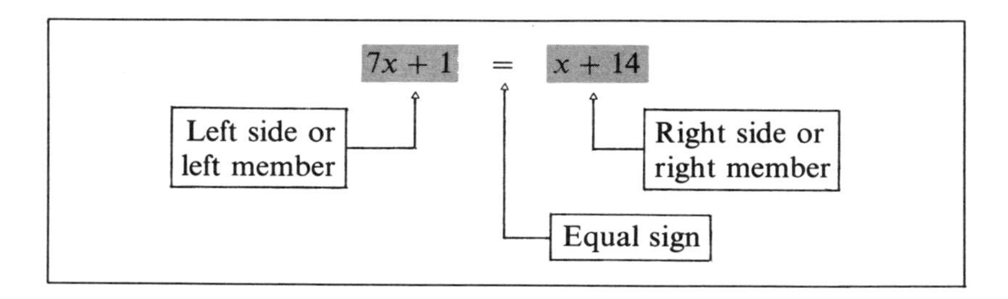

Letters used in equations are called **variables** or **unknowns**. They are placeholders for certain numbers to be substituted. The **solution set** of an equation is the set of values which upon substitution for the letters make the equation a true statement. To **solve an equation** is to find its solution set. The elements of the solution set are the **roots** or **solutions** of the equation.

A polynomial equation is a polynomial in one or more variables, set equal to zero. The degree of a polynomial equation is the degree of the polynomial.

EXAMPLE 1 Examples of equations using x and y as variables.

a. $x + 2 = 0$ First-degree polynomial equation in x.

b. $x^2 + 3x + 2 = 0$ Second-degree polynomial equation in x.

c. $\sqrt{x + 2} = 3$ Radical equation in x.

d. $x + y = 6$ First-degree polynomial equation in x and y.

e. $y + 5 = x^2 + 4x$ First-degree polynomial equation in y or a second-degree polynomial equation in x.

Equations are of two general types, **identities** and **conditional equations**. A conditional equation is true only for certain values of the variables. For example, the equation $x + 2 = 5$ is true only when $x = 3$. When $x = 4$, $x + 2 = 5$ is not true. We call 3 the **solution**, or **root**, of the equation $x + 2 = 5$.

EXAMPLE 2 Examples of conditional equations.

a. $x - 4 = 0$ Solution set: $\{4\}$
b. $x^2 + 5x + 6 = 0$ Solution set: $\{-2, -3\}$

CHECK FOR $x = -2$	CHECK FOR $x = -3$
$x^2 + 5x + 6 = 0$	$x^2 + 5x + 6 = 0$
$(-2)^2 + 5(-2) + 6 = 0$	$(-3)^2 + 5(-3) + 6 = 0$
$4 - 10 + 6 = 0$	$9 - 15 + 6 = 0$
$0 = 0$	$0 = 0$

An **identity** is a statement of equality which is true for all values of the variables, except those values of the variables for which any terms of the equality do not have meaning. If one side of an equation can be converted to the same form as the other side, then the equation is an identity.

EXAMPLE 3 Examples of identities.

a. $x^2 - 4 = (x + 2)(x - 2)$ True for all values of x.

b. $\dfrac{1}{x^2 - 9} = \dfrac{1}{(x - 3)(x + 3)}$

True for all values of x except for $x = \pm 3$ (which would make the denominators zero).

Two or more equations are said to be **equivalent** if they have the same solution set (see Example 4).

EXAMPLE 4 Equivalent equations.

$$x = 2, \qquad x - 2 = 0, \qquad 5x - 7 = x + 1$$

The three equations in Example 4 are equivalent because they all have the same solution set, namely $\{2\}$.

2 Solving elementary equations

To solve an equation, we complete a series of steps—each of which reduces the equation to an equivalent equation—until the solution set is determined. You should already be familiar with the solution of equations by adding or subtracting expressions to or from both sides of the equations, as well as by multiplying or dividing both sides by any nonzero real number.

EXAMPLE 1 Solve $3x - 8 = 7$.

SOLUTION $3x - 8 = 7$

$$3x = 15$$

$$x = 5$$

CHECK Always check solutions in the original equation.

$$3x - 8 = 7$$
$$3(5) - 8 = 7$$
$$15 - 8 = 7$$
$$7 = 7$$

Therefore, $\{5\}$ is the solution set for the equation since 5 makes the statement $3x - 8 = 7$ true. We say 5 **satisfies** the equation.

SOLVING EQUATIONS BY FACTORING

THEOREM If p and q are real numbers and $pq = 0$, then $p = 0$ or $q = 0$.

PROOF If $p \neq 0$,

$$\frac{1}{p}(pq) = \frac{1}{p}(0)$$
$$\left(\frac{1}{\not p} \cdot \frac{\not p}{1}\right) q = 0$$
$$1 \cdot q = 0$$
$$q = 0.$$

The same argument follows to show $p = 0$ when $q \neq 0$.

EXAMPLE 2 Solve $x^2 - x = 20$ by factoring.

SOLUTION

$$x^2 - x = 20$$
$$x^2 - x - 20 = 0$$
$$(x - 5)(x + 4) = 0$$
$$x - 5 = 0 \quad \Big| \quad x + 4 = 0$$
$$x = 5 \quad \Big| \quad x = -4$$

Since both 5 and -4 check, the solutions are 5 and -4. In the case of applied problems, both or only one of the roots may be a solution. A solution set such as this one is often written $\{5, -4\}$.

When we multiply an equation by an expression containing the unknown or raise both sides of the equation to a power, we often produce a derived equation whose solution set contains elements not in the solution set of the original equation. Such derived equations are called **redundant equations** and the additional roots are called **extraneous roots**.

EXAMPLE 3 Solve

$$\frac{2x+1}{x-1} - \frac{x+2}{x+1} = \frac{6x}{x^2-1}.$$

SOLUTION To clear fractions, we multiply each term by the LCD $(x+1)(x-1)$.

$$\boxed{\frac{(x+1)\cancel{(x-1)}}{1}}\left(\frac{2x+1}{\cancel{x-1}}\right) + \boxed{\frac{\cancel{(x+1)}(x-1)}{1}}\left(\frac{-(x+2)}{\cancel{x+1}}\right)$$

$$= \boxed{\frac{\cancel{(x+1)}\cancel{(x-1)}}{1}}\left(\frac{6x}{\cancel{x^2-1}}\right)$$

$$(x+1)(2x+1) + (x-1)(-x-2) = 6x$$

$$2x^2 + 3x + 1 - x^2 - x + 2 = 6x$$

$$x^2 - 4x + 3 = 0$$

$$(x-3)(x-1) = 0$$

We factor, then set each factor to zero.

$x - 3 = 0$	$x - 1 = 0$
$x = 3$	$x = 1$

The number 3 satisfies the original equation, but the number 1 does not. When 1 is substituted in the original equation, we have the following.

$$\frac{2(1)+1}{(1)-1} - \frac{(1)+2}{(1)+1} = \frac{6(1)}{(1)^2-1}$$

$$\frac{3}{0} + \frac{3}{2} = \frac{6}{0}$$

Undefined

Therefore, 3 is the only solution of the equation. The extraneous root 1 was introduced when we multiplied by $(x+1)(x-1)$.

EXAMPLE 4 Solve $x - 1 = \sqrt{x+1}$.

SOLUTION Square both sides of the equation to eliminate the radical.

$$(x-1)^2 = (\sqrt{x+1})^2$$
$$x^2 - 2x + 1 = x + 1$$
$$x^2 - 3x = 0$$
$$x(x-3) = 0$$

We factor then equate each factor to zero to find roots.

$$x = 0 \quad \Big| \quad x - 3 = 0, \quad x = 3$$

CHECK $x = 0$

$$x - 1 = \sqrt{x+1}$$
$$(0) - 1 \neq \sqrt{(0)+1}$$
$$-1 \neq +1$$

CHECK $x = 3$

$$x - 1 = \sqrt{x+1}$$
$$(3) - 1 = \sqrt{(3)+1}$$
$$2 = 2$$

Hence 0 is not a root. This extraneous root was introduced when we squared both sides of the equation. Therefore, 3 is the only solution.

A **defective equation** is one that has fewer roots than an equation from which it was derived. Roots may be lost, for example, by dividing both members of an equation by an expression containing the variable (see Example 5).

EXAMPLE 5 Solve $x^2 - 2x = 0$.

CORRECT SOLUTION

$$x^2 - 2x = 0$$
$$x(x-2) = 0$$
$$x = 0 \quad \Big| \quad x - 2 = 0, \quad x = 2$$

INCORRECT SOLUTION

$$x^2 - 2x = 0$$
$$x^2 = 2x$$
$$\frac{x^2}{x} = \frac{2x}{x}$$
$$x = 2$$

In the correct solution, both 0 and 2 satisfy the original equation and are, therefore, solutions. In dividing by x, we obtained a defective equation and lost the root 0.

A defective equation is also obtained by taking the square roots of both sides of an equation and using only the plus sign on the right side when the square root was taken (see Example 6).

EXAMPLE 6 Solve $x^2 = 4$.

FIRST CORRECT SOLUTION

$x^2 = 4$ Given equation.

$x = \pm\sqrt{4}$

$x = \pm 2$

INCORRECT SOLUTION

$x^2 = 4$

$\sqrt{x^2} = \sqrt{4}$ Should use $\pm$.

$x = 2$

The -2 root was lost.

SECOND CORRECT SOLUTION

$$x^2 = 4$$

$$x^2 - 4 = 0$$

$$(x + 2)(x - 2) = 0$$

$$x + 2 = 0 \quad \Big| \quad x - 2 = 0$$

$$x = -2 \quad \Big| \quad x = 2$$

EXAMPLE 7 Solve $(x - 2)^2 = 9$.

FIRST CORRECT SOLUTION Subtract 9 from both sides and factor the difference of the two squares.

$$(x - 2)^2 = 9$$

$$(x - 2)^2 - 9 = 0$$

$$[(x - 2) - 3][(x - 2) + 3] = 0$$

$$(x - 5)(x + 1) = 0$$

We now set each factor equal to 0.

$$x - 5 = 0 \quad \Big| \quad x + 1 = 0$$

$$x = 5 \quad \Big| \quad x = -1$$

Both 5 and -1 are roots of the equation.

SECOND CORRECT SOLUTION

$(x - 2)^2 = 9$

Both + and − signs must be used.

$\sqrt{(x - 2)^2} = \pm\sqrt{9}$ Take the square root of both sides of the equation.

$x - 2 = \pm 3$

$$x - 2 = 3 \quad \Big| \quad x - 2 = -3$$

$$x = 5 \quad \Big| \quad x = -1$$

INCORRECT SOLUTION

$$(x-2)^2 = 9$$

Using only the positive square root leads to a defective equation.

$$\sqrt{(x-2)^2} = \sqrt{9}$$

$x - 2 = 3$ Defective equation

$x = 5$ The root -1 was lost.

EXAMPLE 8 Solve $x^2 - 6x + 9 = 0$ by factoring.

SOLUTION $x^2 - 6x + 9 = 0$

$$(x-3)(x-3) = 0$$

Equate each factor to zero.

$x - 3 = 0$	$x - 3 = 0$
$x = 3$	$x = 3$

Since the factor $x - 3$ appears twice in the solution of Example 8, 3 is called a **double root**, or root of multiplicity two.

EXAMPLE 9 Solve $x^{3/2} = 27$.

SOLUTION $x^{3/2} = 27$

$$(x^{3/2})^{2/3} = 27^{2/3}$$

$$x = (\sqrt[3]{27})^2$$

$$= 3^2 = 9$$

Exercises 2

A In Exercises 1 and 2, show that each equation is an identity.

1. $3u - [2v - (2u - 5v) + 3v] = 5u - 10v$
2. $(5v - w)(25v^2 + 5vw + w^2) = 125v^3 - w^3$

In Exercises 3–7, solve each equation.

3. $4x - 2[5 - (x - 2) - (3 - x)] = 0$
4. $\dfrac{3z - 7}{4} - \dfrac{z - 11}{3} = 4$
5. $(2y + 7)(5 - 2y) = 0$
6. $(3x - 2)(x + 4) = (5 - 3x)(3 - x)$
7. $(2x - 5)(x - 6)(3 - x) = 0$

In Exercises 8–13, solve by factoring.

8. $5x^2 + 15x = 0$ **9.** $y^2 - y = 30$

10. $7x^2 = 20x + 3$ **11.** $x^3 - 25x = 0$

12. $2x^3 + 5x^2 = 3x$ **13.** $(x - 3)^2 - 16 = 0$

In Exercises 14–20, solve and check. Beware of possible extraneous roots, no solutions, and roots that may have been lost in the process of solution.

14. $\dfrac{v}{v-2} - \dfrac{v+1}{v-4} = -2$ **15.** $\dfrac{5}{x+1} - \dfrac{3}{x+2} = \dfrac{25-4x}{x^2+3x+2}$

16. $\sqrt{z-3} = 3 - z$ **17.** $(5x - 8)^2 = 4$

18. $(3x - 4)^2 = x^2$ **19.** $\sqrt{4x-1} = 1 - 4x$

20. $\sqrt{4u+1} - \sqrt{u-2} = \sqrt{u+3}$

Solve Exercises 21 and 22 by factoring. Use synthetic division as an aid in the factoring.

21. $x^3 - 7x + 6 = 0$ **22.** $x^4 - x^3 - 7x^2 + x + 6 = 0$

23. $x^{3/4} = 8$

B In Exercises 1 and 2, show that each equation is an identity.

1. $x - 2y - [3y - (2x - y) - (-x + 2y)] = 2x - 4y$

2. $(3u + 2v)(9u^2 - 6uv + 4v^2) = 27u^3 + 8v^3$

In Exercises 3–7, solve each equation.

3. $4x - 5 - [(x + 2) - (2x - 3)] = 0$

4. $\dfrac{4z-1}{3} - \dfrac{z+1}{5} = 4$

5. $(x - 6)(3x + 2) = 0$

6. $(x + 5)(2x - 3) = (3 - x)(1 - 2x)$

7. $(2x - 3)(x + 5)(2 - x) = 0$

In Exercises 8–13, solve by factoring.

8. $2x^2 - 10x = 0$ **9.** $y^2 - y = 12$

10. $15x^2 + 4x = 4$ **11.** $x^3 - x = 0$

12. $x^3 + 3x^2 - 10x = 0$ **13.** $(x + 2)^2 - 9 = 0$

In Exercises 14–20, solve and check. Beware of possible extraneous roots, no solutions, and roots that may have been lost in the process of solution.

14. $\dfrac{3u+4}{u-1} - 2 = \dfrac{5u-6}{u-1}$ **15.** $\dfrac{x+1}{x+4} - \dfrac{x+3}{x-7} = \dfrac{1-8x}{x^2-3x-28}$

16. $\sqrt{z-2} = 2 - z$ **17.** $(3x-5)^2 = 16$
18. $(2x+5)^2 = x^2$ **19.** $\sqrt{2x-1} = 1 - 2x$
20. $\sqrt{2-v} + \sqrt{v+3} = \sqrt{7+2v}$

Solve Exercises 21 and 22 by factoring. Use synthetic division as an aid in the factoring.

21. $x^3 - 7x - 6 = 0$ **22.** $x^4 + x^3 - 6x^2 - 4x + 8 = 0$

23. $x^{4/3} = 16$

3 Quadratic equations

QUADRATIC EQUATION IN x

$$ax^2 + bx + c = 0, \quad a \neq 0, \quad a, b, c \in R$$

The equation $3x^2 + 2x - 5 = 0$ is a quadratic equation in x, with $a = 3$, $b = 2$, and $c = -5$.

A quadratic equation is said to be in **standard form** when it is written in the form $ax^2 + bx + c = 0$. It is often necessary to simplify a quadratic equation in order to write it in standard form.

EXAMPLE 1 Convert the quadratic equation $3 + 2x(5x-2) = (2x+3)(3x-1)$ to standard form and then determine the values of a, b, and c.

SOLUTION $3 + 2x(5x-2) = (2x+3)(3x-1)$

$$3 + 10x^2 - 4x = 6x^2 + 7x - 3$$

$$4x^2 - 11x + 6 = 0$$

Therefore, $a = 4$, $b = -11$, and $c = 6$.

In Section 2 of this chapter we considered only those quadratic equations that could be solved by factoring or by taking the square root of both sides of the equation. In this section, we will discuss methods which can be used to solve all kinds of quadratic equations.

SOLVING A QUADRATIC EQUATION BY COMPLETING THE SQUARE

In the last section, we solved a quadratic equation of the type $(x-2)^2 = 3$ as follows.

$$(x-2)^2 = 3$$

$$\sqrt{(x-2)^2} = \pm\sqrt{3}$$

$$x - 2 = \pm\sqrt{3}$$

$$x = 2 \pm \sqrt{3}$$

The method of **completing the square** makes use of the same reasoning. (See Section 5 of Chapter 2.)

EXAMPLE 2 Solve $x^2 - 6x - 5 = 0$ by completing the square.

SOLUTION

$$x^2 - 6x - 5 = 0$$

$$x^2 - 6x = 5$$

$$\frac{1}{2}(-6) = -3 \qquad \text{Take } \tfrac{1}{2} \text{ of the } x\text{-coefficient.}$$

$$x^2 - 6x + (-3)^2 = (-3)^2 + 5$$

Add $(-3)^2$ to both sides to complete the square in x.

$$x^2 - 6x + 9 = 9 + 5$$

Perfect square trinomial.

$$(x - 3)^2 = 14$$

Factor the left side and add 9 and 5 on the right side.

$$\sqrt{(x-3)^2} = \pm\sqrt{14}$$

Take the square root of both sides.

$$x - 3 = \pm\sqrt{14}$$

$$x = 3 \pm \sqrt{14}$$

EXAMPLE 3 Solve $2x(3x - 5) = (x + 1)(x - 5)$.

SOLUTION First, write the quadratic equation in standard form.

$$2x(3x - 5) = (x + 1)(x - 5)$$

$$6x^2 - 10x = x^2 - 4x - 5$$

$$5x^2 - 6x + 5 = 0$$

Standard form.

$$\frac{5x^2}{5} - \frac{6x}{5} + \frac{5}{5} = 0$$

Divide by 5 to reduce the coefficient of x^2 to one. This simplifies the work.

$$x^2 - \frac{6}{5}x = -1$$

$$\frac{1}{2}\left(-\frac{6}{5}\right) = -\frac{3}{5}$$

Take $\frac{1}{2}$ of $-\frac{6}{5}$.

$$x^2 - \frac{6}{5}x + \left(-\frac{3}{5}\right)^2 = \left(-\frac{3}{5}\right)^2 - 1$$

Add $(-\frac{3}{5})^2$ to both sides to complete the square in x.

$$x^2 - \frac{6}{5}x + \frac{9}{25} = \frac{9}{25} - \frac{25}{25}$$

$$\left(x - \frac{3}{5}\right)^2 = -\frac{16}{25}$$

Factor the left side and combine fractions on the right side.

$$\sqrt{\left(x - \frac{3}{5}\right)^2} = \pm\sqrt{-\frac{16}{25}}$$

Take the square root of both sides.

$$x - \frac{3}{5} = \pm\frac{4}{5}i$$

$$x = \frac{3}{5} \pm \frac{4}{5}i, \text{ or } \frac{3 \pm 4i}{5}$$

THE QUADRATIC FORMULA

$$ax^2 + bx + c = 0, \quad \text{where } a, b, c \in R, a \neq 0$$

The roots are

$$x = \frac{-b \pm \sqrt{b^2 - 4ac}}{2a}.$$

DERIVATION OF THE QUADRATIC FORMULA BY COMPLETING THE SQUARE

$$ax^2 + bx + c = 0$$

General form.

$$ax^2 + bx = -c$$

Subtract c from both sides.

$$\frac{ax^2}{a} + \frac{b}{a}x = \frac{-c}{a}$$

Divide by a.

$$\frac{1}{2}\left(\frac{b}{a}\right) = \frac{b}{2a}$$

Take $\frac{1}{2}$ of b/a.

$$x^2 + \frac{b}{a}x + \left(\frac{b}{2a}\right)^2 = \left(\frac{b}{2a}\right)^2 - \frac{c}{a}$$

Add $(b/2a)^2$ to both sides to complete the square in x.

$$\left(x + \frac{b}{2a}\right)^2 = \frac{b^2}{4a^2} - \frac{4a}{4a} \cdot \frac{c}{a}$$

Factor the left side and prepare the fractions on the right side for addition.

$$\left(x + \frac{b}{2a}\right)^2 = \frac{b^2 - 4ac}{4a^2}$$

Add fractions.

$$\sqrt{\left(x + \frac{b}{2a}\right)^2} = \pm\sqrt{\frac{b^2 - 4ac}{4a^2}}$$

Take the square roots of both sides.

$$x + \frac{b}{2a} = \pm\frac{\sqrt{b^2 - 4ac}}{2a}$$

Simplify.

$$x = -\frac{b}{2a} \pm \frac{\sqrt{b^2 - 4ac}}{2a}$$ Subtract $b/2a$ from both sides.

$$x = \frac{-b \pm \sqrt{b^2 - 4ac}}{2a}$$ Quadratic formula.

EXAMPLE 4 Use the quadratic formula to solve

$$x(6x - 4) = (x + 2)(2x - 1).$$

SOLUTION First, write the quadratic equation in standard form and determine a, b, and c.

$$x(6x - 4) = (x + 2)(2x - 1)$$

$$6x^2 - 4x = 2x^2 + 3x - 2$$

$$4x^2 - 7x + 2 = 0$$

$a = 4$, $b = -7$, and $c = 2$

$$x = \frac{-b \pm \sqrt{b^2 - 4ac}}{2a}$$

$$= \frac{-(-7) \pm \sqrt{(-7)^2 - 4(4)(2)}}{2(4)}$$

$$= \frac{7 \pm \sqrt{49 - 32}}{8} = \frac{7 \pm \sqrt{17}}{8}$$

Written in set notation, the solution to Example 4 is

$$\left\{\frac{7 + \sqrt{17}}{8}, \frac{7 - \sqrt{17}}{8}\right\}.$$

If the solution is to be given in decimal notation, use Table 1 at the end of the book or a hand calculator. Written as decimals, rounded to two decimal places, the solution set is $\{1.39, 0.36\}$. Note that these numbers will come close, but they will not satisfy the original equation. They are not exact solutions.

Exercises 3

A In Exercises 1–3, use the method of completing the square to solve each equation.

1. $x^2 - 2x - 2 = 0$ **2.** $4x^2 - 8x + 1 = 0$

3. $9x^2 - 12x + 5 = 0$

In Exercises 4–8, use the quadratic formula to solve each equation. Check solutions.

4. $x^2 - 6x + 6 = 0$ **5.** $x^2 - 10x + 23 = 0$

6. $x^2 - 4x + 6 = 0$ **7.** $x^2 - 3x + 1 = 0$

8. $x^2 + 10x + 26 = 0$

In Exercises 9 and 10, write each quadratic equation in standard form. Use the quadratic formula to solve, and then write the roots as decimals rounded to two decimal places. Check for extraneous roots.

9. $\dfrac{3}{x-1} + \dfrac{1}{x} = \dfrac{2}{x-2}$ **10.** $\dfrac{2}{x-1} - \dfrac{1}{x} + \dfrac{3}{2x+1} = 0$

B In Exercises 1–3, use the method of completing the square to solve each equation.

1. $x^2 - 6x + 4 = 0$ **2.** $2x^2 - 10x + 11 = 0$
3. $3x^2 + 4x + 3 = 0$

In Exercises 4–8, use the quadratic formula to solve each equation. Check solutions.

4. $x^2 - 4x - 1 = 0$ **5.** $x^2 - 6x + 7 = 0$
6. $x^2 - 5x + 8 = 0$ **7.** $9x^2 + 6x - 4 = 0$
8. $x^2 + 4x + 5 = 0$

In Exercises 9 and 10, write each quadratic equation in standard form. Use the quadratic formula to solve, and then find the roots as decimals rounded to two decimal places. Check for extraneous roots.

9. $\dfrac{1}{x-1} + \dfrac{2}{x} + \dfrac{3}{x+1} = 0$

10. $\dfrac{2}{(x-2)(x+1)} + \dfrac{x+2}{x+1} - \dfrac{2}{x-2} = 0$

4 The discriminant of a quadratic equation; forming equations when the roots are known

In the last section, it was shown that the roots of the quadratic equation

$$ax^2 + bx + c = 0, \quad \text{where } a, b, c \in R \text{ and } a \neq 0, \tag{1}$$

are

$$x = \frac{-b \pm \sqrt{b^2 - 4ac}}{2a}.$$

If x_1 and x_2 are the two roots of the equation, then

$$x_1 = -\frac{b}{2a} + \frac{\sqrt{b^2 - 4ac}}{2a} \quad \text{and} \quad x_2 = -\frac{b}{2a} - \frac{\sqrt{b^2 - 4ac}}{2a}. \tag{2}$$

The number $\boldsymbol{b^2 - 4ac}$, which appears under the radical sign in the formula, is called the **discriminant** of the quadratic equation. Useful facts about the roots of a quadratic equation can be determined by computing the value of its discriminant, $b^2 - 4ac$.

If $b^2 - 4ac$ is positive, then $\sqrt{b^2 - 4ac}$ is a real number and the two roots x_1 and x_2, are real and unequal, as can be seen in Equations **2**.

If $b^2 - 4ac$ is zero, it follows from Equations **2** that both x_1 and x_2 equal $-b/2a$. In this case, x_1 and x_2 are real and equal.

If $b^2 - 4ac$ is negative, then $\sqrt{b^2 - 4ac}$ is an imaginary number and x_1 and x_2 are two unequal, imaginary numbers. In this case, x_1 and x_2 are complex conjugate numbers.

DISCRIMINANT TESTS

Given the quadratic equation

$$ax^2 + bx + c = 0, \quad \text{where } a, b, c \in R \text{ and } a \neq 0.$$

	Nature of roots
1. $b^2 - 4ac > 0$ and a perfect square	Real, unequal, and rational.
2. $b^2 - 4ac > 0$ and **not** a perfect square	Real, unequal, and irrational.
3. $b^2 - 4ac = 0$	Real and equal (root of multiplicity two).
4. $b^2 - 4ac < 0$	Conjugate imaginaries.

If $b^2 - 4ac$ is a perfect square, the equation is factorable.

EXAMPLE 1 In each of the following equations, we use the value of the discriminant to determine the nature of the roots.

a. $9x^2 - 12x + 1 = 0$

$$\begin{aligned} b^2 - 4ac &= (-12)^2 - 4(9)(1) \\ &= 144 - 36 = 108 > 0 \end{aligned}$$

Therefore, the roots are real, unequal, and irrational.

b. $x^2 - 10x + 25 = 0$

$$\begin{aligned} b^2 - 4ac &= (-10)^2 - 4(1)(25) \\ &= 100 - 100 = 0 \end{aligned}$$

Therefore, the roots are real and equal.

c. $3x^2 - 2x + 4 = 0$

$$\begin{aligned} b^2 - 4ac &= (-2)^2 - 4(3)(4) \\ &= 4 - 48 < 0 \end{aligned}$$

Therefore, the roots are conjugate complex numbers.

FORMING EQUATIONS WITH SPECIFIED ROOTS

To form an equation having the specified roots r_1 and r_2, we use the fact that $(x - r_1)$ and $(x - r_2)$ are factors of the equation. The required equation is of the form $a(x - r_1)(x - r_2) = 0$, where a is any nonzero constant. The constant a is chosen to eliminate fractions in the equation.

This method of building equations from known roots can be extended to any number of roots merely by having a factor for each required root.

EXAMPLE 2 Form a quadratic equation having the given sets as roots.

a. $\left\{\frac{1}{3}, -\frac{1}{2}\right\}$ **b.** $\{1+\sqrt{6}, 1-\sqrt{6}\}$ **c.** $\left\{\frac{2+3i}{5}, \frac{2-3i}{5}\right\}$

SOLUTION **a.** The required equation is of the form

$$a\left(x-\frac{1}{3}\right)\left(x+\frac{1}{2}\right)=0.$$

We will use $a = 6$, the product of the denominators 2 and 3, to clear the fractions.

$$6\left(x-\frac{1}{3}\right)\left(x+\frac{1}{2}\right)=0$$

$$\frac{6}{1}\left(\frac{3x-1}{3}\right)\left(\frac{2x+1}{2}\right)=0$$

$$(3x-1)(2x+1)=0$$ Equation in factored form.

Therefore, $6x^2 + x - 1 = 0$ is the required quadratic equation.

CHECK $6x^2 + x - 1 = 0$

$$(3x-1)(2x+1)=0$$

$3x - 1 = 0$ | $2x + 1 = 0$

$x = \frac{1}{3}$ | $x = -\frac{1}{2}$

b. Because there are no fractions, we choose $a = 1$.

$$[x-(1+\sqrt{6})][x-(1-\sqrt{6})]=0$$

$$[x-1-\sqrt{6}][x-1+\sqrt{6}]=0$$

$$[(x-1)-\sqrt{6}][(x-1)+\sqrt{6}]=0$$

$$(x-1)^2-(\sqrt{6})^2=0$$ Product of the sum and difference of two quantities.

$$x^2-2x+1-6=0$$

Therefore, $x^2 - 2x - 5 = 0$ is the required equation.

CHECK Use the quadratic formula to solve $x^2 - 2x - 5 = 0$.

$$x = \frac{-(-2) \pm \sqrt{(-2)^2 - 4(1)(-5)}}{2(1)}$$

$$= \frac{2 \pm \sqrt{24}}{2}$$

$$= \frac{2 \pm \sqrt{4 \cdot 6}}{2}$$

$$= \frac{2 \pm 2\sqrt{6}}{2}$$

$$= 1 \pm \sqrt{6}$$

c. We choose $a = 25$ to clear the fractions.

$$25\left[x - \frac{2 + 3i}{5}\right]\left[x - \frac{2 - 3i}{5}\right] = 0$$

$$\frac{\cancel{25}}{1}\left[\frac{5x - 2 - 3i}{\cancel{5}}\right]\left[\frac{5x - 2 + 3i}{\cancel{5}}\right] = 0$$

$$[(5x - 2) - 3i][(5x - 2) + 3i] = 0$$

Product of the sum and difference of two quantities.

$$(5x - 2)^2 - (3i)^2 = 0$$

$$25x^2 - 20x + 4 - 9i^2 = 0$$

Therefore, $25x^2 - 20x + 13 = 0$ is the required equation. The check is left for the student.

EXAMPLE 3 Find a fourth-degree equation whose solution set is $\{0, \frac{2}{3}, 1, -4\}$.

SOLUTION In order for the equation to have the given roots, it must have factors of $x - 0$, $x - \frac{2}{3}$, $x - 1$, and $x - (-4)$. An equation, in factored form, with these roots is

$$a(x)\left(x - \frac{2}{3}\right)(x - 1)(x + 4) = 0.$$

To clear fractions, we choose $a = 3$.

$$\cancel{3}(x)\left(\frac{3x - 2}{\cancel{3}}\right)(x - 1)(x + 4) = 0$$

$$3x^4 + 7x^3 - 18x^2 + 8x = 0$$

Therefore, $3x^4 + 7x^3 - 18x^2 + 8x = 0$ is the required equation.

CHECK For $x = 0$: $0 + 0 - 0 + 0 = 0$

We will use synthetic division to check the other roots.

		3	7	−18	8	
			3	10	−8	Remainder zero shows $x - 1$ is a factor and 1 is a root
For $x = 1$:	1	3	10	−8	(0)	
			−12	8		Remainder zero shows $x + 4$ is a factor and -4 is a root
For $x = -4$:	−4	3	−2	(0)		
			2			Remainder zero shows $x - \frac{2}{3}$ is a factor and $\frac{2}{3}$ is a root
For $x = \frac{2}{3}$:	$\frac{2}{3}$	3	(0)			

Exercises 4

A In Exercises 1–5, use the discriminant to determine the nature of the roots. Do not solve the equations.

1. $x^2 + 2x - 15 = 0$ **2.** $3x^2 - 2x - 4 = 0$

3. $25x^2 - 20x + 4 = 0$ **4.** $10x^2 + 3x - 1 = 0$

5. $\frac{3}{2}x^2 + \frac{4}{5}x + \frac{2}{5} = 0$

6. Find the value of k for which the roots of the equation $k^2x^2 - 6x + 4 = 0$ are equal.

In Exercises 7–10, find a quadratic equation having the given solution set.

7. $\{4, -3\}$ **8.** $\{3\sqrt{2}, -3\sqrt{2}\}$

9. $\left\{\frac{1 - 3\sqrt{3}}{2}, \frac{1 + 3\sqrt{3}}{2}\right\}$ **10.** $\left\{\frac{3 + i\sqrt{2}}{3}, \frac{3 - i\sqrt{2}}{3}\right\}$

11. Find a cubic equation with the solution set $\{3i, -3i, 1\}$.

12. Find a fifth-degree equation with the solution set $\{\sqrt{2}, -\sqrt{2}, 5i, -5i, 2\}$.

B In Exercises 1–5, use the discriminant to determine the nature of the roots. Do not solve the equations.

1. $x^2 - 5x - 14 = 0$ **2.** $2x^2 - 3x - 4 = 0$

3. $4x^2 - 12x + 9 = 0$ **4.** $3x^2 - 7x + 4 = 0$

5. $\frac{2}{3}x^2 + \frac{5}{4}x + \frac{3}{5} = 0$

6. Find the value of k for which the roots of the equation $2x^2 - 3x + k = 0$ are equal.

In Exercises 7–10, find a quadratic equation with the given solution set.

7. $\{3, -2\}$ **8.** $\{2\sqrt{5}, -2\sqrt{5}\}$

9. $\left\{\frac{5-2\sqrt{3}}{3}, \frac{5+2\sqrt{3}}{3}\right\}$ **10.** $\left\{\frac{1-i\sqrt{3}}{2}, \frac{1+i\sqrt{3}}{2}\right\}$

11. Find a cubic equation with the solution set $\{i, -i, 2\}$.

12. Find a fifth-degree equation with the solution set $\{\sqrt{3}, -\sqrt{3}, 2i, -2i, -1\}$.

5 Equations in quadratic form

Some equations which are not quadratic in a particular variable can be solved by the methods of solving quadratic equations. We will illustrate this with several examples.

EXAMPLE 1 Solve $(\sqrt[3]{x})^2 - \sqrt[3]{x} - 2 = 0$.

SOLUTION

$$x^{2/3} - x^{1/3} - 2 = 0$$

$$(x^{1/3})^2 - (x^{1/3}) - 2 = 0 \qquad \text{Equivalent form.}$$

$$u^2 - u - 2 = 0 \qquad \text{Let } u = x^{1/3}.$$

$$(u-2)(u+1) = 0 \qquad \text{Factor.}$$

$u - 2 = 0$	$u + 1 = 0$
$u = 2$	$u = -1$
$x^{1/3} = 2$	$x^{1/3} = -1$
$(x^{1/3})^3 = 2^3$	$(x^{1/3})^3 = (-1)^3$
$x = 8$	$x = -1$

Replace u by $x^{1/3}$. Cube both sides.

Therefore, the solution set is $\{8, -1\}$.

CHECK $x = 8$

$$x^{2/3} - x^{1/3} - 2 = 0$$
$$8^{2/3} - 8^{1/3} - 2 = 0$$
$$4 - 2 - 2 = 0$$
$$0 = 0$$

CHECK $x = -1$

$$x^{2/3} - x^{1/3} - 2 = 0$$
$$(-1)^{2/3} - (-1)^{1/3} - 2 = 0$$
$$1 - (-1) - 2 = 0$$
$$1 + 1 - 2 = 0$$
$$0 = 0$$

EXAMPLE 2 Solve $x^4 - 3x^2 - 4 = 0$.

SOLUTION

$$x^4 - 3x^2 - 4 = 0$$

$$(x^2)^2 - 3(x^2) - 4 = 0 \qquad \text{Equivalent form.}$$

$$u^2 - 3u - 4 = 0 \qquad \text{Let } u = x^2.$$

$$(u - 4)(u + 1) = 0 \qquad \text{Factor.}$$

$u - 4 = 0$	$u + 1 = 0$
$u = 4$	$u = -1$
$x^2 = 4$	$x^2 = -1$
$\sqrt{x^2} = \pm\sqrt{4}$	$\sqrt{x^2} = \pm\sqrt{-1}$
$x = \pm 2$	$x = \pm i$

Replace u by x^2.

Therefore, the solution set is $\{-2, 2, -i, i\}$.

EXAMPLE 3 Solve

$$\left(x - \frac{1}{x}\right)^2 - 4\left(x - \frac{1}{x}\right) + 3 = 0.$$

SOLUTION

$$\left(x - \frac{1}{x}\right)^2 - 4\left(x - \frac{1}{x}\right) + 3 = 0$$

$$u^2 - 4u + 3 = 0 \qquad \text{Let } u = x - \frac{1}{x}.$$

$$(u - 3)(u - 1) = 0 \qquad \text{Factor.}$$

$u - 3 = 0$	$u - 1 = 0$
$u = 3$	$u = 1$
$x - \frac{1}{x} = 3$	$x - \frac{1}{x} = 1$
$x^2 - 1 = 3x$	$x^2 - 1 = x$
$x^2 - 3x - 1 = 0$	$x^2 - x - 1 = 0$
$x = \frac{-(-3) \pm \sqrt{9 + 4}}{2(1)}$	$x = \frac{-(-1) \pm \sqrt{1 + 4}}{2}$
$= \frac{3 \pm \sqrt{13}}{2}$	$= \frac{1 \pm \sqrt{5}}{2}$

Replace u by $x - \frac{1}{x}$.

Use the quadratic formula.

Therefore, the solution set is

$$\left\{\frac{3+\sqrt{13}}{2}, \frac{3-\sqrt{13}}{2}, \frac{1+\sqrt{5}}{2}, \frac{1-\sqrt{5}}{2}\right\}.$$

Exercises 5

Solve the following equations.

A

1. $x^4 - 5x^2 + 4 = 0$
2. $2(x+1) - 5\sqrt{x+1} - 3 = 0$
3. $x^6 - 28x^3 + 27 = 0$
4. $3x^{2/3} + 5x^{1/3} - 2 = 0$
5. $\dfrac{6(x+2)^2}{x^2} - \dfrac{7(x+2)}{x} = 3$

6. Find the roots of $6x^2 - 4 - \sqrt{3x^2 - 2} = 6$ to two decimal places.

B

1. $x^4 - 10x^2 + 9 = 0$
2. $x - 7\sqrt{x} + 12 = 0$
3. $8x^6 + 7x^3 - 1 = 0$
4. $2x^{2/3} + 5x^{1/3} - 3 = 0$
5. $\dfrac{2x^2}{(x-1)^2} - \dfrac{x}{x-1} = 3$

6. Find the roots of $2(x^2 - 3x)^2 + 7x^2 - 21x = 4$ to two decimal places.

6 Applications

We will show three examples where quadratic equations are used to solve word problems.

EXAMPLE 1 Find the dimensions of the right triangle shown in Figure 1.

SOLUTION We make use of the fact that the sum of the squares on the two legs of a right triangle is equal to the square on its hypotenuse (Pythagorean Theorem).

$$(x+1)^2 + x^2 = 3^2$$

$$x^2 + 2x + 1 + x^2 = 9$$

$$2x^2 + 2x - 8 = 0$$

$$x^2 + x - 4 = 0 \qquad \text{Divide both sides by 2.}$$

Then

$$x = \frac{-1 \pm \sqrt{1+16}}{2(1)} = \frac{-1 \pm \sqrt{17}}{2} \doteq 1.562 \text{ or } -2.562.$$

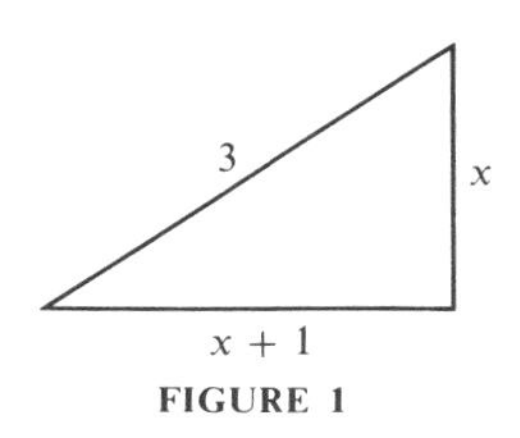

FIGURE 1

The negative dimension, -2.562, does not have meaning in this geometric problem.

Therefore, the height is about 1.562 and the base is about 2.562, rounded to three decimal places.

EXAMPLE 2 A box with no top is to be formed from a rectangular sheet of metal by cutting 2-inch squares from the corners and folding up the sides. The length of the box is to be 3 inches more than its width, and its volume is to be 80 cubic inches.

a. Find the dimensions of the sheet of metal.
b. Find the dimensions of the box.

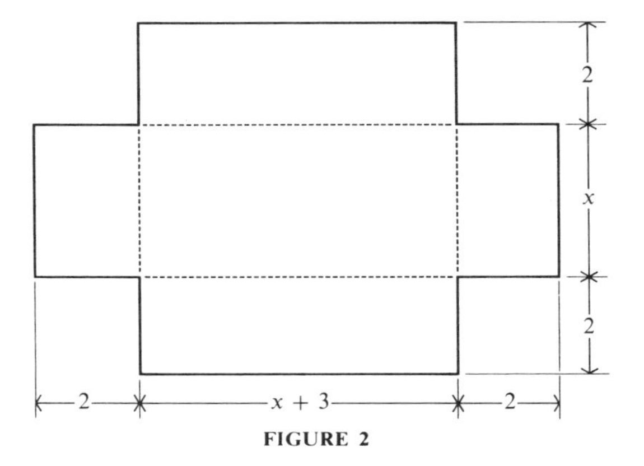

FIGURE 2

SOLUTION

$$(\text{Length})(\text{Width})(\text{Depth}) = \text{Volume}$$
$$(x+3)(x)(2) = 80$$
$$(x+3)(x) = 40$$
$$x^2 + 3x - 40 = 0$$
$$(x+8)(x-5) = 0$$

$x + 8 = 0$	$x - 5 = 0$
$x = -8$	$x = 5$ Width
	$x + 3 = 8$ Length
-8 has no meaning in this problem.	

a. The dimensions of the metal sheet are 9 by 12 inches.
b. The dimensions of the box are:

Depth = 2 inches,

Width = 5 inches,

Length = 8 inches.

CHECK Volume = (Length)(Width)(Height)

$$= (8)(5)(2)$$

$$= 80 \text{ cubic inches}$$

EXAMPLE 3 A projectile is fired straight up with an initial velocity of 320 feet per second. The distance S (in feet) traveled in t (seconds) is given by the formula $S = -16t^2 + 320t$.

a. How many seconds will it take for the projectile to reach a height of 1,200 feet?

b. How long will it be before the projectile returns to the ground?

SOLUTION

a.

$$S = -16t^2 + 320t$$

$$1{,}200 = -16t^2 + 320t \qquad S = 1{,}200.$$

$$16t^2 - 320t + 1{,}200 = 0$$

$$t^2 - 20t + 75 = 0 \qquad \text{Divide both sides by 16.}$$

$$(t - 5)(t - 15) = 0$$

$$t - 5 = 0 \quad \Big| \quad t - 15 = 0$$

$$t = 5 \quad \Big| \quad t = 15$$

The projectile will be at a height of 1,200 feet in 5 seconds and again in 15 seconds (see Figure 3).

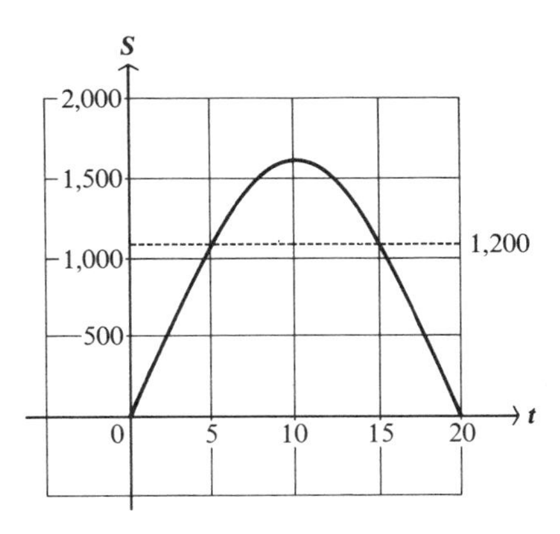

FIGURE 3

b. When $S = 0$, the projectile is on the ground.

$$S = -16t^2 + 320t$$
$$0 = -16t^2 + 320t$$
$$0 = -16t(t - 20)$$
$$t = 0 \quad \Big| \quad t - 20 = 0$$
$$t = 20$$

The projectile was at ground level when $t = 0$ and it returned in 20 seconds ($t = 20$).

Exercises 6

A

1. Find the value of the constant k for which the roots of the equation $9x^2 + kx + 1 = 0$ are equal.
2. If k is a constant and one root of the equation $2x^2 - 5x + k = 0$ is 5, what is the other root?
3. Find the dimensions of a rectangle whose length is 2 inches more than twice its width and which has a 13-inch diagonal.
4. A box with no top is to be formed from a rectangular sheet of metal by cutting 4-inch squares from the corners and folding up the sides. The length of the box is to be 2 inches more than its width, and its volume is to be 32 cubic inches.
   - **a.** Find the dimensions of the sheet of metal.
   - **b.** Find the dimensions of the box.
5. A baseball is thrown upward with an initial velocity of 48 feet per second. The height S (in feet) traveled in t (seconds) is given by the formula $S = -16t^2 + 48t$.
   - **a.** How many seconds will it take for the ball to reach a height of 32 feet?
   - **b.** How long will it be before the ball returns to the ground?

B

1. Find the value of the constant k for which the roots of the equation $3x^2 - 2x + k = 0$ are equal.
2. If k is a constant and one root of the equation $kx^2 - kx = 6$ is 3, what is the other root?
3. Find the dimensions of the right triangle shown in the figure below.

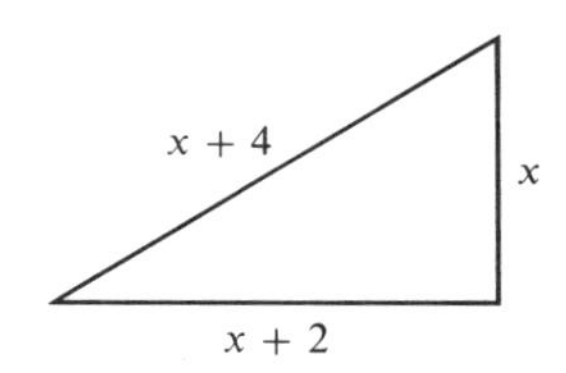

4. A box with no top is formed from a square sheet of metal by cutting 3-inch squares from the corners and folding up the sides. The box is to have a volume of 12 cubic inches.
   a. Find the dimensions of the sheet of metal.
   b. Find the dimensions of the box.
5. A baseball is thrown upward with an initial velocity of 64 feet per second. The height S (in feet) reached in t (seconds) is given by the formula $S = -16t^2 + 64t$.
   a. How many seconds will it take for the ball to reach a height of 48 feet?
   b. How long will it be before the ball returns to the ground?

7 First-degree inequalities with one unknown

An **inequality** is a statement that one mathematical quantity is less than (or greater than) another. Inequalities, as well as equations, are used extensively in the applications of mathematics.

In this section, we will assume that all variables represent real numbers.

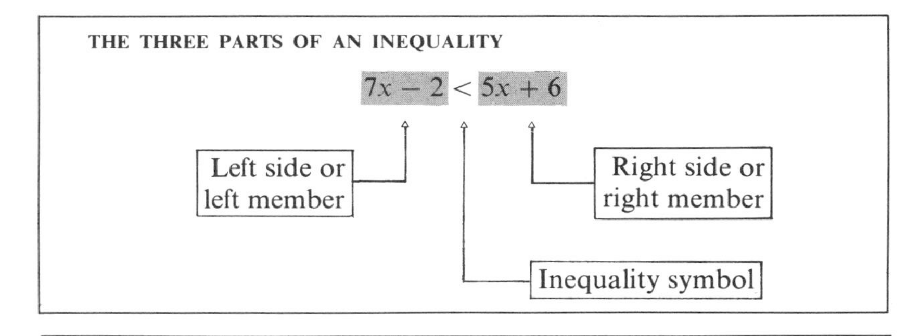

INEQUALITY SYMBOLS USED IN THIS CHAPTER

$<$	Is less than.
$\leq$	Is less than or equal to.
$>$	Is greater than.
$\geq$	Is greater than or equal to.

The **solution** set of an inequality is the set of all numbers which, when substituted for the variable (or variables) in the inequality, make the inequality a true statement. To **solve** an inequality is to find its solution set. As was true with equations, inequalities are **equivalent** if they have the same solution set.

Inequalities are of two types, **conditional**, and **unconditional.** A **conditional inequality** is an inequality which is only true for certain values of the variable(s). For example, the inequality $x + 1 > 4$ is a conditional inequality because it is only true for x greater than 3.

An **unconditional inequality** (or **absolute inequality**) is an inequality which is true for all values of the variable(s), or which contains no variables. For example, the inequality $x^2 \geq 0$ is true for all real values of x. Also, $2 < 3$ is always true and has no variables.

Two inequalities in which the inequality signs point in the same direction are said to have the **same sense**; if the signs point in opposite directions, the inequalities are **opposite in sense**. Thus, the statements $a < b$ and $c < d$ have the same sense; $a > b$ and $c < d$ are opposite in sense.

An **interval** of real numbers is the set of all real numbers between two given numbers (the **endpoints** of the interval) and one, both, or neither endpoint. An interval that contains none of its endpoints is an **open interval**. When used in the discussion of intervals, the symbol (a, b) represents the open interval with endpoints a and b, and $a < b$. The symbol $[a, b]$ represents the **closed interval**, which includes the endpoints a and b and $a < b$. Intervals that contain only one of the endpoints, such as $2 \leq x < 3$, $x \leq 4$, or $x \geq -1$, are sometimes called **half-open** or **half-closed** intervals and are represented by such symbols as $[a, b)$, $(-\infty, b]$, $[a, \infty)$.

Continued inequalities are inequalities of the form $a < x < c$. The continued inequality $a < x < c$ is the intersection of $a < x$ and $x < c$, which can be written $\{x \mid a < x\} \cap \{x \mid x < c\}$. The union $\{x \mid x < -2\} \cup \{x \mid x > 3\}$ *should not* be written $3 < x < -2$. It could be written $x < -2$ or $x > 3$.

The following examples are illustrations of intervals.

EXAMPLE 1 Open interval.

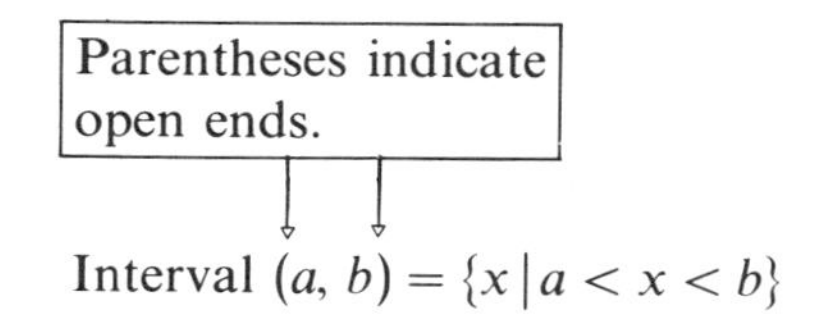

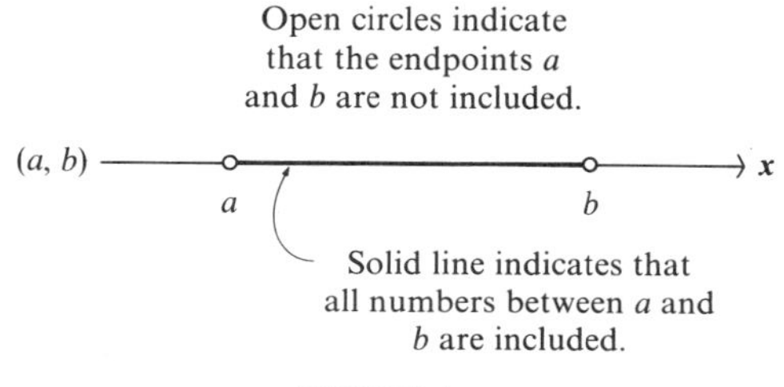

FIGURE 4

EXAMPLE 2 Half-open interval.

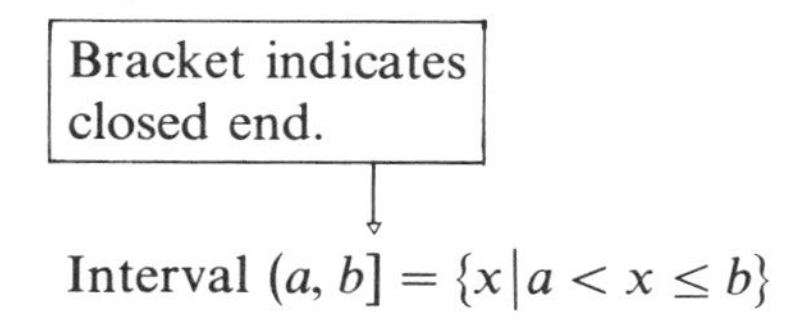

Interval $(a, b] = \{x \mid a < x \leq b\}$

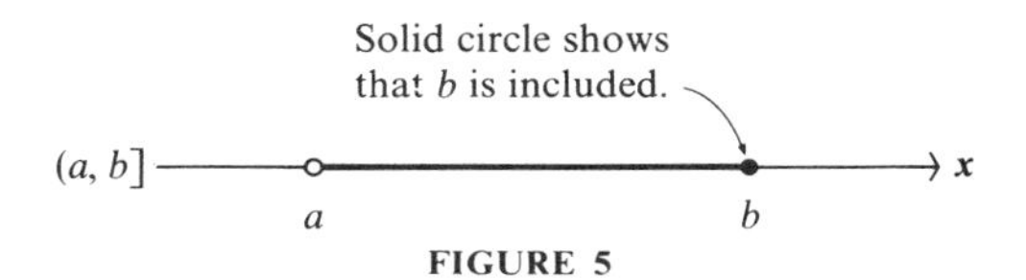

FIGURE 5

EXAMPLE 3 Closed interval.

Interval $[a, b] = \{x \mid a \leq x \leq b\}$

Solid circles show that
a and b are included.

FIGURE 6

EXAMPLE 4 Half-open interval.

Interval $[a, \infty) = \{x \mid x \geq a\}$

Read "infinity." This means numbers increase in size without end.

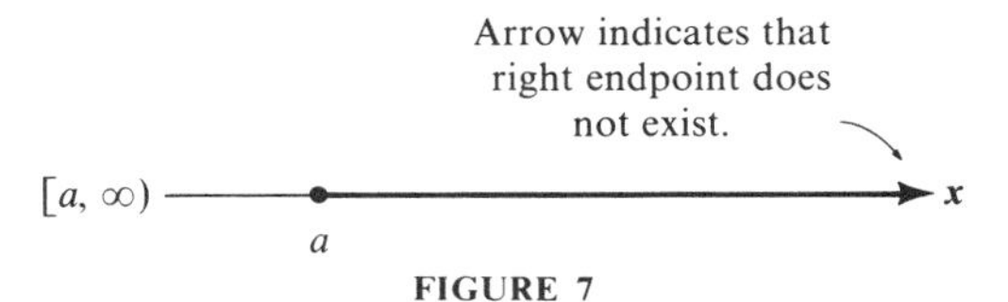

FIGURE 7

EXAMPLE 5 Half-open interval.

Interval $(-\infty, b] = \{x \mid x \leq b\}$

Read "negative infinity." This means numbers decrease in size without end.

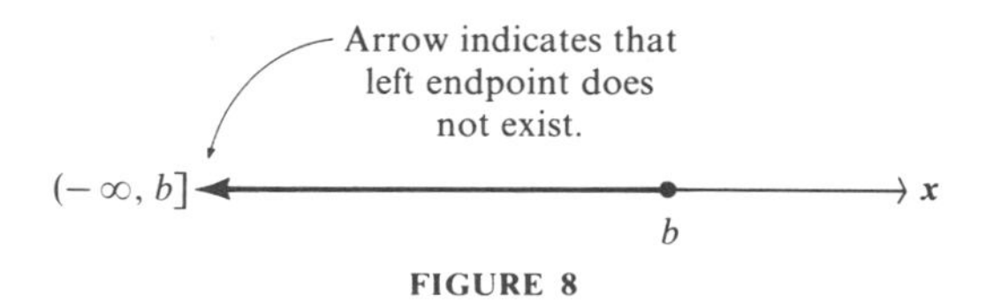

FIGURE 8

EXAMPLE 6 Graph $[-2, 0] \cup [3, 4]$ and write the solution set.

SOLUTION Solution set $\{x \mid -2 \leq x \leq 0\} \cup \{x \mid 3 \leq x \leq 4\}$.

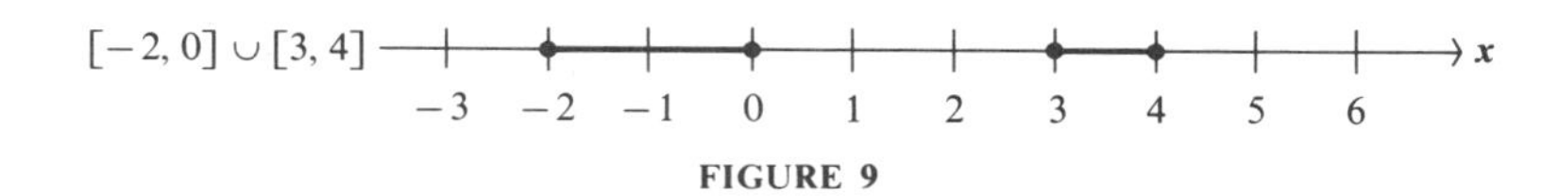

FIGURE 9

EXAMPLE 7 Graph $(-\infty, 4] \cap [2, \infty)$ and write the solution set. Also write the solution using inequality symbols.

SOLUTION $\{x \mid x \leq 4\} \cap \{x \mid x \geq 2\}$

$$2 \leq x \leq 4$$

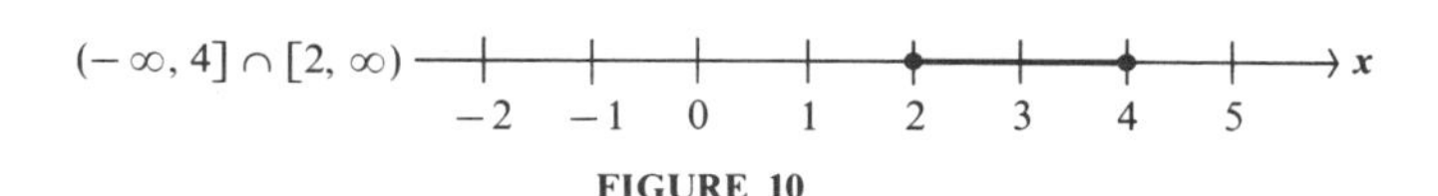

FIGURE 10

The methods used to solve inequalities are, in many ways, similar to those used to solve equations. Before we start solving inequalities, we will give some of the theorems used.

THEOREM 1 The addition of the same real number to, or the subtraction of the same real number from, both sides of an inequality leaves the sense of the inequality unchanged.

EXAMPLE 8

a.	$5 < 7$	Given inequality.
	$5 + 3 < 7 + 3$	Add 3 to both sides.
	$8 < 10$	Same sense.
b.	$5 < 7$	Given inequality.
	$5 - 3 < 7 - 3$	Subtract 3 from both sides.
	$2 < 4$	Same sense.

THEOREM 2 The multiplication or division of both sides of an inequality by the same *positive* number leaves the sense of the inequality unchanged.

EXAMPLE 9

a.	$4 > 3$	Given inequality.
	$4 \times 2 > 3 \times 2$	Multiply both sides by 2.
	$8 > 6$	Same sense.
b.	$8 < 10$	Given inequality.
	$\frac{8}{2} < \frac{10}{2}$	Divide both sides by 2.
	$4 < 5$	Same sense.

THEOREM 3 If both sides of an inequality are multiplied or divided by a negative number, a new inequality of opposite sense is obtained.

EXAMPLE 10

a.	$6 > 4$	Given inequality.
	$6 \times (-2) < 4 \times (-2)$	Multiply both sides by -2.
	$-12 < -8$	Opposite in sense.
b.	$6 > 4$	Given inequality.
	$\frac{6}{-2} < \frac{4}{-2}$	Divide both sides by -2.
	$-3 < -2$	Opposite in sense.

The proof of the above theorems is based on the following axiom.

> **AXIOM** If $a > b$, then $a - b = p$ is a positive number, and conversely.

We will show the proof for the first theorem and leave the proofs of the other two theorems as exercises for the student.

PROOF FOR THEOREM 1

Read "implies" (pointing to $\Rightarrow$)

$$a > b \Rightarrow a - b = p \text{ is a positive number.}$$

For any real number m, it follows that:

$$\begin{aligned} a - b &= a - b + m - m \\ &= a + m - b - m \\ &= (a + m) - (b + m) \end{aligned}$$

But $a - b = p$ is a positive number. Then by substitution:

$$(a + m) - (b + m) = p \qquad \text{A positive number.}$$

and, by the above axiom,

$$a + m > b + m.$$

Since m may be positive or negative, Theorem 1 is proved.

To solve an inequality, we go through a series of steps, each of which reduces the inequality to an equivalent inequality, until the values of the variables are determined.

EXAMPLE 11 Solve $2x - 3 < x + 1$ and graph the solution on a line graph.

SOLUTION $2x - 3 < x + 1$

$$x - 3 < 1 \qquad \text{Add } (-x) \text{ to both sides.}$$

$$x < 4 \qquad \text{Add 3 to both sides.}$$

The solution is $\{x \mid x < 4\}$, or the interval $(-\infty, 4)$.

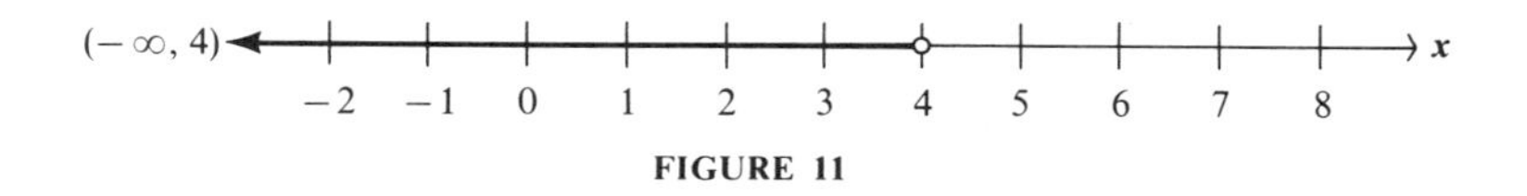

FIGURE 11

EXAMPLE 12 Solve $7x - 2 \leq 5x + 6$ and graph the solution set on a line graph.

SOLUTION Our objective is to end the chain of steps with x on one side of the inequality symbol and a number on the other side.

$$\begin{array}{rcl} 7x - 2 & \leq & 5x + 6 \\ -5x + 2 & & -5x + 2 \\ \hline 2x & \leq & 8 \end{array} \qquad \text{Add } (-5x + 2) \text{ to both sides.}$$

$$\frac{2x}{2} \leq \frac{8}{2} \qquad \text{Divide both sides by 2.}$$

$$x \leq 4 \qquad \text{Solution.}$$

The solution is $\{x \mid x \leq 4\}$, or the half-open interval $(-\infty, 4]$.

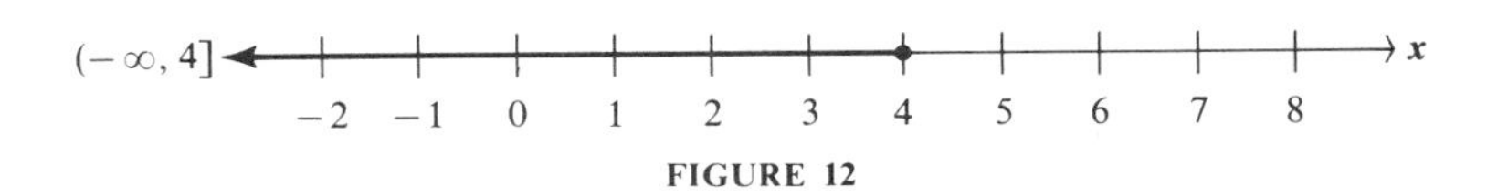

FIGURE 12

EXAMPLE 13 Solve

$$\left\{x \left| \frac{x-1}{5} < \frac{x}{4} - \frac{1}{10}, \quad x \in J\right.\right\}$$

and graph its solution set.

SOLUTION

$$\frac{x-1}{5} < \frac{x}{4} - \frac{1}{10}$$

$$\frac{20}{1} \cdot \frac{x-1}{5} < \frac{20}{1} \cdot \frac{x}{4} - \frac{20}{1} \cdot \frac{1}{10} \qquad \text{Multiply both sides by 20, the LCD.}$$

$$4x - 4 < 5x - 2$$

$$\begin{array}{lcr} -5x + 4 & & -5x + 4 \\ \hline -x & < & 2 \end{array} \qquad \text{Add } -5x + 4 \text{ to both sides.}$$

$$x > -2 \qquad \text{Multiply both sides by } -1\text{; change the sense of the inequality.}$$

The solution is $\{x \mid x > -2, x \in J\}$.

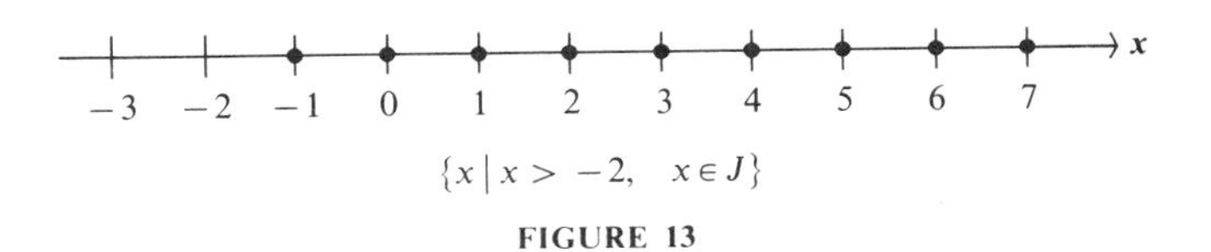

FIGURE 13

THEOREM 4 If $|x| \le b$, then $-b \le x \le b$.

PROOF By the definition of absolute value.

Case I When

$$x \ge 0, \qquad |x| = x \le b.$$

Then

$$x \ge 0 \quad \text{and} \quad x \le b.$$

The intersection of $x \ge 0$ and $x \le b$ is $\{x \mid 0 \le x \le b\}$.

Case II When

$$x < 0, \qquad |x| = -x \le b.$$

Then

$$x < 0 \quad \text{and} \quad x \ge -b.$$

The intersection of $x < 0$ and $x \ge -b$ is $\{x \mid -b \le x < 0\}$.

The union of the solutions of Cases I and II is

$$\{x \mid -b \le x \le b\}.$$

This is shown on the number line in Figure 14.

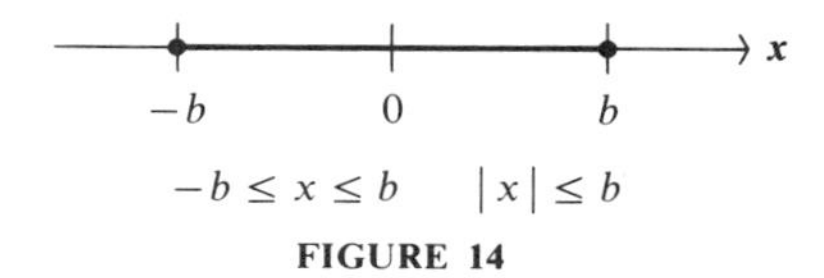

FIGURE 14

In mathematics, expressions such as $|x - a| \le b$ are often used to represent a **neighborhood of** a. The expression $|x - a| \le b$ represents the set of numbers extending from the point $x = a - b$ to the point $x = a + b$ on the x-axis, with midpoint $x = a$. Since $x - a$ could be positive or negative, we use Theorem 4 to solve $|x - a| \le b$.

$$\begin{array}{cccl} & |x - a| & \le b & \\ -b & \le x - a & \le b & \text{By Theorem 4.} \\ +a & +a & +a & \\ \hline a - b & \le x & \le a + b & \text{Which is the neighborhood shown in Figure 15.} \end{array}$$

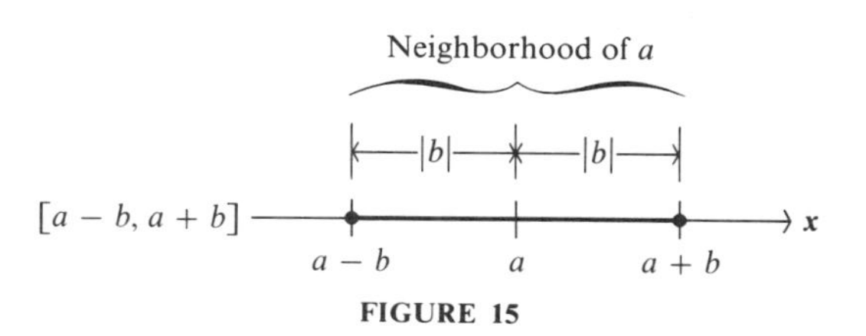

FIGURE 15

EXAMPLE 14 Examples of intervals determined by inspection and graphed on a number line.

a. $|x - 2| \le 1$; x is any number from 0 to 1 unit from 2.

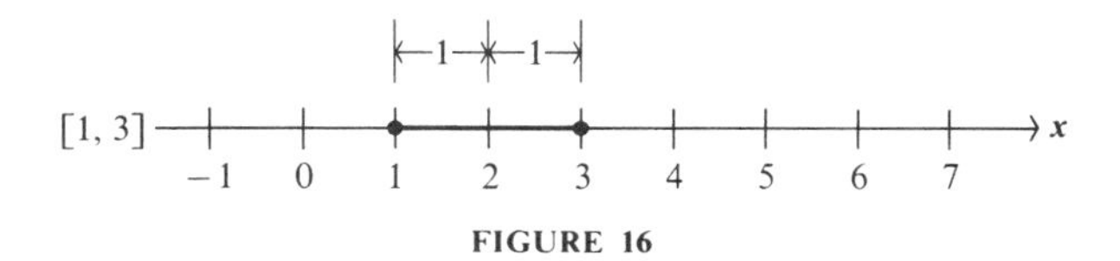

FIGURE 16

b. $|x + 3| < 2$

$$|x + 3| = |x - (-3)|$$

x is any number from 0 to, but not including, 2 units from -3.

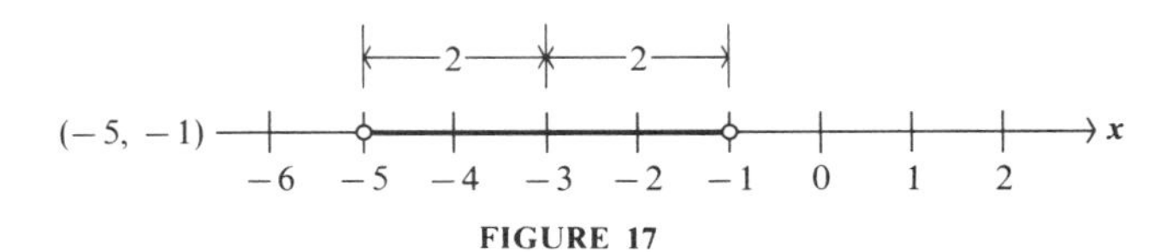

FIGURE 17

THEOREM 5 $|x - a| \geq b = \{x | x \geq a + b\} \cup \{x | x \leq a - b\}$

PROOF

Case I If $x - a \geq 0$, $|x - a| = x - a$. Therefore, we must have the intersection of

$$x - a \geq 0 \quad \text{and} \quad x - a \geq b$$

$$x \geq a \quad \text{and} \quad x \geq a + b$$

On the number line:

Intersection

Case II If $x - a < 0$, $|x - a| = -(x - a)$. Therefore, we must have the intersection of

$$x - a < 0 \quad \text{and} \quad -(x - a) \geq b$$

$$x - a \leq -b$$

$$x < a \quad \text{and} \quad x \leq a - b$$

On the number line:

$a - b$ x

Intersection

The union of the solutions of Cases I and II is

$$\{x \,|\, x \geq a + b\} \cup \{x \,|\, x \leq a - b\}.$$

EXAMPLE 15 $|x + 2| \geq 1$

$$|x + 2| \geq 1 = |x - (-2)| \geq 1$$

x is any number equal to, or greater than, 1 unit from -2. This is also an application of Theorem 5.

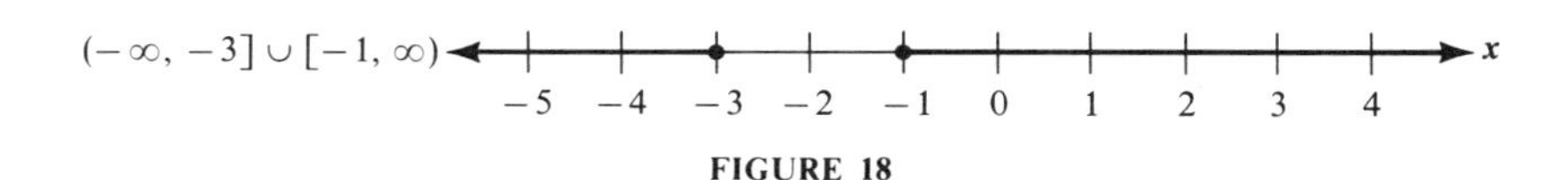

FIGURE 18

These intervals are the union of two sets.

$$\{x \,|\, x \leq -3\} \cup \{x \,|\, x \geq -1\}$$

EXAMPLE 16 Solve

$$\left|\frac{2x - 3}{2}\right| \geq 1$$

and graph its solution set.

SOLUTION Write $|(2x - 3)/2| \geq 1$ in the form $|x - (3/2)| \geq 1$. Then using the definition of a neighborhood, x represents all the points equal to or greater than 1 unit from 3/2.
Therefore, the solution is $\{x \,|\, x \leq \frac{1}{2}\} \cup \{x \,|\, x \geq \frac{5}{2}\}$.

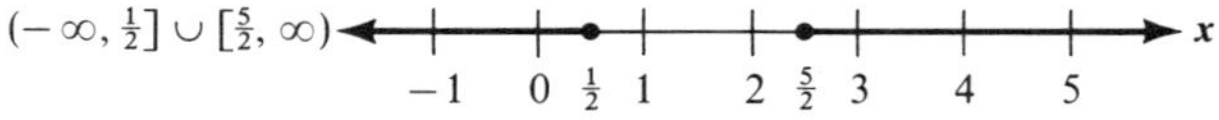

$(-\infty, \frac{1}{2}] \cup [\frac{5}{2}, \infty)$

$-1 \quad 0 \quad \frac{1}{2} \quad 1 \quad 2 \quad \frac{5}{2} \quad 3 \quad 4 \quad 5$ x

FIGURE 19

EXAMPLE 17 Solve

$$\frac{3x + 1}{x - 1} < 1$$

and graph its solution set.

SOLUTION We need to multiply both sides by $x - 1$, but we do not know whether $x - 1$ is positive or negative. Thus, we must consider two cases.

Case I

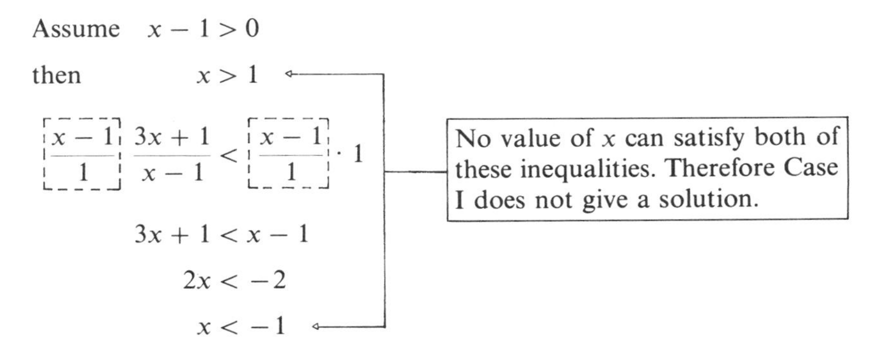

Case II

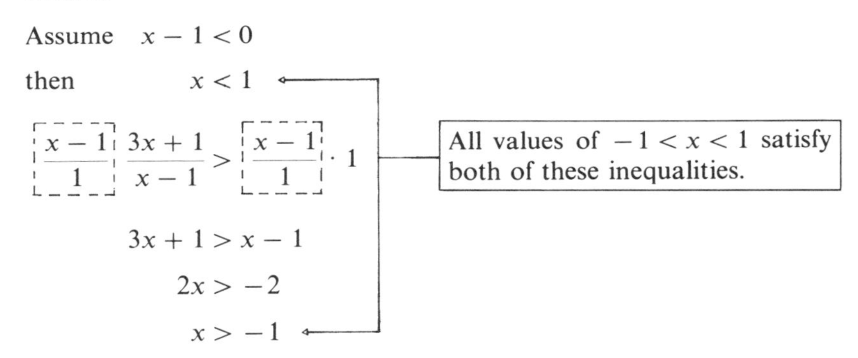

Therefore, the solution set is $\{x \mid -1 < x < 1\}$.

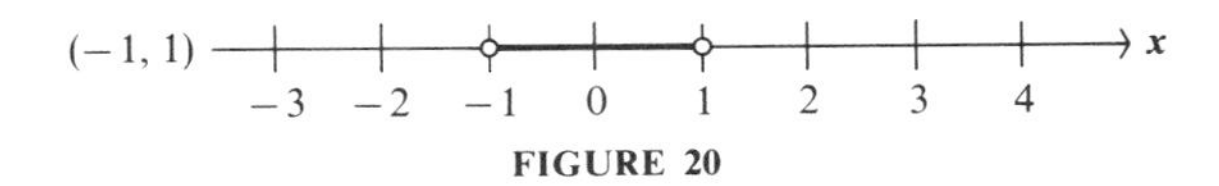

FIGURE 20

EXAMPLE 18 Solve the double inequality $2x - 8 < x - 1 < 2x - 4$.

SOLUTION Our plan is to remove the x-terms from the first and last members of the inequality and the numbers from the center member of the inequality.

$$2x - 8 < x - 1 < 2x - 4$$

$$-7 < -x < -3 \qquad \text{Add } -2x + 1.$$

$$7 > x > 3 \qquad \text{Multiply by } -1; \text{ change the sense.}$$

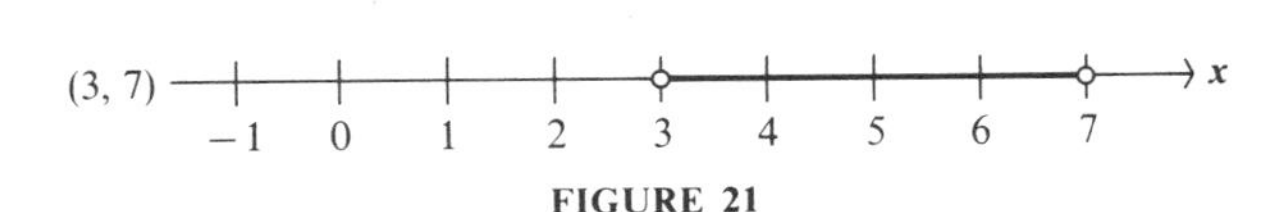

FIGURE 21

EXAMPLE 19 Solve the double inequality $2x - 3 < x + 5 < 3x - 4$.

SOLUTION This is a continued inequality and the solution is the intersection of two sets. We must solve as two separate problems.

$\{2x - 3 < x + 5\}$		$\cap$	$\{x + 5 < 3x - 4\}$	
$x < 8$	Add $-x + 3$.		$9 < 2x$	Add $-x + 4$.
			$\frac{9}{2} < x$	Divide by 2.

Therefore, the solution set is the intersection $\{x \mid x < 8\} \cap \{x \mid x > \frac{9}{2}\} = \{x \mid 4\frac{1}{2} < x < 8\}$.

FIGURE 22

EQUATIONS CONTAINING ABSOLUTE VALUE SYMBOLS

EXAMPLE 20 Solve $|x| = 3$ and graph the solution set.

SOLUTION $|x| = 3$ implies $x = \pm 3$. That is, $|x| = 3$ represents two points each 3 units from 0. Therefore, the solution set is $\{-3, 3\}$.

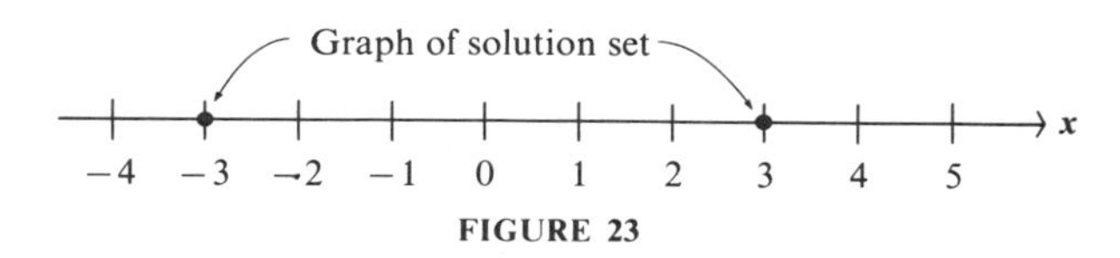

FIGURE 23

EXAMPLE 21 Solve $|x - 1| = |2x - 3|$ and graph the solution set.

SOLUTION $|x - 1| = |2x - 3|$ is equivalent to two statements.

$x - 1 = +(2x - 3)$	$x - 1 = -(2x - 3)$
$x - 1 = 2x - 3$	$x - 1 = -2x + 3$
$2 = x$	$3x = 4$
	$x = \frac{4}{3}$

Therefore, the solution set is $\{\frac{4}{3}, 2\}$.

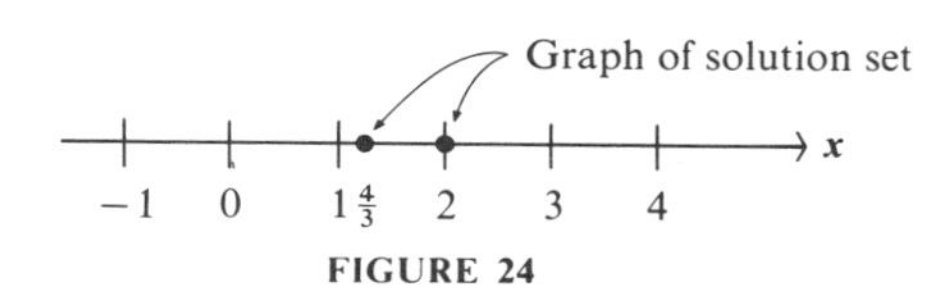

FIGURE 24

EXAMPLE 22 Solve $x^2 - 16 < 0$ and graph the solution set.

SOLUTION

$$x^2 - 16 < 0$$
$$x^2 < 16$$
$$\sqrt{x^2} < \sqrt{16}$$
$$|x| < 4$$
$$-4 < x < 4 \qquad \text{Theorem 4.}$$

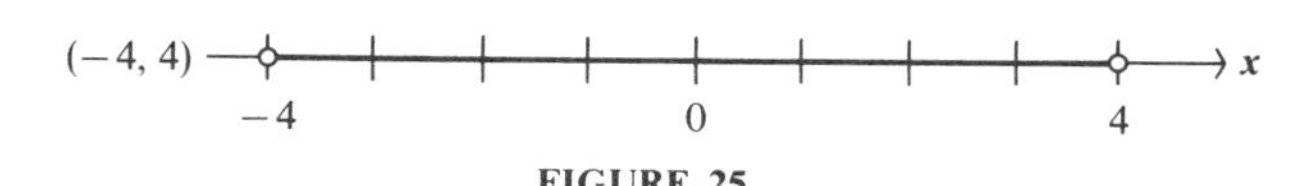

FIGURE 25

Exercises 7

A

1. Write > or < to make each expression true.
   a. -5 __ -2 **b.** 3 __ $|-4|$
   c. $|1 - \sqrt{13}|$ __ 0

In Exercises 2–10, solve and then graph the solution set.

2. $2x - 3 < x - 4$
3. $|x| = 2$
4. $4x - 3 \geq 2x + 5$
5. $|2x - 1| = 7$
6. $3 < 5x - 2 < 8$
7. $|2x - 3| = |x - 1|$
8. $\dfrac{2x - 3}{5} \geq \dfrac{x}{2} - 1$
9. $5x + 1 \leq x - 3 \leq 3x + 5$
10. $x^2 - 9 < 0$

In Exercises 11–14, sketch the interval or intervals represented by each statement and write an equivalent statement using interval notation.

11. $|x - 2| < 1$
12. $|x + 1| < \dfrac{1}{2}$
13. $|2 - x| > 1$
14. $\left|\dfrac{x + 1}{2}\right| \leq 2$
15. If a and b are both positive numbers and $a > b$, show that $a^2 > b^2$.

B

1. Write > or < to make each expression true.
   a. -3 __ -1 **b.** $|-5|$ __ -4
   c. $|1 - \sqrt{15}|$ __ 0

In Exercises 2–10, solve and then graph the solution set.

2. $3x + 1 < x - 3$
3. $|x| = 3$
4. $2x - 1 \geq 5x + 2$
5. $|2x - 1| = 3$
6. $3 < 2x - 3 < 5$
7. $|3x - 4| = |x + 2|$
8. $\frac{3x - 1}{2} \geq \frac{5x}{3} - 1$
9. $2x + 1 > x + 4 > 3x - 4$
10. $x^2 < 4$

In Exercises 11–14, sketch the interval or intervals represented by each statement and write an equivalent statement using interval notation.

11. $|x - 1| < 2$
12. $|x + 1| \leq 1$
13. $|3 - x| > 2$
14. $\left|\frac{2x + 1}{3}\right| \geq 1$
15. If $a + b > 1$ and $a > b$, show that $a^2 - a > b^2 - b$.

Summary

Equations are of two types, **identities** or **conditional equations**. A **conditional equation** is true only for certain values of the variables. A **solution** (or **root**) of a conditional equation is a value of the variable for which the equation is a true statement. The **solution set** of an equation is the set of values of the variables that make the equation a true statement. An **identity** is a statement of equality that is true for all values of the variables, with the exception of those values for which any member of the equality does not have meaning.

When we multiply an equation by an expression containing the variable, or when we raise both sides to a power, we may introduce new roots (called **extraneous** roots) that do not satisfy the original equation. Such equations are called **redundant equations**.

A defective equation is one that has fewer roots than an equation from which it was obtained. Defective equations may be obtained by dividing both sides of an equation by an expression containing the variable.

A **quadratic equation** is an equation that can be written in the form $ax^2 + bx + c = 0$, where a, b, and c are real numbers and $a \neq 0$. The quadratic formula for finding the roots of the quadratic equation is

$$x = \frac{-b \pm \sqrt{b^2 - 4ac}}{2a}.$$

The radicand, $b^2 - 4ac$, which appears under the radical sign in the quadratic formula is called the **discriminant** of the quadratic equation.

DISCRIMINANT TESTS

Given the quadratic equation

$$ax^2 + bx + c = 0, \quad \text{where } a, b, c \in R \text{ and } a \neq 0.$$

	Nature of roots
1. $b^2 - 4ac > 0$ and a perfect square	Real, unequal, and rational
2. $b^2 - 4ac > 0$ and **not** a perfect square	Real, unequal, and irrational.
3. $b^2 - 4ac = 0$	Real and equal (root of multiplicity two).
4. $b^2 - 4ac < 0$	Conjugate imaginaries.

An **inequality** is a statement that one mathematical quantity is less than (or greater than) another. The **solution set** of an inequality is the set of all numbers which, when substituted for the variable (or variables), make the inequality a true statement. Inequalities are of two types, **conditional** and **unconditional**. A conditional inequality is an inequality which is true only for certain values of the variable(s). An unconditional inequality (or **absolute inequality**) is an inequality which is true for all values of the variable(s), or which contains no variables.

INEQUALITY SYMBOLS USED IN THIS CHAPTER

$<$	Is less than.
$>$	Is greater than.
$\leq$	Is less than or equal to.
$\geq$	Is greater than or equal to.

If the inequality signs point in the same direction, the inequalities are said to have the **same sense**; if the signs point in opposite directions, the inequalities are **opposite in sense**.

An interval of real numbers is the set of all real numbers between two given numbers (the endpoints of the interval) and one, both, or neither endpoint.

INTERVAL NOTATION

1. (a, b): **Open interval.** All real numbers between, but not including a and b, $a < b$.
2. $[a, b]$: **Closed interval.** All real numbers between and including a and b, $a < b$.
3. $(a, b]$, $[a, b)$, $(-\infty, b]$, $[a, \infty)$: **Half-open** or **half-closed intervals**. In each case the interval contains one endpoint but not the other.

THEOREM 1 The addition of the same real number to, or the subtraction of the same real number from, both sides of an inequality leaves the sense of the inequality unchanged.

THEOREM 2 The multiplication or division of both sides of an inequality by the same *positive* number leaves the sense of the inequality unchanged.

THEOREM 3 If both sides of an inequality are multiplied or divided by a negative number, a new inequality of opposite sense is obtained.

The expression $|x - a| < b$ is frequently referred to as a **neighborhood of *a***. The expression $|x - a| < b$ represents the open interval $(a - b, a + b)$, as shown in Figure 15.

Review exercises

A

In Exercises 1 and 2, determine whether the equation is an identity or a conditional equation. If conditional, give the solution set.

1. $2x(x - 5) + (2x + 7)(3x + 2) = 8x^2 + 15x + 14$
2. $2[x - x(x + 2) - 1] = 1 - (3x - 1)(x + 2) - 4(x + 1)$
3. Graph each of the indicated intervals.
   a. $[-3, 1)$ **b.** $[-1, 2] \cup [3, 5]$

In Exercises 4–7, solve, then graph and write the solution sets.

4. $|x - 2| \geq 1$
5. $3x - 1 > x + 2$
6. $|5x - 1| = |2x - 3|$
7. $2x - 3 < 3x + 1 \leq 5x - 3$

In Exercises 8–13, solve the equations.

8. $15x^2 - x = 2$

9. $x^3 + 5x^2 = 14x$

10. $\dfrac{5}{x+2} + \dfrac{1}{x-2} = \dfrac{10}{2x-1}$

11. $4x^2 - 8x + 13 = 0$

12. $x^4 + 5x^2 = 36$

13. $2\sqrt{2x+1} = \sqrt{x+5} + 3$

14. Form a third-degree equation with solution set

$$\left\{3, \frac{-3 + 3i\sqrt{3}}{2}, \frac{-3 - 3i\sqrt{3}}{2}\right\}.$$

15. Use the discriminant to determine the nature of the roots of the equation $5x^2 - 2x - 7 = 0$.

16. Find the dimensions of the right triangle shown in the figure below.

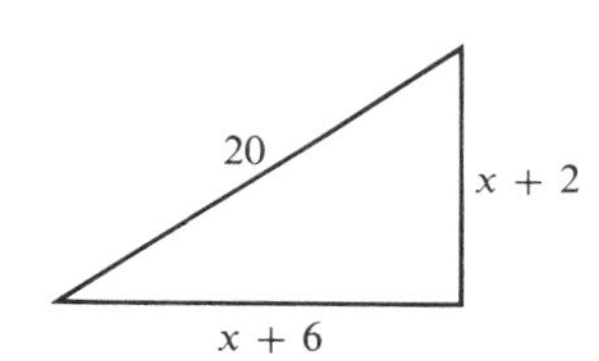

B In Exercises 1 and 2, determine whether the equation is an identity or a conditional equation. If conditional, give the solution set.

1. $2x(4x + 5) - 3 = (x + 1)(7x - 3)$

2. $(x - 1)(3x + 5) + 4 = (3x - 1)(x + 1)$

3. Graph each of the indicated intervals.

a. $(-1, 3]$ **b.** $[-2, 0] \cup [1, 4)$

In Exercises 4–7, solve, then graph and write the solution sets.

4. $|x - 3| < 1$

5. $2x + 1 > 4x - 3$

6. $|5x - 3| = |x + 1|$

7. $3x - 1 \leq x + 3 \leq 2x + 5$

In Exercises 8–13, solve the equations.

8. $10x^2 - x = 3$

9. $x^3 - 2x^2 - 15x = 0$

10. $\dfrac{2}{x-3} + \dfrac{5}{x+1} = \dfrac{6}{x-2}$

11. $2x^2 - 6x + 5 = 0$

12. $6x^{2/3} - x^{1/3} = 1$

13. $2\sqrt{3x-2} = \sqrt{4x+1} + 1$

14. Form a fourth-degree equation with solution set $\{-1, 2, 3 + i, 3 - i\}$.

15. Use the discriminant to determine the nature of the roots of the equation $4x^2 - 15x + 17 = 0$.
16. A box with no top is to be formed from a rectangular sheet of metal by cutting 3-inch squares from the corners and folding up the sides. The length of the box is to be 4 inches more than its width, and its volume is to be 63 cubic inches.
    a. Find the dimensions of the sheet of metal.
    b. Find the dimensions of the box.

Diagnostic test

The purpose of this test is to see how well you understand the work of this chapter. Allow yourself approximately 55 minutes to do the test. Solutions to the problems, together with section references, are given in the Answers section at the end of the book. We suggest that you study the sections referred to for the problems you do incorrectly.

1. Determine whether the following equation is an identity or a conditional equation. Show your work.

$$3[x - x(x - 2) - 1] = 3x(4 - x) - (3 + 3x)$$

In Exercises 2–4, solve, then graph and write the solution sets.

2. $|x - 3| \leq 1$
3. $4x + 1 > 3x - 2 \geq x + 6$
4. $|3x - 2| = |x + 2|$

In Exercises 5–8, solve the equations.

5. $10x^2 - x = 3$
6. $4x^2 - 4x - 1 = 0$
7. $\dfrac{4}{x - 1} - \dfrac{1}{x + 1} = \dfrac{4 - 2x^2}{x^2 - 1}$
8. $\sqrt{2x - 1} = \sqrt{x - 1} + 1$
9. Solve $2x^{2/3} - 5x^{1/3} = 3$.
10. Form a fourth-degree equation with the solution set $\{3, -4, 2i, -2i\}$.
11. Use the discriminant to determine the nature of the roots of the equation $6x^2 - 7x + 3 = 0$.
12. A box with no top is to be formed from a rectangular sheet of metal by cutting 4-inch squares from the corners and folding up the sides. The length of the box is to be 2 inches more than its width, and its volume is to be 252 cubic inches.
    a. Find the dimensions of the sheet of metal.
    b. Find the dimensions of the box.

6 Functions, linear equations and inequalities, and graphs

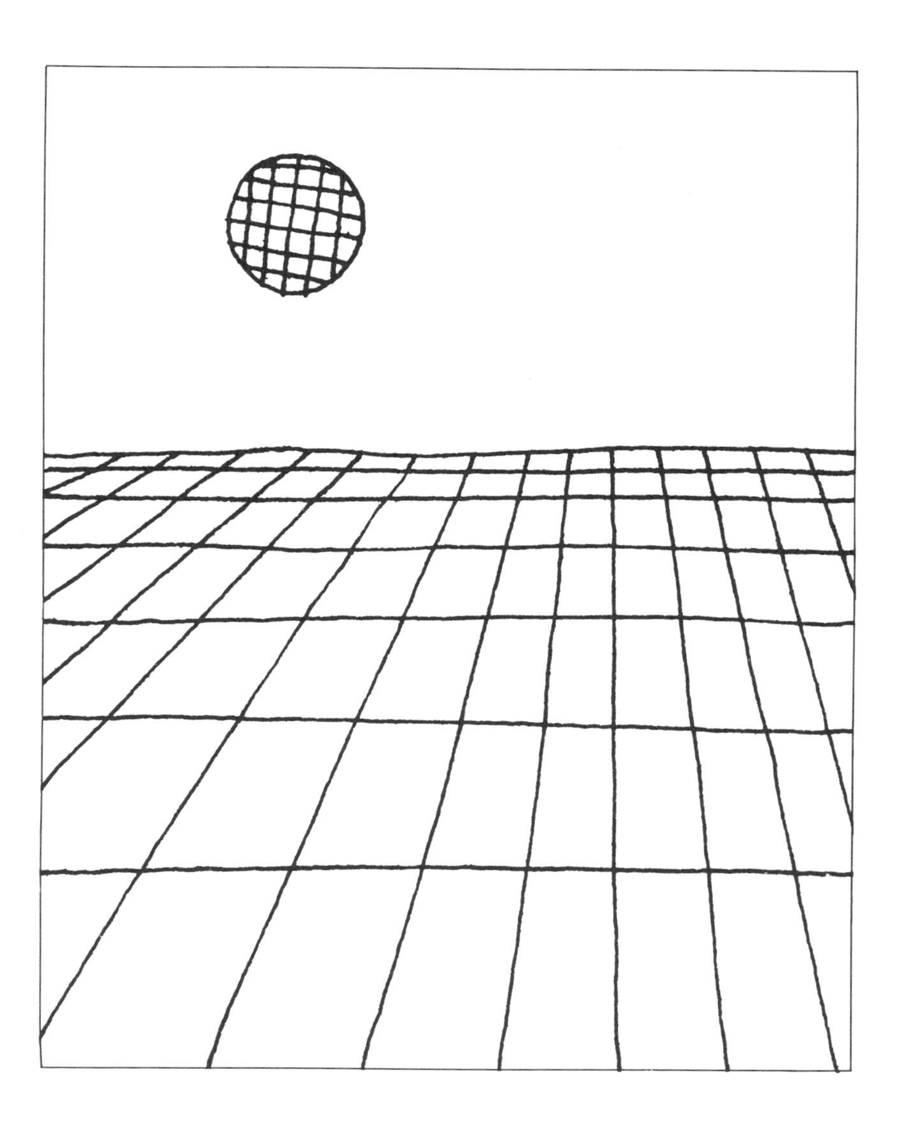

Relations and **functions** are fundamental parts of mathematics. They appear in many kinds of equations and formulas. As an example, the formula for the circumference C of a circle in terms of its diameter d, $C = \pi d$, defines C as a function of d. The expression $y = 2x - 1$ defines y as a function of x, and so on.

Before giving formal definitions of a relation and a function, we review our rectangular coordinate system.

1 The rectangular coordinate system

In Section 7 of Chapter 5, we plotted points and sets of real numbers on a number line. This is sometimes called **one-space**. In this chapter, we will also be working in **two-space**. The two-space graph that we use is a **rectangular coordinate system**.

For a review of the terms used in the rectangular coordinate system, see Figures 1 and 2.

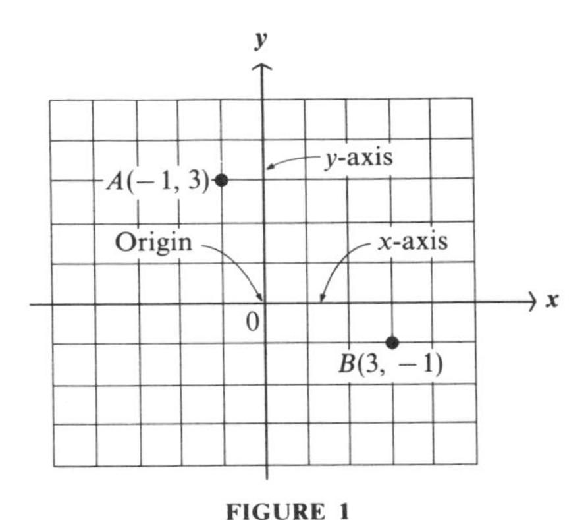

FIGURE 1

The *xy*-plane

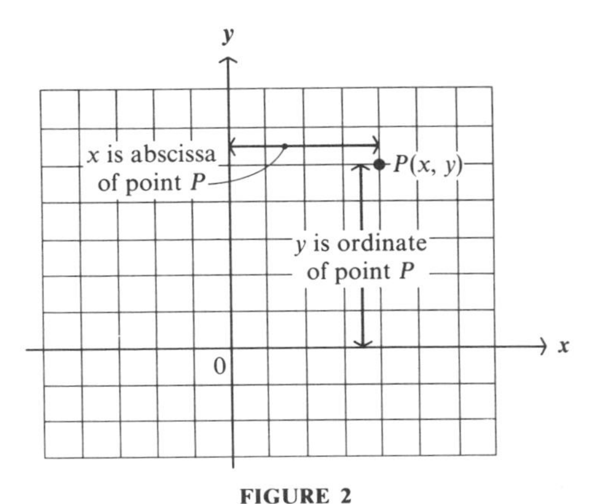

FIGURE 2

Ordinate and abscissa of a point

Points on the ***xy*-plane** are located by means of **ordered pairs**. As shown in Figure 1, the pairs $(-1, 3)$ and $(3, -1)$ represent two entirely different points, A and B. An ordered pair of **coordinates** is a set in which the order of the elements is significant.

In the ordered pair $(-1, 3)$, -1 is called the **first coordinate** (or **abscissa**) of point A and 3 is called the **second coordinate** (or **ordinate**) of point A. Together they are called the **coordinates** of A.

2 Relations and functions

RELATION

A relation is a set of ordered pairs of mathematical quantities.

EXAMPLE 1 Examples of relations.

a. $\{(1, 2), (3, 4), (1, 3), (5, 8)\}$
b. $\{(a, b), (a, c), (c, d), (e, f)\}$
c. $\{(x, y) \mid x < 3 \quad \text{and} \quad y > -1\}$

The domain of a relation is the set of first coordinates of the ordered pairs of that relation. The domain in part a of Example 1 is $\{1, 3, 5\}$.

The range of a relation is the set of second coordinates of the ordered pairs of that relation. The range in part a of Example 1 is $\{2, 3, 4, 8\}$.

EXAMPLE 2 Graph the relation and state the domain and range of $\{(2, 0), (4, 2), (2, 3), (-2, 3)\ (-2, -2)\}$.

SOLUTION

Domain: $\{2, 4, -2\}$
Range: $\{0, 2, 3, -2\}$

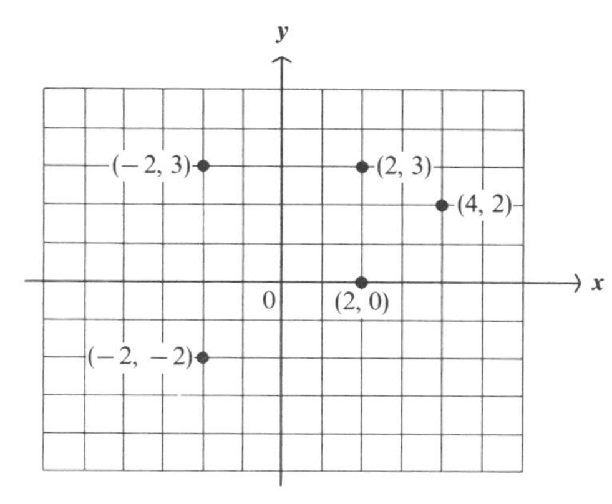

FIGURE 3

> **DEFINITION OF FUNCTION**
>
> A function f is a relation such that no two ordered pairs of the set have different second coordinates corresponding to the same first coordinate. The set of all first coordinates of the ordered pairs is the **domain of f**. The set of all second coordinates of the ordered pairs is the **range of f**.

By examining the graph for Example 2, you can see that the ordered pairs (2, 0) and (2, 3), and (−2, 3) and (−2, −2), have the same first coordinates but different second coordinates. Because of this fact, the set of points in Example 2 does not represent a function. It is a relation.

EXAMPLE 3 Which of the following are graphs of functions?

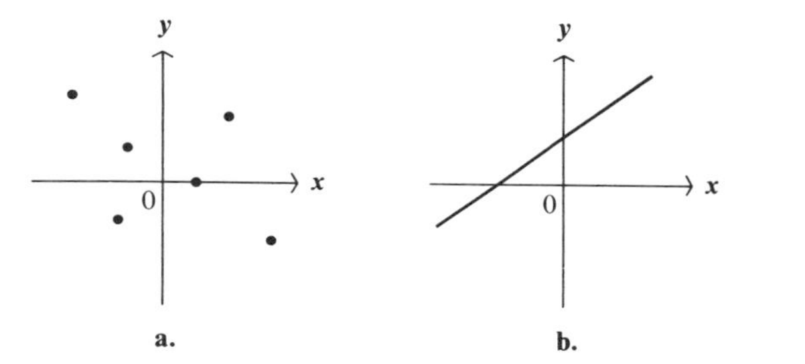

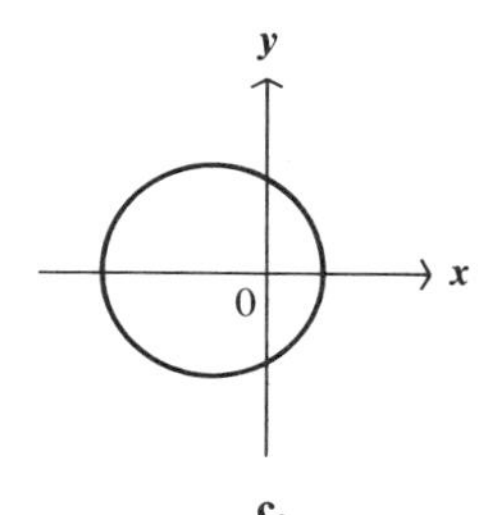

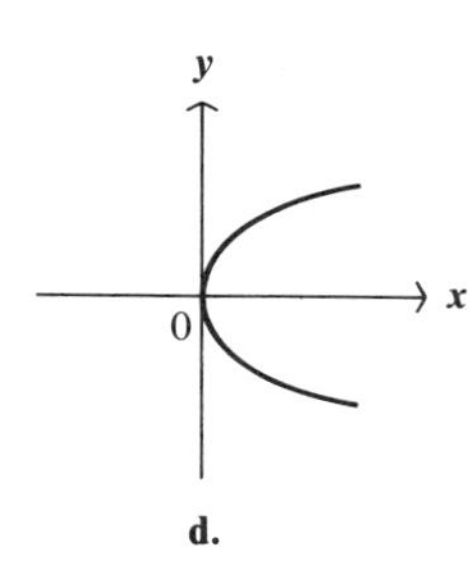

FIGURE 4

SOLUTION To be a function, no two ordered pairs of the set can have the same first coordinates and different second coordinates. This means that no vertical line can intersect the graph in more than one point. By applying this test, we see that **a** and **b** are graphs of functions. Graphs **c** and **d** are relations (they are not functions).

If we have a rule or formula giving y in terms of x and there is no more than one value of y for each value of x, then y is said to be a function of x. This can be written $y = f(x)$ and is read "y is a function of x."

EXAMPLE 4 Given $f(x) = 2x - 1$, find:

a. $f(2)$ **b.** $f(0)$

c. $\dfrac{f(3) - f(2)}{f(4)}$ **d.** $\dfrac{f(x+h) - f(x)}{h}, \quad h \neq 0$

SOLUTION **a.** To find $f(2)$, merely replace x with 2 and then evaluate the resulting expression.

$$f(x) = 2x - 1$$
$$f(2) = 2(2) - 1 = 3$$

b.

$$f(x) = 2x - 1$$
$$f(0) = 2(0) - 1 = -1$$

c.

$$f(x) = 2x - 1$$
$$f(3) = 2(3) - 1 = 5$$
$$f(2) = 2(2) - 1 = 3$$
$$f(4) = 2(4) - 1 = 7$$

Therefore,

$$\frac{f(3) - f(2)}{f(4)} = \frac{5 - 3}{7} = \frac{2}{7}$$

d.

$$f(x) = 2x - 1$$
$$f(x + h) = 2(x + h) - 1$$
$$= 2x + 2h - 1$$

Therefore,

$$\frac{f(x + h) - f(x)}{h} = \frac{2x + 2h - 1 - (2x - 1)}{h}$$
$$= \frac{2x + 2h - 1 - 2x + 1}{h}$$
$$= \frac{2h}{h} = 2$$

Other names, such as $P(x)$, $Q(x)$, $g(x)$, or simply f, are used to represent a function of x. If the function is given in terms of the variable z or θ, we represent the function by $f(z)$, or $f(\theta)$. Of course, we could also use $P(z)$ or $P(\theta)$. When more than one function is used in a discussion, such as $y = 2x + 3$ and $y = -x + 2$, we could, for example, name one function $f(x)$ and the other $g(x)$. Notation such as $f(x)$, $g(x)$, and $P(\theta)$ is used to name particular functions in a given discussion.

EXAMPLE 5 Let $f(x) = 2x + 3$ and $g(x) = -x + 5$. Find $f(4) \div g(3)$.

SOLUTION

$$\frac{f(4)}{g(3)} = \frac{2(4) + 3}{-(3) + 5} = \frac{11}{2}$$

EXAMPLE 6 Given the function $\{(-1, -3), (0, -1), (1, 1), (2, 3)\}$.

a. Write its domain.
b. Write its range.
c. Graph the function.

SOLUTION

a. The domain is the set of all first coordinates of the ordered pairs of the function.

$$\text{Domain} = \{-1, 0, 1, 2\}.$$

b. The range is the set of all second coordinates of the ordered pairs of the function.

$$\text{Range} = \{-3, -1, 1, 3\}.$$

c. The graph of the function consists of the four points corresponding to the ordered pairs (see Figure 5).

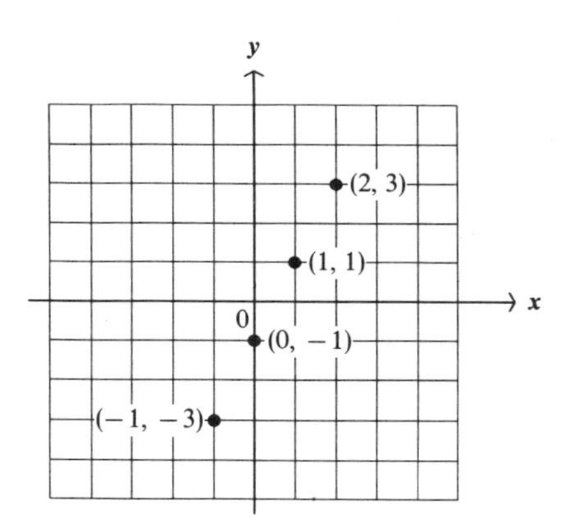

FIGURE 5

When a function is written in the form $y = f(x)$, as in $y = 2x + 1$, x is called the **independent variable** and y the **dependent variable**. We say y **depends** for its value on the number assigned to x. We also say that y is expressed **explicitly** in terms of x, or that y is a function of x. The function $2x - y + 1 = 0$ is equivalent to the function $y = 2x + 1$. In the form $2x - y + 1 = 0$, the function is called an **implicit function**. An implicit function is a function not expressed in terms of a single variable. We can solve either form of the above function for x as a function of y.

$$2x - y = -1$$

$$x = \frac{y-1}{2} \quad \text{or} \quad f(y) = \frac{y-1}{2} \quad \text{or} \quad x = f(y)$$

In this latter form, x is expressed explicitly in terms of y. This makes y the independent variable and x the dependent variable.

IMPLICIT AND EXPLICIT FUNCTIONS

	Equivalent functions	Type of function
1.	$2x - y + 1 = 0$	Implicit.
2.	$y = 2x + 1$	Explicit.
	or $f(x) = 2x + 1$	Explicit.
3.	$x = \frac{y - 1}{2}$	Explicit.
	or $f(y) = \frac{y - 1}{2}$	Explicit.

EXAMPLE 7 Solve the implicit function $2x - 3y + 12 = 0$ for y as a function of x.

SOLUTION

$$2x - 3y + 12 = 0$$

$$-3y = -2x - 12$$

$$\frac{-3y}{-3} = \frac{-2x}{-3} - \frac{12}{-3}$$

Therefore,

$$y = \frac{2}{3}x + 4$$

or

$$f(x) = \frac{2}{3}x + 4$$

EXAMPLE 8 Give the domain and range of the function $f(x) = \sqrt{x + 3}$.

SOLUTION In order that $f(x)$ be a real number, $x + 3$ must be greater than or equal to zero. That is,

$$x + 3 \geq 0$$

$$x \geq -3$$

Therefore, the domain is $\{x \mid x \geq -3\}$. The expression $\sqrt{x + 3}$ is never negative (see principal roots, Section 7 of Chapter 1). Therefore, the range is $\{f(x) \mid f(x) \geq 0\}$.

Exercises 2

A

1. Determine if each set is a function. Give a reason for each answer.
   a. $\{(-1, 2), (0, 3), (1, 4), (2, 5)\}$
   b. $\{(-2, 1), (1, 2), (0, -3), (0, 2)\}$
2. Give the range and the domain of the function
$$\{(-5, -2), (-3, -1), (-1, 1), (1, 3), (2, 5)\}.$$
3. Graph the relation $\{(-2, -3), (-1, 2), (-2, 4), (2, 3)\}$.
4. Which of the graphs in Figure 6 represent functions?

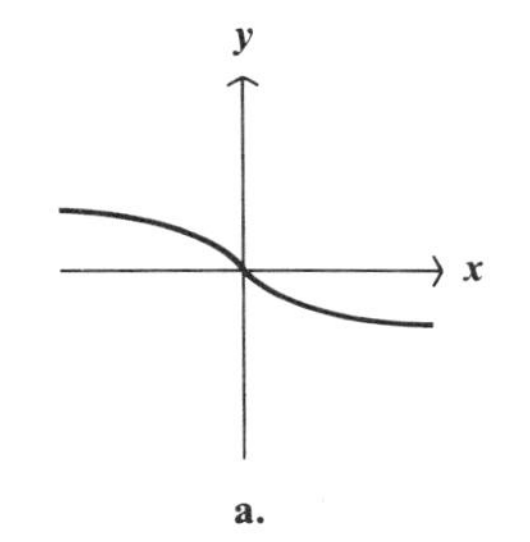

a.

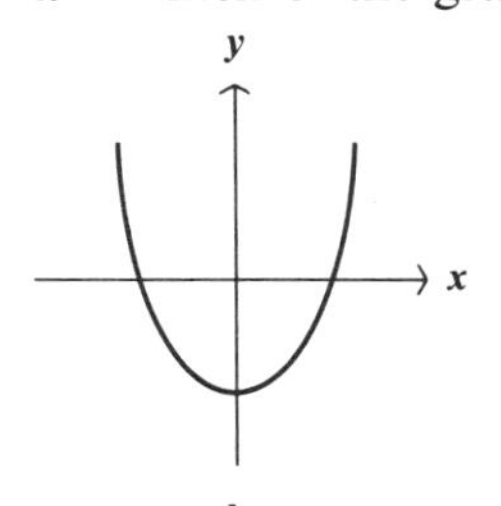

b.

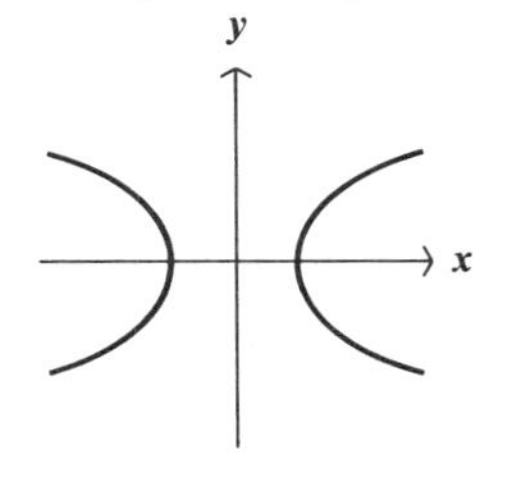

c.

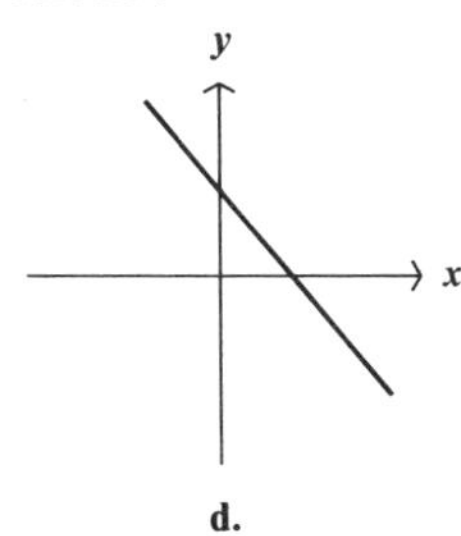

d.

FIGURE 6

5. Name the independent variable of the function $x = 5z + 2$.
6. Solve the implicit function $4x + 3y + 12 = 0$ for y as a function of x.
7. Given $f(x) = 5x - 2$, find each of the following.
   a. $f(0)$ **b.** $f(-3)$ **c.** $\dfrac{f(x+h) - f(x)}{h}$
8. Give the domain and range of the function $f(x) = \sqrt{2x + 1}$.

B

1. Determine if each set is a function. Give a reason for each answer.
   a. $\{(-3, 4), (1, 2), (2, 2), (3, 1)\}$
   b. $\{(1, 3), (2, 4), (5, 3), (1, 4)\}$
2. Give the range and the domain of the function
$$\{(-2, 0), (1, 2), (3, 4), (5, -2)\}.$$
3. Graph the relation $\{(1, -2), (3, 1), (3, 3), (-3, -2)\}$.
4. Which of the graphs in Figure 7 represent functions?

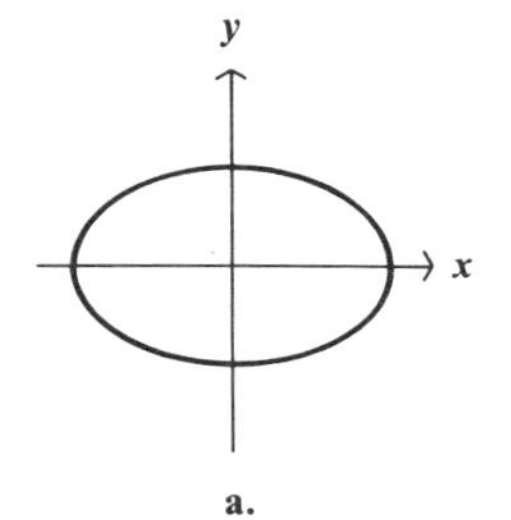

a.

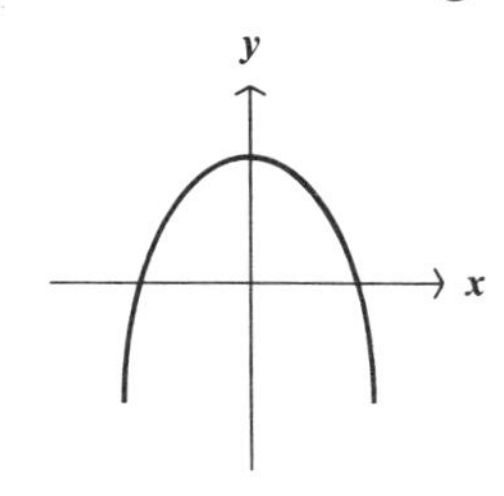

b.

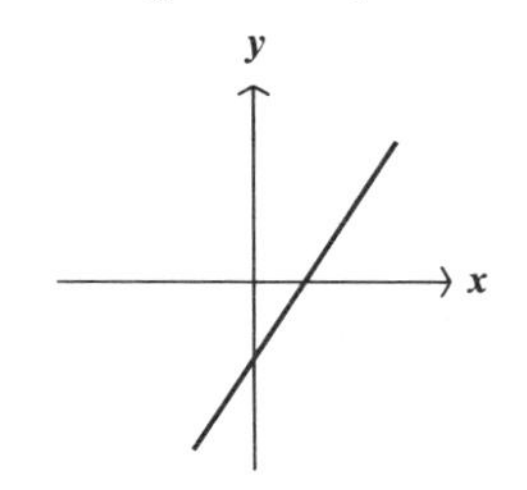

c.

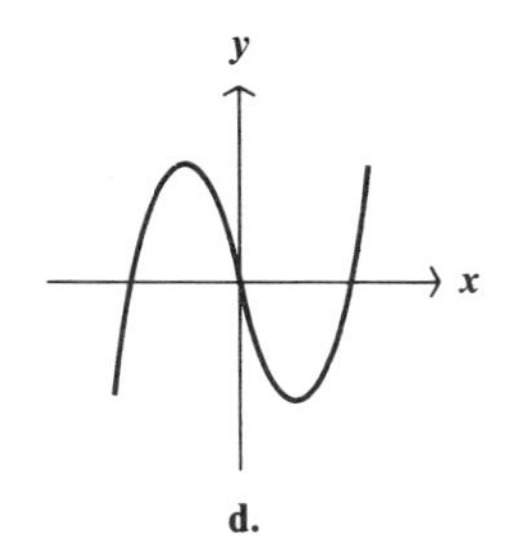

d.

FIGURE 7

5. Name the independent variable of the function $y = 3x - 5$.
6. Solve the implicit function $3x + 2y - 6 = 0$ for y as a function of x.
7. Given $f(x) = 3x - 4$, find each of the following.

 a. $f(0)$ **b.** $f(2)$ **c.** $\dfrac{f(x+h) - f(x)}{h}$

8. Give the domain and range of the function $f(x) = \sqrt{x - 2}$.

3 Equations of lines

When the term *line* is used in mathematics, it is understood to be a straight line with no endpoints. Equations which define straight lines are called **linear equations**. They are first-degree equations. Linear equations whose graphs are vertical lines are **linear relations**. For example, $x = 2$. Linear equations whose graphs are not vertical lines are **linear functions**. For example, $f(x) = 2x + 3$.

GENERAL FORM OF LINEAR EQUATIONS

$$Ax + By + C = 0, \quad \text{where } A, B, C \in R, \quad A \text{ and } B \text{ not both zero.} \tag{1}$$

When the coordinates of a point on a line and its steepness, or **slope**, are known, the line is determined and we can write its equation. Recall from geometry that two points determine a line. Thus, knowing the coordinates of two points on a line enables us to write its equation. Before deriving the equations of lines, we will develop the concept of slope of a line.

SLOPE OF A LINE

If (x_1, y_1) and (x_2, y_2) are the coordinates of two distinct points on a nonvertical line, then the slope of the line is the number

$$m = \frac{y_2 - y_1}{x_2 - x_1}.$$

This value m is a ratio of differences of corresponding coordinates of the given distinct points, and it can be interpreted as the change in ordinate divided by the change in abscissa between the points. The ratio m is the same for any two distinct points selected on the line, and as such it gives the rate of change of ordinate with respect to abscissa for a point moving along the line. We illustrate with Figure 8.

Let $P_1(x_1, y_1)$ and $P_2(x_2, y_2)$ be two points on line P_1P_2. The letter m is used to represent the slope.

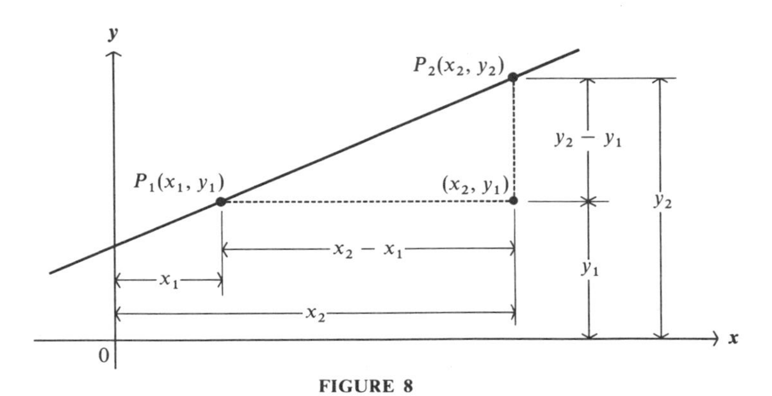

FIGURE 8

> **SLOPE m OF A LINE**
>
> $$m = \frac{y_2 - y_1}{x_2 - x_1}, \qquad x_2 \neq x_1$$

Two lines are parallel if and only if they have the same slope.

EXAMPLE 1 Find the slope of the line that passes through the points $A(-2, -1)$ and $B(4, 3)$.

SOLUTION We choose $P_1(x_1, y_1) = A(-2, -1)$ and $P_2(x_2, y_2) = B(4, 3)$. (See Figure 9.)

$$m = \frac{y_2 - y_1}{x_2 - x_1} = \frac{3 - (-1)}{4 - (-2)} = \frac{4}{6} = \frac{2}{3}$$

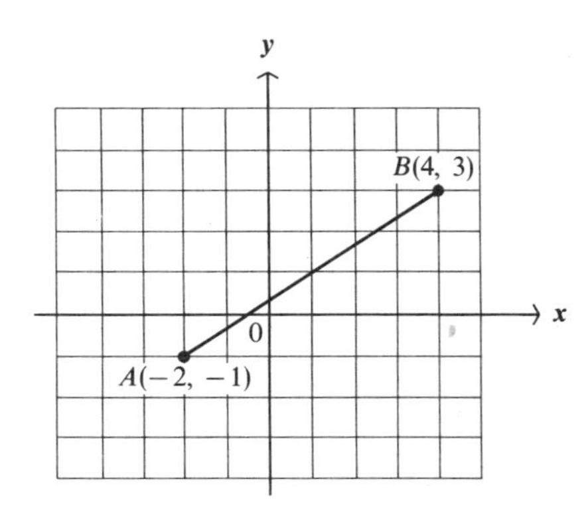

FIGURE 9

The slope is not changed if the points are interchanged in the formula. That is, if we choose $P_1(x_1, y_1) = B(4, 3)$ and $P_2(x_2, y_2) = A(-2, -1)$, we have

$$m = \frac{y_2 - y_1}{x_2 - x_1} = \frac{(-1) - (3)}{(-2) - (4)} = \frac{-4}{-6} = \frac{2}{3}.$$

The slope of a line is positive if a point moving along the line in a positive x-direction rises. The slope is negative if a point moving along the line in a positive x-direction falls. For horizontal lines, the ordinate never changes, therefore, a point moving along the line in a positive x-direction stays at the same level. Horizontal lines have a slope of zero, as shown by the expression

$$m = \frac{y_2 - y_1}{x_2 - x_1} = \frac{0}{x_2 - x_1} = 0.$$

The slope of a vertical line is undefined, as shown by the expression

$$m = \frac{y_2 - y_1}{x_2 - x_1} = \frac{y_2 - y_1}{0},$$

which is undefined. All points on vertical lines have the same x-coordinate. Therefore, $x_2 = x_1$ and $x_2 - x_1 = 0$.

Figure 10 shows lines having positive slope, negative slope, zero slope, and no slope (slope undefined).

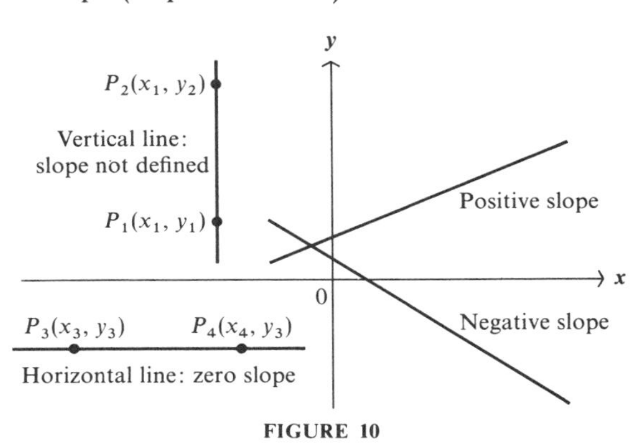

FIGURE 10

POINT-SLOPE FORM OF THE EQUATION OF A LINE

One of the simplest ways in which a line is determined occurs when the slope of the line and the coordinates of a point on the line are known.

Let $P_1(x_1, y_1)$ be a given point on a line and let m be the slope of the line. We let $P(x, y)$ be any point, different from P_1, on the given line (see Figure 11). By the slope formula, the slope of the line through $P_1 P$ is

$$\frac{y - y_1}{x - x_1} = m.$$

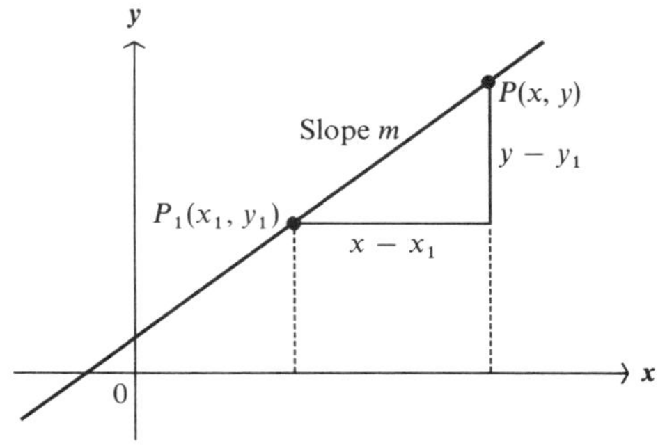

FIGURE 11

Multiplying both sides by $(x - x_1)$, we obtain

$$y - y_1 = m(x - x_1) \qquad \text{Point-slope form.}$$

POINT-SLOPE FORM OF THE EQUATION OF A LINE

Given a line containing the point $P_1(x_1, y_1)$ and with slope m, the point-slope equation of the line is

$$y - y_1 = m(x - x_1). \qquad \textbf{(2)}$$

Note: This equation is satisfied by the point $P_1(x_1, y_1)$.

EXAMPLE 2 Find an equation of the line that passes through the point $(3, -1)$ and has a slope of -2.

SOLUTION Let $(x_1, y_1) = (3, -1)$ and $m = -2$.

$$y - y_1 = m(x - x_1)$$
$$y - (-1) = (-2)(x - 3)$$
$$y + 1 = -2x + 6$$

Transforming this equation to the $Ax + By + C = 0$ form, we have

$$2x + y - 5 = 0.$$

INTERCEPTS

The **intercepts of a line** are the coordinates of the points where the line intersects the two axes (see Figure 12). To find the x-intercept, let $y = 0$ in the equation and solve for x. Similarly, to find the y-intercept, let $x = 0$ and solve for y.

LINES PARALLEL TO THE COORDINATE AXES

If a line is parallel to the y-axis and has a as its x-intercept, it follows that the abscissa of every point on this line is $x = a$. Conversely, we see that every point with abscissa $x = a$ lies on the given line (Figure 13). Therefore, the equation of any line parallel to the y-axis is of the form

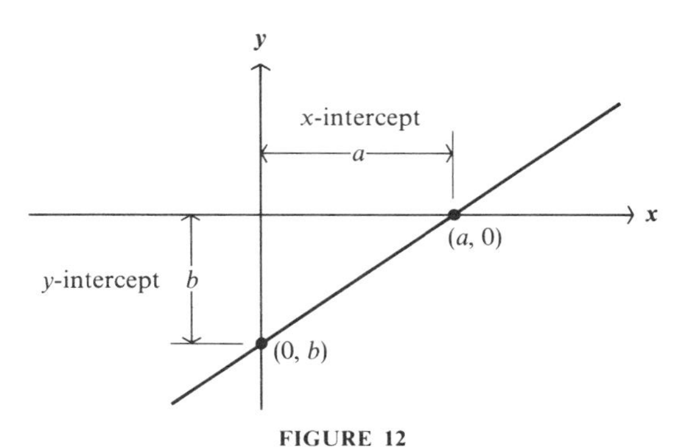

FIGURE 12

x- and y-intercepts

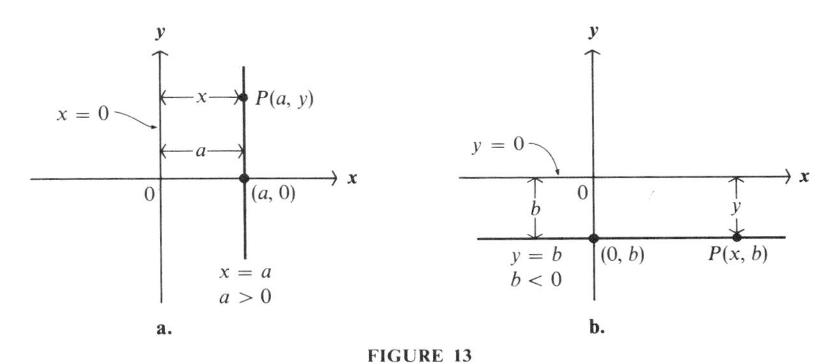

FIGURE 13

Line parallel to coordinate axis

$x = a$. In a similar way, it can be seen that the equation of a line parallel to the x-axis with the y-intercept b is of the form $y = b$.

LINES PARALLEL TO A COORDINATE AXIS

An equation of a line parallel to the y-axis is

$$x = a, \quad \text{where } a \text{ is the } x\text{-intercept of the line.} \tag{3}$$

An equation of a line parallel to the x-axis is

$$y = b, \quad \text{where } b \text{ is the } y\text{-intercept of the line.} \tag{4}$$

Note: $x = 0$ is the equation of the y-axis and $y = 0$ is the equation of the x-axis.

EXAMPLE 3 Find the x- and y-intercepts, and draw the graph of the equation $3x - 4y = 12$.

SOLUTION To find the x-intercept, let $y = 0$ and solve for x.

$$3x - 4y = 12$$
$$3x - 4(0) = 12$$
$$3x = 12$$
$$x = 4 \qquad x\text{-intercept.}$$

Thus the line intersects the y-axis at the point $(0, -3)$.

To find the y-intercept, let $x = 0$ and solve for y.

$$3x - 4y = 12$$
$$3(0) - 4y = 12$$
$$-4y = 12$$
$$y = -3 \qquad y\text{-intercept.}$$

Thus the line intersects the y-axis at the point $(0, -3)$.

This shows one easy way to graph a straight line: find and plot its x- and y-intercepts and then drawn the line through these points. This method does not work when the line passes through the origin. In that case, plot other points.

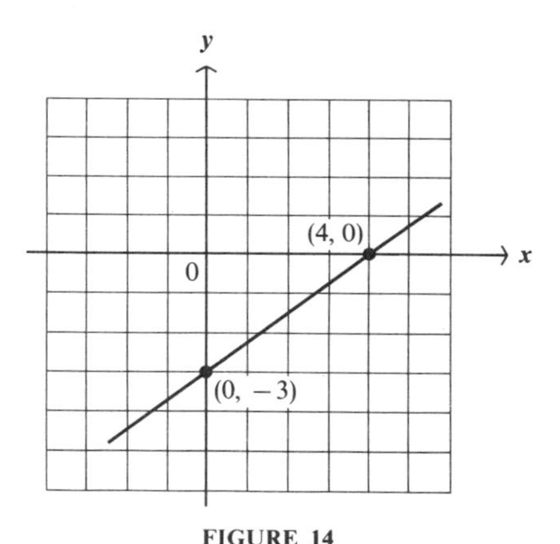

FIGURE 14

SLOPE-INTERCEPT FORM OF THE EQUATION OF A LINE

The slope-intercept form of the equation of a line is a special case of the point-slope form of the equation of a line. The given point in the slope-intercept form is a point on the y-axis, the y-intercept.

Let $(0, b)$ be the given point and m be the given slope. We then use the point-slope form.

$$y - y_1 = m(x - x_1)$$
$$y - b = m(x - 0)$$
$$y - b = mx$$
$$y = mx + b \qquad \text{Slope-intercept form.}$$

SLOPE-INTERCEPT FORM OF THE EQUATION OF A LINE

Given a line with slope m and y-intercept b, the *slope-intercept* equation of the line is

$$y = mx + b. \tag{5}$$

EXAMPLE 4 Transform the equation $3x - 4y - 8 = 0$ to the slope-intercept form and sketch its graph.

SOLUTION

$$3x - 4y - 8 = 0$$
$$-4y = -3x + 8$$
$$y = \frac{3}{4}x - 2$$

Comparing this to $y = mx + b$, we see that $m = \frac{3}{4}$ and $b = -2$. To graph the line, start at the point $(0, -2)$, then move 4 units to the right and up 3 units. This locates a second point and gives a slope of $\frac{3}{4}$.

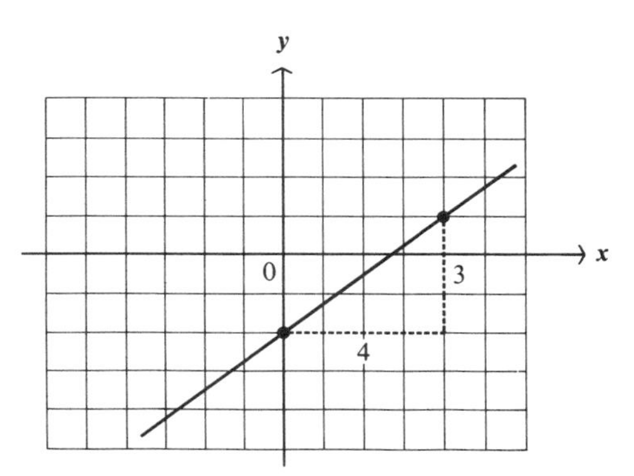

FIGURE 15

REDUCING THE $Ax + By + C = 0$ FORM OF A LINEAR EQUATION TO THE SLOPE-INTERCEPT FORM

We transform $Ax + By + C = 0$, $B \neq 0$, to the form $y = mx + b$.

$$Ax + By + C = 0$$
$$By = -Ax - C$$
$$y = \boxed{-\frac{A}{B}}x + \boxed{\frac{-C}{B}}$$

Comparing this to $\qquad y = \quad mx \quad + \quad b,$

we see that $-A/B$ of the equation $Ax + By + C = 0$ is the slope m and $-C/B$ is the y-intercept. Thus, in an equation such as $5x + 2y - 3 = 0$, the slope is $-\frac{5}{2}$ and the y-intercept is $b = \frac{3}{2}$. If $A = 0$, then $y = b$ and the line has slope 0.

INTERCEPT FORM OF THE EQUATION OF A LINE

When both the x- and y-intercepts of a line are known and are not zero, we can use the intercept form of the equation of a line.

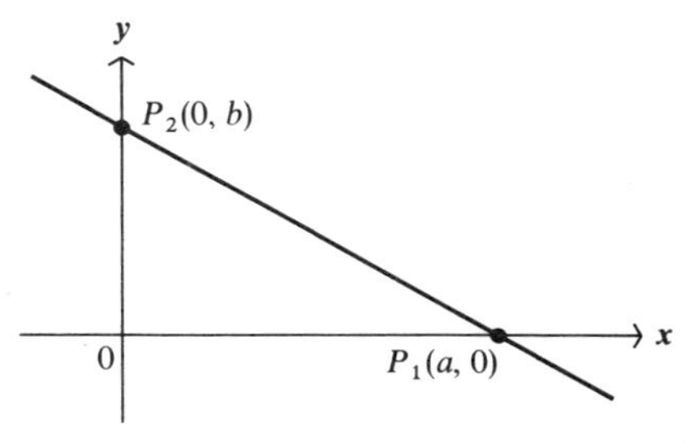

FIGURE 16

> **INTERCEPT FORM OF THE EQUATION OF A LINE**
>
> Given a line with intercepts $(a, 0)$ and $(0, b)$, the *intercept* equation of the line is
>
> $$\frac{x}{a} + \frac{y}{b} = 1, \qquad a \text{ and } b \neq 0. \qquad \textbf{(6)}$$

This form is derived as follows. Let $P_1(a, 0)$ and $P_2(0, b)$ be the given x- and y-intercepts. Then

$$m = \frac{y_2 - y_1}{x_2 - x_1} = \frac{b - 0}{0 - a} = -\frac{b}{a}.$$

Then, substituting $(0, b)$ for (x_1, y_1) and $-b/a$ for m in the point-slope form of the equation of a line, we have the following.

$$y - y_1 = m(x - x_1)$$

$$y - b = -\frac{b}{a}(x - 0)$$

$$\frac{b}{a}x + y = b$$

$$\boxed{\frac{1}{b}}\frac{b}{a}x + \boxed{\frac{1}{b}}\frac{y}{1} = \boxed{\frac{1}{b}}\frac{b}{1}$$

$$\frac{x}{a} + \frac{y}{b} = 1 \qquad \text{Intercept form.}$$

EXAMPLE 5 Find the equation of the line that intersects the x-axis at $(-4, 0)$ and the y-axis at $(0, 5)$.

SOLUTION The x- and y-intercepts are, respectively, -4 and 5.

$$\frac{x}{a} + \frac{y}{b} = 1$$

$$\frac{x}{-4} + \frac{y}{5} = 1$$

$$-\frac{x}{4} + \frac{y}{5} = 1$$

$$5x - 4y + 20 = 0 \qquad \text{General form.}$$

TWO-POINT FORM OF THE EQUATION OF A LINE

If $P_1(x_1, y_1)$ and $P_2(x_2, y_2)$ are two given points on a line and if $x_1 \neq x_2$, then the slope of the line through these two points is

$$m = \frac{y_2 - y_1}{x_2 - x_1}, \text{ as we have seen before.}$$

If we substitute this form of the slope and either of the two given points into the point-slope form of the equation of a line, we get the **two-point form**.

$$y - y_1 = m(x - x_1) \qquad \text{Point-slope form.}$$

$$y - y_1 = \frac{y_2 - y_1}{x_2 - x_1}(x - x_1) \qquad \text{(Two-point form)}$$

In practice, this formula can be omitted. If we know two points on a line, we can find the slope. Then, using the slope and either of the given points in the formula $y - y_1 = m(x - x_1)$, we can find the equation of the line.

EXAMPLE 6 Find an equation of the line that contains the points $A(-2, 3)$ and $B(4, -1)$.

SOLUTION It makes no difference whether we choose A as P_1 and B as P_2, or B as P_1 and A as P_2. Suppose we choose A as P_1 and B as P_2. Then $(x_1, y_1) = (-2, 3)$ and $(x_2, y_2) = (4, -1)$.

$$y - y_1 = \frac{y_2 - y_1}{x_2 - x_1}(x - x_1)$$

$$y - 3 = \frac{-1 - (3)}{4 - (-2)}[x - (-2)]$$

$$y - 3 = \frac{-2}{3}(x + 2)$$

$$3y - 9 = -2x - 4$$

The required equation in general form is $2x + 3y - 5 = 0$.

When we write equations in general form, that is, the form $Ax + By + C = 0$, it is assumed that $A > 0$ and A, B, and C are integers as small as possible. For example, the equation

$$-\frac{2}{3}x = \frac{3}{2}y + 1$$

would be changed to

$$4x + 9y + 6 = 0.$$

Exercises 3

A In Exercises 1–4, find the slope of the line that passes through the given points.

1. $(-3, -1)$ and $(4, 2)$ **2.** $(-3, 2)$ and $(4, -1)$

3. $(-5, -6)$ and $(1, 7)$ **4.** $(-3, -1)$ and $(2, -1)$

In Exercises 5–8, find the slope of each line.

5. $y = 3x + 2$ **6.** $2x + 5y = 10$

7. $\frac{x}{4} - \frac{y}{5} = 1$ **8.** $y + 3 = 0$

In Exercises 9–12, graph the line that has the given slope and passes through the given point.

9. $P(-4, -1)$; $m = 1$ **10.** $P(3, -5)$; $m = -2$

11. $P(1, 3)$; $m = \frac{3}{4}$ **12.** $P(-5, 2)$; $m = -\frac{2}{3}$

In Exercises 13–17, find the equation of the line that satisfies each of the following conditions. Express your equation in the form $Ax + By + C = 0$.

13. $m = \frac{2}{5}$; passes through $(6, -1)$

14. $m = -2$; x-intercept -3

15. x-intercept -4; y-intercept 3

16. Passes through $(-4, -1)$ and $(5, -2)$

17. Parallel to x-axis; y-intercept 5

In Exercises 18–23, graph each equation.

18. $2x - 6y = 12$ **19.** $\frac{x}{5} - \frac{y}{2} = 1$

20. $x + 4 = 0$ **21.** $y + 5 = 0$

22. $6x + 3 = \frac{3y}{2}$ **23.** $2(2x + y) = 3(3x + 4y)$

24. The right triangle ABC has vertices of $A(-1, -1)$, $B(3, -2)$, and $C(4, 2)$.

a. Find the slope of line AB.

b. Find the slope of line BC.

c. What relationship can you discover between the slope of AB and the slope of BC in this special case?

B In Exercises 1–4, find the slope of the line that passes through the given points.

1. $(-2, -3)$ and $(1, 4)$ **2.** $(-2, 1)$ and $(3, -1)$

3. $(-4, -5)$ and $(2, 4)$ **4.** $(-1, -2)$ and $(1, -2)$

In Exercises 5–8, find the slope of each line.

5. $y = 2x + 1$ **6.** $3y + 4x = 12$

7. $\frac{x}{2} + \frac{y}{3} = 1$ **8.** $y = 5$

In Exercises 9–12, graph the line that has the given slope and passes through the given point.

9. $P(-1, -3);\ m = 2$ **10.** $P(2, -4);\ m = -1$

11. $P(3, 4);\ m = \frac{2}{3}$ **12.** $P(-4, 5);\ m = -\frac{3}{5}$

In Exercises 13–17, find the equation of the line that satisfies each of the following conditions. Express your equation in the form $Ax + By + C = 0$.

13. $m = \frac{2}{3}$; passes through $(3, 1)$

14. $m = -1$; y-intercept 5

15. x-intercept 3; y-intercept 2

16. Passes through $(-2, -3)$ and $(3, -1)$

17. Parallel to y-axis; x-intercept 4

In Exercises 18–23, graph each equation.

18. $2x + 3y = 6$ **19.** $\frac{x}{4} - \frac{y}{5} = 1$

20. $x - 5 = 0$ **21.** $y + 6 = 0$

22. $x + 2 = \frac{2y}{3}$ **23.** $3(x - y) = 2(3x - 2y)$

24. Show that the points $(3, 4)$, $(-1, 2)$, $(-3, -4)$, and $(1, -2)$ are the vertices of a parallelogram. (A parallelogram is a four-sided figure with opposite sides that are parallel.)

4 Distance between two points

Let two points P_1 and P_2 have coordinates (x_1, y_1) and (x_2, y_2), and let d be the length of the line segment joining them (see Figure 17). These points can be located anywhere in the xy-plane. If the two points lie on a horizontal line such as $P_1(x_1, y_1)$ and $P_3(x_2, y_1)$ in Figure 17, the distance between them is $|x_2 - x_1|$. If the two points lie on a vertical line such as $P_3(x_2, y_1)$ and $P_2(x_2, y_2)$ in Figure 17, the distance between them is $|y_2 - y_1|$. Draw a horizontal line through P_1 and a vertical line through P_2. These two lines intersect, forming a right angle at the point

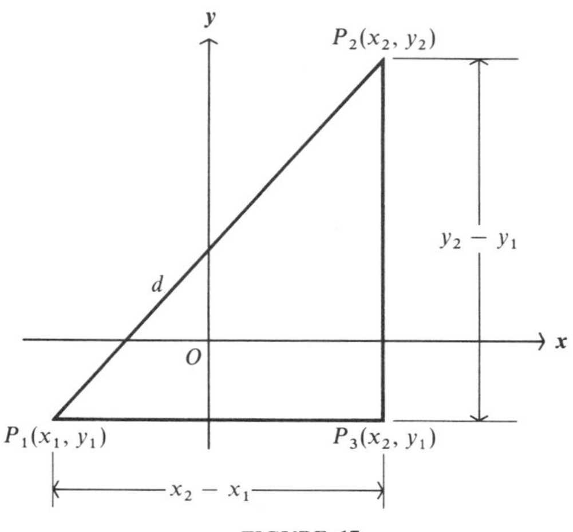

FIGURE 17

$P_3(x_2, y_1)$. Draw the line segment $\overline{P_1P_2}$, which represents the distance d between points P_1 and P_2. Then, by the Pythagorean Theorem,

$$d^2 = (x_2 - x_1)^2 + (y_2 - y_1)^2$$
$$d = \sqrt{(x_2 - x_1)^2 + (y_2 - y_1)^2}.$$

DISTANCE BETWEEN TWO POINTS

The distance d between two points $P_1(x_1, y_1)$ and $P_2(x_2, y_2)$ is

$$d = \sqrt{(x_2 - x_1)^2 + (y_2 - y_1)^2}.$$

If the coordinates of points P_1 and P_2 were interchanged in the formula, the results would be the same. That is:

$$d = \sqrt{(x_2 - x_1)^2 + (y_2 - y_1)^2} = \sqrt{(x_1 - x_2)^2 + (y_1 - y_2)^2}.$$

EXAMPLE 1 Find the distance between the points $A(-3, -2)$ and $B(9, 3)$.

SOLUTION

$$\begin{aligned} d &= \sqrt{(x_2 - x_1)^2 + (y_2 - y_1)^2} \\ &= \sqrt{[9 - (-3)]^2 + [3 - (-2)]^2} \\ &= \sqrt{(9 + 3)^2 + (3 + 2)^2} \\ &= \sqrt{144 + 25} = \sqrt{169} = 13 \end{aligned}$$

EXAMPLE 2 Find the distance between the points $A(-3, 1)$ and $B(0, -2)$.

SOLUTION

$$\begin{aligned} d &= \sqrt{(x_2 - x_1)^2 + (y_2 - y_1)^2} \\ &= \sqrt{[0 - (-3)]^2 + [(-2) - (+1)]^2} \\ &= \sqrt{3^2 + (-3)^2} = \sqrt{18} = \sqrt{(9)(2)} = \sqrt{9}\sqrt{2} = 3\sqrt{2} \end{aligned}$$

If the distance is to be given in decimal notation, use Table 1 (in the back of the book) or a hand calculator. Written as a decimal and rounded to three decimal places, $\sqrt{18} \doteq 4.243$.

EXAMPLE 3 Show that the triangle whose vertices are $(-1, 2)$, $(4, -3)$, and $(5, 3)$ is an isosceles triangle.

SOLUTION An isosceles triangle is a triangle with two equal sides. Graph the triangle and show that two of its sides have the same length.

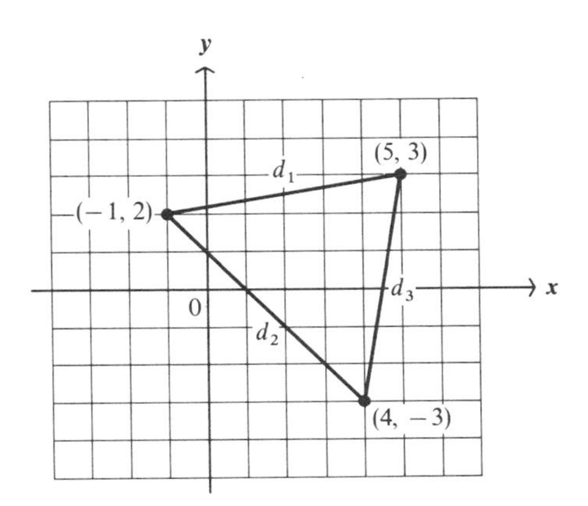

FIGURE 18

We will show that d_1 and d_3 have the same length.

$$\begin{aligned} d &= \sqrt{(x_2 - x_1)^2 + (y_2 - y_1)^2} \\ d_1 &= \sqrt{[5 - (-1)]^2 + (3 - 2)^2} \\ &= \sqrt{(6)^2 + (1)^2} \\ &= \sqrt{36 + 1} \\ &= \sqrt{37} \end{aligned}$$

← d_1 and d_3 have same length

$$d_2 = \sqrt{[4-(-1)]^2 + (-3-2)^2}$$
$$= \sqrt{(5)^2 + (-5)^2}$$
$$= \sqrt{25 + 25}$$
$$= \sqrt{50}$$
$$d_3 = \sqrt{(5-4)^2 + [3-(-3)]^2}$$
$$= \sqrt{(1)^2 + (6)^2}$$
$$= \sqrt{1 + 36}$$
$$= \sqrt{37}$$ ← d_1 and d_3 have same length

Since $d_1 = d_3$, the triangle is an isosceles triangle.

This is an example of an analytic proof or solution. An **analytic proof** consists essentially of algebraic (rather than geometric) methods.

MIDPOINT OF LINE SEGMENT

Let $M(\bar{x}, \bar{y})$ be the midpoint of the line segment $\overline{P_1 P_2}$, where the coordinates of P_1 and P_2 are (x_1, y_1) and (x_2, y_2), respectively (see Figure 19). Let horizontal lines through P_1 and M intersect the vertical lines

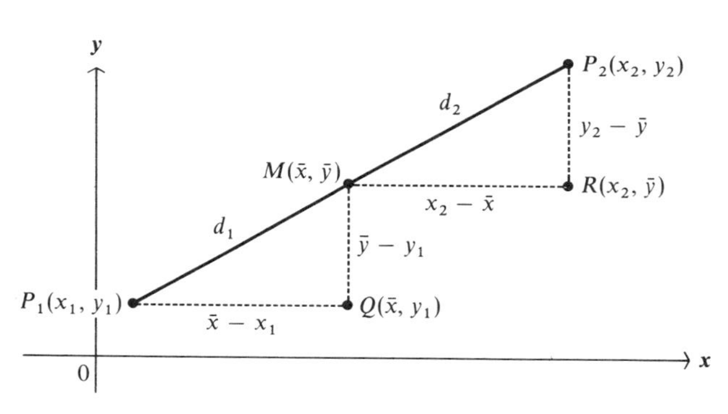

FIGURE 19

through P_2 and M at R and Q, respectively. Then because $d_1 = d_2$ (M is the midpoint of $\overline{P_1 P_2}$) and angles MP_1Q and P_2MR are equal, right triangles P_1QM and MRP_2 are congruent. Their corresponding sides are therefore equal.

$$\bar{x} - x_1 = x_2 - \bar{x} \quad \text{and} \quad \bar{y} - y_1 = y_2 - \bar{y}$$
$$2\bar{x} = x_2 + x_1 \qquad 2\bar{y} = y_2 + y_1$$
$$\bar{x} = \frac{x_2 + x_1}{2} \qquad \bar{y} = \frac{y_2 + y_1}{2}$$

These $\bar{x}$ and $\bar{y}$ values are the coordinates of the midpoint of the line segment $\overline{P_1 P_2}$.

MIDPOINT OF A LINE SEGMENT

The coordinates of the midpoint $M(\bar{x}, \bar{y})$ of the line segment with endpoints (x_1, y_1) and (x_2, y_2) are

$$\bar{x} = \frac{x_2 + x_1}{2} \quad \text{and} \quad \bar{y} = \frac{y_2 + y_1}{2}.$$

EXAMPLE 4 The vertices of triangle ABC are $A(3, -6)$, $B(5, 4)$, and $C(-7, -3)$.

a. Find the midpoint of side $\overline{AB}$.

b. Find the midpoint of side $\overline{BC}$.

SOLUTION

$$\text{Midpoint of a segment} = (\bar{x}, \bar{y}) = \left(\frac{x_2 + x_1}{2}, \frac{y_2 + y_1}{2}\right).$$

a. For $\overline{AB}$, let $(x_1, y_1) = (3, -6)$ and $(x_2, y_2) = (5, 4)$. Then

$$(\bar{x}, \bar{y}) = \left(\frac{5 + 3}{2}, \frac{4 + (-6)}{2}\right) = (4, -1).$$

b. For $\overline{BC}$, let $(x_1, y_1) = (5, 4)$ and $(x_2, y_2) = (-7, -3)$. Then

$$(\bar{x}, \bar{y}) = \left(\frac{-7 + 5}{2}, \frac{-3 + 4}{2}\right) = \left(-1, \frac{1}{2}\right).$$

Exercises 4

A

1. Find the distance between each of the following pairs of points. Leave your answers in simplified radical form.

a. $(-4, 2)$ and $(3, -1)$ **b.** $(2, 7)$ and $(-5, -2)$

2. Find the distance between the points $(3, -5)$ and $(-4, 2)$, correct to 3 decimal places.

3. Find the coordinates of the midpoint of the line segment joining each of the following pairs of points.

a. $(5, -4)$ and $(-7, 10)$ **b.** $(-8, 3)$ and $(14, -7)$

4. Find the distance from the origin to the point $(8, -6)$.

5. The midpoint of a line segment is $(2, -1)$ and one endpoint of the segment is $(-1, -3)$. Find the coordinates of the other endpoint of the segment.

6. Find y, given that $(4, y)$ is equidistant from $(5, -3)$ and $(-3, 1)$.

7. Show that the triangle whose vertices are $(3, 1)$, $(1, 6)$, and $(-2, -1)$ is a right triangle. (Hint: If a, b, and c are sides of a right triangle, then $a^2 + b^2 = c^2$.)
8. Given right triangle OAB with O at $(0, 0)$, A at $(a, 0)$, and B at (a, b). M is the midpoint of the hypotenuse $\overline{OB}$ of the triangle. Prove analytically, that M is equidistant from the vertices of the triangle.

B

1. Find the distance between each of the following pairs of points. Leave your answers in simplified radical form.
   a. $(-3, 2)$ and $(5, -3)$ **b.** $(4, -2)$ and $(7, 5)$
2. Find the distance between the points $(6, 2)$ and $(-2, -3)$, correct to 3 decimal places.
3. Find the coordinates of the midpoint of the line segment joining each of the following pairs of points.
   a. $(-2, -3)$ and $(4, 7)$ **b.** $(-3, 2)$ and $(5, -6)$
4. Find the distance from the origin to the point $(12, 5)$.
5. The midpoint of a line segment is $(-1, 2)$ and one endpoint of the segment is $(-5, -3)$. Find the coordinates of the other endpoint of the segment.
6. Find x, given that $(x, 1)$ is equidistant from $(1, 6)$ and $(9, 4)$.
7. Show that the triangle whose vertices are $(1, -1)$, $(-1, 7)$, and $(-3, -2)$ is a right triangle.
8. Given isosceles triangle OAB with $\overline{OB} = \overline{AB}$; D is the midpoint of $\overline{OB}$ and C is the midpoint of $\overline{AB}$. The coordinates of O are $(0, 0)$; of A are $(a, 0)$; and of B are $(a/2, b)$. Prove analytically that the medians $\overline{OC}$ and $\overline{AD}$ are equal.

5 Graphs of first-degree inequalities

In this section, we are mainly concerned with graphing first-degree inequalities, but we will also be graphing first-degree equations. Before discussing graphing, we will define the terms that will be used.

A line in a plane divides the plane into two **half-planes** (see Figure 20). The line that divides the plane is called the **boundary** of both half-planes.

The graph of a single inequality, in no more than two variables, is a half-plane. To find the boundary line, replace the inequality symbol with an equals sign and then graph the resulting equation. *If the inequality symbol includes the equality symbol ($\leq$ or $\geq$), the boundary line is a part of the solution and is drawn as a solid line. If the inequality symbol does not include the equality symbol ($<$ or $>$), the boundary line is* ***not*** *a part of the solution and is drawn as a dashed line.*

EXAMPLE 1 Graph $2x - 5y \geq 10$.

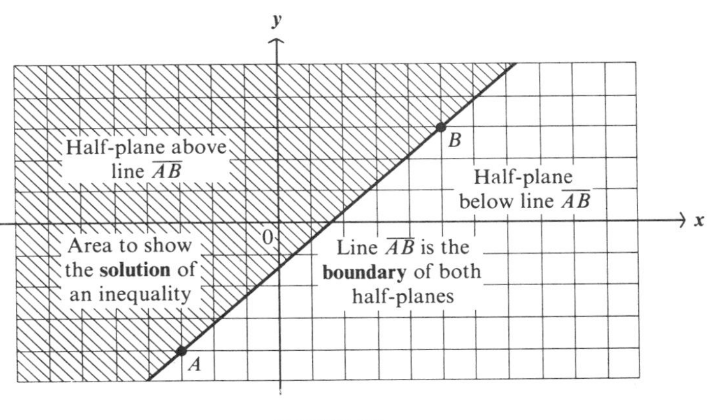

FIGURE 20

SOLUTION Change the symbol $\geq$ of the inequality to $=$ in order to find the boundary line.

$$2x - 5y = 10$$

Graph this equation by finding the x- and y-intercepts. To find the x-intercepts, let $y = 0$. To find the y-intercept, let $x = 0$.

$$2x - 5(0) = 10$$
$$x = 5$$
$$2(0) - 5y = 10$$
$$y = -2$$

x	y
5	0
0	-2

Therefore, the boundary line goes through (5, 0) and (0, -2). The boundary line is solid because the equality symbol is included in the given inequality.

When we substitute the coordinates of the origin, (0, 0), into the original inequality we get an untrue statement.

$$2x - 5y \geq 10$$
$$2(0) - 5(0) \geq 10$$
$$0 \geq 10 \text{ Not true.}$$

Therefore, the solution is not the half-plane that includes the origin. Thus the solution is the half-plane not containing the origin (see Figure 21). One test point is all that is needed to locate the correct half-plane. Unless the boundary line passes through the origin, the origin is always an easy test point to use.

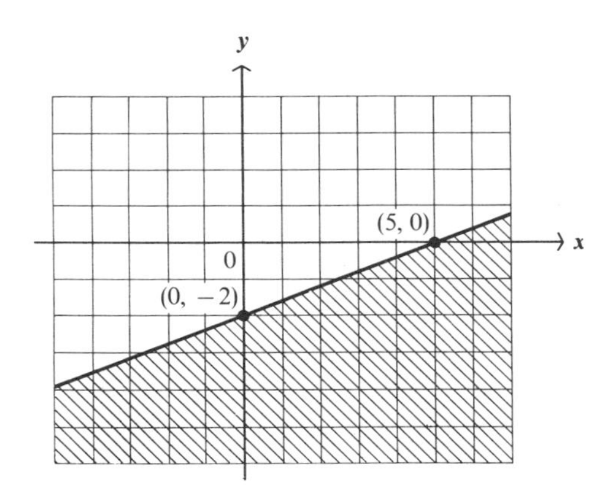

FIGURE 21

EXAMPLE 2 Graph $x + 3y < 3$.

SOLUTION Change the symbol $<$ of the inequality to $=$ in order to find the boundary line.

$$x + 3y = 3$$

Find the x-intercept. $\quad x + 3(0) = 3$

$$x = 3$$

Find the y-intercept. $\quad 0 + 3y = 3$

$$y = 1$$

x	y
3	0
0	1

Therefore, the boundary line goes through (3, 0) and (0, 1). The boundary line is dashed because the equality symbol is not included in the given inequality. When we substitute the coordinates of the origin, (0, 0), into the original inequality we get a true statement.

$$x + 3y < 3$$

$$0 + 3(0) < 3$$

$$0 < 3 \text{ True.}$$

Therefore, the solution is the half-plane that includes the origin (see Figure 22).

EXAMPLE 3 Graph in a plane all points with coordinates that satisfy $|x| \leq 3$.

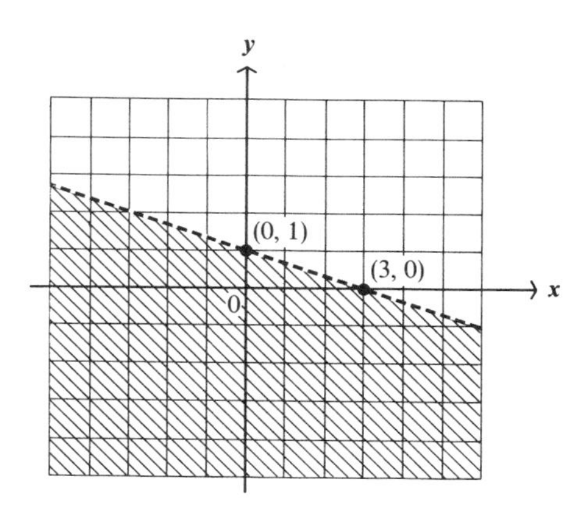

FIGURE 22

SOLUTION From the definition of absolute value, we have

$x \leq 3$	$-x \leq 3$
	$x \geq -3$
One boundary line is	One boundary line is
$x = 3.$	$x = -3.$

That is, $-3 \leq x \leq 3$ is equivalent to $|x| \leq 3$. Using set notation, we have $\{x \mid x \geq -3\} \cap \{x \mid x \leq 3\}$ (see Figure 23).

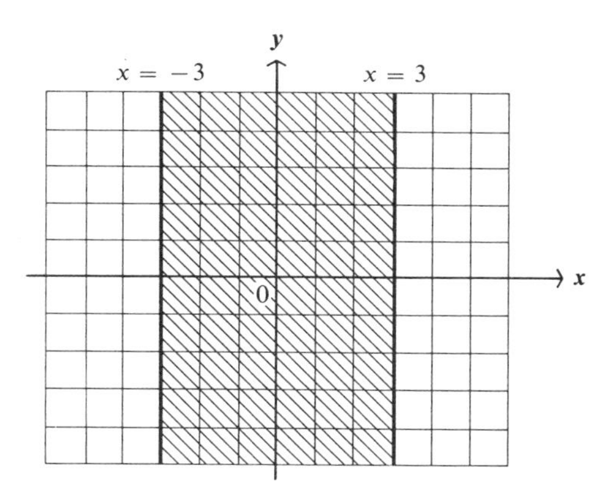

FIGURE 23

If we were to graph the inequality in Example 3 in one-space, as we did in Section 7 of Chapter 5, it would appear as in Figure 24.

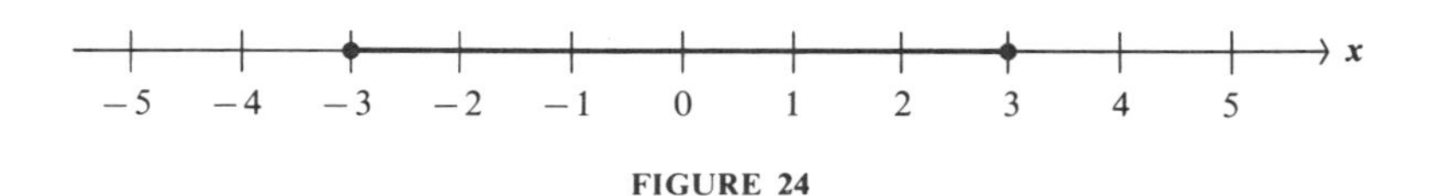

FIGURE 24

In this chapter we are graphing in **two-space**. When we are in two-space, if we have an equation or an inequality with only x's in it, we automatically assume the statement is true for all y.

EXAMPLE 4 Graph the solution set defined by the following system of inequalities.

1. $3x + 4y \leq 12$

2. $y \geq 0$

3. $0 \leq x \leq 3$

SOLUTION The solution set for the system is the set of points that satisfies all three of the inequalities simultaneously. The solution is the intersection of the three graphs. We will graph them one at a time. Figure 25 shows the complete graph.

1. $3x + 4y \leq 12$
$3x + 4y = 12$

x	y
0	3
4	0

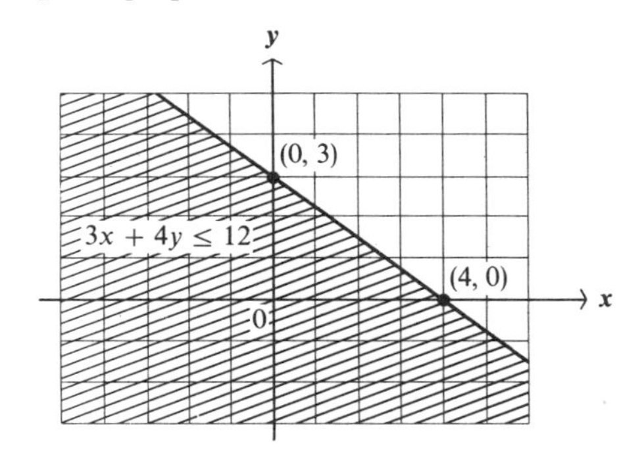

2. $y \geq 0$

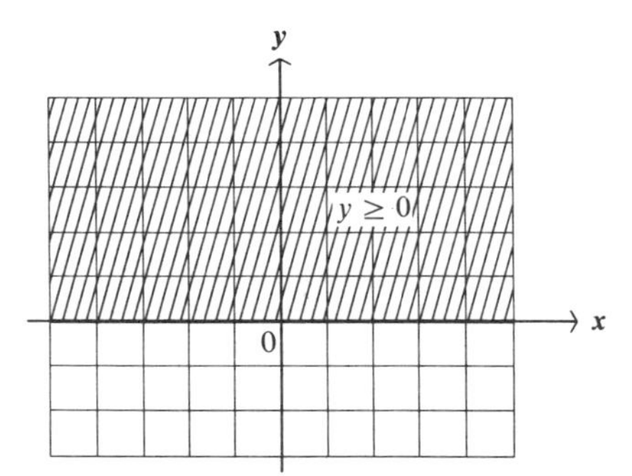

3. $0 \leq x \leq 3$

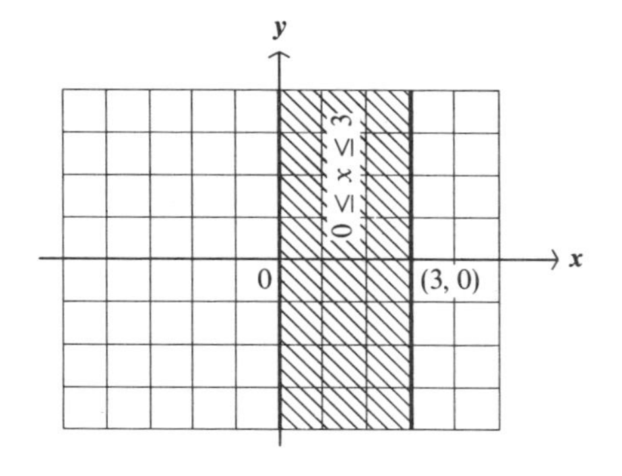

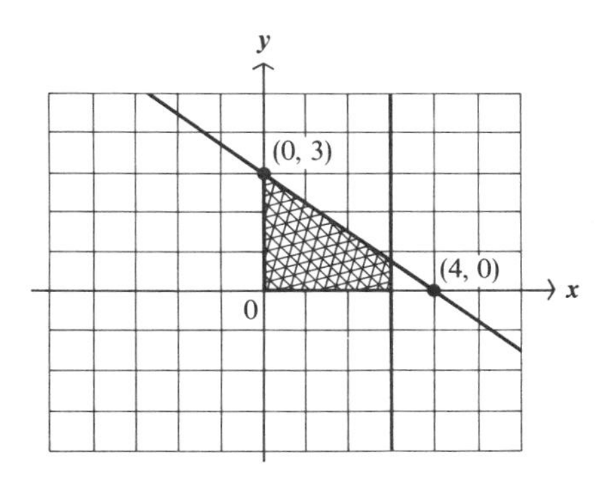

FIGURE 25

EXAMPLE 5 Graph the solution set for

$$2x + 1 < 3x + 2 < 5x - 4.$$

SOLUTION We will write two separate inequalities.

$2x + 1 < 3x + 2$	$3x + 2 < 5x - 4$
$-1 < x$	$6 < 2x$
	$3 < x$

Therefore, the solution set is the intersection of $x > -1$ and $x > 3$, which is $x > 3$ for all y (see Figure 26).

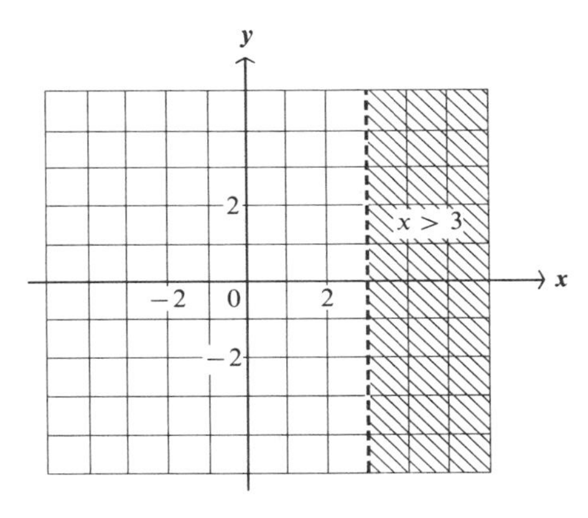

FIGURE 26

EXAMPLE 6 Graph

$$\frac{2x}{3} - y < 0.$$

SOLUTION

$$\frac{2}{3}x - y < 0$$

$$-y < -\frac{2}{3}x$$

$$y > \frac{2}{3}x$$

Substitute = for > to find the boundary line.

$$y = \frac{2}{3}x$$

This equation is in the form $y = mx + b$, so the slope is $\frac{2}{3}$ and the y-intercept is 0. We next draw the dashed boundary line. Because the boundary line goes through the origin, we must choose some other point to determine which half-plane is the solution. We can choose any point not on the boundary line. We will choose a point for which the given inequality is easy to evaluate, such as (1, 0).

$$\frac{2x}{3} - y < 0$$

$$\frac{2(1)}{3} - (0) < 0$$

$$\frac{2}{3} < 0 \text{ Not true.}$$

Therefore, the top half-plane is the solution.

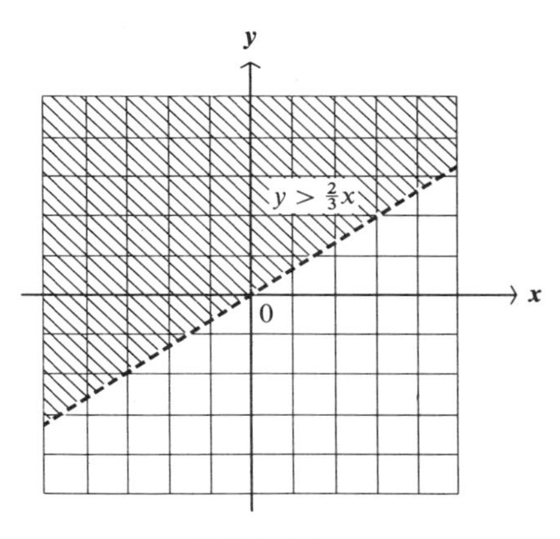

FIGURE 27

EXAMPLE 7 Graph

$$\frac{2x-1}{x+1} < 3.$$

SOLUTION We must clear fractions by multiplying through by $x + 1$. However, since we do not know the value of x, we do not know whether $x + 1$ is positive or negative. Since multiplying the inequality by a negative number would change the sense of the inequality, we will consider the two possibilities in Cases I and II.

Case I If $x + 1$ is positive, then $x + 1 > 0$ and $x > -1$.

Then $$\frac{2x-1}{x+1} < 3$$

becomes $$2x - 1 < 3x + 3,$$

and $$-4 < x.$$

The intersection of $x > -4$ and $x > -1$ is $x > -1$. Therefore, one of the boundaries is the line $x = -1$.

Case II If $x + 1$ is negative, then $x + 1 < 0$ and $x < -1$.

Then $$\frac{2x-1}{x+1} < 3$$

Note opposite sense.

becomes $$2x - 1 > 3x + 3,$$

and $$-4 > x.$$

The intersection of $x < -4$ and $x < -1$ is $x < -4$, so the other boundary line is $x = -4$. The union of Cases I and II is $x > -1$ or $x < -4$ (see Figure 28).

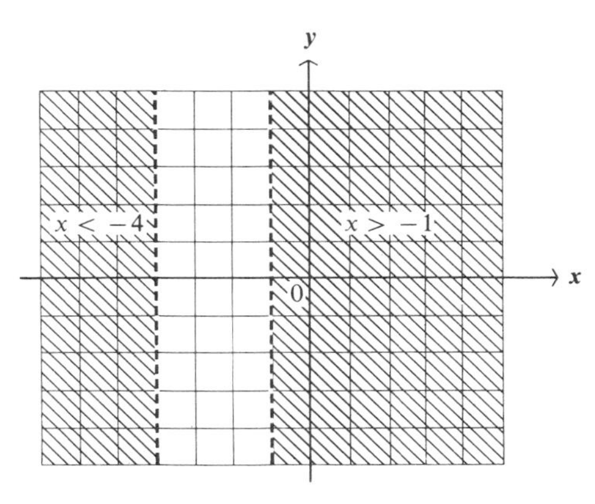

FIGURE 28

Exercises 5

A Graph the following inequalities.

1. $4x - 6y \leq -24$

2. $5x + 2 > 3x - 4$

3. $|x| \leq 5$

4. $5x - 2y > 0$

5. $2x - 2 < 3x - 4 < 5x + 2$

6. $\dfrac{3x - 1}{x - 1} > 2$

7. Graph the solution set defined by the following system of inequalities.

$$\begin{cases} -5x + 6y \leq 30 \\ x \leq 0 \\ y \geq 2 \end{cases}$$

B Graph the following inequalities.

1. $5x - 4y \leq 20$

2. $x + 4 > 2x + 1$

3. $|x| \geq 2$

4. $4y - 3x < 0$

5. $x - 2 > 3x + 2 > 4x + 6$

6. $\dfrac{3x + 2}{x + 2} \leq 1$

7. Graph the solution set defined by the following system of inequalities.

$$\begin{cases} 5x - 4y \leq 20 \\ x \geq 0 \\ y \geq -3 \end{cases}$$

Summary

The pair (x, y), used to locate a point on the rectangular coordinate xy-plane, is an example of an **ordered pair**. The ordered pair (a, b) is different from the ordered pair (b, a). We call x the **first coordinate** (or **abscissa**) and y the **second coordinate** (or **ordinate**) of the point.

A **relation** is a set of ordered pairs of mathematical quantities. The **domain** of the relation is the set of all first coordinates of the ordered pairs of the relation. The **range** of a relation is the set of all the second coordinates of the ordered pairs of the relation.

A **function** f is a relation such that no two ordered pairs of the set have different second coordinates for the same first coordinate. Functions are represented by expressions such as $P(x)$, $Q(x)$, $f(t)$, $f(\theta)$, or just f. When a function is written in the form $y = 3x - 5$, x is called the **independent variable** and y is called the **dependent variable**. For this same function, y is said to be expressed **explicitly** in terms of x. An equation is an **implicit equation** if no variable is solved for explicitly in terms of the other variables.

Equations that define straight lines are called **linear equations**. First-degree equations are linear equations.

SLOPE (m) OF A LINE

For a line through the points $P_1(x_1, y_1)$ and $P_2(x_2, y_2)$:

$$m = \frac{y_2 - y_1}{x_2 - x_1}, \qquad x_2 \neq x_1.$$

DISTANCE BETWEEN TWO POINTS $P_1(x_1, y_1)$ AND $P_2(x_2, y_2)$

$$d = \sqrt{(x_2 - x_1)^2 + (y_2 - y_1)^2}$$

If $P_1(x_1, y_1)$ and $P_2(x_2, y_1)$ lie on a horizontal line, the distance between them is $|x_2 - x_1|$. If $P_1(x_1, y_1)$ and $P_2(x_1, y_2)$ lie on a vertical line, the distance between them is $|y_2 - y_1|$.

EQUATIONS OF A LINE

General form of linear equation

$$Ax + By + C = 0, \qquad A \text{ and } B \text{ not both zero}$$

Point-slope form

$$y - y_1 = m(x - x_1)$$

Slope-intercept form

$$y = mx + b \qquad b \text{ is the } y\text{-intercept.}$$

Intercept form

$$\frac{x}{a} + \frac{y}{b} = 1, \qquad a \neq 0, \qquad b \neq 0$$

a is the x-intercept,
b is the y-intercept.

Two-point form

$$y - y_1 = \frac{y_2 - y_1}{x_2 - x_1}(x - x_1) \qquad x_2 \neq x_1$$

Parallel to y-axis

$$x = a \qquad a \text{ is the } x\text{-intercept.}$$

Parallel to x-axis

$$y = b \qquad b \text{ is the } y\text{-intercept.}$$

MIDPOINT OF LINE SEGMENT CONNECTING POINTS $P_1(x_1, y_1)$, AND $P_2(x_2, y_2)$

$$\bar{x} = \frac{x_2 + x_1}{2} \qquad \bar{y} = \frac{y_2 + y_1}{2}$$

Solving an equation (or inequality) with an absolute value sign is equivalent to solving two equations (or inequalities). In particular, $|x| < a$ is replaced with $-a < x < a$ and $a < |x|$ is replaced with $x < -a$ or $a < x$.

$$|x| = a \quad \text{implies} \quad x = -a, \quad x = a$$

Review exercises

Write all equations of lines in the general form.

A

1. Find the distance from the origin to the point $(4, -3)$.
2. Given the function $S = 16t^2 - 48t$.
   a. Which variable is the dependent variable?
   b. Which variable is the independent variable?
3. Find the slope of the line $3x - 4y + 7 = 0$.
4. Graph the line through the point $(0, -4)$ with slope of $\frac{2}{3}$.

In Exercises 5–7, graph each relation in two-space.

5. $|x + 3| \leq 2$ 6. $7x + 2y > 14$ 7. $\dfrac{2x}{3} + \dfrac{y}{2} < 0$

8. Given $f(x) = 5x - 3$ find $\dfrac{f(x + h) - f(x)}{h}$.

9. Parallelogram $OABC$ has vertices $O(0, 0)$, $A(5, 0)$, $B(8, 6)$, and $C(3, 6)$.
   a. Find the slope of $\overline{AB}$.
   b. Find an equation of the line through C parallel to $\overline{AB}$.
   c. Find the midpoint of diagonal $\overline{AC}$.
   d. Write an equation of the line through A and C.
   e. Prove that the diagonals $\overline{AC}$ and $\overline{OB}$ bisect each other.

B

1. Find the distance from the origin to the point $(-6, 8)$.
2. Given the function $y = 3x + 4$.
   a. Which variable is the dependent variable?
   b. Which variable is the independent variable?
3. Find the slope of the line $5x + 8y - 3 = 0$.
4. Graph the line through the point $(-2, 4)$ with slope of $-\frac{3}{5}$.

In Exercises 5–7, graph each relation in two-space.

5. $|x - 1| \leq 3$ 6. $2x - 7y > 14$ 7. $x - y \geq 0$

8. Given $f(x) = 3x - 4$, find $\dfrac{f(x + h) - f(x)}{h}$.

9. Triangle ABC has vertices $A(1, -4)$, $B(5, 2)$, and $C(-6, 5)$.
   a. Find the slope of $\overline{AB}$.
   b. Find an equation of the line through C parallel to $\overline{AB}$.
   c. Find the midpoint of $\overline{AB}$.
   d. Write an equation of the line through B and C.
   e. Prove that the triangle is an isosceles triangle.

Diagnostic test

The purpose of this test is to see how well you understand the work in this chapter. Allow yourself approximately 60 minutes to do the test. Solutions to the problems, together with section references, are given in the Answers section at the end of the book. We suggest that you study the sections referred to for the problems you do incorrectly.

1. Which of the following are graphs of functions?

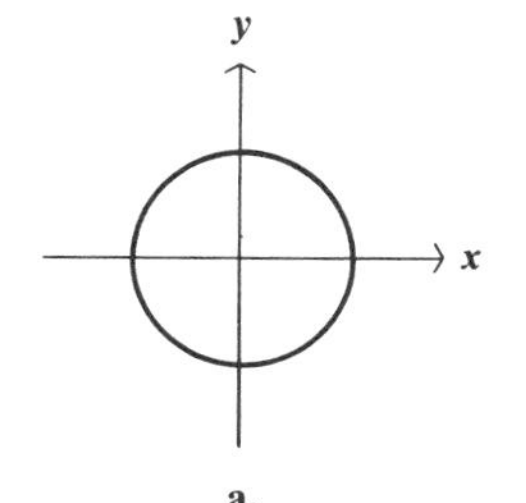

a.

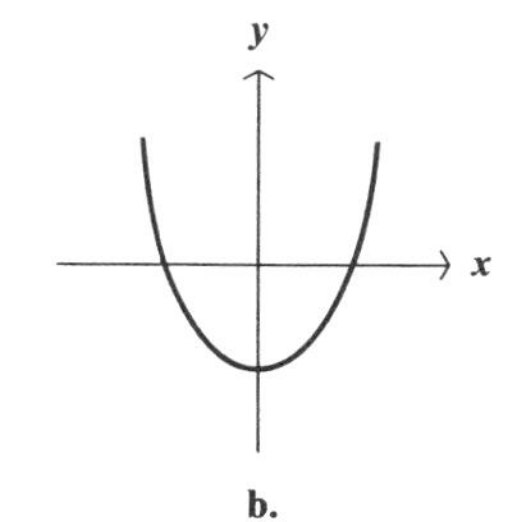

b.

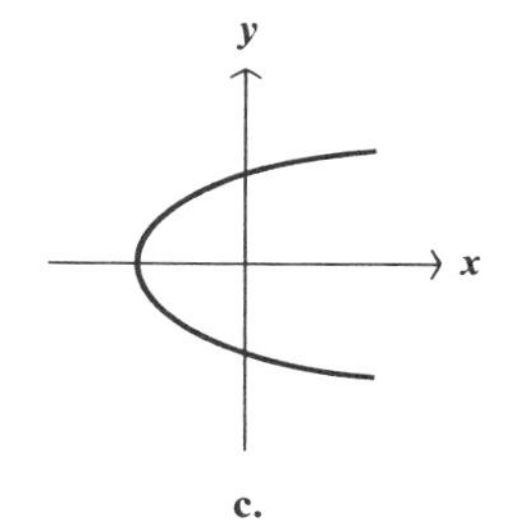

c.

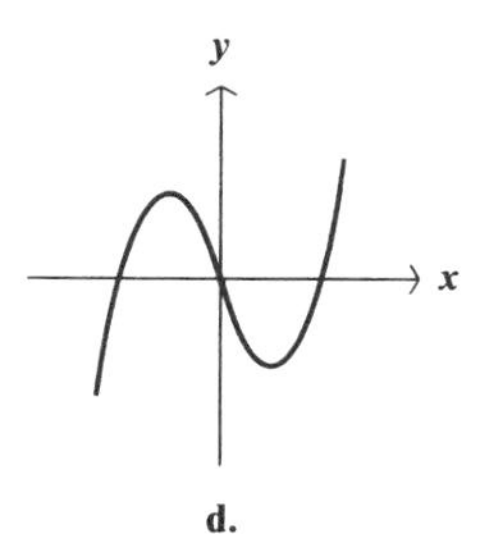

d.

2. Is the equation $x^2 + y^2 + 4x = 5$ an implicit or explicit equation?
3. Name the independent variable of the function $y = 3x - 4$.
4. Graph each of the following relations in two-space.

   a. $|2x - 3| \geq 5$ b. $\frac{2x}{3} - \frac{y}{2} < 0$

5. The quadrilateral $ABCD$ has vertices $A(-1, -4)$, $B(3, -2)$, $C(1, 6)$, and $D(-5, 2)$.
   a. Find the length of $\overline{AB}$.
   b. Write the equation of the line through B parallel to side $\overline{AD}$.
   c. Let E be the midpoint of $\overline{AB}$, F the midpoint of $\overline{BC}$, G the midpoint of $\overline{CD}$, and H the midpoint of $\overline{AD}$. Draw lines to form the quadrilateral $EFGH$. Prove analytically that the quadrilateral $EFGH$ is a parallelogram.
6. Determine the domain and range of the function $f(x) = \sqrt{5x - 1}$.

7 Exponential and logarithmic functions

Many applications of mathematics make use of exponential and logarithmic functions. For example, growth and decay of population, bacteria, or radium, the Richter scale for earthquakes, compound interest, continuous compound amount of principal and interest, probability theory, and many other disciplines make use of exponential functions.

Logarithms were invented in the seventeenth century and, until recent years, they were an important tool in all kinds of arithmetic calculations. Now, however, with the influx of inexpensive hand calculators, this application of logarithms is less important. But the principle of logarithms and their applications in expressions and equations has not diminished in importance.

In Chapter 5, we worked with algebraic functions. An **algebraic function** is a function made up of algebraic expressions. For a definition of algebraic expressions, see Section 1 of Chapter 2.

In this chapter, we shall be working with **transcendental functions**. *A transcendental function is any function which is not an algebraic function.* Trigonometric functions and logarithmic functions are examples of transcendental functions.

1 Exponential functions

An **exponential function** is a function in which an unknown quantity appears in one or more exponents (see Examples 1 and 2).

EXAMPLE 1 Graph $f(x) = 2^x$ and $g(x) = 2^{-x}$ on the same axes.

SOLUTION We will make a brief table of values for each function.

$f(x) = 2^x$

x	-3	-2	-1	0	1	2	3	As $x \to \infty$
$f(x)$	$\frac{1}{8}$	$\frac{1}{4}$	$\frac{1}{2}$	1	2	4	8	$f(x) \to \infty$

$g(x) = 2^{-x}$

x	-3	-2	-1	0	1	2	3	As $x \to \infty$
$g(x)$	8	4	2	1	$\frac{1}{2}$	$\frac{1}{4}$	$\frac{1}{8}$	$g(x) \to 0$

Notice that as x decreases through negative values, $f(x) = 2^x$ approaches, but never intersects, the x-axis. Also, as x increases through positive values, $g(x) = 2^{-x}$ approaches, but never intersects, the x-axis. The x-axis is, therefore, a horizontal asymptote for both graphs. The interval of the domain of both functions is $(-\infty, \infty)$, and the interval of the range of both functions is $(0, \infty)$ (see Figure 1).

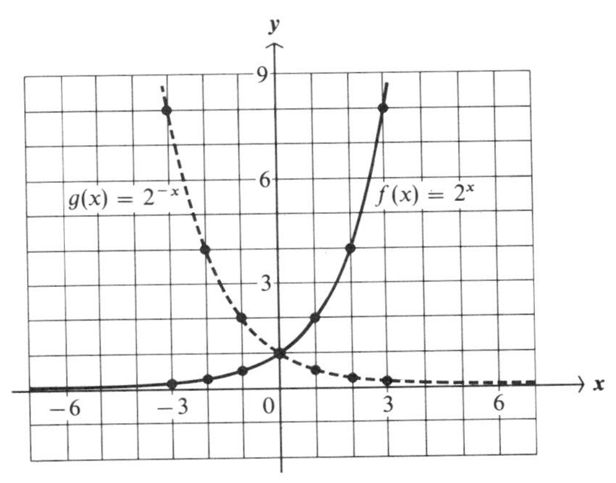

FIGURE 1

EXAMPLE 2 Graph $f(x) = 2^{-x^2}$.

SOLUTION Since

$$f(x) = 2^{-x^2} = \left(\frac{1}{2}\right)^{x^2},$$

we observe that as the absolute value of x increases, $f(x)$ approaches the x-axis. Therefore, the x-axis is a horizontal asymptote of the graph. Notice that the greatest value of $f(x)$ occurs when $x = 0$. With the aid of a calculator, we will make a brief table of values and then plot the graph.

x	−2	−1	−0.8	−0.5	−0.2	0	0.2	0.5	0.8	1	2
$f(x)$	0.06	0.5	0.64	0.84	0.97	1	0.97	0.84	0.64	0.5	0.06

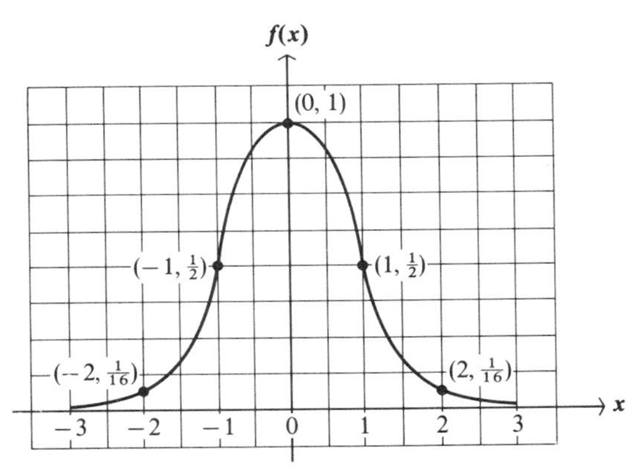

FIGURE 2

Functions of this type occur in the study of probability.

Functions such as the one shown in Figure 2, are said to be **symmetric to the $f(x)$-axis**. *Two points, $(x, f(x))$ and $(-x, f(x))$, whose $f(x)$-coordinates are equal and whose x-coordinates differ only in sign, are said to be **symmetric** with respect to the $f(x)$-axis.* The interval of the domain is $(-\infty, \infty)$, and the interval of the range is $(0, 1]$.

Exercises 1

Graph the following functions. Make use of symmetric points when possible.

A

1. $f(x) = 4^x$ 2. $f(x) = 3^{x+1}$ 3. $f(x) = 2^{-x}$
4. $f(x) = 3^{|x|}$ 5. $f(x) = 3^{1-x}$ 6. $f(x) = 2^x + 2^{-x}$
7. $f(x) = \left(\frac{2}{3}\right)^{x2}$

B

1. $f(x) = 3^x$ 2. $f(x) = 2^{x+1}$ 3. $f(x) = 4^{-x}$
4. $f(x) = 2^{|x|}$ 5. $f(x) = 2^{1-x}$ 6. $f(x) = \left(\frac{1}{2}\right)^x + \left(\frac{1}{2}\right)^{-x}$
7. $f(x) = \left(\frac{1}{4}\right)^{x2}$

2 Laws of exponents

In this section, we will recall certain laws of exponents. These laws of exponents (discussed in Chapter 4) hold for all types of real exponents for a positive real base.

LAWS OF EXPONENTS

Let a be any positive number and x and y be any two rational numbers.

1. $a^x \cdot a^y = a^{x+y}$ *Example:* $2^3 \cdot 2^4 = 2^7$
2. $\dfrac{a^x}{a^y} = a^{x-y}$ *Example:* $\dfrac{2^7}{2^2} = 2^5$
3. $(a^x)^y = a^{xy}$ *Example:* $(2^3)^4 = 2^{12}$
4. $\sqrt[y]{a^x} = a^{x/y}$ $y \in J > 1$ *Example:* $\sqrt[3]{2^5} = 2^{5/3}$
5. $a^0 = 1$ *Example:* $2^0 = 1$
6. $a^{-x} = \dfrac{1}{a^x}$ *Example:* $2^{-5} = \dfrac{1}{2^5} = \dfrac{1}{32}$
7. If a is any number greater than zero and not 1 and if y is positive, it is always possible to find a real exponent x such that
$$a^x = y.$$
But if y is negative or zero, no real value of x exists satisfying this relation.

3 Definition of a logarithm

If b, x, and N are three numbers such that

$$N = b^x \tag{1}$$

then we say that x is the logarithm of N to the base b, and write

$$x = \log_b N. \tag{2}$$

DEFINITION The **logarithm** of a positive number to a given base is the exponent to which the base must be raised to give the number.

Since $10^2 = 100$, 2 is the logarithm of 100 to the base 10, written $\log_{10} 100 = 2$. In like manner, $10^{-2} = 1/10^2 = 0.01$. Therefore -2 is the logarithm of 0.01 to the base 10, written $\log_{10} 0.01 = -2$.

Equations 1 and 2 are different ways of expressing the same relation among the three numbers b (the *base*), N (the *number*), and x (the *exponent* or *logarithm*). Equation 1 states the relation in *exponential form* and Equation 2 says the same thing in *logarithmic form*. *If x is the logarithm of N to a given base, then N is called the* ***antilogarithm*** *of x to that base.*

We assume in this book that $\log_b N$ is used only for $N > 0$ and $b > 1$, which implies that $\log_b N$ is a real number.

The following values are important and should be memorized. Since $b^1 = b$ and $b^0 = 1$, for any base b,

$$\log_b b = 1 \tag{3}$$

$$\log_b 1 = 0. \tag{4}$$

A **logarithmic function** is a function containing the logarithm of the variable (see Example 3).

In each of the parts of Example 1, we express the same relation between the three respective numbers, first as an exponential statement and then as a logarithmic statement.

EXAMPLE 1

	Exponential statement	Logarithmic statement
a.	$4^3 = 64$ 4 is the base. 3 is the exponent of the power. 4^3 is the third power of 4, and 64 is the third power of 4.	$\log_4 64 = 3$ 4 is the base. 3 is the logarithm of 64 to the base 4.
b.	$4^{2.5} = 32$ 4 is the base. 2.5 is the exponent of the power. $4^{2.5}$ is the 2.5 power of 4, and 32 is the 2.5 power of 4.	$\log_4 32 = 2.5$ 4 is the base. 2.5 is the logarithm of 32 to the base 4.

c. $e^2 = 7.39$
e is the base.
2 is the exponent of the power.
e^2 is the second power of e, and 7.39 is the second power of e.

$\log_e 7.39 = 2$
e is the base.
2 is the logarithm of 7.39 to the base e.

EXAMPLE 2 Find the value of $\log_{32} 8$.

SOLUTION Let $x = \log_{32} 8$. Next, write in exponential form: $32^x = 8$. Express 32 and 8 as powers of 2 to obtain an exponential equation with 2 as a base on both sides.

$$(2^5)^x = 2^3$$
$$2^{5x} = 2^3$$

Now, since base numbers are equal, exponents must be equal:

$$5x = 3$$
$$x = \frac{3}{5}.$$

Therefore $\log_{32} 8 = \frac{3}{5}$.

EXAMPLE 3 Find the number N if $\log_{81} N = \frac{3}{4}$.

SOLUTION

Write in exponential form: $N = (81)^{3/4}$

Write 81 as a power of 3: $N = (3^4)^{3/4}$

Apply Law 3 of Exponents: $N = 3^3 = 27$

Exercises 3

A In Exercises 1–6, express each statement in logarithmic form.

1. $3^2 = 9$ **2.** $10^{-3} = 0.001$ **3.** $16^{1/2} = 4$

4. $27^{-2/3} = \frac{1}{9}$ **5.** $3^0 = 1$ **6.** $3^1 = 3$

In Exercises 7–10, express each statement in exponential form.

7. $\log_4 64 = 3$ **8.** $\log_{16} 4 = \frac{1}{2}$

9. $\log_{10} 0.01 = -2$ **10.** $\log_b b = 1$

In Exercises 11–13, find the value of each of the logarithms.

11. $\log_4 64$ **12.** $\log_5 1$ **13.** $\log_8 \frac{1}{64}$

In Exercises 14–16, find the value of b, N, or x.

14. $\log_3 N = 2$ **15.** $\log_b 32 = 2.5$ **16.** $3^{\log_3 2} = x$

B In Exercises 1–6, express each statement in logarithmic form.

1. $2^3 = 8$ **2.** $10^{-1} = 0.1$ **3.** $25^{1/2} = 5$

4. $8^{-2/3} = \frac{1}{4}$ **5.** $7^0 = 1$ **6.** $5^1 = 5$

In Exercises 7–10, express each statement in exponential form.

7. $\log_8 64 = 2$ **8.** $\log_5 125 = 3$

9. $\log_{10} 10 = 1$ **10.** $\log_{10} 100 = 2$

In Exercises 11–13, find the value of each of the logarithms.

11. $\log_3 27$ **12.** $\log_{27} 9$ **13.** $\log_9 \frac{1}{3}$

In Exercises 14–16, find the value of N or x.

14. $\log_2 N = 3$ **15.** $\log_{10} 10{,}000 = x$ **16.** $10^{\log_{10} 3} = x$

4 Fundamental laws of logarithms

Since logarithms are exponents, the laws of exponents enable us to derive certain laws of logarithms that are used in solving logarithmic equations and in computation. Any base b greater than 1 may be used. We shall assume that M and N are positive numbers.

1. MULTIPLICATION

The logarithm of the product of two or more numbers is equal to the sum of the logarithms of the numbers.

$$\log_b (M \cdot N) = \log_b M + \log_b N$$

$$\log_b (M \cdot N \cdot P \cdot \cdots \cdot R) = \log_b M + \log_b N + \log_b P + \cdots + \log_b R$$

PROOF Let $x = \log_b M$ and $y = \log_b N$.

Write in exponential form: $M = b^x$ and $N = b^y$

Multiply M by N: $MN = b^x \cdot b^y = b^{x+y}$

Write in logarithmic form: $\log_b MN = \log_b b^{x+y} = x + y$

Replace x and y by their values: $\log_b MN = \log_b M + \log_b N$

EXAMPLE 1 $\log_{10}(289 \times 356) = \log_{10} 289 + \log_{10} 356$

> **2. DIVISION**
>
> The logarithm of the quotient of two numbers is equal to the logarithm of the dividend minus the logarithm of the divisor.
>
> $$\log_b\left(\frac{M}{N}\right) = \log_b M - \log_b N$$

PROOF Let $x = \log_b M$ and $y = \log_b N$.

Write in exponential form: $M = b^x$ and $N = b^y$

Divide M by N: $\dfrac{M}{N} = \dfrac{b^x}{b^y} = b^{x-y}$

Write in logarithmic form: $\log_b\left(\dfrac{M}{N}\right) = x - y$

Replace x and y by their values: $\log_b\left(\dfrac{M}{N}\right) = \log_b M - \log_b N$

EXAMPLE 2 $\log_{10}\left(\dfrac{125}{346}\right) = \log_{10} 125 - \log_{10} 346$

> **3. POWER OF A NUMBER**
>
> The logarithm of a power of a number is equal to the logarithm of the number multiplied by the exponent of the power.
>
> $$\log_b(M^n) = n \log_b M$$

PROOF Let $x = \log_b M$.

Write in exponential form: $M = b^x$

Raise to the nth power: $M^n = (b^x)^n = b^{nx}$

Write in logarithmic form: $\log_b M^n = nx$

Replace x by its value: $\log_b M^n = n \log_b M$

EXAMPLE 3 $\log_{10}(3.51)^7 = 7 \log_{10} 3.51$.

> **4. ROOT OF A NUMBER**
>
> The logarithm of a root of a number is equal to the logarithm of the number divided by the index of the root.
>
> $$\log_b(\sqrt[n]{M}) = \frac{\log_b M}{n}$$

PROOF Since, by Law 4 of Exponents, $\sqrt[n]{M} = M^{1/n}$, by the above law of logarithms we have

$$\log_b (\sqrt[n]{M}) = \log_b (M^{1/n}) = \frac{1}{n} \log_b M.$$

EXAMPLE 4 $\log_{10} \sqrt[5]{146} = \log_{10} (146)^{1/5} = \frac{1}{5} \log_{10} 146$

EXAMPLE 5

a. $\log_{10} \dfrac{345 \times 561}{284} = \log_{10} 345 + \log_{10} 561 - \log_{10} 284$

b. $\log_{10} \dfrac{156 \times (4.15)^3}{(1.05)^{12}} = \log_{10} 156 + 3 \log_{10} 4.15 - 12 \log_{10} 1.05$

c. $\log_{10} \dfrac{68.4 \times \sqrt{7.61}}{\sqrt[4]{3.48}} = \log_{10} 68.4 + \dfrac{1}{2} \log_{10} 7.61 - \dfrac{1}{4} \log_{10} 3.48$

d. $\log_{10} \dfrac{\overset{\overset{4}{\cancel{8}}}{\cancel{32}}}{\underset{7}{\cancel{28}} \times \underset{31}{\cancel{62}}} = \log_{10} 4 - \log_{10} 7 - \log_{10} 31$

EXAMPLE 6 Given $\log_{10} 2 = 0.301$, $\log_{10} 3 = 0.477$, and $\log_{10} 7 = 0.845$, find $\log_{10} \frac{54}{7}$.

SOLUTION

$$\begin{aligned} \log_{10} \frac{54}{7} &= \log_{10} \frac{2 \times 3^3}{7} \\ &= \log_{10} 2 + 3 \log_{10} 3 - \log_{10} 7 \\ &= 0.301 + 3(0.477) - 0.845 \\ &= 0.887 \end{aligned}$$

EXAMPLE 7 Express $3 \log_b x + 5 \log_b y - \frac{1}{2} \log_b z$ as a single logarithm.

SOLUTION

$$\begin{aligned} 3 \log_b x + 5 \log_b y - \frac{1}{2} \log_b z &= \log_b x^3 + \log_b y^5 - \log_b z^{1/2} \\ &= \log_b \frac{x^3 y^5}{\sqrt{z}} \end{aligned}$$

A WORD OF CAUTION

a. $\log (M + N) \neq \log M + \log N$
because $\log M + \log N = \log MN$.

b. $\log (M - N) \neq \log M - \log N$

because $\log M - \log N = \log \frac{M}{N}$.

Exercises 4

A Given $\log_{10} 2 = 0.301$, $\log_{10} 3 = 0.477$, and $\log_{10} 7 = 0.845$, find the following logarithms. (Recall that $\log_{10} 10 = 1$.)

1. $\log_{10} 14$ **2.** $\log_{10} \frac{18}{7}$ **3.** $\log_{10} 50$

4. $\log_{10} \sqrt{21}$ **5.** $\log_{10} 60$ **6.** $\log_{10} \sqrt[10]{2}$

In Exercises 7–10, write each expression as a single logarithm. (Assume that all of the logarithms have the same base.)

7. $\log x + \log y$ **8.** $2 \log x - 3 \log y$

9. $\frac{1}{2} \log x + \log y - \log z$ **10.** $\log (x^2 - y^2) - \log (x + y)$

In Exercises 11–13, solve for y in terms of x.

11. $\log_2 y = 3x$ **12.** $\log_{10} y = 3 \log_{10} x$

13. $\frac{1}{3} \log_{10} y = \log_{10} x^4 - 3 \log_{10} x$

14. Solve for x. $\log_{10} (x - 5) + \log_{10} 5 = \log_{10} 10 + \log_{10} 3$

B Given $\log_{10} 2 = 0.301$, $\log_{10} 3 = 0.477$, and $\log_{10} 7 = 0.845$, find the following logarithms. (Recall that $\log_{10} 10 = 1$.)

1. $\log_{10} 6$ **2.** $\log_{10} \frac{8}{7}$ **3.** $\log_{10} 25$

4. $\log_{10} \sqrt{14}$ **5.** $\log_{10} 72$ **6.** $\log_{10} \sqrt[7]{2}$

In Exercises 7–10, write each expression as a single logarithm. (Assume that all of the logarithms have the same base.)

7. $\log u - \log v$ **8.** $4 \log x - 2 \log y$

9. $\frac{1}{3} \log x - \log y + \log z$ **10.** $\log (x^4 - y^4) - \log (x^2 - y^2)$

In Exercises 11–13, solve for y in terms of x.

11. $\log_4 y = 2x$ **12.** $\log_{10} y = -3 \log_{10} x$

13. $\frac{1}{3} \log_{10} y = 3 \log_{10} x - \log_{10} x^2$

14. Solve for x. $\log_{10} (x + 3) + \log_{10} 4 = \log_{10} 100 + \log_{10} 2$

5 Systems of logarithms

There are two important systems of logarithms. The **natural** or **Napierian** system uses the base e, where e is approximately 2.71828. This system is used in most advanced applications of mathematics. For this system, the notation $\ln x$ is often used instead of $\log_e x$. The system generally used for computation is the **common** or **Briggs** system which uses the base 10. Hereafter, when we speak of the logarithm of a number without specifying the base, it is to be understood that the base of the logarithm is 10. Instead of writing $\log_{10} N$, we shall write $\log N$.

6 Logarithms to the base 10

The logarithms of integral powers of 10 ($\ldots 10^{-3}, 10^{-2}, 10^{-1}, 10^0, 10^1, 10^2, 10^3, \ldots$) are integers.

EXAMPLE 1

	Exponential form	Logarithmic form
a.	$10^3 = 1{,}000$	$\log 1{,}000 = 3$
b.	$10^2 = 100$	$\log 100 = 2$
c.	$10^1 = 10$	$\log 10 = 1$
d.	$10^0 = 1$	$\log 1 = 0$
e.	$10^{-1} = 0.10$	$\log 0.10 = -1$
f.	$10^{-2} = 0.01$	$\log 0.01 = -2$

Most of the logarithms we use are irrational numbers that are given with an accuracy specified to a certain number of decimal places. A collection of such numbers is given in Table 2.

EXAMPLE 2 The following logarithms are given to four decimal accuracy.

	Exponential form	Logarithmic form
a.	$10^{2.3010} = 200$	$\log 200 = 2.3010$
b.	$10^{1.6990} = 50$	$\log 50 = 1.6990$
c.	$10^{0.3010} = 2$	$\log 2 = 0.3010$

To find the values of logarithms of numbers, such as those in Example 2, we use a table or a calculator. The values in a table are calculated by methods beyond the level of this course

Any positive number written in ordinary decimal notation can be written in scientific notation; that is, as the product of a number greater than or equal to 1 and less than 10 and a power of 10. (See Section 10 of Chapter 1 for a discussion of how to write a number in scientific notation.)

EXAMPLE 3 Writing numbers in scientific notation.

a. $2{,}714 = 2.714 \times 10^3$
b. $315.6 = 3.156 \times 10^2$
c. $21.59 = 2.159 \times 10^1$
d. $7.187 = 7.187 \times 10^0$
e. $0.8570 = 8.570 \times 10^{-1}$
f. $0.01444 = 1.444 \times 10^{-2}$
g. $0.00009889 = 9.889 \times 10^{-5}$

In scientific notation, two numbers that have the same sequence of digits can differ only in the power of 10 that is used.

EXAMPLE 4

a. $314 = 3.14 \times 10^2$
b. $31.4 = 3.14 \times 10^1$
c. $0.314 = 3.14 \times 10^{-1}$
d. $0.00000314 = 3.14 \times 10^{-6}$

We recall that $\log 1 = 0$, $\log 10 = 1$, and that $\log x_1 > \log x_2$ if $x_1 > x_2$; therefore it follows that the value of $\log N$, where $0 < N < 10$, is a number greater than 0 and less than 1.

EXAMPLE 5 The logarithms of numbers between 1 and 10 are numbers between 0 and 1.

a. $\log 1 = 0$
b. $\log 1.10 = 0.0414$
c. $\log 2.00 = 0.3010$
d. $\log 3.14 = 0.4969$
e. $\log 9.99 = 0.9996$
f. $\log 10.0 = 1.0000$

The logarithms of the numbers in Example 4 can be expressed as follows.

EXAMPLE 6

a.
$$\begin{aligned}\log 314 &= \log (3.14 \times 10^2)\\ &= \log 3.14 + \log 10^2\\ &= \log 3.14 + 2 \log 10\\ &= \log 3.14 + 2(1)\\ &= 0.4969 + 2 = 2.4969\end{aligned}$$

b. $\log 31.4 = \log(3.14 \times 10^1)$
$= \log 3.14 + \log 10^1$
$= 0.4969 + 1 = 1.4969$

c. $\log 0.314 = \log(3.14 \times 10^{-1})$
$= \log 3.14 + \log 10^{-1}$
$= \log 3.14 + (-1) \log 10$
$= 0.4969 - 1$

By adding 10 and then subtracting 10 from the number $0.4969 - 1$, we can write

$$\begin{aligned} 0.4969 - 1 &= 0.4969 - 1 + 10 - 10 \\ &= 0.4969 + 9 - 10 \\ &= 9.4969 - 10. \end{aligned}$$

This is done to simplify the calculation, as you will see in Section 9.

d. $\log 0.00000314 = \log(3.14 \times 10^{-6})$
$= \log 3.14 + \log 10^{-6}$
$= 0.4969 - 6$
$= 4.4969 - 10$

Hence, the common logarithm of any positive number can be written as the sum of two parts, (1) a nonnegative decimal less than 1 and (2) the exponent applied to 10 when the number is written in scientific notation.

The decimal part of the common logarithm of a number is called the **mantissa** of the logarithm. As we saw in Example 6, the *mantissa depends only on the particular sequence of digits in the number.* It is independent of the position of the decimal point.

The exponent of 10, which is the whole part of the logarithm, is called the **characteristic**. *The characteristic depends only on the position of the decimal point in the number.* It is independent of the sequence of digits.

7 Rule for the characteristic of a common logarithm

As we stated in the previous section, the characteristic of the logarithm of a given number is equal to the exponent of ten when that number is written in scientific notation.

EXAMPLE 1 Notice in the table at the top of page 372 that the characteristics (listed in the last column) and the exponents of 10 (listed in the second column) are the same.

	Number	The number in scientific notation	Characteristic of the logarithm of the number
a.	2,045	2.045×10^{3}	3
b.	204.5	2.045×10^{2}	2
c.	20.45	2.045×10^{1}	1
d.	2.045	2.045×10^{0}	0
e.	0.2045	2.045×10^{-1}	−1
f.	0.02045	2.045×10^{-2}	−2
g.	0.002045	2.045×10^{-3}	−3
h.	0.0002045	2.045×10^{-4}	−4

You may find the $\wedge$ symbol, called a **caret**, to be an aid in converting a number to its equivalent form in scientific notation. The caret shows the position where the decimal point of the transformed number is to be placed. Some examples will show the use of the caret.

EXAMPLE 2

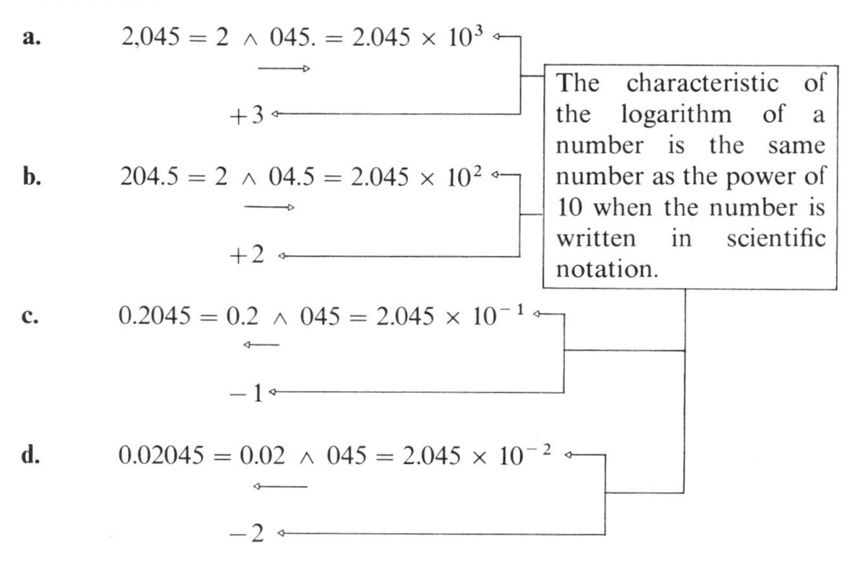

Notice that the caret is placed immediately to the right of the first nonzero digit of the number. Then the number is rewritten with the decimal point in the position where the caret was used. This equivalent number is then multiplied by 10^n, where n is the number of places of the decimal point to the right or left of the caret; n is positive when the decimal point is to the right of the caret, as in Examples 2a and 2b above, and negative when the decimal point is to the left of the caret, as in Examples 2c and 2d.

For computational purposes, it is better to write negative characteristics as the difference of two numbers, the second of which is a multiple of 10. For example,

the characteristic	-1	is written	$9 - 10$,
the characteristic	-2	is written	$8 - 10$,
the characteristic	-4	is written	$6 - 10$,
the characteristic	-12	is written	$8 - 20$.

Exercises 7

A In Exercises 1–9, write the characteristic of the logarithm of each number.

1. 367 **2.** 584,000 **3.** 0.0844
4. 0.00078 **5.** 8.443 **6.** 0.0088
7. 45,000 **8.** 4.54×10^8 **9.** 4.75×10^{-3}

In Exercises 10–15, rewrite each number and place its decimal point according to the given characteristic of its logarithm.

	Number	Characteristic		Number	Characteristic
10.	286	2	**11.**	245	1
12.	841	0	**13.**	745	-1
14.	547	-3	**15.**	207	$7 - 10$

B In Exercises 1–9, write the characteristic of the logarithm of each number.

1. 485 **2.** 186,300 **3.** 0.0714
4. 0.00054 **5.** 5.863 **6.** 0.0078
7. 98,000 **8.** 5.88×10^{12} **9.** 2.81×10^{-2}

In Exercises 10–15, rewrite each number and place its decimal point according to the given characteristic of its logarithm.

	Number	Characteristic		Number	Characteristic
10.	184	1	**11.**	366	3
12.	545	0	**13.**	275	-1
14.	709	-3	**15.**	234	$8 - 10$

8 Interpolation

Interpolation is the process of approximating a number that is not an entry in the table but lies between two consecutive entries.

The logarithms of numbers with three significant digits are read directly from Table 2. By interpolation, we can find the logarithm of a number with four significant digits. If we wish to find the logarithm of a number with more than four significant digits, we usually round the number to four significant digits before using the table as the table is not designed to give greater accuracy than that. For greater accuracy, you can use your calculator or a five-place table of logarithms. We have not included a five-place table because of the added space it requires.

EXAMPLE 1 Find log 78.46.

SOLUTION The characteristic is 1 and the logarithm of 78.46 must lie between the log of 78.40 and log 78.50. We use Table 2.

$$10\left\{\begin{array}{l} 6\left\{\begin{array}{l}\log 78.40 = 1.8943\\ \log 78.46 = \ ?\end{array}\right\} d \\ \log 78.50 = 1.8949\end{array}\right\} 6$$

$$\frac{d}{6} = \frac{6}{10} = \frac{3}{5}$$

$$d = \frac{3}{5}(6) = \frac{18}{5} \doteq 4, \qquad \text{in last digit}$$

Therefore, log 78.46 = 1.8943 + 0.0004 = 1.8947, accurate to four decimal places.

EXAMPLE 2 Given $\log N = 8.7045 - 10$, find N.

SOLUTION We first disregard the characteristic and look for the succession of digits in the number N which correspond to the mantissa 0.7045. We find that this mantissa lies between the two table entries 0.7042 and 0.7050.

Ignoring decimal points

$$10\left\{\begin{array}{l} d\left\{\begin{array}{l}\log 5060 = 7042\\ \log N = 7045\end{array}\right\} 3 \\ \log 5070 = 7050\end{array}\right\} 8$$

$$\frac{d}{10} = \frac{3}{8}$$

$$d = \frac{3}{8}(10) \doteq 4, \qquad \text{to one digit}$$

Using decimal points

$$10\left\{\begin{array}{l} d\left\{\begin{array}{l}\log 0.05060 = 8.7042 - 10 \\ \log \quad N \quad = 8.7045 - 10\end{array}\right\}3 \\ \log 0.05070 = 8.7050 - 10\end{array}\right\}8$$

Therefore, $N = 0.05064$, to four significant digits.

EXAMPLE 3 Use your calculator to find log 78.46.

SOLUTION

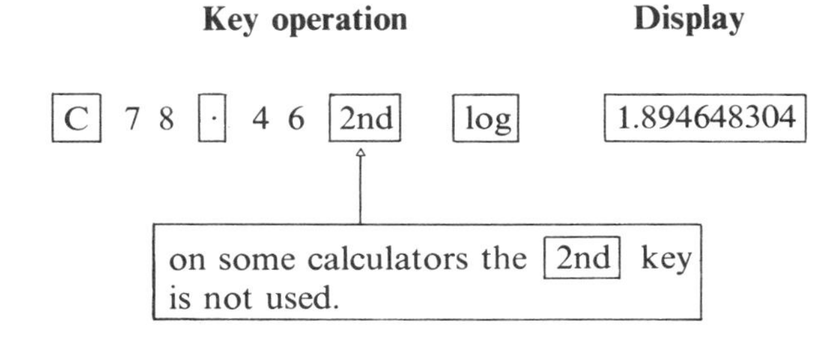

EXAMPLE 4 Use your calculator to find N when $\log N = 2.5612$.

SOLUTION

Key operation	Display
[C] 1 0 [y^x] 2 [·] 5 6 1 2 [=]	[364.0826638]

Rounded to four significant digits, we have 364.1.

EXAMPLE 5 Use your calculator to find N when $\log N = 8.4615 - 10$.

SOLUTION $\log N = 8.4615 - 10 = -1.5385$

Key operation

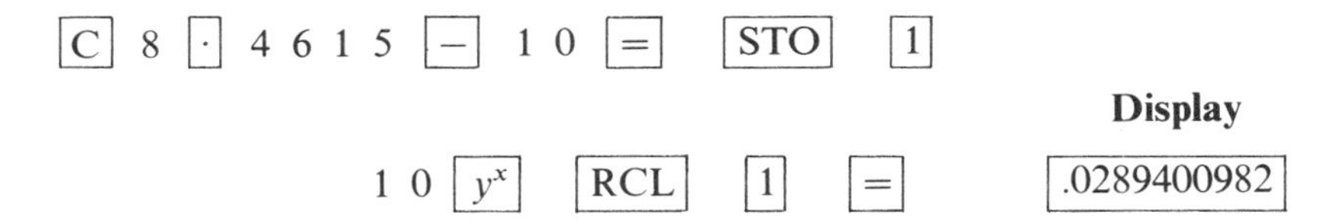

Rounded to four significant digits, we have 0.0289.

Exercises 8

Use Table 2 to solve these exercises. Assume the numbers given in the exercises are exact and round answers to the accuracy of the table.

A In Exercises 1–5, find the logarithm of each number.

1. 58.6 **2.** 384.2 **3.** 0.0055
4. 0.08004 **5.** 2.14×10^2

In Exercises 6–10, let each of the given numbers be log N and find N.

6. 0.8457 **7.** 0.4236 **8.** 2.7033
9. $9.5521 - 10$ **10.** $7.8671 - 10$

B In Exercises 1–5, find the logarithm of each number.

1. 25.4 **2.** ·434.6 **3.** 0.0072
4. 0.09003 **5.** 6.85×10^3

In Exercises 6–10, let each of the given numbers, be log N and find N.

6. 0.7589 **7.** 0.8776 **8.** 3.5441
9. $9.4084 - 10$ **10.** $7.9332 - 10$

9 Computations with logarithms

Before any computation is done in a problem, it is important to analyze the problem and make an outline of the procedure to be followed. Without an outline, you are apt to waste time and possibly omit some important detail. This is especially true in solving triangles by the use of logarithms. The characteristics of the logarithms of each given number should be written in the proper location in the outline before going to the tables for the mantissas.

In the five following examples, the numbers in the examples are assumed to be exact, and the answer for each example is given correct to four significant digits.

EXAMPLE 1 Find the value of $x = 21.7 \times 3.14 \times 0.896$.

SOLUTION By Law 1, Section 4:

$$\log x = \log 21.7 + \log 3.14 + \log 0.896$$

The blank outline, with the characteristics of the numbers written in, is shown at the left. The same outline is also shown at the right after it has been filled in from the tables.

$\log 21.7 = 1.$	$\log 21.7 = 1.3365$
$\log 3.14 = 0.$	$\log 3.14 = 0.4969$
$\log 0.896 = \underline{9.\qquad\quad - 10(+)}$	$\log 0.896 = \underline{9.9523 - 10(+)}$
$\log x =$	$\log x = 11.7857 - 10$
$x =$	$x = 61.06$

EXAMPLE 2 Find the value of $x = 9.04/0.0654$.

SOLUTION Take the logarithms of both sides of the equation.

$$\log x = \log 9.04 - \log 0.0654$$

$$\begin{aligned} \log 9.04 &= 0. \\ \log 0.0654 &= \underline{8. \qquad\qquad - 10(-)} \\ \log x &= \\ x &= \end{aligned} \qquad\qquad \begin{aligned} \log 9.04 &= 10.9562 - 10 \\ \log 0.0654 &= \underline{8.8156 - 10(-)} \\ \log x &= 2.1406 \\ x &= 138.2 \end{aligned}$$

Notice that 10 was added to and then subtracted from the logarithm of 9.04 in order to keep the decimal part of log x positive.

EXAMPLE 3 Find the value of $x = \dfrac{(1.500)^4}{(3.461)(0.8960)}$.

SOLUTION Take the logarithm of both sides of the equation.

$$\begin{aligned} \log x &= 4 \log 1.500 - [\log 3.461 + \log 0.8960] \\ &= 4(0. \qquad) - [(0. \qquad) + (9. \quad - 10)] \end{aligned}$$

We let the above be our outline and proceed to find the mantissas from the tables.

$$\begin{aligned} \log x &= 4(0.1761) - [0.5392 + (9.9523 - 10)] \\ &= 0.7044 - [10.4915 - 10] \\ &= 0.2129 \\ x &= 1.633 \end{aligned}$$

EXAMPLE 4 Find the value of $x = \sqrt[3]{0.7165}$.

SOLUTION Rewrite the equation in the form $x = (0.7165)^{1/3}$. Then take the logarithm of both sides of this equation.

$$\begin{aligned} \log x &= \log (0.7165)^{1/3} \\ &= \frac{1}{3} \log 0.7165 \\ &= \frac{1}{3}(9.8552 - 10) \end{aligned}$$

We are about to divide the logarithm (9.8552 − 10) by 3. In order to make the negative part of the characteristic evenly divisible by 3, we both add and subtract 20 in the characteristic. Thus we rewrite the logarithm (9.8552 − 10) as

$$\begin{aligned} 9.8552 - 10 &= 9.8552 + 20 - 20 - 10 \\ &= 29.8552 - 30. \end{aligned}$$

Then $$\log x = \frac{1}{3}(29.8552 - 30)$$

$$= 9.9517 - 10.$$

Therefore $$x = 0.8948.$$

EXAMPLE 5 Find the value of $x = \dfrac{75 \times (-3.85)}{5.79}$.

SOLUTION As stated in Section 3, we are restricted to the logarithms of positive numbers, but the factor -3.85 is negative. By inspection we observe that

$$x = \frac{75 \times (-3.85)}{5.79} = -\frac{75 \times 3.85}{5.79} = -F,$$

where $F = \dfrac{75 \times 3.85}{5.79}$. Then we proceed as follows.

$\log 75 = 1.$	$\log 75 = 1.8751$
$\log 3.85 = \underline{0.\qquad (+)}$	$\log 3.85 = \underline{0.5855(+)}$
	2.4606
$\log 5.79 = \underline{0.\qquad (-)}$	$\log 5.79 = \underline{0.7627(-)}$
$\log F =$	$\log F = 1.6979$
$F =$	$F = 49.88$

Therefore $x = -F = -49.88$.

Exercises 9

In the following exercises, use logarithms (Table 2) to obtain the results correct to four significant digits. Assume the given numbers to be exact numbers. We suggest that you use your hand calculator to verify the results obtained by the logarithmic calculations of the following exercises. (The calculations obtained by using your calculator will be more accurate, because the entries in Table 2 have been rounded to four significant digits.)

A

1. 29.47×3.058 **2.** $711.6 \div 49.1$ **3.** $(1.16)^5$

4. $\sqrt[5]{80.9}$ **5.** $(18.5)^{2/3}$ **6.** $\sqrt{\dfrac{850}{1.46 \times 7.84}}$

7. $7.86 + \log 14.96$ **8.** $38.6 \times \log 187.4$ **9.** $\dfrac{85(-6.75)}{38.4}$

10. $(2.81)^{-2.5}$ 11. $\dfrac{\log 18.5}{1.17}$ 12. $\dfrac{\log 18.5}{\log 1.17}$

13. $\sqrt[3]{\dfrac{75(3.5)^4}{15.6 \times 2.45}}$ 14. $(\log 297)(\log 0.075)$

15. Calculate the volume of a sphere of radius $r = 2.64$, using the formula $V = \frac{4}{3}\pi r^3$ (use $\pi = 3.14$).

B

1. 47.64×7.056 2. $609.1 \div 15.4$ 3. $(1.84)^5$

4. $\sqrt[4]{75.6}$ 5. $(14.6)^{2/3}$ 6. $\sqrt{\dfrac{944}{1.85 \times 79.4}}$

7. $1.467 + \log 27.36$ 8. $27.1 \times \log 27.1$ 9. $\dfrac{32(-3.47)}{28.4}$

10. $(3.51)^{-2.4}$ 11. $\dfrac{\log 186}{47.1}$ 12. $\dfrac{\log 58.6}{\log 6.95}$

13. $\sqrt[3]{\dfrac{42(5.6)^4}{12.9 \times 3.77}}$ 14. $(\log 486)(\log 7.41)$

15. Calculate the volume of a right-circular cylinder whose altitude is $h = 7.25$ and whose base radius is $r = 1.75$, using the formula $V = \pi r^2 h$ (use $\pi = 3.14$).

10 Graphs of logarithmic functions

For a given value of the base b, we can obtain the graph of the function

$$y = \log_b x$$

by plotting positive values of x as the abscissas and the corresponding values of $\log_b x$ as the ordinates of points on the graph.

We show graphs for the functions $y = \log_{10} x$ and $y = \log_2 x$ (see Figure 3). Coordinates of points on the graph of $y = \log_{10} x$ may be obtained either from a table of logarithms or from the exponential form of $y = \log_{10} x$, namely, $x = 10^y$. In the exponential form, we give y values and solve for x. The coordinates of some of these points are given below.

x	0.01	0.1	1	10	100
y	-2	-1	0	1	2

To find coordinates of points on the graph of the function $y = \log_2 x$, we change the function to exponential form, give values to y, and solve for x.

$$y = \log_2 x$$

$$x = 2^y$$

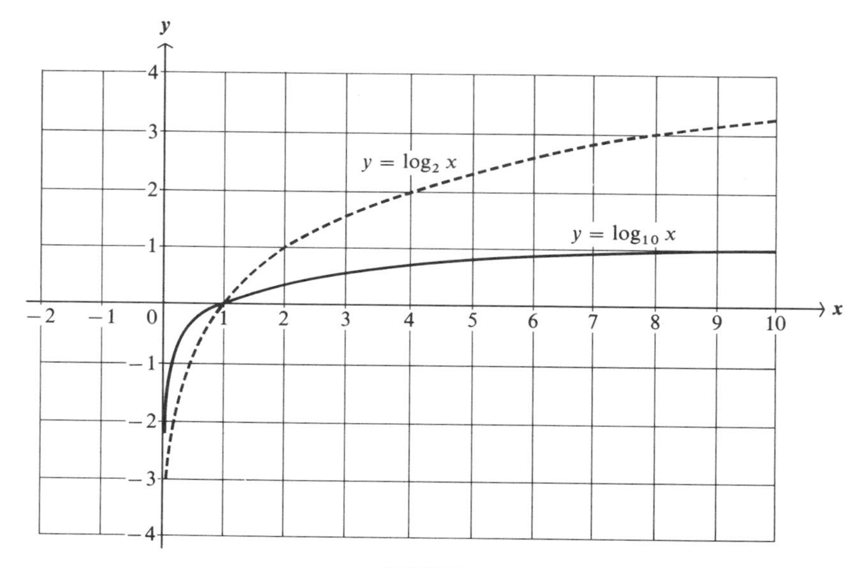

FIGURE 3

The coordinates of some of the points on the graph of the function $y = \log_2 x$ are given below.

x	$\frac{1}{4}$	$\frac{1}{2}$	1	2	4	8
y	-2	-1	0	1	2	3

PROPERTIES OF THE LOGARITHM FUNCTION

Several important properties of the logarithm are illustrated by its graph.

1. The logarithm of a number greater than 1 is positive.
2. The logarithm of 1 is zero.
3. The logarithms of numbers between 0 and 1 are negative.
4. As x approaches 0 as a limit from the positive side of 0, $\log x$ decreases through negative values, and its absolute value increases beyond all bounds.
5. The logarithm of 0 does not exist.
6. The logarithm of a negative number is not on the graph of real numbers (our definition of the logarithm excludes the logarithm of a negative number).
7. As x increases through positive values, $\log x$ increases more and more slowly.
8. The domain is the interval $(0, \infty)$. The range is the interval $(-\infty, \infty)$.

11 Exponential and logarithmic equations and inequalities

An **exponential equation** is an equation in which an unknown quantity appears in one or more exponents, as illustrated in Examples 1 and 2.

A **logarithmic equation** is an equation containing the logarithm of the variable (see Example 3).

Generally, simple exponential equations can be solved by taking logarithms of both members of the equation, equating them, and then solving the resulting algebraic equation.

EXAMPLE 1 Solve the equation $3^x = 50$.

SOLUTION Take the logarithm of both members of the equation and equate them.

$$\log 3^x = \log 50$$

$$x \log 3 = \log 50$$

$$x = \frac{\log 50}{\log 3} = \frac{1.6990}{0.4771} \doteq 3.562$$

In Example 1, we assumed 3 and 50 to be exact numbers and expressed the value of x to four significant figures, as Table 2 is a four-decimal-place table.

A WORD OF CAUTION You are warned that to divide log 50 by log 3, you do not subtract the logarithm of 3 from the logarithm of 50. That would be dividing 50 by 3 and not log 50 by log 3.

EXAMPLE 2 Solve the equation $(34.7)^{2x} = (12.5)^{3x-2}$.

SOLUTION Take the logarithm of both sides of the equation and equate the two. Then solve the resulting algebraic equation.

$$\log (34.7)^{2x} = \log (12.5)^{3x-2}$$

$$2x \log 34.7 = (3x - 2) \log 12.5$$

$$2x(1.5403) \doteq (3x - 2)(1.0969)$$

$$3.0806x \doteq 3.2907x - 2.1938$$

$$0.2101x \doteq 2.1938$$

$$x \doteq 10.44$$

Certain equations in which the unknown does not appear in an exponent can be solved by first reducing the equations to exponential form.

EXAMPLE 3 Solve for x: $2 \log (x + 3) = \log (7x + 1) + \log 2$.

SOLUTION Rearrange terms and combine logarithms.

$$2 \log (x + 3) - \log (7x + 1) = \log 2$$

$$\log (x + 3)^2 - \log (7x + 1) = \log 2$$

$$\log \frac{(x + 3)^2}{7x + 1} = \log 2$$

Therefore

$$\frac{(x + 3)^2}{7x + 1} = 2.$$

We now solve for x.

$$(x + 3)^2 = 2(7x + 1)$$

$$x^2 + 6x + 9 = 14x + 2$$

$$x^2 - 8x + 7 = 0$$

$$(x - 1)(x - 7) = 0$$

Therefore $x = 1$, and 7. Both roots satisfy the original equation.

EXAMPLE 4 Find $\log_3 35.7$.

SOLUTION Let $x = \log_3 35.7$, and then write the equation in exponential form.

$$3^x = 35.7$$

Take the logarithm of both members to the base 10, and then solve the resulting equation for x.

$$\log 3^x = \log 35.7$$

$$x \log 3 = \log 35.7$$

$$x = \frac{\log 35.7}{\log 3} = \frac{1.5527}{0.4771} \doteq 3.255$$

Therefore $\log_3 35.7 = 3.255$.

EXAMPLE 5 Solve the inequality $(0.6)^x > 2$.

SOLUTION Take the logarithm of both sides of the inequality then solve the resulting inequality.

$$\log (0.6)^x > \log 2$$

$$x \log (0.6) > \log 2$$

When we divide both sides of the inequality by log 0.6, which is negative, the sense of the inequality is reversed.

$$x < \frac{\log 2}{\log 0.6} \doteq \frac{0.3010}{9.7782 - 10} \doteq \frac{0.3010}{-0.2218} \doteq -1.357$$

Therefore, the approximate solution is $\{x \mid x < -1.357, x \in R\}$.

Exercises 11

A In Exercises 1–7, solve each equation. Assume that the given numbers are exact and carry out the work as far as possible with Table 2.

1. $2^x = 3$ **2.** $(6.4)^{x+1} = 100$

3. $5^x = (3.7)^{2x+1}$ **4.** $(1.07)^{-x} = 0.486$

5. $\dfrac{(1.06)^x - 1}{0.06} = 20$ **6.** $\log x + \log(11 - x) = 1$

7. $\log(x + 4) - \log 10 = \log 6 - \log x$

In Exercises 8 and 9, find each of the specified logarithms.

8. $\log_7 784$ **9.** $\log_8 365$

10. Graph $y = \log_4 x$.

11. Solve the inequality $(0.7)^x > 3$.

12. Find the monthly payment on a 25-year, \$40,000 loan at 9% interest.

The formula for finding the monthly payment on a homeowner's mortgage is

$$R = \frac{Ai(1 + i)^n}{(1 + i)^n - 1}.$$

R = Monthly payment

i = Interest rate per month expressed as a decimal

n = Number of months

A = Original amount of mortgage

B In Exercises 1–7, solve each equation. Assume that the given numbers are exact and carry out the work as far as possible with Table 2.

1. $3^x = 4$ **2.** $(4.6)^{x+1} = 100$

3. $3^x = (5)^{2x+1}$ **4.** $(1.04)^{-x} = 0.725$

5. $\dfrac{(1.05)^x - 1}{0.05} = 30$

6. $\log(5x - 7) = \log(2x - 3) + \log 3$
7. $\log(2x + 1) = \log 1 + \log(x + 2)$

In Exercises 8 and 9, find each of the specified logarithms.

8. $\log_5 75$ 9. $\log_2 0.135$
10. Graph $y = \log_3 x$.
11. Solve the inequality $(0.4)^x < 2$.

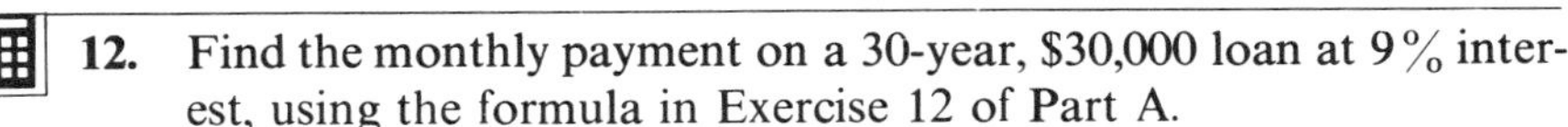

12. Find the monthly payment on a 30-year, \$30,000 loan at 9% interest, using the formula in Exercise 12 of Part A.

Summary

A **transcendental function** is any function which is not an algebraic function. Logarithmic functions and trigonometric functions are examples of transcendental functions.

An **exponential function** is a function in which an unknown quantity appears in one or more exponents.

A **logarithmic function** is a function containing the logarithm of the variable.

The **logarithm of a positive number** to a given base is the exponent to which the base must be raised to give the number.

If x is the logarithm of N to a given base, then N is called the **antilogarithm** of x to that base.

LAWS OF LOGARITHMS

1. $\log_b (M \cdot N) = \log_b M + \log_b N$
2. $\log_b \left(\frac{M}{N}\right) = \log_b M - \log_b N$
3. $\log_b (M^n) = n \log_b M$
4. $\log_b (\sqrt[n]{M}) = \frac{\log_b M}{n}$

There are **two** important systems of logarithms. The **natural** or **Napierian** system uses the base e (approximately 2.71828), and the **common** or **Briggs** system uses the base 10. The logarithm of a number is made up of two parts. The first is the integer part, called the **characteristic**, and the second is the decimal part, called the **mantissa**. The characteristic is the exponent of 10 when the number is written in scientific notation. The mantissa is the logarithm of the number between 1 and 10 when the number is written in scientific notation. The mantissa is found in Table 2.

IMPORTANT PROPERTIES OF LOGARITHMS

1. The logarithm of a number greater than 1 is positive.
2. The logarithm of 1 is zero.
3. The logarithms of numbers between 0 and 1 are negative.
4. As x approaches 0 as a limit from the positive side of 0, $\log x$ decreases through negative values, and its absolute value increases beyond all bounds.
5. The logarithm of 0 does not exist.
6. The logarithm of a negative number is not on the graph of real numbers (our definition of the logarithm excludes the logarithm of a negative number).
7. As x increases through positive values, $\log x$ increases more and more slowly.
8. The domain is the interval $(0, \infty)$. The range is the interval $(-\infty, \infty)$.

Review exercises

A

1. Write the exponential statement $4^3 = 64$ in logarithmic form.

In Exercises 2–4, find the unknown, b, N, or x.

2. $\log_3 \dfrac{1}{9} = x$ 3. $\log_{10} N = -2$ 4. $\log_b 8 = \dfrac{3}{2}$

Given $\log 2 = 0.3010$ and $\log 3 = 0.4771$, find the value of each of the logarithms listed in Exercises 5–7.

5. $\log 2^{10}$ 6. $\log \sqrt{9}$ 7. $\log 48$
8. Transform the expression $\log k^3 - \log (5/\sqrt[3]{k^2}) + \log 5\sqrt[3]{k}$ into a single logarithm.
9. Graph $f(x) = 3^{1/2x}$.

In Exercises 10 and 11, find the solution set.

10. $\log_{16} (2x - 1) = \dfrac{1}{2}$ 11. $\left(\dfrac{1}{4}\right)^x < 4$

Use logarithms to work Exercises 12–14. Assume the given numbers to be exact and carry out the work as far as possible with Table 2.

12. $\sqrt[5]{0.756}$ 13. $\dfrac{(0.844)\sqrt[3]{185}}{3.57}$
14. Solve $(7.6)^x = 0.488$.

B

1. Write the exponential statement $3^4 = 81$ in logarithmic form.

In Exercises 2–4, find the unknown, b, N, or x.

2. $\log_2 \dfrac{1}{4} = x$ 3. $\log_{10} N = -3$ 4. $\log_b 9 = \dfrac{2}{3}$

Given $\log 2 = 0.3010$ and $\log 3 = 0.4771$, find the value of each of the logarithms listed in Exercises 5–7.

5. $\log 2^3$ 6. $\log \sqrt{2}$ 7. $\log 72$
8. Transform the expression $\log a^4 - \log (6/\sqrt[3]{a}) + \log 6\sqrt[3]{a^2}$ into a single logarithm.
9. Graph $f(x) = \left(\dfrac{1}{2}\right)^{2x}$.

In Exercises 10 and 11, find the solution set.

10. $\log_9 (x + 1) = \dfrac{1}{2}$ 11. $\left(\dfrac{1}{3}\right)^x < 3$

Use logarithms to work Exercises 12–14. Assume the given numbers to be exact and carry out the work as far as possible with Table 2.

12. $\sqrt[5]{0.617}$ 13. $\dfrac{(0.56)\sqrt[3]{75}}{2.83}$
14. Solve $(5.8)^x = 0.961$.

Diagnostic test

The purpose of this test is to see how well you understand the material given in this chapter. Allow yourself approximately 55 minutes to do the test. Solutions to the problems, together with section references, are given in the Answers section at the end of this book. We suggest that you study the sections referred to for the problems you do incorrectly.

1. Write the logarithmic function $\log_{25} \frac{1}{5} = -\frac{1}{2}$ in exponential form.

In Problems 2–6, find the unknown, b, N, or x.

2. $\log_7 \sqrt{7} = x$ 3. $\log_b 2 = \dfrac{1}{2}$
4. $\log_3 N = -4$ 5. $\left(\dfrac{1}{6}\right)^x < 6$
6. $\log (4x^2 - 1) - \log (2x + 1) = \log (x^2 - 1) - \log (x - 1)$
7. Given $\log 2 = 0.3010$, find each of the following.
   a. $\log \sqrt[5]{2}$ b. $\log 16$

8. Write the expression $\log \sqrt{h^5} - 2 \log \sqrt[4]{h} + \log h^4$ as a single, simplified term.
9. Graph $f(x) = 2^{x-1}$.

In Problems 10–12, use logarithms (Table 2) to find answers correct to four significant digits. Assume the given numbers to be exact numbers.

10. $\dfrac{\log 2}{3.45}$ 11. $(69.4)^{2/5}$ 12. Solve $(6.31)^x = 7.67$.

8 Polar coordinates and complex numbers

In Chapter 4 (Sections 3 and 4), we defined complex numbers and showed how to add, subtract, multiply, and divide complex numbers.

In this chapter, we show (1) how to plot a complex number on a complex plane, (2) how a complex number can be written in **polar** or **trigonometric form**; and (3) how to multiply, divide, and find powers and roots of complex numbers when they are written in polar form.

1 Graphical representation of complex numbers

In working with complex numbers, it has been found helpful to represent them graphically by points on a **complex plane**. A complex plane is a coordinate plane on which the horizontal axis is the **real axis** and the vertical axis is the **imaginary axis**. Each complex number $x + yi$ determines a unique ordered pair (x, y). The corresponding point $P(x, y)$ in the complex plane is the geometric representation of $x + yi$.

EXAMPLE 1 In Figure 1 we have indicated the geometric representation of several complex numbers. Note that the conjugate numbers

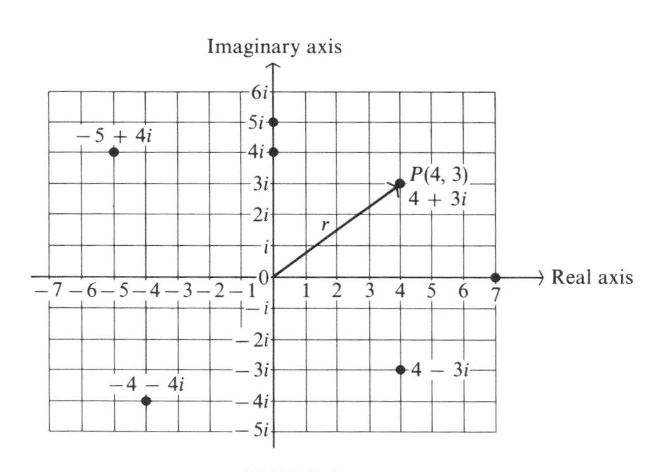

FIGURE 1

Complex plane

$4 + 3i$ and $4 - 3i$ are symmetric to the real axis. A complex number such as $4 + 3i$, $P(4, 3)$, is often represented by a line from the origin to the point. The length of this line, called r, is the absolute value of the complex number. This length r of the complex number $4 + 3i$ is

$$\begin{aligned} r &= |4 + 3i| \\ &= \sqrt{4^2 + 3^2} \\ &= 5. \end{aligned}$$

Exercises 1

Represent graphically on the complex plane each of the following complex numbers and their conjugates.

A

1. $2 + 3i$
2. $3 - 2i$
3. $-4i$
4. 3
5. $-2 - 3i$

B

1. $4 + 2i$
2. $3 - 4i$
3. $5i$
4. -2
5. $-3 - 2i$

2 Polar coordinates

The position of a point in a plane is determined if its rectangular coordinates are known. An alternative method of fixing the position of a point is to give its **polar coordinates**. These are r, the distance of the point from the origin or **pole**, and θ, an angle which the line between the point and the pole makes with the positive x-axis or **polar axis**. The coordinates are written (r, θ); r is called the **radius** of the point; θ is its **angle**.

The radius is expressed in any convenient unit of length, and the angle is commonly measured in radians, though degree measure may be used. In previous chapters, the radius of a point has always been taken as positive. In connection with polar coordinates, however, it is often desirable to let r be negative. To plot a point when its polar coordinates (r, θ) are given, we first place the angle θ in standard position, that is, with its vertex at the pole and its initial side on the polar axis. Then, if r is positive, the point (r, θ) lies on the terminal side of θ and at the distance r from the pole; if r is negative, (r, θ) lies at the distance $|r|$ measured from the pole in the opposite direction, that is, along the line made by extending the terminal side of θ through the pole. Thus, in Figure 2b, the two points P_1: $(4, \frac{2}{3}\pi)$ and P_2: $(-4, \frac{2}{3}\pi)$ are shown. If $r = 0$, the point (r, θ) is the pole, regardless of the value of θ. Unless there is some contrary indication, we shall assume that $r \neq 0$.

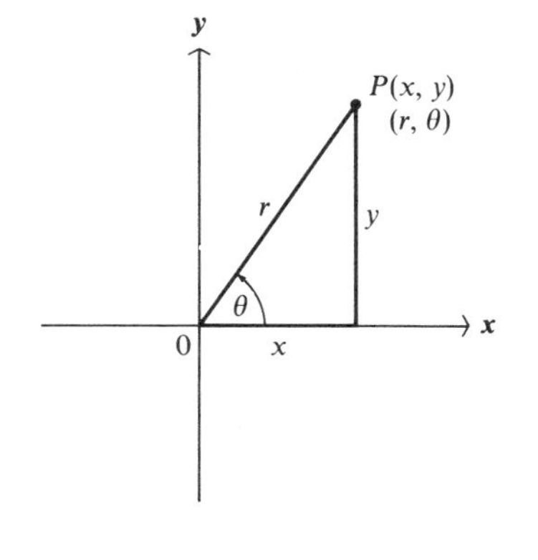

a.

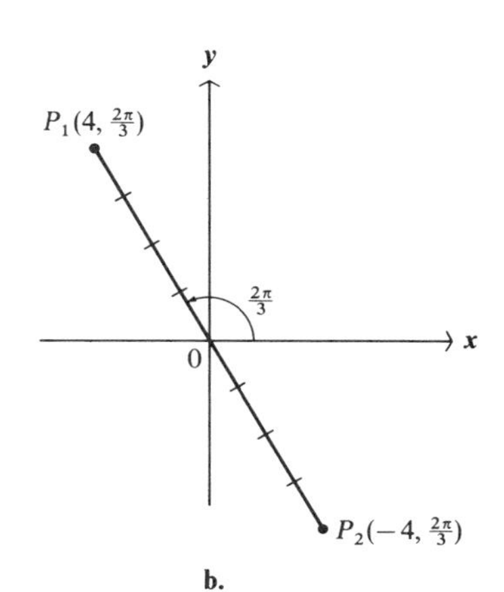

b.

FIGURE 2

It is often desirable to use rectangular and polar coordinates together, the polar axis of the polar system coinciding with the positive x-axis, and x, y, and r measured in the same units. We can always convert from one system to the other by means of the following relations (see Figure 2a).

$$x = r\cos\theta \qquad y = r\sin\theta \tag{1}$$

$$r = \sqrt{x^2 + y^2} \qquad \theta = \arctan\frac{y}{x} \tag{2}$$

In Equations 2, r is taken as positive, and θ is any one of the infinitely many values of arctan (y/x) whose terminal side lies in the same quadrant (or on the same axis) as the point (x, y). If we wish to do so, we may make $r = -\sqrt{x^2 + y^2}$ and take θ in the opposite quadrant from the one containing (x, y).

Polar coordinates have many uses, only one of which is the application to complex numbers.

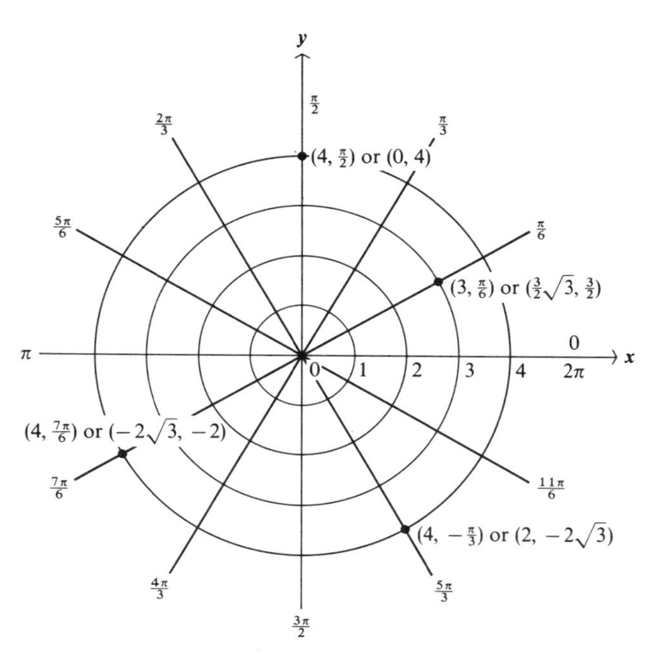

FIGURE 3

EXAMPLE 1 Find the rectangular coordinates of the point whose polar coordinates are $(3, \frac{1}{6}\pi)$ (see Figure 3; this is the real-number plane).

SOLUTION From Equation 1, we have

$$x = r\cos\theta = 3\cos\frac{1}{6}\pi = 3\left(\frac{1}{2}\sqrt{3}\right) = \frac{3}{2}\sqrt{3},$$

$$y = r\sin\theta = 3\sin\frac{1}{6}\pi = 3\left(\frac{1}{2}\right) = \frac{3}{2}.$$

The required coordinates are $(\frac{3}{2}\sqrt{3}, \frac{3}{2})$.

EXAMPLE 2 Find the polar coordinates of the point whose rectangular coordinates are $(2, -2\sqrt{3})$. (See Figure 3.)

SOLUTION From Equation 2, we can take

$$r = \sqrt{2^2 + (-2\sqrt{3})^2} = \sqrt{4 + 12} = 4,$$

$$\theta = \text{Arctan}\,(-\sqrt{3}) = -\frac{1}{3}\pi.$$

The required coordinates can thus be taken as $(4, -\frac{1}{3}\pi)$. We could have taken $r = -4$ and $\theta = \frac{2}{3}\pi$. There are, in fact, infinitely many sets of polar coordinates for every point of the plane. However, it is usually convenient to take r positive and to give θ a numerically small correct value.

Exercises 2

A In Exercises 1–5, plot the point whose polar coordinates are given, and find its rectangular coordinates. Tables may be used when necessary.

1. $\left(2, \frac{\pi}{3}\right)$ **2.** $\left(8, -\frac{\pi}{2}\right)$ **3.** $(10, 100^\circ)$

4. $\left(4, \frac{3\pi}{4}\right)$ **5.** $\left(5, -\frac{5\pi}{6}\right)$

In Exercises 6–9, the rectangular coordinates are given. Find one set of polar coordinates and plot the complex number for the point.

6. $(-\sqrt{3}, 1)$ **7.** $(-5, 0)$ **8.** $(0, 2)$ **9.** $(-2, -3)$

B In Exercises 1–5, plot the point whose polar coordinates are given and find its rectangular coordinates. Tables may be used when necessary.

1. $\left(4, \frac{\pi}{6}\right)$ **2.** $(-2, 2\pi)$ **3.** $(10, 200^\circ)$

4. $\left(6, \frac{5\pi}{2}\right)$ **5.** $\left(6, -\frac{13}{6}\pi\right)$

In Exercises 6–9, the rectangular coordinates are given. Find one set of polar coordinates and plot the complex number for the point.

6. $(2, 2)$ **7.** $(0, 2)$ **8.** $(\sqrt{2}, -\sqrt{2})$ **9.** $\left(\frac{1}{2}, -\frac{1}{2}\sqrt{3}\right)$

3 Polar or trigonometric forms of a complex number

Expressing a complex number in **polar form** (sometimes called **trigonometric form**) gives us a powerful tool for performing certain kinds of calculations. For example, to find

$$\left(\frac{1+i}{\sqrt{2}}\right)^{100}$$

by repeated multiplication would be too much of a challenge for most of us. However changing this expression to polar form greatly simplifies the calculation. We show this calculation using polar coordinates in Example 2 of Section 5.

Suppose that (r, θ) are the polar coordinates of the point $x + yi$ on the complex plane as shown in Figure 4. This number $x + yi$ determines

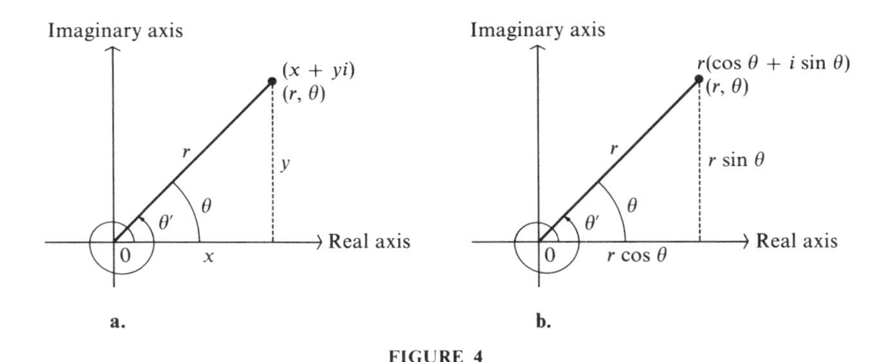

FIGURE 4

both a point on the plane and also the length of a line segment from the origin to the point. The positive (or zero) length r of this line, or *radius*, is called the **absolute value** or **modulus** of the complex number.

> ABSOLUTE VALUE OR MODULUS OF THE COMPLEX NUMBER $(x + yi)$
>
> $$|x + yi| = r = \sqrt{x^2 + y^2} \qquad (1)$$

The **amplitude** or **argument** (θ) of the complex number $x + yi$ is any angle whose terminal side passes through the point (x, y), where the angle is placed in standard position on the axes. It may be chosen as any one of infinitely many values that differ by multiples of 2π.

AMPLITUDE OR ARGUMENT OF THE COMPLEX NUMBER ($x + yi$)

$$\theta = \arctan \frac{y}{x} \qquad \textbf{(2)}$$

In Equation 2, the angle θ must be chosen to put the point representing the complex number in the proper quadrant. This means that θ must satisfy the two equations $x = r \cos \theta$ and $y = r \sin \theta$.

Since $x = r \cos \theta$ and $y = r \sin \theta$, we can write

$$x + yi = r \cos \theta + i(r \sin \theta),$$

or

$$x + yi = r(\cos \theta + i \sin \theta).$$

RECTANGULAR FORM TO POLAR FORM OF COMPLEX NUMBER

$$x + yi = r(\cos \theta + i \sin \theta) \qquad \textbf{(3)}$$

The expression $\cos \theta + i \sin \theta$ is sometimes abbreviated to the more compact symbol cis θ.

EXAMPLE 1 Write the complex number $\sqrt{3} + i$ in polar form and find its absolute value and amplitude.

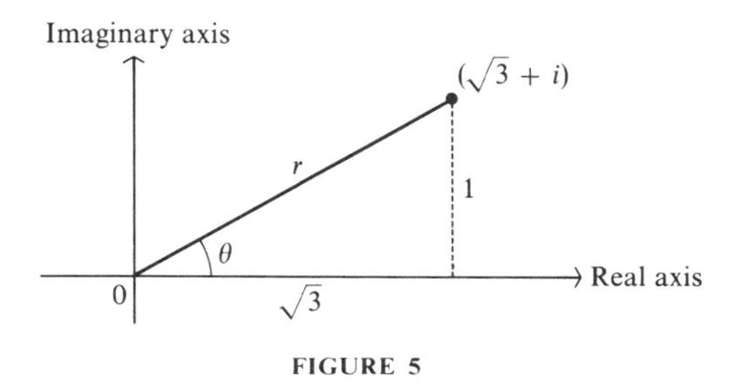

FIGURE 5

SOLUTION Plot the point $(\sqrt{3}, 1)$ which represents the given complex number; draw the radius and sketch the reference triangle (see Figure 5). Find r and θ as follows.

$$r = \sqrt{x^2 + y^2} = \sqrt{3 + 1} = 2 \qquad \text{Absolute value}$$

The amplitude is $\theta = \arctan 1/\sqrt{3} = \frac{1}{6}\pi$. Applying Equation 3

$$x + yi = r(\cos \theta + i \sin \theta),$$

we get

$$\sqrt{3} + i = 2\left(\cos \frac{1}{6}\pi + i \sin \frac{1}{6}\pi\right).$$

We check the results as follows.

$$\begin{aligned} 2\left(\cos \frac{1}{6}\pi + i \sin \frac{1}{6}\pi\right) &= 2\left(\frac{1}{2}\sqrt{3} + i \cdot \frac{1}{2}\right) \\ &= \sqrt{3} + i \end{aligned}$$

EXAMPLE 2 Write the complex number $4i$ in polar form and give several values of its argument.

SOLUTION The point that represents the complex number $4i$ is shown in Figure 1. Here we see that $4i$ is 4 units from the origin, making $r = 4$. The argument is

$$\theta = \frac{1}{2}\pi \qquad \text{or} \qquad \frac{1}{2}\pi + 2n\pi.$$

Therefore $4i = 4(\cos \frac{1}{2}\pi + i \sin \frac{1}{2}\pi)$. A more general polar form for this complex number is

$$4i = 4\left[\cos\left(\frac{1}{2}\pi + 2n\pi\right) + i \sin\left(\frac{1}{2}\pi + 2n\pi\right)\right].$$

EXAMPLE 3 The particular real and pure imaginary numbers 3, -5, and $-2i$ may be written in the polar form as follows.

$$\begin{aligned} 3 &= 3(\cos 0 + i \sin 0) \\ -5 &= 5(\cos \pi + i \sin \pi) \\ -2i &= 2\left(\cos \frac{3}{2}\pi + i \sin \frac{3}{2}\pi\right) \end{aligned}$$

EXAMPLE 4 Express the complex number $8(\cos 120^\circ + i \sin 120^\circ)$ in the rectangular form $x + yi$.

SOLUTION Since $\cos 120^\circ = -\frac{1}{2}$ and $\sin 120^\circ = \frac{1}{2}\sqrt{3}$, we have

$$\begin{aligned} 8(\cos 120^\circ + i \sin 120^\circ) &= 8\left(-\frac{1}{2} + \frac{1}{2}\sqrt{3}\, i\right) \\ &= -4 + 4\sqrt{3}\, i \end{aligned}$$

EXAMPLE 5 Plot the point corresponding to the complex number $2 \operatorname{cis} \frac{5}{6}\pi$ and express the number in polar and rectangular forms.

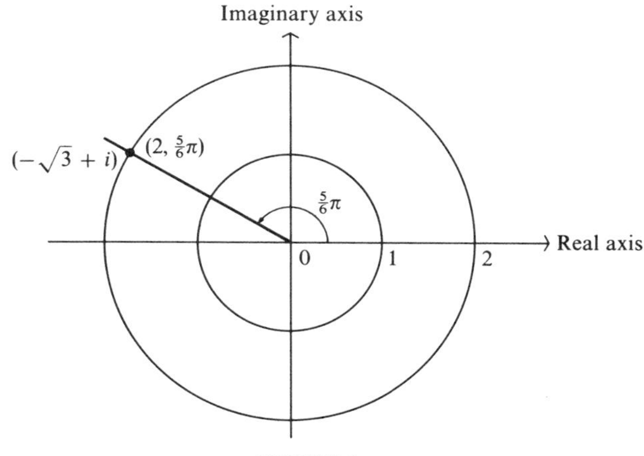

FIGURE 6

SOLUTION The polar coordinates of a point are understood to be indicated by (r, θ). Thus for the complex number $2 \text{ cis } \frac{5}{6}\pi$, we have $r = 2$ and $\theta = \frac{5}{6}\pi$. Place $\theta = \frac{5}{6}\pi$ in standard position, and then locate a point on the terminal side of θ two units from the origin (see Figure 6). In polar form, this complex number is written

$$2\left(\cos \frac{5}{6}\pi + i \sin \frac{5}{6}\pi\right).$$

We can express this complex number in rectangular form as follows.

$$\begin{aligned} 2\left(\cos \frac{5}{6}\pi + i \sin \frac{5}{6}\pi\right) &= 2\left(-\frac{1}{2}\sqrt{3} + i\frac{1}{2}\right) \\ &= -\sqrt{3} + i \end{aligned}$$

Exercises 3

A In Exercises 1–5, represent each complex number graphically on a complex plane. Then find the absolute value and the numerically smallest value of the amplitude.

1. $-1 + i\sqrt{3}$ **2.** $1 + i$ **3.** $-\sqrt{2} + i\sqrt{2}$

4. $-5i$ **5.** 7

In Exercises 6–10, represent each complex number graphically and write the number in the rectangular form $x + yi$.

6. $2(\cos \pi + i \sin \pi)$ **7.** $\cos \pi + i \sin \pi$

8. $8(\cos 270^\circ + i \sin 270^\circ)$ **9.** $2 \text{ cis } \frac{3\pi}{2}$

10. $3 \text{ cis } \left(-\frac{\pi}{4}\right)$

In Exercises 11–13, write each complex number in both polar and rectangular form, corresponding to the given set of polar coordinates (r, θ), and plot the corresponding point.

11. $\left(4, \frac{\pi}{3}\right)$ **12.** $(4, 2\pi)$ **13.** $\left(3, -\frac{2\pi}{3}\right)$

14. Prove that $|\cos\theta + i\sin\theta| = 1$.

B In Exercises 1–5, represent each complex number graphically on a complex plane. Then find the absolute value and the numerically smallest value of the amplitude.

1. $2 - 2i$ **2.** $1 - i$ **3.** $-3 - 3i$

4. $3i$ **5.** -5

In Exercises 6–10, represent each complex number graphically and write the number in the rectangular form $x + yi$.

6. $3(\cos 60^\circ + i\sin 60^\circ)$ **7.** $6\left(\cos\frac{5\pi}{6} + i\sin\frac{5\pi}{6}\right)$

8. $2(\cos 4\pi + i\sin 4\pi)$ **9.** $3 \text{ cis } \frac{\pi}{2}$

10. $5 \text{ cis}\left(-\frac{5\pi}{3}\right)$

In Exercises 11–13, write each complex number in both polar and rectangular form, corresponding to the given set of polar coordinates (r, θ), and plot the corresponding point.

11. $\left(2, \frac{\pi}{6}\right)$ **12.** $\left(3, \frac{3\pi}{2}\right)$ **13.** $\left(2, -\frac{\pi}{3}\right)$

14. Prove that $(\cos\theta + i\sin\theta)^2 = \text{cis } 2\theta$.

4 Multiplication and division of complex numbers in polar form

THEOREM 1 The product of two complex numbers is a number whose absolute value is the **product** of their absolute values and whose amplitude is the **sum** of the amplitudes of the two factors. Thus if

$z_1 = r_1(\cos\theta_1 + i\sin\theta_1)$ and $z_2 = r_2(\cos\theta_2 + i\sin\theta_2)$

then

$$z_1 z_2 = r_1 r_2[\cos(\theta_1 + \theta_2) + i\sin(\theta_1 + \theta_2)]. \quad \textbf{(1)}$$

PROOF By direct multiplication, we have

$$\begin{aligned} z_1 z_2 &= [r_1(\cos\theta_1 + i\sin\theta_1)][r_2(\cos\theta_2 + i\sin\theta_2)] \\ &= (r_1 r_2)[(\cos\theta_1\cos\theta_2 - \sin\theta_1\sin\theta_2) \\ &\quad + i(\sin\theta_1\cos\theta_2 + \cos\theta_1\sin\theta_2)]. \end{aligned}$$

But from trigonometry,

$$\cos\theta_1\cos\theta_2 - \sin\theta_1\sin\theta_2 = \cos(\theta_1 + \theta_2),$$

and

$$\sin\theta_1\cos\theta_2 + \cos\theta_1\sin\theta_2 = \sin(\theta_1 + \theta_2).$$

Making these substitutions, we get

$$z_1 z_2 = (r_1 r_2)[\cos(\theta_1 + \theta_2) + i\sin(\theta_1 + \theta_2)].$$

EXAMPLE 1 Multiply $\sqrt{2} - i\sqrt{2}$ and $1 + i\sqrt{3}$.

SOLUTION The absolute value of the first number is $\sqrt{2+2} = 2$, and its amplitude is Arctan $(-1) = -45°$. The absolute value of the second number is $\sqrt{1+3} = 2$, and its amplitude is Arctan $(\sqrt{3}) = 60°$. Therefore, the required product is

$$\begin{aligned} &2[\cos(-45°) + i\sin(-45°)] \cdot 2[\cos 60° + i\sin 60°] \\ &\qquad = 4(\cos 15° + i\sin 15°). \end{aligned}$$

THEOREM 2 If one complex number is divided by a second number, the quotient is a number whose absolute value is the quotient of the absolute value of the first divided by the absolute value of the second and whose amplitude is the amplitude of the first minus the amplitude of the second. Thus if

$$z_1 = r_1(\cos\theta_1 + i\sin\theta_1)$$

and

$$z_2 = r_2(\cos\theta_2 + i\sin\theta_2),$$

then

$$\frac{z_1}{z_2} = \frac{r_1}{r_2}[\cos(\theta_1 - \theta_2) + i\sin(\theta_1 - \theta_2)]. \qquad \textbf{(2)}$$

PROOF Let $z_1 = r_1(\cos\theta_1 + i\sin\theta_1)$ and

$$z_2 = r_2(\cos\theta_2 + i\sin\theta_2) \neq 0$$

be any two given complex numbers. Then by multiplying both numerator and denominator of the fraction by the conjugate of $\cos\theta_2 + i\sin\theta_2$, we get

$$\frac{z_1}{z_2} = \frac{r_1(\cos\theta_1 + i\sin\theta_1)}{r_2(\cos\theta_2 + i\sin\theta_2)} = \frac{r_1}{r_2}\frac{(\cos\theta_1 + i\sin\theta_1)(\cos\theta_2 - i\sin\theta_2)}{(\cos\theta_2 + i\sin\theta_2)(\cos\theta_2 - i\sin\theta_2)}$$

$$= \frac{r_1}{r_2}\frac{(\cos\theta_1 + i\sin\theta_1)(\cos\theta_2 - i\sin\theta_2)}{\cos^2\theta_2 + \sin^2\theta_2}.$$

Since $\cos(-\theta_2) = \cos\theta_2$ and $\sin(-\theta_2) = -\sin\theta_2$, the factor $(\cos\theta_2 - i\sin\theta_2)$ can be written $[\cos(-\theta_2) + i\sin(-\theta_2)]$, and since $\cos^2\theta_2 + \sin^2\theta_2 = 1$, the right-hand member of the above equation reduces to

$$\frac{r_1}{r_2}(\cos\theta_1 + i\sin\theta_1)[\cos(-\theta_2) + i\sin(-\theta_2)].$$

Then, by applying Theorem 1, we get

$$\frac{z_1}{z_2} = \frac{r_1}{r_2}[\cos(\theta_1 - \theta_2) + i\sin(\theta_1 - \theta_2)].$$

EXAMPLE 2 Divide $6(\cos 170° + i\sin 170°)$ by $3(\cos 110° + i\sin 110°)$.

SOLUTION We apply Theorem 2.

$$[6(\cos 170° + i\sin 170°)] \div [3(\cos 110° + i\sin 110°)]$$

$$= \frac{6}{3}[\cos(170° - 110°) + i\sin(170° - 110°)]$$

$$= 2(\cos 60° + i\sin 60°)$$

$$= 2\left(\frac{1}{2} + i\frac{1}{2}\sqrt{3}\right)$$

$$= 1 + i\sqrt{3}$$

5 Powers of complex numbers; De Moivre's theorem

As a consequence of Theorem 1 in Section 4, when n is a positive integer,

$$z_1 z_2 \cdots z_n$$

$$= (r_1 r_2 \cdots r_n)[\cos(\theta_1 + \theta_2 + \cdots + \theta_n) + i\sin(\theta_1 + \theta_2 + \cdots + \theta_n)].$$

DE MOIVRE'S THEOREM

If $z = r(\cos \theta + i \sin \theta)$ and if n is a positive integer, then

$$z^n = r^n(\cos n\theta + i \sin n\theta). \quad \textbf{(1)}$$

When $r = 1$, this formula reduces to **De Moivre's Theorem** for positive integral exponents.

$$(\cos \theta + i \sin \theta)^n = \cos n\theta + i \sin n\theta \quad \textbf{(2)}$$

Abraham De Moivre developed this theorem in 1722.

As a special case of Theorem 2 in Section 4, it follows that

$$\frac{1}{z} = \frac{\cos 0 + i \sin 0}{r(\cos \theta + i \sin \theta)} = \frac{1}{r}[\cos (0 - \theta) + i \sin (0 - \theta)]$$

$$= \frac{1}{r}(\cos \theta - i \sin \theta).$$

In applying Equation 1, we have

$$z^{-n} = \frac{1}{z^n} = \frac{1}{r^n}[\cos (-n\theta) + i \sin (-n\theta)]. \quad \textbf{(3)}$$

Therefore, both Equations 1 and 2 are true when the exponent is any positive or negative integer.

EXAMPLE 1 Find $(1 + i)^{10}$.

SOLUTION We express $1 + i$ in polar form and apply Equation 1.

$$\begin{aligned}(1 + i)^{10} &= [\sqrt{2}(\cos 45^\circ + i \sin 45^\circ)]^{10} \\ &= (\sqrt{2})^{10}(\cos 450^\circ + i \sin 450^\circ) \\ &= 32(\cos 90^\circ + i \sin 90^\circ) \\ &= 32(0 + i) \\ &= 32i\end{aligned}$$

As an alternate solution to Example 1, we could use the binomial theorem (Section 2, Chapter 12) to expand $(1 + i)^{10}$. After simplifying and collecting terms, we should have only the term $32i$ remaining. This latter solution would be long and certainly is not recommended. The next example makes a stronger case for the importance of De Moivre's Theorem.

EXAMPLE 2 Find the value of $z = \{(1 + i)/\sqrt{2}\}^{100}$.

SOLUTION We express $(1 + i)/\sqrt{2}$ in polar form and then apply Equation 1.

$$\left(\frac{1+i}{\sqrt{2}}\right)^{100} = \left(\frac{1}{\sqrt{2}} + \frac{1}{\sqrt{2}}i\right)^{100} = [1(\cos 45^\circ + i \sin 45^\circ)]^{100}$$

$$= 1^{100}(\cos 4{,}500^\circ + i \sin 4{,}500^\circ)$$

But $4{,}500^\circ/360^\circ = 12.5$, which means $4{,}500^\circ = 12.5(360^\circ) = \frac{1}{2}(360^\circ) + 12(360^\circ) = 180^\circ + 12(360^\circ)$. Then because $\cos(\theta + n360^\circ) = \cos\theta$ and $\sin(\theta + n360^\circ) = \sin\theta$,

$$\cos[180^\circ + 12(360^\circ)] = \cos 180^\circ$$

and

$$\sin[180^\circ + 12(360^\circ)] = \sin 180^\circ.$$

Therefore,

$$\left(\frac{1+i}{\sqrt{2}}\right)^{100} = 1(\cos 180^\circ + i \sin 180^\circ)$$

$$= 1[(-1) + i(0)]$$

$$= -1$$

(Do you believe -1 is the correct value? If not, expand $\{(1 + i)/\sqrt{2}\}^{100}$ by using the binomial theorem! See Section 2, Chapter 12.)

EXAMPLE 3 Find the value of $z = \dfrac{128}{(1 + i\sqrt{3})^6}$.

SOLUTION Let $z = \dfrac{128}{(1 + i\sqrt{3})^6} = 128(1 + i\sqrt{3})^{-6}$.

We express $1 + i\sqrt{3}$ in polar form.

$$1 + i\sqrt{3} = 2(\cos 60^\circ + i \sin 60^\circ)$$

Then, by applying Equation 3, we have

$$[2(\cos 60^\circ + i \sin 60^\circ)]^{-6} = 2^{-6}[\cos(-360^\circ) + i \sin(-360^\circ)]$$

$$= \frac{1}{64}(1 + 0).$$

Therefore,

$$z = \frac{128}{(1 + i\sqrt{3})^6} = 128\left(\frac{1}{64}\right) = 2.$$

EXAMPLE 4 Derive identities for $\cos 3\theta$ and $\sin 3\theta$ by expanding

$$(\cos\theta + i \sin\theta)^3$$

using De Moivre's Theorem and the rule $(a+b)^3 = a^3 + 3a^2b + 3ab^2 + b^3$, and then equating the real parts and the imaginary parts.

SOLUTION By De Moivre's Theorem

$$(\cos\theta + i\sin\theta)^3 = \cos 3\theta + i\sin 3\theta. \qquad \textbf{(4)}$$

We use the rule for expanding $(a+b)^3$, replace i^2 by -1 and i^3 by $-i$, and arrange in the form $a+bi$.

$$\begin{aligned}(\cos\theta + i\sin\theta)^3 &= \cos^3\theta + 3(\cos\theta)^2(i\sin\theta)\\ &\quad + 3(\cos\theta)(i\sin\theta)^2 + (i\sin\theta)^3\\ &= \cos^3\theta + (3\cos^2\theta\sin\theta)i\\ &\quad - 3\cos\theta\sin^2\theta - (\sin^3\theta)i\\ &= (\cos^3\theta - 3\cos\theta\sin^2\theta)\\ &\quad + (3\cos^2\theta\sin\theta - \sin^3\theta)i\end{aligned}$$

By replacing $\sin^2\theta$ by $(1-\cos^2\theta)$ in the first term and $\cos^2\theta$ by $(1-\sin^2\theta)$ in the second term, we reduce the expression as follows.

$$\begin{aligned}&[\cos^3\theta - 3\cos\theta(1-\cos^2\theta)] + [3(1-\sin^2\theta)\sin\theta - \sin^3\theta]i\\ &\qquad = (4\cos^3\theta - 3\cos\theta) + (3\sin\theta - 4\sin^3\theta)i\end{aligned}$$

Replacing $(\cos\theta + i\sin\theta)^3$ of Equation 4 by this last expression, we have

$$(4\cos^3\theta - 3\cos\theta) + (3\sin\theta - 4\sin^3\theta)i = \cos 3\theta + (\sin 3\theta)i.$$

By applying the definition of equality, we have the desired identities.

$$\cos 3\theta = 4\cos^3\theta - 3\cos\theta$$

$$\sin 3\theta = 3\sin\theta - 4\sin^3\theta$$

Exercises 5

A In Exercises 1–8, perform each indicated operation, giving each result in both polar and rectangular form; in the latter form, express the terms either exactly or to four decimal places.

1. $2(\cos 50^\circ + i\sin 50^\circ)\cdot 3(\cos 40^\circ + i\sin 40^\circ)$
2. $[8(\cos 305^\circ + i\sin 305^\circ)] \div [4(\cos 65^\circ + i\sin 65^\circ)]$
3. $5(\text{cis } 15^\circ)\cdot 2(\text{cis } 125^\circ)$
4. $(\text{cis } 15^\circ)^{10}$
5. $[2(\cos 12^\circ + i\sin 12^\circ)]^{-5}$
6. $[3(-1+i)]^4$
7. $\dfrac{8}{(\sqrt{3}-i)^2}$
8. $(1+i\sqrt{3})(1+i)(1-i)$
9. Derive identities for $\cos 4\theta$ and $\sin 4\theta$ by expanding $(\cos\theta + i\sin\theta)^2$ using De Moivre's Theorem and by squaring the binomial; then equate the real and imaginary parts.

B In Exercises 1–8, perform each indicated operation, giving each result in both polar and rectangular form; in the latter form, express the terms either exactly or to four decimal places.

1. $4(\cos 70^\circ + i \sin 70^\circ) \cdot 2(\cos 20^\circ + i \sin 20^\circ)$
2. $[10(\cos 287^\circ + i \sin 287^\circ)] \div [2(\cos 107^\circ + i \sin 107^\circ)]$
3. $5(\text{cis } 23^\circ) \cdot 3(\text{cis } 47^\circ)$
4. $(\text{cis } 9^\circ)^{10}$
5. $[2(\cos 15^\circ + i \sin 15^\circ)]^{-4}$
6. $[2(1 - i)]^6$
7. $\dfrac{16}{(\sqrt{3} + i)^2}$
8. $(1 - i)(\sqrt{3} + i)(\sqrt{3} - i)$
9. Derive identities for $\cos 2\theta$ and $\sin 2\theta$ by expanding $(\cos \theta + i \sin \theta)^2$ using De Moivre's Theorem and by squaring the binomial; then equate the real and imaginary parts.

6 Roots of complex numbers

The statement $\sqrt{9} = x$ implies $x^2 = 9$. The statement $\sqrt[n]{z} = w$ implies $w^n = z$. Thus the problem of extracting the nth roots of a complex number z is one of solving the equation

$$w^n = z \tag{1}$$

for w, when z and the positive integer n are given.

We express z in polar form,

$$z = r(\cos \theta + i \sin \theta),$$

and set

$$w = R(\cos \phi + i \sin \phi),$$

where R and ϕ are presently unknown. Then Equation 1 becomes

$$R^n(\cos n\phi + i \sin n\phi) = r(\cos \theta + i \sin \theta). \tag{2}$$

Since points that represent equal complex numbers coincide, it follows that their absolute values are equal and their amplitudes differ only by integral multiples of 360°. From Equation 2, we then obtain

$$R^n = r \qquad \text{and} \qquad n\phi = \theta + k \cdot 360^\circ \tag{3}$$

where k is an integer, or

$$R = \sqrt[n]{r} \qquad \text{and} \qquad \phi = \frac{\theta}{n} + \frac{k360^\circ}{n} \tag{4}$$

where $\sqrt[n]{r}$ denotes the principal nth root of the positive number r. Substituting these values of R and ϕ in $w = R(\cos \phi + i \sin \phi)$, we obtain nth roots of z.

$$w_k = \sqrt[n]{r}\left[\cos\left(\frac{\theta}{n} + \frac{k \cdot 360^\circ}{n}\right) + i \sin\left(\frac{\theta}{n} + \frac{k \cdot 360^\circ}{n}\right)\right] \tag{5}$$

If we give k the n values $0, 1, 2, \ldots, n-1$, we obtain n distinct complex numbers w_k, which are all nth roots of z. If we give k any other integral value, we find that we obtain one of the values w_k previously obtained.

> **ROOTS OF A COMPLEX NUMBER**
>
> If n is a positive integer and $z \neq 0$, a complex number $z = r(\cos\theta + i\sin\theta)$, real or imaginary, has n and only n distinct nth roots, which are given by the formula
>
> $$w_k = \sqrt[n]{r}\left[\cos\left(\frac{\theta}{n} + \frac{k \cdot 360^\circ}{n}\right) + i\sin\left(\frac{\theta}{n} + \frac{k \cdot 360^\circ}{n}\right)\right], \tag{6}$$
>
> where k takes the values $0, 1, 2, \ldots, n-1$.

EXAMPLE 1 Find all the fourth roots of $-8 - 8i\sqrt{3}$ and plot the points which represent these roots on the complex plane.

SOLUTION In polar form,

$$-8 - 8i\sqrt{3} = 16(\cos 240^\circ + i\sin 240^\circ).$$

Therefore $r = 16$ and $\theta = 240^\circ$. Substituting these values for r and θ in Equation 6, we have

$$w_k = \sqrt[4]{16}\left[\cos\left(\frac{240^\circ}{4} + \frac{k \cdot 360^\circ}{4}\right) + i\sin\left(\frac{240^\circ}{4} + \frac{k \cdot 360^\circ}{4}\right)\right],$$

or

$$w_k = 2\cos(60^\circ + k\,90^\circ) + i\sin(60^\circ + k\,90^\circ).$$

The values of k are 0, 1, 2, and 3, giving the following four roots.

$$w_0 = 2(\cos 60^\circ + i\sin 60^\circ) = 2\left(\frac{1}{2} + i\frac{1}{2}\sqrt{3}\right) = 1 + i\sqrt{3}$$

$$w_1 = 2(\cos 150^\circ + i\sin 150^\circ) = 2\left(-\frac{1}{2}\sqrt{3} + i\frac{1}{2}\right) = -\sqrt{3} + i$$

$$w_2 = 2(\cos 240^\circ + i\sin 240^\circ) = 2\left[-\frac{1}{2} + i\left(-\frac{1}{2}\sqrt{3}\right)\right] = -1 - i\sqrt{3}$$

$$w_3 = 2(\cos 330^\circ + i\sin 330^\circ) = 2\left(\frac{1}{2}\sqrt{3} - i\frac{1}{2}\right) = \sqrt{3} - i$$

Since the roots have equal amplitudes, they can be represented by points on a circle of radius 2 on the complex plane (see Figure 7).

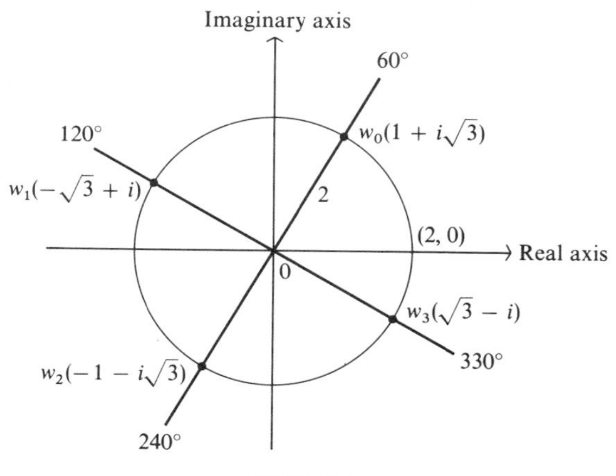

FIGURE 7

EXAMPLE 2 Find all of the tenth roots of 1.

SOLUTION In polar form,

$$1 = \cos 0^\circ + i \sin 0^\circ.$$

Therefore $r = 1$ and $\theta = 0^\circ$. Substituting these values for r and θ in Equation 6, we have

$$w_k = \sqrt[10]{1}\left[\cos\left(\frac{0^\circ}{10} + \frac{k \cdot 360^\circ}{10}\right) + i \sin\left(\frac{0^\circ}{10} + \frac{k \cdot 360^\circ}{10}\right)\right]$$

or

$$w_k = \cos k36^\circ + i \sin k36^\circ.$$

We assign the values 0, 1, 2, ..., 9 for k, and obtain the following roots.

$$w_0 = \cos 0^\circ + i \sin 0^\circ = 1$$
$$w_1 = \cos 36^\circ + i \sin 36^\circ = 0.8090 + 0.5878i$$
$$w_2 = \cos 72^\circ + i \sin 72^\circ = 0.3090 + 0.9511i$$
$$w_3 = \cos 108^\circ + i \sin 108^\circ = -0.3090 + 0.9511i$$
$$w_4 = \cos 144^\circ + i \sin 144^\circ = -0.8090 + 0.5878i$$
$$w_5 = \cos 180^\circ + i \sin 180^\circ = -1$$
$$w_6 = \cos 216^\circ + i \sin 216^\circ = -0.8090 - 0.5878i$$
$$w_7 = \cos 252^\circ + i \sin 252^\circ = -0.3090 - 0.9511i$$
$$w_8 = \cos 288^\circ + i \sin 288^\circ = 0.3090 - 0.9511i$$
$$w_9 = \cos 324^\circ + i \sin 324^\circ = 0.8090 - 0.5878i$$

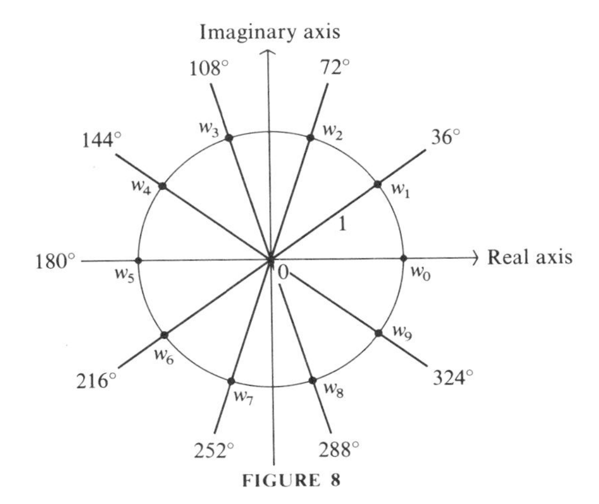

FIGURE 8

We could have used the compact symbol cis θ (see Section 3) to express the above roots as follows.

$$w_0 = \text{cis } 0^\circ, \quad w_1 = \text{cis } 36^\circ, \quad w_2 = \text{cis } 72^\circ, \quad \text{etc.}$$

These roots are represented by points, spaced at equal intervals around a unit circle, on the complex plane (see Figure 8).

Exercises 6

A In Exercises 1–6, find all of the indicated roots in the rectangular form $a + bi$. Express the roots either exactly or to four decimal places. Draw a figure, plotting each of the required roots.

1. Cube roots of $-i$

2. Sixth roots of 1

3. Fifth roots of $4 - 4i$

4. Cube roots of 1

5. Square roots of $-1 + i\sqrt{3}$

6. Square roots of $-5 + 12i$

In Exercises 7 and 8, solve each equation completely. Express the roots in rectangular form.

7. $x^3 + 27 = 0$

8. $x^4 + 81 = 0$

B In Exercises 1–6, find all of the indicated roots in the rectangular form $a + bi$. Express the roots either exactly or to four decimal places. Draw a figure, plotting each of the required roots.

1. Cube roots of -1

2. Fourth roots of i

3. Fifth roots of $-4 - 4i$

4. Sixth roots of 64

5. Cube roots of $-4\sqrt{2} + i4\sqrt{2}$

6. Fourth roots of $8 - i8\sqrt{3}$

In Exercises 7 and 8, solve each equation completely. Express the roots in rectangular form.

7. $x^6 + 64 = 0$ **8.** $x^4 + 4x^2 + 16 = 0$

Summary

The **complex number** $x + yi$ can be represented by the point (x, y) on the complex plane.

An alternative method of fixing the position of a point is to give its **polar coordinates**. The polar coordinates are r, the distance of the point from the origin or **pole**, and θ, the angle that the line between the point and the pole makes with the positive x-axis or **polar axis**. The fixed point, O (in Figure 9), is the **pole**; the distance, $OP = r$, from the pole to

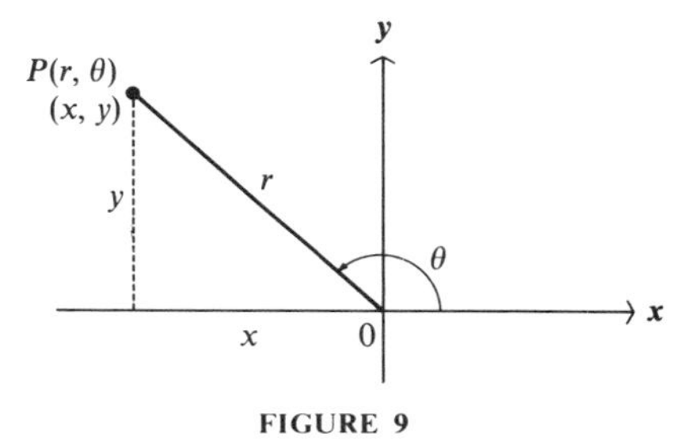

FIGURE 9

the given point is the **radius**; the angle θ (positive when measured in a counterclockwise direction) is the **polar angle**. The polar coordinates of the point P are written (r, θ). The **polar angle** is sometimes called the **amplitude** or **argument** of the point. From Figure 9, it can be seen that the relations between rectangular and polar coordinates are as follows.

$$x = r \cos \theta \qquad r = \sqrt{x^2 + y^2}$$

$$y = r \sin \theta \qquad \theta = \arctan \frac{y}{x}$$

The **absolute value** or **modulus** of the complex number $(x + yi)$ is

$$|x + yi| = r = \sqrt{x^2 + y^2}.$$

The **amplitude** or **argument** of the complex number $x + yi$ is

$$\theta = \arctan \frac{y}{x}.$$

The **rectangular form to polar form** of a complex number is

$$x + yi = r(\cos \theta + i \sin \theta) = r \text{ cis } \theta.$$

If $z_1 = r_1(\cos\theta_1 + i\sin\theta_1)$ and $z_2 = r_2(\cos\theta_2 + i\sin\theta_2)$, then

$$z_1z_2 = r_1r_2[\cos(\theta_1 + \theta_2) + i\sin(\theta_1 + \theta_2)]$$

and

$$\frac{z_1}{z_2} = \frac{r_1}{r_2}[\cos(\theta_1 - \theta_2) + i\sin(\theta_1 - \theta_2)].$$

If $z = r(\cos\theta + i\sin\theta)$, then

$$z^n = r^n(\cos n\theta + i\sin n\theta)$$

and

$$z^{-n} = \frac{1}{z^n} = \frac{1}{r^n}[\cos(-n\theta) + i\sin(-n\theta)].$$

When $r = 1$,

$$(\cos\theta + i\sin\theta)^n = \cos n\theta + i\sin n\theta.$$

If n is a positive integer and $z \neq 0$, a complex number $z = r(\cos\theta + i\sin\theta)$, real or imaginary, has n and only n distinct nth roots, which are given by the formula

$$w_k = \sqrt[n]{r}\left[\cos\left(\frac{\theta}{n} + \frac{k\cdot 360^\circ}{n}\right) + i\sin\left(\frac{\theta}{n} + \frac{k\cdot 360^\circ}{n}\right)\right],$$

where k takes the values $0, 1, 2, \ldots, n-1$.

Review exercises

A

1. a. Graph the complex number $12 + 5i$ on the complex plane.
   b. Find the absolute value of $12 + 5i$.
   c. Find the amplitude of $12 + 5i$.
2. Write $3 - 3i$ in polar form.
3. Find two square roots of -1. Express the roots in polar form and plot them on the complex plane.

In Exercises 4–7, perform the indicated operations. Write each answer in both polar and rectangular form.

4. $5(\cos 146^\circ + i\sin 146^\circ)\cdot 3(\cos 34^\circ + i\sin 34^\circ)$
5. $[10(\cos 475^\circ + i\sin 475^\circ)] \div [5(\cos 115^\circ + i\sin 115^\circ)]$
6. $[2(\cos 12^\circ + i\sin 12^\circ)]^5$
7. $\dfrac{16}{(1+i)^6}$

8. Find all the fifth roots of $-16 - i16\sqrt{3}$. Express the roots in polar form and plot the points that represent the roots on the complex plane.
9. Solve $x^4 + 4x^2 + 3 = 0$. Express the roots in rectangular form.

B

1. **a.** Graph the complex number $-3 + 4i$ on the complex plane.
   b. Find the absolute value of $-3 + 4i$.
   c. Find the amplitude of $-3 + 4i$.
2. Write $2 + i2\sqrt{3}$ in polar form.
3. Find the two square roots of i. Express the roots in polar form and plot them on the complex plane.

In Exercises 4–7, perform the indicated operations. Write each answer in both polar and rectangular form.

4. $3(\cos 174^\circ + i \sin 174^\circ) \cdot 2(\cos 96^\circ + i \sin 96^\circ)$
5. $[12(\cos 195^\circ + i \sin 195^\circ)] \div [4(\cos 15^\circ + i \sin 15^\circ)]$
6. $[2(\cos 15^\circ + i \sin 15^\circ)]^6$
7. $\dfrac{32}{(1 + i\sqrt{3})^4}$
8. Find all the fifth roots of $16 + i16\sqrt{3}$. Express the roots in polar form and plot the points that represent the roots on the complex plane.
9. Solve $x^4 + 5x^2 + 4 = 0$. Express the roots in rectangular form.

Diagnostic test

The purpose of this test is to see how well you understand the work in this chapter. Allow yourself approximately 50 minutes to do the test. Solutions for all the problems together with section references, are given in the Answers section at the end of the book. We suggest that you study the sections referred to for the problems you do incorrectly.

1. Write the complex number $-\sqrt{3} + i$ in polar form and graph it on the complex plane.
2. Write the complex number $2 \text{ cis } 120^\circ$ in rectangular form.
3. Find the rectangular coordinates of the point whose polar coordinates are $(2, 120^\circ)$.

In Problems 4–7, perform the indicated operations and write each answer in both polar and rectangular form.

4. $[16(\cos 250^\circ + i \sin 250^\circ)] \div [4(\cos 70^\circ + i \sin 70^\circ)]$
5. $7(\cos 47^\circ + i \sin 47^\circ) \cdot 2(\cos 13^\circ + i \sin 13^\circ)$
6. $[2(\cos 135^\circ + i \sin 135^\circ)]^4$

7. $\dfrac{(1 + i\sqrt{3})^{10}}{2}$

8. Simplify $\dfrac{8}{(-1 + i)^4}$.

9. Find the three cube roots of the imaginary unit i. Express the roots in polar form and graph them on a circle in a complex plane.

10. Solve $x^4 + 3x^2 - 4 = 0$.

9 Analytic geometry

1 Introduction

In our discussion of analytic geometry in this chapter, our chief concerns are: (1) given a condition for a set of points, to write the equation that contains those points; and (2) given a second-degree equation, to identify its type and graph it.

Analytic geometry is a subject in which algebra, geometry, and trigonometry are studied together in such a way that each is helpful in simplifying and clarifying the solution of problems arising in the others.

The invention of analytic geometry is usually attributed to René Descartes, a French mathematician and philosopher, whose *La Geometrie*, which appeared in 1637, was the source from which the mathematical world learned the value of the methods of analytic geometry.

Some analytic geometry has already been discussed in Chapter 6. In that chapter, we found the slopes of straight lines, wrote the equations of straight lines, graphed straight lines, and found x- and y-intercepts, midpoints of lines and the distance between two points, all of which are considered to be part of analytic geometry.

ANALYTIC GEOMETRY COVERED IN CHAPTER 6

Slope of a line

$$m = \frac{y_2 - y_1}{x_2 - x_1}, \qquad x_2 \neq x_1$$

Equations of lines

General form

$$Ax + By + C = 0, \qquad A \text{ and } B \text{ not both zero}$$

Point-slope form

$$y - y_1 = m(x - x_1), \qquad m \text{ the slope}$$

Slope-intercept form

$$y = mx + b, \qquad b \text{ the } y\text{-intercept, } m \text{ the slope}$$

Intercept form

$$\frac{x}{a} + \frac{y}{b} = 1, \qquad a \text{ the } x\text{-intercept and } b \text{ the } y\text{-intercept}$$

Two-point form

$$y - y_1 = \frac{y_2 - y_1}{x_2 - x_1}(x - x_1), \qquad x_2 \neq x_1$$

Parallel to y-axis

$$x = a, \qquad a \text{ the } x\text{-intercept}$$

Parallel to x-axis

$$y = b, \qquad b \text{ the } y\text{-intercept}$$

Midpoint of a line segment

$$\bar{x} = \frac{x_2 + x_1}{2} \qquad \text{and} \qquad \bar{y} = \frac{y_2 + y_1}{2}$$

Distance between two points (x_1, y_1) *and* (x_2, y_2)

$$d = \sqrt{(x_2 - x_1)^2 + (y_2 - y_1)^2}$$

2 Inclination and slope of a straight line

The direction of a straight line may be indicated by the angle that it makes with a reference line, such as the x-axis.

The **inclination** of a straight line not parallel to the x-axis is the least angle α measured counterclockwise from the positive x-axis to the line; the inclination of a line parallel to the x-axis is 0°. The inclination is a positive angle less than 180°, or is 0°.

EXAMPLE 1 In Figure 1, the inclination of line l_1 is $\alpha = 60°$, and the inclination of the line l_2 is $\alpha = 120°$.

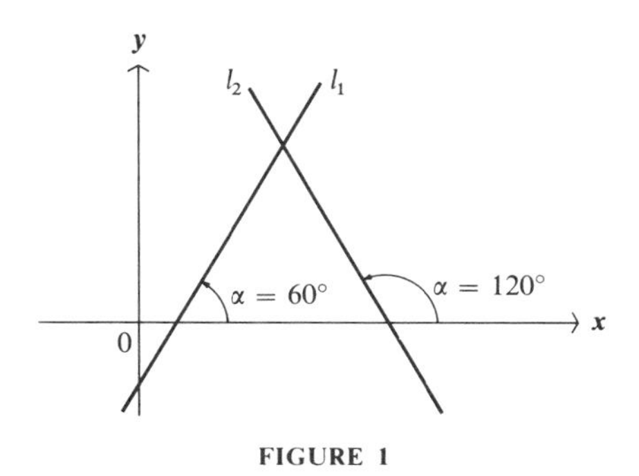

FIGURE 1

For many purposes, the direction of a line is more conveniently expressed by the tangent of the angle of inclination.

The **slope of a straight line** is defined as the tangent of its inclination α.

$$m = \tan \alpha \qquad \textbf{(1)}$$

It follows from the definition of slope that a line parallel to the x-axis has slope 0, since its inclination is 0°. It also follows that the slope of a line parallel to the y-axis is undefined, since its inclination is 90° and $\tan 90^\circ$ does not exist. However, a line whose inclination is close to 90° has a slope, which is numerically large.

ANOTHER WAY OF FINDING THE SLOPE OF A LINE

When the equation of the line

$$Ax + By + C = 0$$

is written in the slope-intercept form, we have

$$Ax + By = -C$$

$$By = -Ax - C$$

$$y = \boxed{-\frac{A}{B}}\, x - \frac{C}{B}.$$

Comparing with $\qquad y = \boxed{m}\, x + b \qquad$ Slope-intercept form

we see that:

> The slope m of $Ax + By = -C$ is
>
> $$m = -\frac{A}{B} \tag{2}$$
>
> where $B \neq 0$.

Equation 2 gives us an easy way of determining the slope of a line.

EXAMPLE 2 Use Equation 2, $m = -A/B$, to find the slope of a line.

a. The slope of $3x + 4y = 1$ is $-\frac{3}{4}$.

b. The slope of $2x - y = 3$ is $-\dfrac{2}{-1} = 2$.

c. The slope of

$$\frac{3}{4}x - \frac{1}{2}y = 1$$

is

$$-\frac{\frac{3}{4}}{-\frac{1}{2}} = \frac{3}{2}.$$

The slope of a line measures the steepness of a line; the greater the absolute value of the slope, the steeper is the line. The steepness of a road or hillside is often given by its slope expressed as a percentage and called its **grade**. For example, a grade of 4% means a slope of 0.04 or 4/100, that is, a rise of 4 feet in 100 feet of horizontal distance.

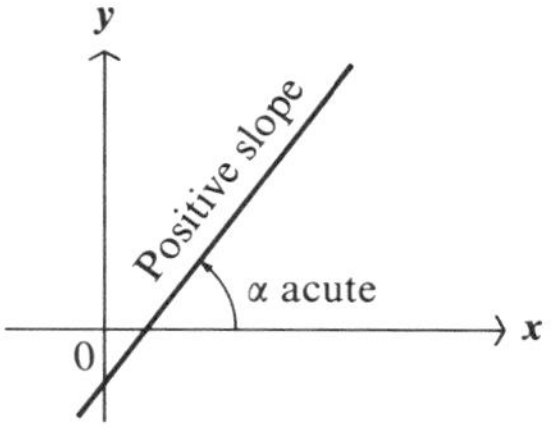

Positive slope
α acute
a.

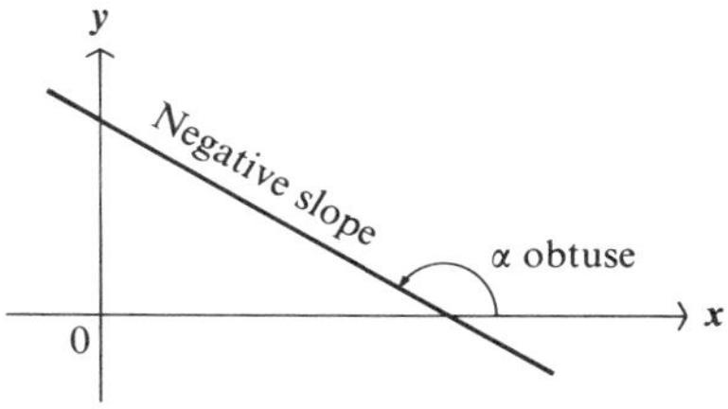

Negative slope
α obtuse
b.

FIGURE 2

If a line rises from left to right (see Figure 2a), its inclination is an acute angle, the tangent is positive and the slope is positive. If the line falls from left to right (see Figure 2b), its inclination is an obtuse angle, the tangent is negative, and the slope is negative. The slope of a line may be any real number, positive, negative, or zero.

SLOPES OF PERPENDICULAR LINES

Let l_1 and l_2 be any two perpendicular lines not parallel to either of the coordinate axes (see Figure 3), with inclinations of α_1 and α_2 and slopes

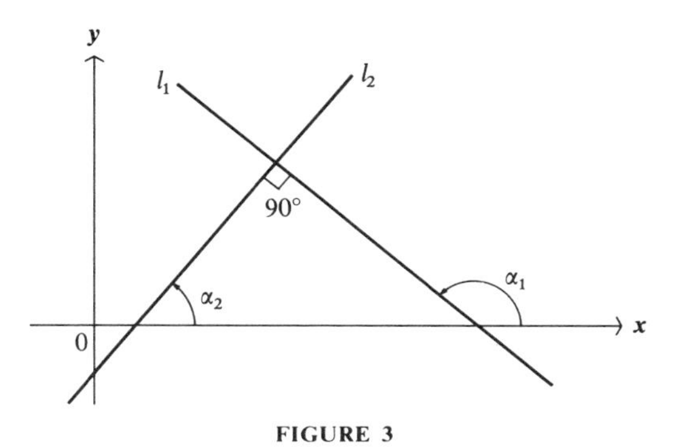

FIGURE 3

of m_1 and m_2, respectively. Assume α_1 is the larger of the two angles of inclination.

By geometry, $\alpha_1 = 90° + \alpha_2$. By trigonometry,

$$m_1 = \tan \alpha_1 = \tan (90° + \alpha_2) = \cot (-\alpha_2) = -\cot \alpha_2$$

$$= -\frac{1}{\tan \alpha_2} = -\frac{1}{m_2}. \qquad \textbf{(3)}$$

This relation $m_1 = -1/m_2$ may also be written $m_2 = -1/m_1$ or $m_1 m_2 = -1$.

Conversely, if l_1 and l_2 are two lines whose slopes m_1 and m_2 satisfy the relation $m_1 = -1/m_2$, it follows that

$$\tan \alpha_1 = -\frac{1}{\tan \alpha_2} = -\cot \alpha_2. \quad \textbf{(4)}$$

By trigonometry, the angles α_1 and α_2 differ by 90°, since the inclination (by definition) is never greater than 180°, and therefore the lines l_1 and l_2 are perpendicular lines.

> Two lines with slopes m_1 and m_2 are perpendicular if and only if their slopes are negative reciprocals of each other, that is, if
>
> $$m_1 = -\frac{1}{m_2}, \quad \text{or} \quad m_2 = -\frac{1}{m_1}, \quad \text{or} \quad m_1 m_2 = -1, \quad \textbf{(5)}$$
>
> provided neither slope is 0. However, if one slope is 0 and the other slope is undefined, the lines are perpendicular.

ANGLE BETWEEN TWO LINES WHOSE SLOPES ARE GIVEN

The angle from a line l_1 to a line l_2 is defined as the smallest, nonnegative (counterclockwise) angle through which l_1 must be rotated in order to make it coincide with l_2 (see Figure 4). We shall call this angle, through which one line must rotate to coincide with the other, ϕ. The angle ϕ is in the interval [0°, 180°).

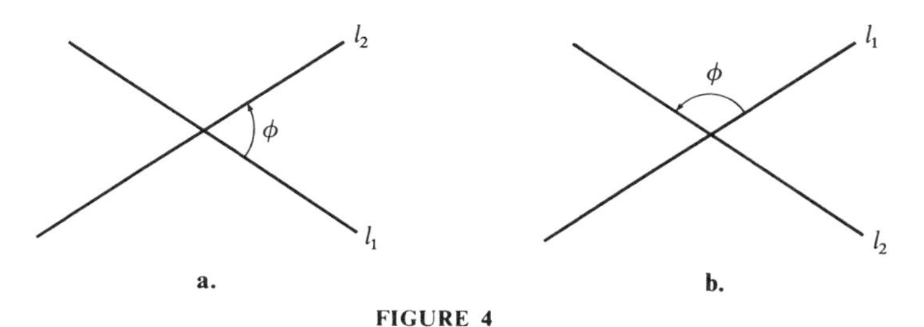

FIGURE 4

Since the direction of a line is frequently given by its slope, it is desirable to have a formula for the angle between lines in terms of their slopes. Let l_1 and l_2 be two lines, neither of which is parallel to the y-axis and which are not perpendicular to each other; let their inclinations be α_1 and α_2 and their slopes be m_1 and m_2, respectively. Let ϕ be the angle from l_1 to l_2.

There are two cases to consider: Case I, in which $\alpha_2 > \alpha_1$, and Case II, in which $\alpha_1 > \alpha_2$ (see Figure 5). In Case I, $\alpha_2 > \alpha_1$ (see Figure 5a). We

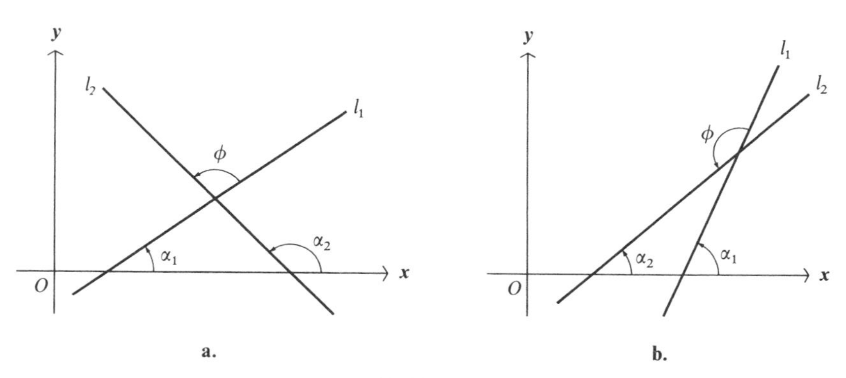

FIGURE 5

have, by geometry, $\alpha_2 = \alpha_1 + \phi$ or $\phi = \alpha_2 - \alpha_1$. Thus $\tan\phi = \tan(\alpha_2 - \alpha_1)$. In Case II, $\alpha_1 > \alpha_2$ (see Figure 5b). We have $\alpha_1 = \alpha_2 + (180° - \phi)$ or $\phi = [180° + (\alpha_2 - \alpha_1)]$. Thus

$$\begin{aligned}\tan\phi &= \tan[180° + (\alpha_2 - \alpha_1)]\\ &= \tan(\alpha_2 - \alpha_1).\end{aligned}$$

In either case, by trigonometry, we get

$$\begin{aligned}\tan\phi &= \tan(\alpha_2 - \alpha_1)\\ &= \frac{\tan\alpha_2 - \tan\alpha_1}{1 + \tan\alpha_2\tan\alpha_1}\\ &= \frac{m_2 - m_1}{1 + m_2 m_1}.\end{aligned}$$

Therefore, we have the following relationships.

> The angle ϕ from a line l_1 with slope m_1 to a line l_2 with slope m_2 is given by the formula
>
> $$\tan\phi = \frac{m_2 - m_1}{1 + m_2 m_1}, \qquad \textbf{(6)}$$
>
> provided neither line is parallel to the y-axis and that the given lines are not perpendicular to each other.

EXAMPLE 3 Find angle A of the triangle whose vertices are $A(-2, -3)$, $B(4, -1)$, and $C(2, 3)$.

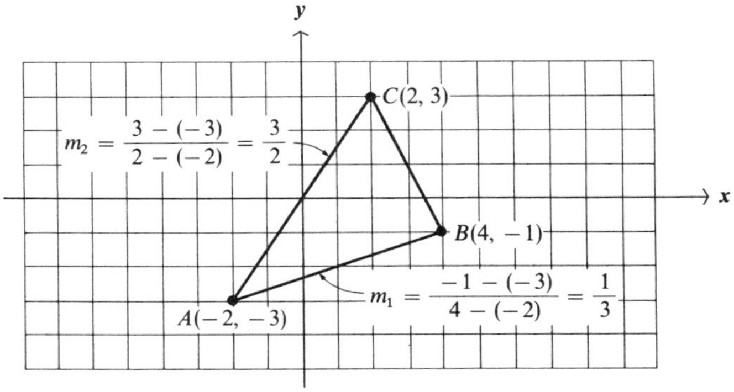

FIGURE 6

SOLUTION Sketch the figure and find the slopes of sides AB and AC (see Figure 6). Then,

$$\tan A = \frac{m_2 - m_1}{1 + m_2 m_1}$$

$$= \frac{\frac{3}{2} - \frac{1}{3}}{1 + \frac{3}{2}(\frac{1}{3})} = \frac{7}{9}$$

$$A = \text{Arctan}\,\frac{7}{9} \doteq 37^\circ\ 50'.$$

Therefore, $A = 37^\circ\ 50'$, to the nearest 10 minutes.

EXAMPLE 4 Use the slopes of the sides of the triangle whose vertices are $A(1, -2)$, $B(5, 1)$, and $C(2, 5)$ to prove that it is a right triangle. Determine which angle is the 90° angle.

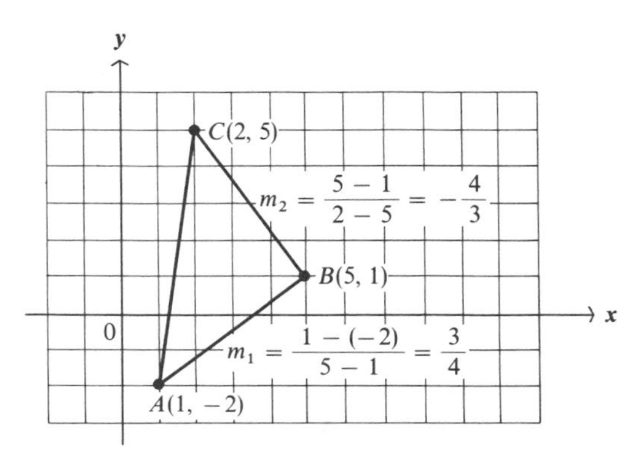

FIGURE 7

SOLUTION Sketch the triangle and find the slopes of its sides (see Figure 7). The angle whose sides have slopes that are negative reciprocals of each other will be the right angle.

$$m_1 m_2 = \left(\frac{3}{4}\right)\left(-\frac{4}{3}\right) = -1$$

Therefore B is a right angle and triangle ABC is a right triangle.

When two nonperpendicular lines intersect, any two adjacent angles at the point of intersection, are supplementary; one angle is acute and the other is obtuse. When asked to find the angle of intersection of two nonperpendicular intersecting lines, unless otherwise specified, we calculate the **acute** angle formed at the point of intersection. To obtain this acute angle, without having to graph the lines, we use absolute value in the formula for $\tan \phi$. That is

$$\tan \phi = \left| \frac{m_2 - m_1}{1 + m_2 m_1} \right|.$$

EXAMPLE 5 Find the acute angle formed by the intersection of the graphs of $3x + 2y = 6$ and $3x - y = 3$.

SOLUTION Let l_1 be line $3x + 2y = 6$ and l_2 be line $3x - y = 3$. We use the fact that when the equation of a line is in the form $Ax + By + C = 0$, its slope is $-A/B$.

$$l_1: \quad 3x + 2y = 6 \qquad\qquad l_2: \; 3x - y = 3$$

$$m_1 = -\frac{3}{2} \qquad\qquad m_2 = -\frac{3}{-1} = 3$$

Then

$$\tan \phi = \left| \frac{3 - (-\frac{3}{2})}{1 + 3(-\frac{3}{2})} \right| = \left| -\frac{9}{7} \right| = \frac{9}{7}.$$

Therefore, the acute angle of intersection is $\phi = \text{Arctan } \frac{9}{7} \doteq 52^\circ\ 08'$, to the nearest minute.

EXAMPLE 6 Write an equation of the line that passes through the point $(2, 1)$ and is perpendicular to the line $3x + 5y = 10$.

SOLUTION First, find the slope of $3x + 5y = 10$.

$$3x + 5y = 10$$

$$m = -\frac{A}{B}$$

$$m = -\frac{3}{5}$$

To find the slope (m_2) of the perpendicular line, we use the formula $m_2 = -1/m_1 = \frac{5}{3}$. Then, by using $m = \frac{5}{3}$ and the point (2, 1), we find

$$y - y_1 = m(x - x_1)$$

$$y - 1 = \frac{5}{3}(x - 2)$$

$$3y - 3 = 5x - 10$$

$$5x - 3y - 7 = 0.$$

The required equation is $5x - 3y - 7 = 0$.

Exercises 2

A

1. Find angle B of the triangle whose vertices are $A(1, -2)$, $B(4, 2)$, and $C(-2, 3)$.
2. Use the slopes of the sides of the triangle whose vertices are $A(-1, -2)$, $B(2, 0)$, and $C(-3, 1)$ to prove that it is a right triangle. Determine which angle is the 90° angle.
3. Find the acute angle formed by the intersection of the graphs of $5x - 3y = 1$ and $3x + 2y = 8$.
4. Write the equation of the line that passes through the point $(-2, 1)$ and is perpendicular to the line $3x - 5y = 15$.
5. Find the slope of the line that makes an angle of 45° with the line $3x + y = 6$.

B

1. Find angle A of the triangle whose vertices are $A(1, -2)$, $B(4, 2)$, and $C(-2, 3)$.
2. Use the slopes of the sides of the triangle whose vertices are $A(-2, -1)$, $B(3, 1)$, and $C(1, 6)$ to prove that it is a right triangle. Determine which angle is the 90° angle.
3. Find the acute angle formed by the intersection of the graphs of $2x - 3y = 1$ and $4x + 2y = 5$.
4. Write an equation of the line that passes through the point $(3, -1)$ and is perpendicular to the line $4x + 2y = 7$.
5. Find the slope of the line that makes an angle of 45° with the line $4x + 2y = 5$.

3 Symmetry

Two points, (x, y) and $(x, -y)$, whose x-coordinates are equal and whose y-coordinates differ only in sign are said to be **symmetric with respect to the x-axis**. Similarly, the points (x, y) and $(-x, y)$ are **symmetric with respect to the y-axis** (see Figure 8). Points B and C are symmetric with respect to the x-axis, as are A and D; A and B are symmetric with respect to the y-axis, as are C and D.

Two points, (x, y) and $(-x, -y)$, whose x-coordinates have equal absolute value but are opposite in sign and whose y-coordinates have

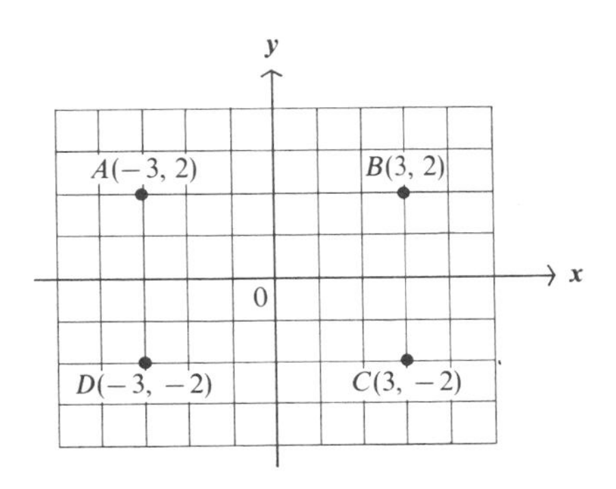

FIGURE 8

equal absolute value but are opposite in sign are said to be **symmetric with respect to the origin**. Thus points A and C in Figure 8 are symmetric with respect to the origin, as are points B and D.

If there are no odd powers of y in a polynomial equation, its graph is symmetric with respect to the x-axis, and if the polynomial equation is free of odd powers of x, then the graph is symmetric with respect to the y-axis.

EXAMPLE 1 Show that the graph of $y^2 = 4x$ is symmetric with respect to the x-axis.

SOLUTION If, when substituting $(x, -y)$ or (x, y) in the equation, there is no change in the equation, then its graph is symmetric with respect to the x-axis. Or, as there are no odd powers of y, the graph is symmetric with respect to the x-axis.

$$y^2 = 4x$$

$$(-y)^2 = 4(x) \qquad \boxed{\text{No change}}$$

$$y^2 = 4x$$

Therefore, $y^2 = 4x$ is symmetric with respect to the x-axis.

EXAMPLE 2 What properties of symmetry does the curve $xy = 4$ have?

SOLUTION When $(-x, -y)$ is substituted for (x, y) in the equation, there is no change.

$$xy = 4$$

$$(-x)(-y) = 4 \qquad \boxed{\text{No change}}$$

$$xy = 4$$

Therefore, the graph of $xy = 4$ is symmetric with respect to the origin.

EXAMPLE 3 What properties of symmetry does the curve $y = \sqrt{x^2 + 4}$ have?

SOLUTION When we substitute $(-x, y)$ for (x, y) in the equation, there is no change in the equation since $(-x)^2 = (+x)^2 = x^2$. Therefore, $y = \sqrt{x^2 + 4}$ is symmetric with respect to the y-axis.

Considerations of symmetry are very helpful in drawing the graphs of given equations. If, for example, the part of a curve that lies above the x-axis has been drawn and if the graph is known to be symmetric with respect to the x-axis, then we can locate as many points as we please on the lower half by choosing points on the part already drawn and plotting the corresponding points symmetric with respect to the x-axis. Similarly, if the curve is symmetric with respect to the y-axis, we first draw the part of it to the right of the y-axis and then locate symmetric points to the left of the y-axis.

Exercises 3

State whether the graph of each equation is symmetric with respect to the x-axis, the y-axis, the origin, or all or none of these.

A

1. $x^2 = y - 4$ **2.** $x^2 + y^2 = 4$
3. $y^2 = 5 - 2x$ **4.** $x^2 - 10x + y^2 = 26$
5. $4x^2 - y^2 + 4x + 4y = 4$ **6.** $xy = 4$

B

1. $x^2 = y + 1$ **2.** $x^2 + y^2 = 1$
3. $y^2 = 4x + 7$ **4.** $x^2 - 4x + y^2 = 5$
5. $x^2 + 4y^2 - 2x + 36y + 36 = 0$ **6.** $xy = 9$

4 Conic sections

In the remaining part of this chapter, our main concern will be writing quadratic equations and drawing the graphs for these equations. The graphs of quadratic equations are called **conic sections**. *Conic sections*

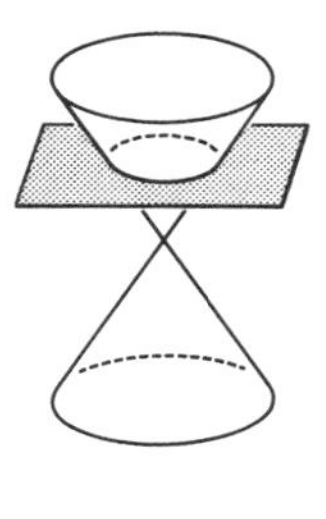

a.

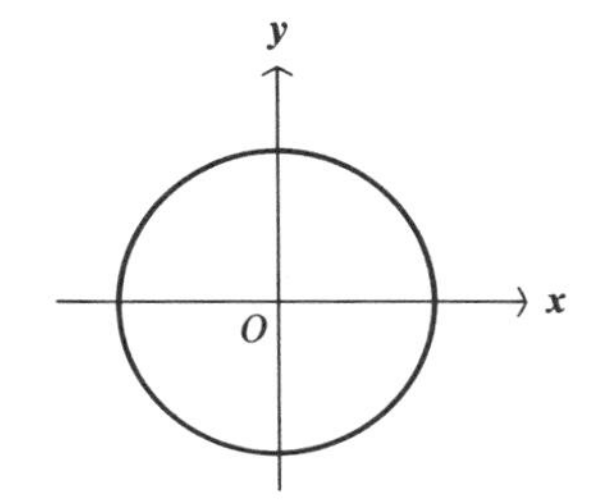

Typical equations of circles
(center at the origin)
$x^2 + y^2 = 4$
$x^2 + y^2 = 25$
$2x^2 + 2y^2 = 7$
[center at (h, k), radius r]
$(x - h)^2 + (y - k)^2 = r^2$

b.

FIGURE 9

Circle

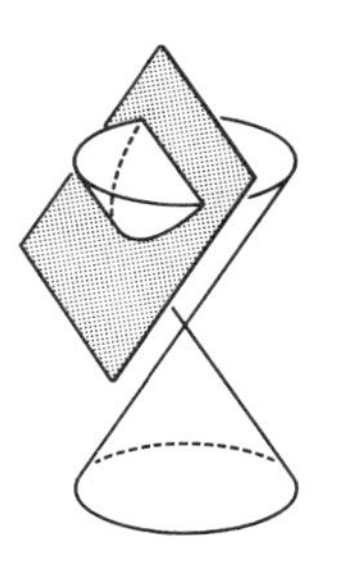

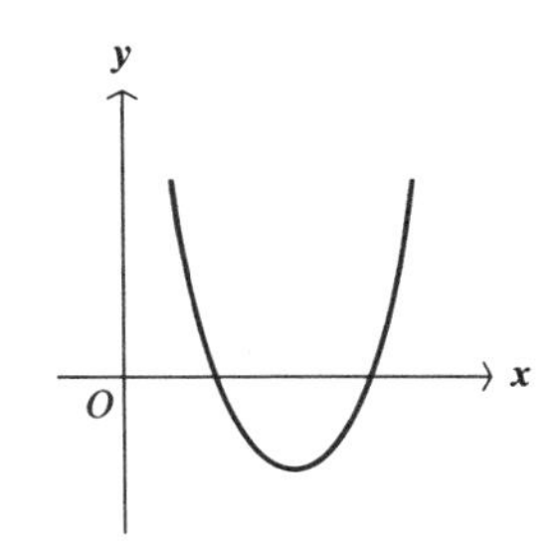

Typical equations of parabolas

$$y = 2x^2 - 5x + 1$$
$$y = 7 - 2x - 5x^2$$
$$(y - k) = 4c(x - h)^2$$
$$(x - h) = 4c(y - k)^2$$
[vertex at (h, k)]

a. b.

FIGURE 10

Parabola

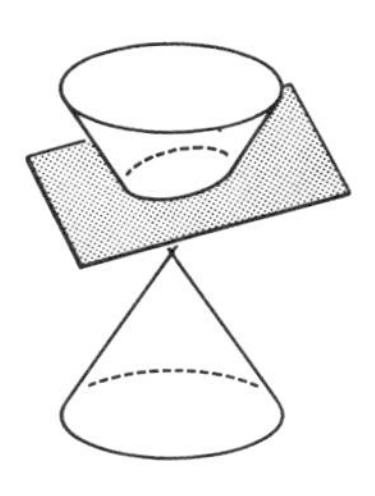

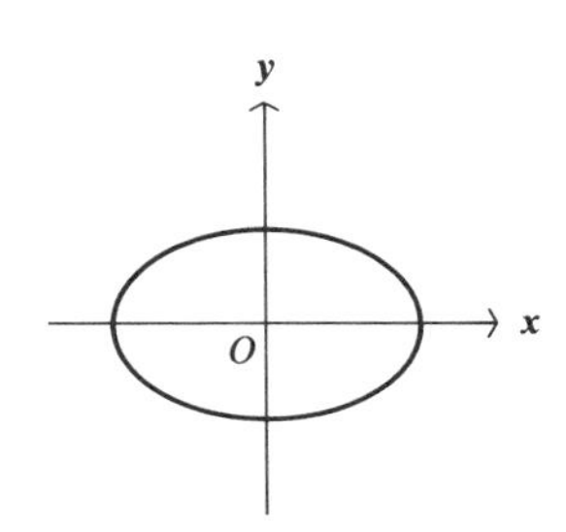

Typical equations of ellipses

$$2x^2 + y^2 = 4$$
$$3x^2 + 4y^2 = 12$$
$$\frac{(x - h)^2}{a^2} + \frac{(y - k)^2}{b^2} = 1$$
[center at (h, k)]

a. b.

FIGURE 11

Ellipse

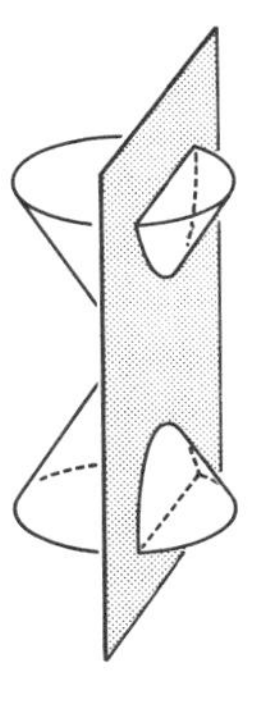

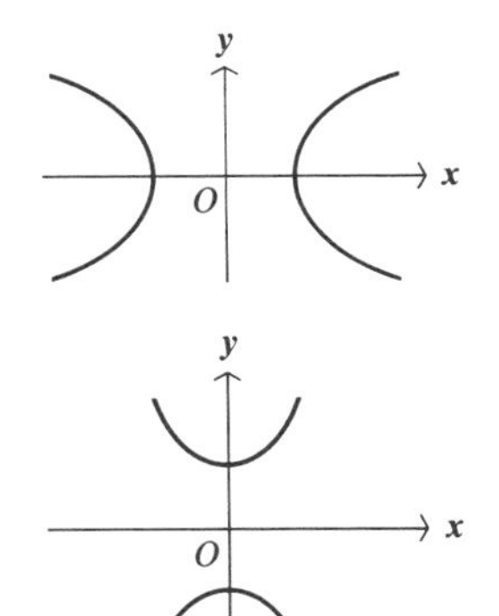

Typical equations of hyperbolas

$$2x^2 - y^2 = 4$$
$$\frac{(x - h)^2}{a^2} - \frac{(y - k)^2}{b^2} = 1$$
[center at (h, k)]

$$3y^2 - 4x^2 = 12$$
$$\frac{(y - k)^2}{a^2} - \frac{(x - h)^2}{b^2} = 1$$
[center at (h, k)]

a. b.

FIGURE 12

Hyperbola

are curves formed when a plane intersects a right circular cone. Examples of conic sections are parabolas, circles, ellipses, hyperbolas, and intersecting straight lines. These conic sections are shown in Figures 9, 10, 11, and 12.

5 The circle

A **circle** is a plane curve consisting of all points at a given distance, called the **radius**, from a fixed point in the plane, called the **center**.

To find the equation of a circle from its definition, we let $C(h, k)$ be the center and let r be the radius (see Figure 13).

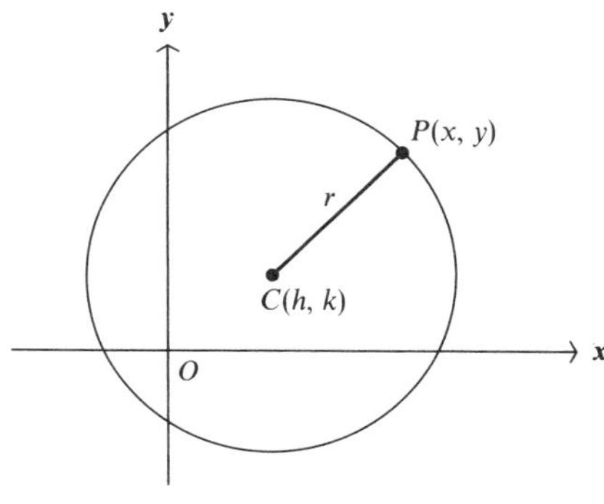

FIGURE 13

If $P(x, y)$ is any point on the circle, the expression for its distance from $C(h, k)$ is

$$\sqrt{(x-h)^2 + (y-k)^2}.$$

Since this distance is equal to r, we have as the equation of the circle,

$$\sqrt{(x-h)^2 + (y-k)^2} = r.$$

Squaring both sides of the equation gives the **standard form** of the equation of a circle.

EQUATION OF A CIRCLE (STANDARD FORM)

$$(x-h)^2 + (y-k)^2 = r^2, \qquad \textbf{(1)}$$

where r is the radius and $C(h, k)$ is the center.

Conversely, if the coordinates of point $P(x, y)$ satisfy Equation 1, then P lies on the circle.

If the center is at the origin, then h and k are both zero and Equation 1 reduces to Equation 2.

> **EQUATION OF CIRCLE (CENTER AT ORIGIN)**
>
> $$x^2 + y^2 = r^2, \tag{2}$$
>
> where r is the radius.

EXAMPLE 1 Find the equation of the circle whose center is at the origin, passing through the point $(3, -2)$.

SOLUTION When the center is at the origin, we use the formula $x^2 + y^2 = r^2$.

$$x^2 + y^2 = r^2$$

$$\begin{aligned} r^2 &= x^2 + y^2 \\ &= (3)^2 + (-2)^2 \\ &= 9 + 4 \\ &= 13 \end{aligned}$$

Therefore, the required equation is $x^2 + y^2 = 13$.

The standard form of the equation of a circle, $(x - h)^2 + (y - k)^2 = r^2$, is convenient in that it shows at a glance the center and radius of the circle. However, another form, called **the general form**, is often the form in which the equation is given. We often change an equation from general to standard form in order to identify the center and radius of the circle.

> **EQUATION OF A CIRCLE (GENERAL FORM)**
>
> $$Ax^2 + Ay^2 + Dx + Ey + F = 0 \tag{3}$$
>
> where $A \neq 0$.

We shall illustrate the standard and general forms of the equation of a circle with an example.

EXAMPLE 2 Write the equation of the circle whose center is at $(3, -1)$ and whose radius is 4, in both standard and general forms.

SOLUTION Written in standard form we have

$$[x - (3)]^2 + [y - (-1)]^2 = (4)^2$$

$$(x - 3)^2 + (y + 1)^2 = 16.$$

To reduce this equation to the general form, we square the binomials and then rearrange terms.

$$(x-3)^2+(y+1)^2=16$$
$$x^2-6x+9+y^2+2y+1=16$$

The general form is $x^2+y^2-6x+2y-6=0$.

We can generalize the above process by starting with the standard form and transforming it to the general form.

$$(x-h)^2+(y-k)^2=r^2$$
$$x^2-2xh+h^2+y^2-2ky+k^2=r^2$$
$$x^2+y^2-2hx-2ky+(h^2+k^2-r^2)=0$$

This is in the form

$$x^2+y^2+D'x+E'y+F'=0,$$

where $D'=-2h$, $E'=-2k$, and $F'=h^2+k^2-r^2$. Multiplying by $A\neq 0$ clears the equation of fractions.

$$Ax^2+Ay^2+Dx+Ey+F=0,$$

which is the general form of the equation of a circle, where $D=AD'$, $E=AE'$, and $F=AF'$.

Notice that the coefficients of x^2 and y^2 are the same in equations of circles.

EXAMPLE 3 Find the equation of the circle whose center is at $(2, -1)$ and which passes through the point $(-2, -4)$. Write the equation in general form.

SOLUTION We start with Equation 1 and then expand and rearrange terms.

$$(x-h)^2+(y-k)^2=r^2, \qquad \text{where } r^2=[2-(-2)]^2$$
$$(x-2)^2+[y-(-1)]^2=25 \qquad +[(-1)-(-4)]^2$$
$$x^2-4x+4+y^2+2y+1=25 \qquad =16+9=25$$
$$x^2+y^2-4x+2y-20=0$$

The required equation is $x^2+y^2-4x+2y-20=0$.

EXAMPLE 4 Graph $x^2+y^2+4x-2y-20=0$.

SOLUTION We transform the equation from general form to standard form so that we can identify the center and radius of the circle.

$$x^2+y^2+4x-2y-20=0$$
$$(x^2+4x+\quad)+(y^2-2y+\quad)=20$$
$$(x^2+4x+4)+(y^2-2y+1)=20+4+1$$

Add 4 and 1 to both sides to complete the squares on x and y.

$$(x + 2)^2 + (y - 1)^2 = 25$$

$$[x - (-2)]^2 + (y - 1)^2 = 5^2$$

Therefore, the center of the circle is at $(-2, 1)$ and its radius is 5 (see Figure 14). (For more discussion on how to complete a square see Example 3 in Section 3 of Chapter 5.)

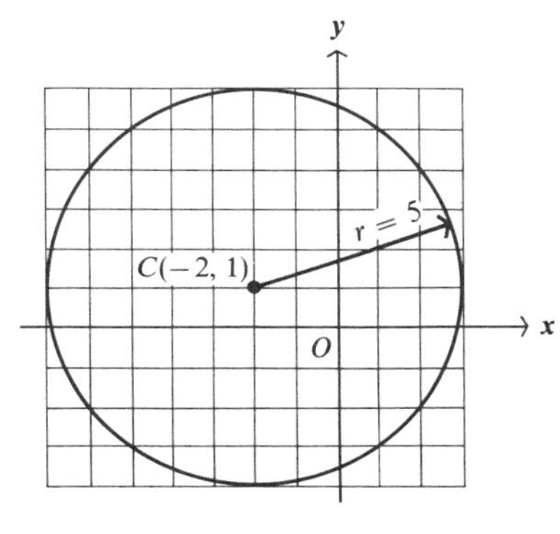

FIGURE 14

EXAMPLE 5 Express $x^2 + y^2 - 2x - 4y + 5 = 0$ in standard form.

SOLUTION

$$x^2 + y^2 - 2x - 4y + 5 = 0$$

$$(x^2 - 2x + \quad) + (y^2 - 4y + \quad) = -5$$

$$(x^2 - 2x + 1) + (y^2 - 4y + 4) = -5 + 1 + 4$$

$$(x - 1)^2 + (y - 2)^2 = 0$$

Since neither term on the left side of the equation can be negative, their sum can be zero only if both terms are zero. This is possible only when $x = 1$ and $y = 2$. Therefore, the point $(1, 2)$ is the only point in the plane that satisfies the original equation. Equations of this kind are sometimes called **point circles** since the radius is zero and only one point satisfies the equation.

EXAMPLE 6 Express $x^2 + y^2 + 2x - 10y + 30 = 0$ in standard form.

SOLUTION

$$x^2 + y^2 + 2x - 10y + 30 = 0$$

$$(x^2 + 2x + \quad) + (y^2 - 10y + \quad) = -30$$

$$(x^2 + 2x + 1) + (y^2 - 10y + 25) = -30 + 26$$

$$(x + 1)^2 + (y - 5)^2 = -4$$

Since neither term on the left side of the equation can be negative, their sum cannot be negative. Therefore, there is no point in the plane that satisfies this equation. The solution is $\varnothing$.

EXAMPLE 7 Reduce $3x^2 + 3y^2 + 4x - 10y - 7 = 0$ to the standard form of the equation of a circle, and find the coordinates of the center and the radius of the circle.

SOLUTION First reduce the coefficients of the second degree terms to 1 by multiplying both sides of the equation by $\frac{1}{3}$.

$$\frac{1}{3}(3x^2 + 3y^2 + 4x - 10y - 7) = \frac{1}{3}(0)$$

$$x^2 + y^2 + \frac{4}{3}x - \frac{10}{3}y - \frac{7}{3} = 0$$

$$\left(x^2 + \frac{4}{3}x + \qquad\right) + \left(y^2 - \frac{10}{3}y + \qquad\right) - \frac{7}{3} = 0$$

$$\left(x^2 + \frac{4}{3}x + \frac{4}{9}\right) + \left(y^2 - \frac{10}{3}y + \frac{25}{9}\right) = \frac{7}{3} + \frac{4}{9} + \frac{25}{9}$$

$$\left(x + \frac{2}{3}\right)^2 + \left(y - \frac{5}{3}\right)^2 = \frac{50}{9}$$

Therefore, the center is at $(-\frac{2}{3}, \frac{5}{3})$ and the radius is $5\sqrt{2}/3$.

Exercises 5

A In Exercises 1–5, write the equation in standard form of each circle described.

1. Center (2, 3); radius 2
2. Center (0, 0); radius 1
3. Center $(-1, 3)$; passes through (3, 1)
4. $(3, -5)$ and $(-1, 3)$ are endpoints of a diameter
5. Center $(-4, 3)$; tangent to y-axis

In Exercises 6–8, write the equation in general form of each circle described.

6. Tangent to both axes at $(-2, 0)$ and (0, 2), respectively
7. Center (3, 7); tangent to y-axis
8. Center $(4, -3)$; passes through origin

In Exercises 9 and 10, express each equation in standard form and then sketch its graph.

9. $x^2 + y^2 + 6x - 16 = 0$
10. $9x^2 + 9y^2 - 12x - 24y - 5 = 0$

B In Exercises 1–5, write the equation in standard form of each circle described.

1. Center $(2, 1)$; radius 3
2. Center $(0, 0)$; radius 2
3. Center $(-3, -1)$; passes through $(1, 1)$
4. $(-2, 3)$ and $(4, -1)$ are endpoints of a diameter
5. Center $(4, 5)$; tangent to x-axis

In Exercises 6–8, write the equation in general form of each circle described.

6. Tangent to both axes at $(3, 0)$ and $(0, -3)$, respectively.
7. Center $(4, 5)$; tangent to y-axis
8. Center $(-3, 4)$; passes through origin

In Exercises 9 and 10, express each equation in standard form and then sketch its graph.

9. $x^2 + y^2 - 2x - 4y + 1 = 0$
10. $9x^2 + 9y^2 - 6x + 18y + 9 = 0$

6 The parabola

A ***parabola*** *is the set of all points in a plane equidistant from a given fixed point and a given fixed line not containing the given point.*

The fixed point is called the **focus** and the fixed line is called the **directrix** of the parabola. The line through the focus perpendicular to the directrix is called the **axis** of the parabola. The **vertex** of the parabola is the point where the parabola intersects its axis. By the definition of the parabola, its vertex is midway between its directrix and focus.

In Figure 15, we let F be the focus of the parabola and $D'D$ the directrix. The set of points P such that

$$\overline{FP} = \overline{RP},$$

where R is the foot of the perpendicular from P to $D'D$, is the parabola. If we plot a number of these points and draw a smooth curve through them, we obtain the graph in Figure 15 representing this curve. The **latus rectum** of a parabola is the chord of the parabola that passes through its focal point and is perpendicular to its axis. The line segment $\overline{KK'}$ is the latus rectum of the parabola shown in Figure 15. Knowing the endpoints of the latus rectum is helpful in drawing the graph of the parabola.

The cross section of high grade reflecting surfaces such as the mirrors used in reflecting telescopes, radar antennas, spot light reflectors, and so on, are parabolas. Some comets have parabolic orbits, some have hyperbolic orbits, while others have elliptical orbits. Comets that have elliptical orbits, such as Halley's comet, reappear periodically. Comets

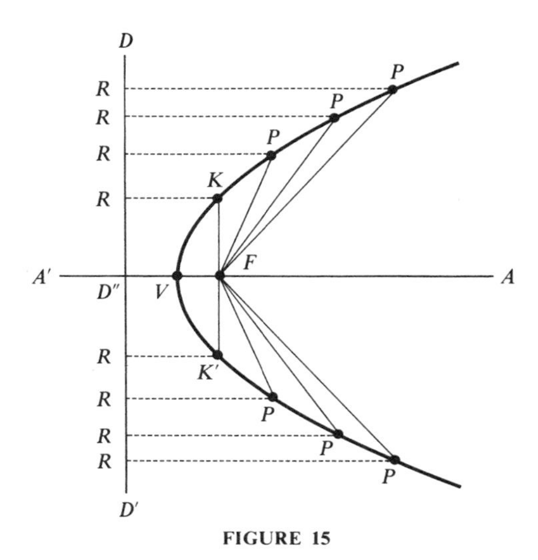

FIGURE 15

that follow parabolic or hyperbolic orbits are seen for a short period of time, but never reappear.

To derive an equation of the parabola, we take the x-axis as the axis of the curve and the vertex as the origin (see Figure 16). This location of the axis gives the simplest form of the equation. Since the vertex is midway between the focus and the directrix, the coordinates of the focus F are $(c, 0)$, and the equation of the directrix DD' is $x = -c$.

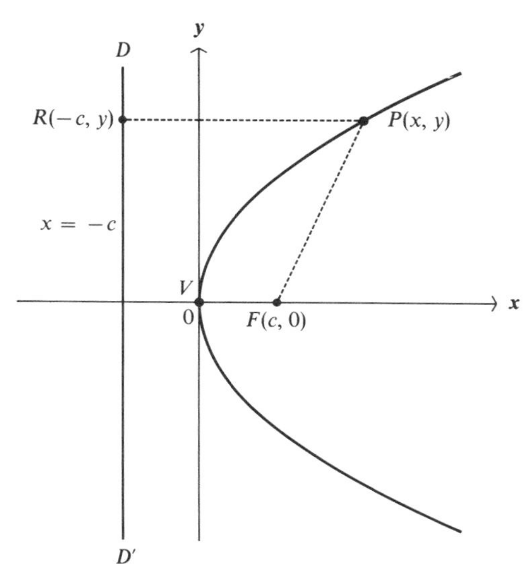

FIGURE 16

Let $P(x, y)$ be any point on the parabola, and let R be the foot of the perpendicular from P to DD'. The coordinates of R are $(-c, y)$. By the definition of the parabola,

$$\overline{FP} = \overline{RP},$$

or

$$(\overline{FP})^2 = (\overline{RP})^2. \tag{1}$$

Then by the distance formula,

$$(x - c)^2 + (y - 0)^2 = [x - (-c)]^2$$
$$x^2 - 2cx + c^2 + y^2 = x^2 + 2cx + c^2. \tag{2}$$

This reduces to

$$y^2 = 4cx \tag{3}$$

Therefore, the coordinates of any point on the parabola satisfy Equation 3.

Conversely, if P is any point whose coordinates satisfy Equation 3, then we obtain Equations 2 and 1 by adding $(x - c)^2$ to both members of Equation 3 and simplifying. Therefore, $\overline{FP} = \overline{RP}$, and the point P lies on the parabola.

When $c > 0$, as in Figure 16, we obtained the equation $y^2 = 4cx$. In a similar manner, if $c < 0$, the focus is to the left of the vertex, the directrix is to the right of the origin, and the parabola opens to the negative part of the x-axis. By interchanging x and y in Equation 3, we obtain

$$x^2 = 4cy. \tag{4}$$

This defines a parabola with focus on the y-axis and directrix $y = -c$.

EQUATIONS OF PARABOLAS

If the vertex of a parabola is at the origin and the focus is on the x-axis, the equation is

$$y^2 = 4cx, \tag{3}$$

where $2c$ is the **directed distance** from the directrix to the focus. The focus is at the point $(c, 0)$ and the directrix is the line $x = -c$. The number c is positive when the focus is to the right of the directrix and negative when the focus is to the left of the directrix.

If the vertex of a parabola is at the origin and the focus is on the y-axis, the equation is

$$x^2 = 4cy, \tag{4}$$

where $2c$ is the directed distance from the directrix to the focus. The focus is at $(0, c)$ and the directrix is $y = -c$. The number c is positive when the focus is above the directrix and negative when the focus is below the directrix.

To find the length of the latus rectum of a parabola (Figure 17), we set $x = c$ (the value of x at the focus) in the equation $y^2 = 4cx$. We find the coordinates of P_1 and P_2 and then find the length of $\overline{P_1 P_2}$.

$$\begin{aligned} y^2 &= 4cx \\ &= 4c(c) \\ &= 4c^2 \end{aligned}$$

Thus $y = \pm 2c$.

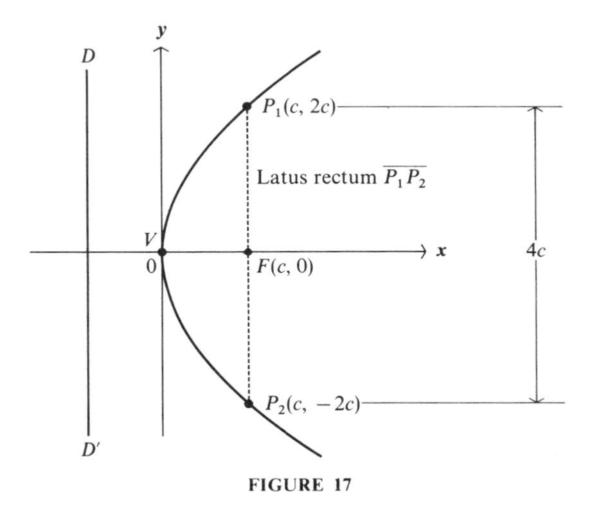

FIGURE 17

Then the coordinates of P_1 are $(c, 2c)$ and the coordinates of P_2 are $(c, -2c)$. Therefore, the length of the latus rectum $\overline{P_1 P_2}$ is $|4c|$. Notice that this length is the absolute value of the coefficient of x in Equation 3 and the coefficient of y in Equation 4.

When the parabola is in one of the standard forms 3 or 4, the following hold.

1. The vertex is always at the origin.
2. The axis of the parabola is the same as the first-degree variable in the equation, that is, the x-axis for Equation 3 and the y-axis for Equation 4.
3. The length of the latus rectum is the absolute value of the coefficient of x in Equation 3 and the coefficient of y in Equation 4.

4. Two points on the parabola can be found by moving half the length of the latus rectum in each direction from the focus along a perpendicular line through the focus.
5. If the coefficient of the first-degree variable in the equation is positive, the parabola opens to the positive part of that axis. If the coefficient of the first-degree variable is negative, the parabola opens to the negative part of that axis.

EXAMPLE 1 Sketch and discuss $y^2 = 4x$.

SOLUTION The equation is that of a parabola in standard form 3, with $c = 1$. Therefore, the vertex is at the origin, the axis is the x-axis, the focus is at $(1, 0)$, and the latus rectum is 4 units long. The length of the latus rectum enables us to find two points $(1, 2)$ and $(1, -2)$, on the parabola. Then, with these two points, the vertex, and the fact that the curve is symmetric to the x-axis, we can make a sketch of the parabola (see Figure 18).

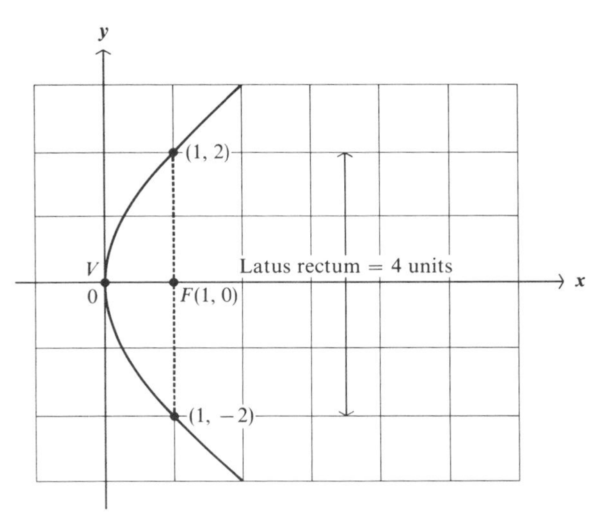

FIGURE 18

EXAMPLE 2 Sketch and discuss $x^2 = -8y$.

SOLUTION The equation is that of a parabola in standard form 4. The vertex is at the origin, the axis is the negative part of the y-axis, and the focus is $(0, -2)$. The length of its latus rectum is $|-8| = 8$. Thus the ends of the latus rectum are at $(4, -2)$ and $(-4, -2)$. With these two points, along with the origin as the vertex and the fact that the

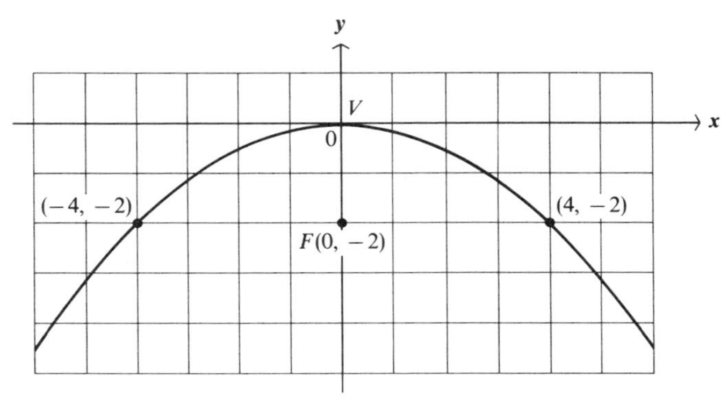

FIGURE 19

curve is symmetric to the y-axis, we can make a sketch of the curve (see Figure 19).

EXAMPLE 3 Find the equation of the parabola having its vertex at the origin, axis along the x-axis, and passing through the point $(-4, -6)$.

SOLUTION Since the parabola is in standard position, its equation has the form $y^2 = 4cx$. The coordinates $(-4, -6)$ of the given point must satisfy the equation $y^2 = 4cx$.

$$y^2 = 4cx$$
$$(-6)^2 = 4c(-4)$$
$$36 = -16c$$

Thus $c = -\frac{9}{4}$.

Therefore, substituting $c = -\frac{9}{4}$ gives the required equation of $y^2 = 4(-\frac{9}{4})x$, or $y^2 = -9x$.

Exercises 6

A In Exercises 1–4, sketch and discuss each parabola.

1. $y^2 = 6x$ **2.** $x^2 = 8y$
3. $y^2 = -4x$ **4.** $3x^2 = -8y$

In Exercises 5–8, find the equation of the parabola satisfying the given conditions.

5. Directrix $x = -2$; focus $(2, 0)$
6. Vertex $(0, 0)$; focus $(4, 0)$.
7. Vertex at origin; axis along x-axis, passing through $(6, 4)$
8. Axis along y-axis; vertex at origin; opening downward; length of latus rectum 8

9. Find the equations of the two lines perpendicular to the latus rectum at its end points for the parabola $x^2 = -8y$.
10. Sketch the parabola $x^2 = 8y$ and draw lines from the endpoints of its latus rectum to its vertex. Find the angle formed by these two lines.

B In Exercises 1–4, sketch and discuss each parabola.

1. $y^2 = 2x$
2. $x^2 = 4y$
3. $y^2 = -12x$
4. $5x^2 = -12y$

In Exercises 5–8, find the equation of the parabola satisfying the given conditions.

5. Directrix $x = -4$; focus $(4, 0)$
6. Vertex $(0, 0)$; focus $(8, 0)$
7. Vertex at origin; axis along x-axis, passing through $(4, 6)$.
8. Axis along y-axis; vertex at origin; opening downward; length of latus rectum 12.
9. Find the equations of the two lines perpendicular to the latus rectum at its endpoints for the parabola $x^2 = -4y$.
10. Sketch the parabola $y^2 = 4x$ and draw lines from the endpoints of its latus rectum to its vertex. Find the angle formed by these two lines.

7 The ellipse

Johannes Kepler, a German astronomer, announced to the world in 1609 that "the planets move around the sun in ellipses having the sun at one of the foci." Our moon moves around the earth in an elliptical orbit and artificial satellites move in elliptical orbits.

> An **ellipse** is the set of all points $P(x, y)$ in a plane such that the sum of the distances from $P(x, y)$ to a pair of fixed points (foci) is a constant.

The simplest form of the equation of an ellipse is obtained by taking the x-axis through the foci and the origin midway between the foci. In Figure 20, let F and F' be the foci and let $\overline{FF'} = 2c$; then the coordinates of F are $(c, 0)$ and those of F' are $(-c, 0)$. Let $P(x, y)$ be any point on the ellipse. Let the sum of the distances of $P(x, y)$ from F and F' be $2a$. Hence $2a > 2c$ and $a > c$. By the definition of the ellipse,

$$\overline{FP} + \overline{F'P} = 2a.$$

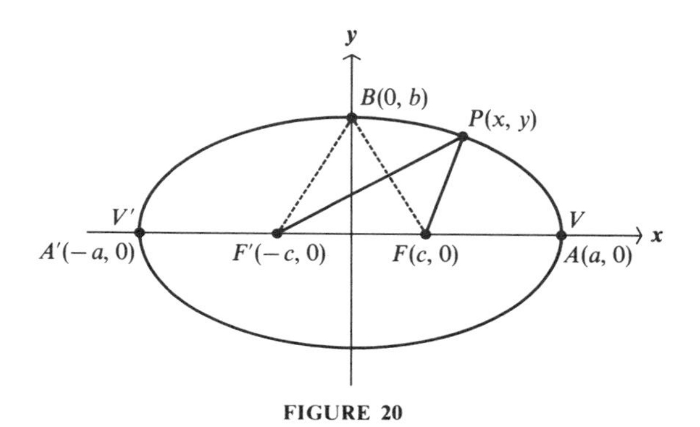

FIGURE 20

Then by using the distance formula, this becomes

$$\sqrt{(x-c)^2+y^2}+\sqrt{(x+c)^2+y^2}=2a.$$

When we rearrange the terms, square both sides, and simplify, we have

$$a\sqrt{(x+c)^2+y^2}=a^2+cx.$$

If we square both sides again and simplify, we obtain

$$(a^2-c^2)x^2+a^2y^2=a^2(a^2-c^2). \qquad \textbf{(1)}$$

But $a > c$, so that $a^2 - c^2$ is positive and is denoted by the number b^2. Substituting b^2 for $a^2 - c^2$ in Equation 1 we have

$$b^2x^2+a^2y^2=a^2b^2. \qquad \textbf{(2)}$$

Dividing by a^2b^2 gives

$$\frac{x^2}{a^2}+\frac{y^2}{b^2}=1. \qquad \textbf{(3)}$$

This equation is satisfied by the coordinates of any point on the ellipse.

Notice that in developing the equation, we squared both sides of the equation at two of the steps. In both cases, both sides of the equation were nonnegative. Thus, no extraneous roots were introduced, and the steps may be reversed. When $y = 0$ in

$$\frac{x^2}{a^2}+\frac{y^2}{b^2}=1,$$

then

$$\frac{x^2}{a^2}+\frac{0}{b^2}=1$$

$$x^2=a^2$$

$$x=\pm a$$ The x-intercepts are $(\pm a, 0)$.

When $x = 0$ in

$$\frac{x^2}{a^2} + \frac{y^2}{b^2} = 1,$$

then

$$\frac{0}{a^2} + \frac{y^2}{b^2} = 1$$

$$y^2 = b^2$$

$$y = \pm b \qquad \text{The } y\text{-intercepts are } (0, \pm b).$$

Observe that there are two axes of symmetry: the x-axis and the y-axis. In both Equations 3 and 4, it is understood that $a > b$. The **major axis is the longer** of the two axes of the ellipse and the **minor axis is the shorter** of the two axes. In Equation 3, the major axis of the ellipse is on the x-axis. In Equation 4, the major axis of the ellipse is on the y-axis. The ends of the major axis are called the **vertices of the ellipse**.

EQUATIONS OF ELLIPSES

If an ellipse has center at (0, 0), ends of major axis at $(\pm a, 0)$, ends of minor axis at $(0, \pm b)$, and foci at $(\pm c, 0)$, its equation is

$$\frac{x^2}{a^2} + \frac{y^2}{b^2} = 1. \tag{3}$$

If an ellipse has center at (0, 0), ends of major axis at $(0, \pm a)$, ends of minor axis at $(\pm b, 0)$, and foci $(0, \pm c)$, its equation is

$$\frac{y^2}{a^2} + \frac{x^2}{b^2} = 1. \tag{4}$$

In each case, $2a$ is the sum of the distances of any point of the ellipse from the foci, and b is defined by

$$b^2 = a^2 - c^2, \tag{5}$$

where $2c$ is the distance between the foci.

The numbers a and b are the lengths of the **semimajor axis** and the **semiminor axis**, respectively. By the definition of the ellipse, the sum of the distances BF and BF' is $2a$ (see Figure 21); since $BF = BF'$, it follows that $BF = a$. The important relation

$$a^2 = b^2 + c^2 \tag{6}$$

connecting the constants a, b, c can be visualized in triangle OBF of Figure 21.

A chord of the ellipse through a focus perpendicular to the major axis (RL in Figure 21) is called the **latus rectum**. By substituting the

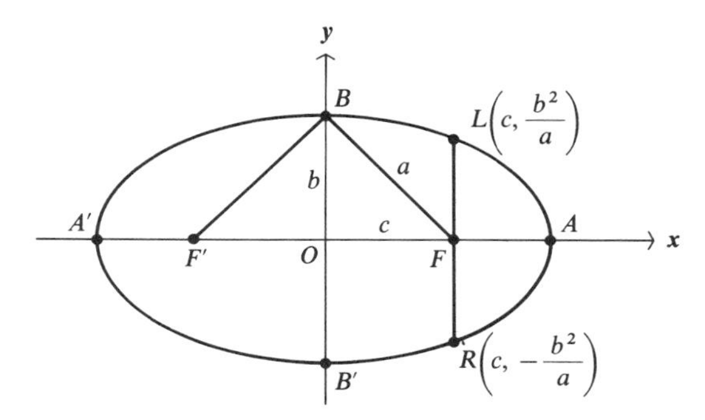

FIGURE 21

coordinates of a focal point in Equation 3 or 4, we find that the length of the latus rectum RL is $2b^2/a$.

The ratio c/a is called the **eccentricity of the ellipse** and it is denoted by e.

$$e = \frac{c}{a} = \frac{\sqrt{a^2 - b^2}}{a} \tag{7}$$

Since $c < a$, the eccentricity of an ellipse is always between 0 and 1. If e is close to 0, then c is small compared with a. Since $c^2 = a^2 - b^2$, it follows that the difference between a and b is small compared with a; therefore, the foci are close together and the major and minor axes are nearly equal, so that the ellipse is very nearly circular. As e increases, the ellipse becomes less nearly circular and more flattened. If e is close to 1, c is nearly equal to a. Since $b^2 = a^2 - c^2$, it follows that b is very small compared to a; therefore, the ellipse is very narrow and elongated. Hence the shape of an ellipse is related to its eccentricity. It is sometimes said that a circle is a special case of an ellipse when the eccentricity is 0. In this case, the foci coincide (at the center).

EXAMPLE 1 Find the vertices, foci, lengths of the semiaxes, eccentricity, and graph of the ellipse whose equation is $4x^2 + 9y^2 = 36$.

SOLUTION First reduce the given equation to standard form.

$$\frac{4x^2}{36} + \frac{9y^2}{36} = \frac{36}{36}$$

$$\frac{x^2}{9} + \frac{y^2}{4} = 1$$

The lengths of the semiaxes are $a = 3$ and $b = 2$; $c = \sqrt{a^2 - b^2} = \sqrt{5} \doteq 2.24$. Hence the vertices are $(\pm 3, 0)$, the minor axis intercepts are $(0, \pm 2)$, the foci are $(\pm\sqrt{5}, 0)$, and the eccentricity is $e = \sqrt{5}/3 \doteq 0.75$ (see Figure 22). The length of the latus rectum is $2b^2/a = 2(2)^2/3 = 8/3$.

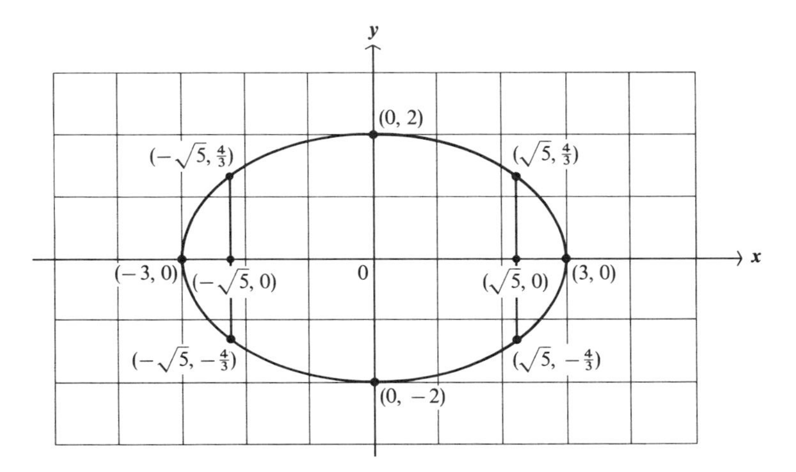

FIGURE 22

EXAMPLE 2 Find the vertices, foci, lengths of the semiaxes, eccentricity, and graph of the ellipse whose equation is $16x^2 + 4y^2 = 1$.

SOLUTION First reduce the given equation to standard form.

$$\frac{16x^2}{64} + \frac{4y^2}{64} = \frac{64}{64}$$

$$\frac{x^2}{4} + \frac{y^2}{16} = 1$$

Because the denominator of the y^2 term is greater than the denominator of the x^2 term, the major axis is on the y-axis. The x-intercepts are $b = \pm\sqrt{4} = \pm 2$, the y-intercepts are $a = \pm\sqrt{16} = \pm 4$, and $c = \sqrt{a^2 - b^2} = \sqrt{16 - 4} = \sqrt{12}$. Thus the foci are at $(0, \pm 2\sqrt{3})$, the eccentricity is $e = \sqrt{12}/4 \doteq 0.87$, and the lengths of the semiaxes are 4 and 2 (see Figure 23).

$$\text{Length of latus rectum} = \frac{2b^2}{a} = \frac{2(4)}{4} = 2$$

EXAMPLE 3 Find the equation of the ellipse with vertices at $(\pm 6, 0)$ and eccentricity $= \frac{1}{2}$.

SOLUTION Since its vertices are $(\pm 6, 0)$, $a = 6$ and the center is at the origin. The major axis is along the x-axis. Since $e = c/a$ and $e = \frac{1}{2}$, $c/6 = \frac{1}{2}$ and $c = 3$.

$$\begin{aligned} b^2 &= a^2 - c^2 \\ &= 36 - 9 = 27 \end{aligned}$$

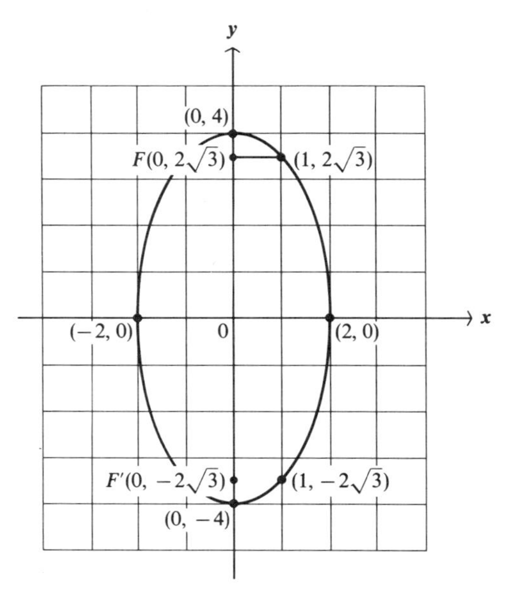

FIGURE 23

We use Equation 3,

$$\frac{x^2}{a^2} + \frac{y^2}{b^2} = 1.$$

Therefore,

$$\frac{x^2}{36} + \frac{y^2}{27} = 1$$

is the required equation.

Exercises 7

A In Exercises 1–3, find the vertices, foci, and lengths of the semiaxes. Graph each curve.

1. $x^2 + 4y^2 = 100$ **2.** $9x^2 + 4y^2 = 36$

3. $81x^2 + 100y^2 = 100$

In Exercises 4–6, find the equation of the ellipse satisfying the given data. Graph each curve.

4. Foci $(\pm 3, 0)$; one vertex at $(5, 0)$

5. Vertices $(\pm 10, 0)$; eccentricity $\frac{2}{5}$

6. One vertex $(6, 0)$; minor axis 6

In Exercises 7 and 8, find the equation of the ellipse with center at the origin satisfying the given data.

7. Minor axis 12; latus rectum $\frac{18}{5}$; foci on x-axis

8. One vertex $(-3, 0)$; eccentricity $\frac{2}{3}$

9. Graph and identify the curve $x = \frac{2}{5}\sqrt{25 - y^2}$.

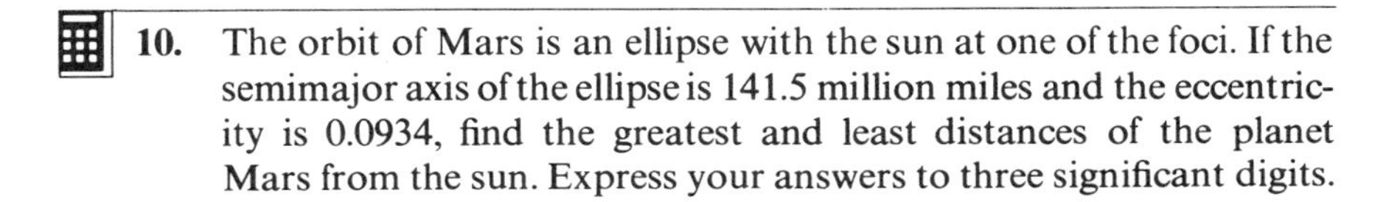

10. The orbit of Mars is an ellipse with the sun at one of the foci. If the semimajor axis of the ellipse is 141.5 million miles and the eccentricity is 0.0934, find the greatest and least distances of the planet Mars from the sun. Express your answers to three significant digits.

B In Exercises 1–3, find the vertices, foci, and lengths of the semiaxes. Graph each curve.

1. $16x^2 + 25y^2 = 400$ **2.** $25x^2 + 4y^2 = 100$

3. $3x^2 + 2y^2 = 6$

In Exercises 4–6, find the equation of the ellipse satisfying the given data. Graph each curve.

4. Vertices $(\pm 13, 0)$; foci $(\pm 8, 0)$

5. Foci $(0, \pm 2)$; $e = \frac{2}{5}$

6. Foci $(0, \pm 3)$; ends of minor axis $(\pm 4, 0)$.

In Exercises 7 and 8, find the equation of the ellipse with center at the origin satisfying the given data.

7. Major axis 6; latus rectum 4; foci on the x-axis

8. One vertex $(0, -4)$; eccentricity $\frac{3}{4}$.

9. Graph and identify the curve $y = -\frac{2}{3}\sqrt{9 - x^2}$.

10. The earth's orbit is an ellipse, with the sun at one of the foci. If the semimajor axis of the ellipse is 92.9 million miles and the eccentricity is 0.0167, find the greatest and least distances of the earth from the sun. Express your answers to three significant digits.

8 The hyperbola

A **hyperbola** is the set of all points $P(x, y)$ in a plane such that the difference between the distances from $P(x, y)$ to a pair of fixed points (foci) is a constant.

The simplest form of the equation of a hyperbola is found by taking the x-axis through the foci and the origin midway between the foci. In Figure 24, we let F and F' be the foci and $FF' = 2c$; then the coordinates

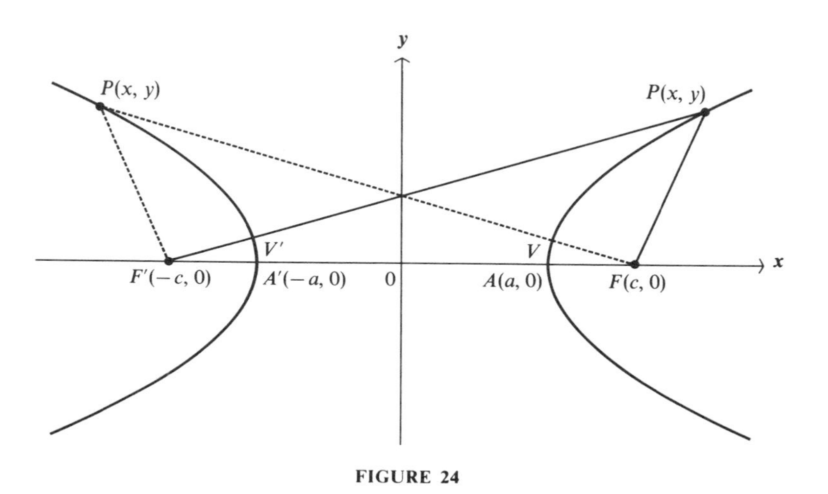

FIGURE 24

of F and F' are $(c, 0)$ and $(-c, 0)$, respectively. Let $P(x, y)$ be any point on the hyperbola. Let the difference of the distances from $P(x, y)$ to F or F' be $2a$. Hence $2a < 2c$ and $a < c$.

By the definition of hyperbola, $|\overline{F'P} - \overline{FP}| = 2a$, so that either

$$\overline{F'P} - \overline{FP} = 2a \quad \text{or} \quad \overline{FP} - \overline{F'P} = 2a. \tag{1}$$

By application of the distance formula, $\overline{F'P} - \overline{FP} = 2a$ becomes

$$\sqrt{(x + c)^2 + y^2} - \sqrt{(x - c)^2 + y^2} = 2a.$$

By rearranging terms, squaring, and simplifying, we get

$$cx - a^2 = a\sqrt{(x - c)^2 + y^2}.$$

Squaring both sides again and simplifying, we have

$$(c^2 - a^2)x^2 - a^2y^2 = a^2(c^2 - a^2). \tag{2}$$

The same result is obtained beginning with $\overline{FP} - \overline{F'P} = 2a$. But $a < c$, so that $c^2 - a^2$ is positive; we denote this positive number by b^2. Equation 2 then becomes

$$b^2x^2 - a^2y^2 = a^2b^2. \tag{3}$$

Dividing by a^2b^2, we get

$$\frac{x^2}{a^2} - \frac{y^2}{b^2} = 1. \tag{4}$$

In developing the equation, we squared both sides of the equation in two of the steps. In both cases, both sides of the equation were nonnegative. Thus we have introduced no extraneous roots, and the steps may be reversed. Therefore, if Equation 4 is satisfied by the coordinates of any point P, then the point P lies on the hyperbola and Equation 4 is the required equation of the hyperbola.

EQUATION OF A HYPERBOLA

If the foci are on the x-axis and the origin is midway between them, then the equation of a hyperbola is

$$\frac{x^2}{a^2} - \frac{y^2}{b^2} = 1, \tag{4}$$

where $2a$ is the absolute value of the difference of the distances of any point on the hyperbola from the foci and b is defined by

$$b^2 = c^2 - a^2. \tag{5}$$

The distance between the foci is $2c$.

If the foci of a hyperbola are on the y-axis, with coordinates $F(0, c)$ and $F'(0, -c)$, if $|\overline{FP} - \overline{F'P}| = 2a$ and if $b = \sqrt{c^2 - a^2}$, then the equation of the hyperbola is

$$\frac{y^2}{a^2} - \frac{x^2}{b^2} = 1. \tag{6}$$

SOME PROPERTIES OF THE HYPERBOLA

Symmetry Because both x and y (in Equation 4) are of second degree, the hyperbola is symmetric to both the x- and y-axes and, therefore, symmetric to the origin.

Intercepts If we set $y = 0$, we get $x = \pm a$; but if $x = 0$, we have $y^2 = -b^2$, so that y is imaginary. There are therefore two points, $A(a, 0)$ and $A'(-a, 0)$, where the curve intersects the x-axis. The curve does not intersect the y-axis. The points A and A' are called the **vertices** of the hyperbola.

Extent By solving Equation 4 for y and for x, we get

$$y = \pm\frac{b}{a}\sqrt{x^2 - a^2} \qquad \text{and} \qquad x = \pm\frac{a}{b}\sqrt{y^2 + b^2}. \tag{7}$$

The first equation shows that if x is numerically less than a, then y is imaginary, so that we must exclude the region between $x = a$ and $x = -a$. From the second equation, we see that x is real for all real values of y, so that there are no excluded values of y in the range. The graph consists of two disconnected curves, called **branches**.

The segment $\overline{AA'}$ joining the vertices is called the **transverse axis** of the hyperbola; it is of length $2a$. Although the hyperbola does not intersect the y-axis, the segment joining the points $B(0, b)$ and $B'(0, -b)$ is called the **conjugate axis**; it is of length $2b$. The origin, which is a center of symmetry of the curve, is the **center** of the hyperbola. The numbers a and b are the lengths of the **semitransverse** axis and the **semiconjugate** axis, respectively.

A chord of the hyperbola through either focus perpendicular to the transverse axis and terminated by the curve is called the **latus rectum**. The ordinate corresponding to a focus is found by letting $x = \pm c$ in Equation 7; we obtain

$$y = \pm \frac{b}{a}\sqrt{c^2 - a^2} = \pm \frac{b^2}{a}, \tag{8}$$

since $c^2 - a^2 = b^2$. Therefore, the length of the latus rectum is $2b^2/a$.

The ratio c/a is called the eccentricity of the hyperbola and is denoted by e.

$$e = \frac{c}{a} \tag{9}$$

Since $c > a$, the eccentricity of a hyperbola is always greater than 1.

EXAMPLE 1 Find the vertices, foci, lengths of the semiaxes, and eccentricity of the hyperbola whose equation is $4x^2 - 9y^2 = 36$.

SOLUTION Reduce the given equation to standard form.

$$\frac{4x^2}{36} - \frac{9y^2}{36} = \frac{36}{36}$$

$$\frac{x^2}{9} - \frac{y^2}{4} = 1$$

$$a^2 = 9, \qquad b^2 = 4$$

$$a = 3, \qquad b = 2$$

$$c = \sqrt{a^2 + b^2} = \sqrt{9 + 4} = \sqrt{13} \doteq 3.6$$

Therefore, vertices are at $(\pm 3, 0)$, foci are at $(\pm\sqrt{13}, 0)$, the length of semitransverse axis is 3; the length of semiconjugate axis is 2; and the eccentricity is $\sqrt{13}/3 \doteq 1.2$.

ASYMPTOTES OF A HYPERBOLA

When we solve the equation of the hyperbola

$$\frac{x^2}{a^2} - \frac{y^2}{b^2} = 1$$

for y in terms of x, we get

$$y = \pm \frac{b}{a}\sqrt{x^2 - a^2} = \pm \frac{bx}{a}\sqrt{1 - \frac{a^2}{x^2}}. \tag{10}$$

When x grows larger and larger as compared to a, the fraction a^2/x^2 gets smaller and smaller. We say a^2/x^2 approaches zero as x grows large without limit. Thus as x grows exceedingly large, the value of $\sqrt{1 - (a^2/x^2)}$ approaches 1, and the graph of the hyperbola comes closer and closer to the straight lines whose equations are $y = \pm(b/a)x$. For this reason, the straight lines

$$y = \pm \frac{b}{a}x \tag{11}$$

are called the **asymptotes** of the hyperbola whose equation is

$$\frac{x^2}{a^2} - \frac{y^2}{b^2} = 1.$$

By a similar argument, the asymptotes of the hyperbola

$$\frac{y^2}{a^2} - \frac{x^2}{b^2} = 1 \qquad \text{are} \quad y = \pm \frac{a}{b}x.$$

An **asymptote** is a line which has the property that the distance from a point P on the curve to the line approaches zero as the distance of P from the origin increases without limits (see Figure 25).

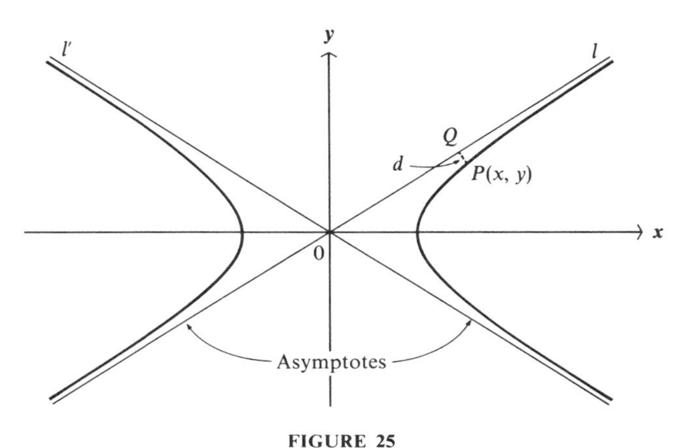

FIGURE 25

Asymptotes of the hyperbola

A convenient way to sketch a hyperbola is outlined in these five steps.

1. Locate the vertices A and A', $(\pm a, 0)$.
2. Locate the ends of the conjugate axis B and B', $(0, \pm b)$.
3. Construct a rectangle with sides parallel to the coordinate axes and through the points A, A', B, and B'.
4. Draw the diagonals of the rectangle of Step 3. The extensions of these diagonal lines are the asymptotes of the hyperbola.
5. Having drawn the asymptotes and plotted the vertices, and knowing that the hyperbola is symmetric with respect to both axes and to the origin, we can make a sketch of the curve (see Example 2). We use the latus rectum to find additional points.

EXAMPLE 2 Sketch the hyperbola $\dfrac{x^2}{9} - \dfrac{y^2}{4} = 1$.

SOLUTION

$a^2 = 9$; $a = 3$, so the vertices A and A' are at $(3, 0)$ and $(-3, 0)$.
$b^2 = 4$; $b = 2$, so B and B' are located at $(0, 2)$ and $(0, -2)$.
$c = \pm\sqrt{a^2 + b^2} = \pm\sqrt{9 + 4} = \pm\sqrt{13}$, so the foci are $(\pm\sqrt{13}, 0)$.

The ends of latus rectum are at $y = \pm(b^2/a) = \pm\frac{4}{3}$. Construct a rectangle with sides parallel to the coordinate axes, through A, A', B, and B'. Draw the diagonals of the rectangle. The extensions of these diagonals are the asymptotes of the hyperbola. Then, having the ver-

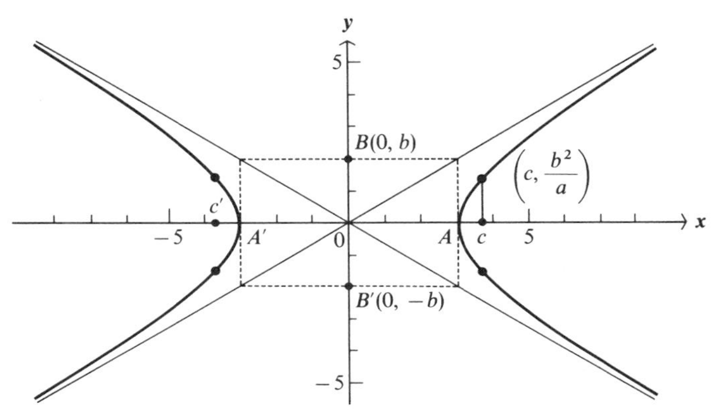

FIGURE 26

tices, the asymptotes, the end points of the latus rectum, and the property that the hyperbola is symmetric with respect to the origin, we can make a fairly accurate sketch of the curve (see Figure 26).

If two hyperbolas are so related that the transverse and conjugate axes of one curve are respectively the conjugate and transverse axes of the other, each hyperbola is called the **conjugate** of the other, and together they are said to form a pair of **conjugate hyperbolas**.

If the equation of one hyperbola is $(x^2/a^2) - (y^2/b^2) = 1$, its conjugate has the equation $(y^2/b^2) - (x^2/a^2) = 1$ (see Figure 27). The foci of

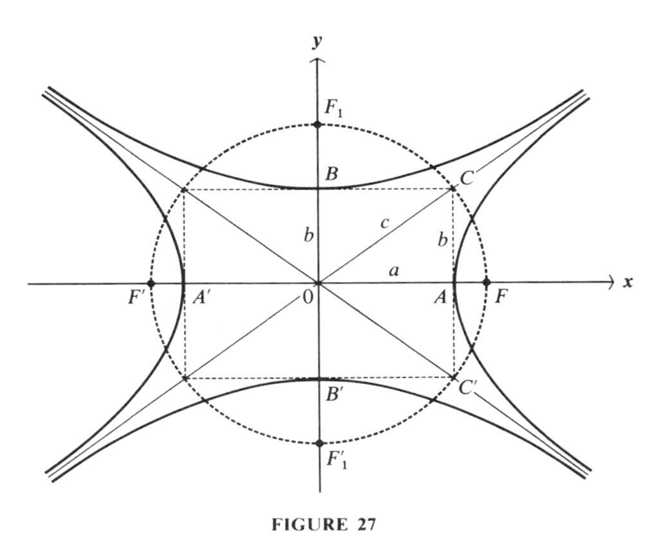

FIGURE 27

Two conjugate hyperbolas

the first hyperbola are on the x-axis and those of the second are on the y-axis.

The foci of two conjugate hyperbolas are equidistant from the center, since $c^2 = a^2 + b^2$ for both curves. Two conjugate hyperbolas have the same asymptotes (see Figure 27).

EXAMPLE 3

a. Find the equation of the hyperbola having foci $(\pm 5, 0)$ and vertices $(\pm 3, 0)$.

b. Find the equation of the hyperbola conjugate to the hyperbola of Part a.

SOLUTION

a. Because the vertices and foci are on the x-axis, the equation is in the form $(x^2/a^2) - (y^2/b^2) = 1$. Since the vertices are $(\pm 3, 0)$, $a = 3$ and $a^2 = 9$. Also, since the foci are $(\pm 5, 0)$, $c = 5$ and $c^2 = 25$.

$$b^2 = c^2 - a^2 = 25 - 9 = 16$$

Therefore, the required equation is $\dfrac{x^2}{9} - \dfrac{y^2}{16} = 1$.

b. The hyperbola conjugate to $(x^2/9) - (y^2/16) = 1$ is

$$\frac{x^2}{9} - \frac{y^2}{16} = -1 \qquad \text{or} \qquad \frac{y^2}{16} - \frac{x^2}{9} = 1.$$

Exercises 8

A In Exercises 1 and 2, find the vertices, foci, eccentricity, length of a latus rectum, the equations of the asymptotes, and sketch the graph of each hyperbola.

1. $5y^2 - 2x^2 = 40$ **2.** $x^2 - y^2 = 16$

In Exercises 3–5, find the equation of each hyperbola.

3. Vertices $(0, \pm 4)$; foci $(0, \pm 8)$

4. Foci $(0, \pm 8)$, ends of conjugate axis $(\pm 4, 0)$

5. Foci $(\pm 6, 0)$; eccentricity $\frac{3}{2}$

6. Find the equation of the hyperbola whose asymptotes are $y = \pm\frac{2}{3}x$ if the vertices are at $(0, \pm 4)$.

7. Write the equation of the hyperbola conjugate to $4y^2 - x^2 = 8$. Then find the equations of the asymptotes and sketch the conjugate hyperbola.

8. Find the equation of the hyperbola which has the foci of the ellipse $4x^2 + 9y^2 = 36$ as vertices and the vertices of the ellipse as foci.

B In Exercises 1 and 2, find the vertices, foci, eccentricity, length of a latus rectum, the equations of the asymptotes, and sketch the graph of each hyperbola.

1. $16x^2 - 25y^2 = 400$ **2.** $y^2 - 4x^2 = 8$

In Exercises 3–5, find the equation of each hyperbola.

3. Vertices $(\pm 6, 0)$; eccentricity $\dfrac{\sqrt{13}}{2}$

4. Foci $(\pm 5, 0)$; vertices $(\pm 3, 0)$

5. One end of conjugate axis $(3, 0)$; eccentricity 2

6. Find the equation of the hyperbola whose asymptotes are $y = \pm\frac{3}{2}x$ and whose vertices are at $(\pm 4, 0)$.
7. Write the equation of the hyperbola conjugate to $4x^2 - 9y^2 = 36$. Then find the equations of the asymptotes and sketch the conjugate hyperbola.
8. Find the equation of the hyperbola with eccentricity 2 which has the same foci as the ellipse $9x^2 + 25y^2 = 225$.

9 Translation of axes

In many cases, the solution of a problem in analytic geometry can be simplified by the use of a different pair of coordinate axes from the one used in the statement of the problem. The process of changing from one pair of coordinate axes to another is called a **transformation of the axes**. If the new axes are parallel, respectively, to the old ones, and if they have the same positive directions, the transformation is called a **translation of axes**. If the origin remains unchanged, and the new axes are obtained by rotating the old axes about the origin through a given angle, then the transformation is a **rotation of axes**. In this section, we shall discuss translation of axes.

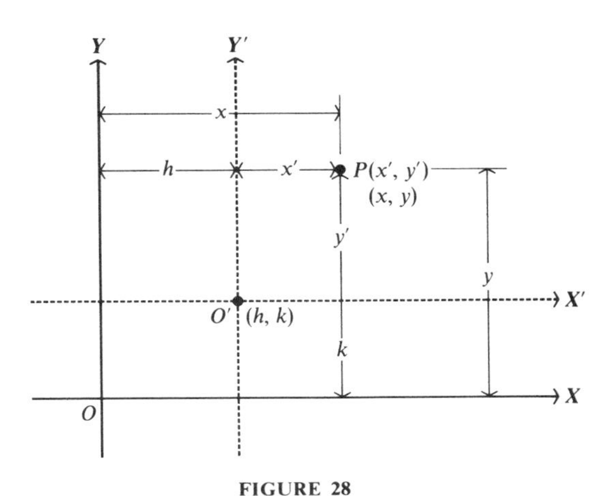

FIGURE 28

Let OX and OY (see Figure 28) be the original axes and let $O'X'$ and $O'Y'$ be the new axes, parallel, respectively, to the old and having the same positive directions. Let the coordinates of O' in reference to OX and OY, be (h, k).

Let P be any point in the plane and let its coordinates, in reference to the old axes, be (x, y) and, in reference to the new axes, be (x', y'). Then from Figure 28, we see that $x = x' + h$ and $y = y' + k$. These equations are called the **equations of translation**.

EQUATIONS OF TRANSLATION

If a translation of axes is made to a new origin O' whose coordinates with respect to the old axes are (h, k), the relation between the old and new coordinates of any point is given by

$$x = x' + h \quad \text{or} \quad x' = x - h$$
$$y = y' + k \quad \text{or} \quad y' = y - k, \tag{1}$$

where x' and y' represent coordinates with respect to the new axes.

EXAMPLE 1 Find the equation of the straight line $2x - 3y + 6 = 0$ when the origin has been translated to the point (3, 4). Graph the original equation on the xy-axes and the transformed equation on the $x'y'$-axes [with origin at (3, 4)] and thus show that both equations represent the same line.

SOLUTION Using the equations of translation and the fact that the origin of the $x'y'$-axes is at (3, 4), we have

$$x = x' + h \qquad y = y' + k$$
$$x = x' + 3 \qquad y = y' + 4$$

Then

$$2x - 3y + 6 = 0$$
$$2(x' + 3) - 3(y' + 4) + 6 = 0 \qquad \text{Substitute } x' + 3 \text{ for } x \text{ and } y' + 4 \text{ for } y.$$
$$2x' + 6 - 3y' - 12 + 6 = 0$$
$$2x' - 3y' = 0 \qquad \text{Transformed equation.}$$

We find some points on each equation.

$2x - 3y + 6 = 0$

x	y
0	2
−3	0
3	4

$2x' - 3y' = 0$

x'	y'
0	0
3	2

The origin of the $x'y'$-axes, (3, 4) on the xy-axes, is on both lines and both lines have the same slope, $\frac{2}{3}$. Therefore both equations represent the same line (see Figure 29).

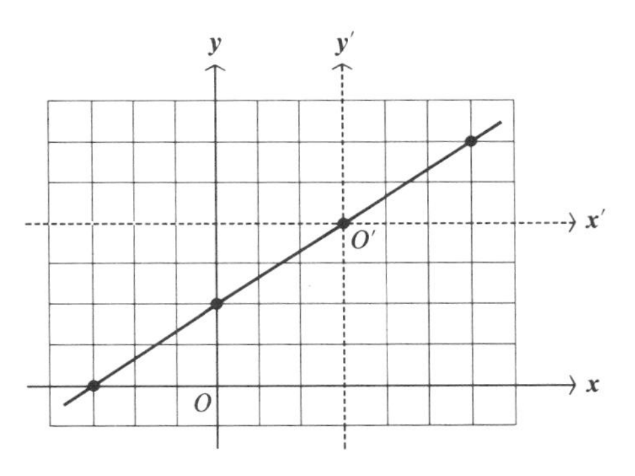

FIGURE 29

EXAMPLE 2 Find a translation of axes that will transform the equations $9x^2 + 4y^2 + 18x - 24y + 9 = 0$ into one in which the coefficients of the first-degree terms are zero.

FIRST SOLUTION If we substitute the values of x and y from Equations 1 in the given equation and collect the coefficients of the various powers of x' and y', we have

$$9x^2 + 4y^2 \qquad + 18x \qquad - 24y \qquad + 9 = 0$$
$$9(x' + h)^2 + 4(y' + k)^2 + 18(x' + h) - 24(y' + k) + 9 = 0.$$

This reduces to

$$9(x')^2 + 4(y')^2 + (18h + 18)x' + (8k - 24)y'$$
$$+ 9h^2 + 4k^2 + 18h - 24k + 9 = 0.$$

Equating the coefficients of x' and y' to zero gives the following.

$$18h + 18 = 0 \qquad \text{and} \qquad 8k - 24 = 0$$
$$h = -1 \qquad \text{and} \qquad k = 3$$

On substituting these values of h and k in the transformed equation, we obtain

$$9x'^2 + 4y'^2 - 36 = 0.$$

This curve is the ellipse

$$\frac{(x')^2}{4} + \frac{(y')^2}{9} = 1,$$

which has its center at the new origin, its major axis on the y'-axis, and semiaxes $a = 3$ and $b = 2$ (see Figure 30).

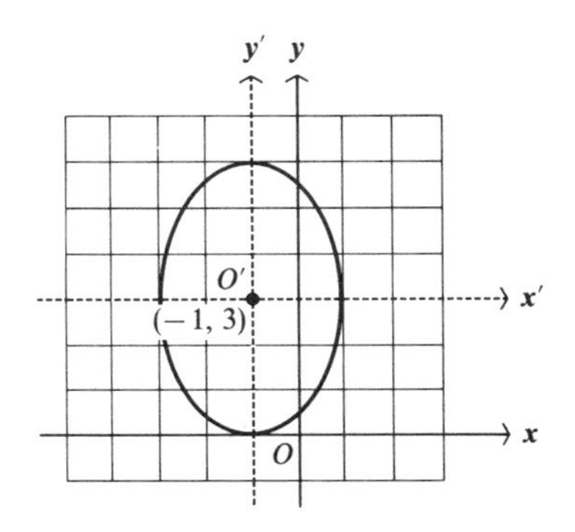

FIGURE 30

SECOND SOLUTION By collecting the terms in x and y and factoring out the coefficients of x^2 and y^2, respectively, we can write the given equation in the form

$$9(x^2 + 2x \qquad) + 4(y^2 - 6y \qquad) = -9.$$

Then by completing the squares within the parentheses, we have

$$9(x^2 + 2x + 1) + 4(y^2 - 6y + 9) = -9 + 9 + 36.$$

Because of the coefficients outside the parentheses, by inserting 1 and 9 inside the parentheses, we actually add 9 and 36, respectively, to the left side of the equation. To preserve the equality, we must add the same numbers to the right side of the equation. The equation further reduces to

$$9(x + 1)^2 + 4(y - 3)^2 = 36.$$

If we now translate the origin by substituting $x + 1 = x'$ and $y - 3 = y'$, we get

$$9(x')^2 + 4(y')^2 = 36.$$

Exercises 9

A

1. Find the new coordinates of the point $(-2, 3)$ when the origin is translated to the point $(3, 1)$.
2. Find the new coordinates of the point $(-3, 4)$ when the origin is translated to the point $(2, -4)$.

In Exercises 3–5, transform the equation by translating the origin to the point indicated. Identify the type of conic section.

3. $x^2 + y^2 - 4x - 6y - 3 = 0$; $(2, 3)$
4. $4x^2 + y^2 + 24x + 2y + 29 = 0$; $(-3, -1)$
5. $4x^2 - y^2 - 12x - 6y + 24 = 0$; $(\frac{3}{2}, -3)$

6. By translation, remove the first-degree term in x and the constant term in the equation $y^2 - 6y + 5 = 4x$ and give the new origin.

B

1. Find the new coordinates of the point $(1, -2)$ when the origin is translated to the point $(1, 4)$
2. Find the new coordinates of the point $(3, -1)$ when the origin is translated to the point $(1, 4)$.

In Exercises 3–5, transform the equation by translating the origin to the point indicated. Identify the type of conic section.

3. $x^2 + y^2 + 4x - 10y + 4 = 0$; $(-2, 5)$
4. $x^2 + 4y^2 + 4x - 8y - 8 = 0$; $(-2, 1)$
5. $2x^2 - 3y^2 - 12x - 12y = 0$; $(3, -2)$
6. By translation, remove the first-degree term in x and the constant term in the equation $x^2 - 4x - 4y - 8 = 0$ and give the new origin.

10 Standard and general forms of the equations of conic sections

The equations of the circle, ellipse, and hyperbola developed in Sections 5, 7, and 8 are called **central conics**. Central conics have one of the coordinate axes as an axis; the origin is the center of the circle, ellipse, and hyperbola.

If a parabola has its vertex at any point (h, k) or if an ellipse or hyperbola has its center at (h, k), and if the principal axis of either curve is parallel to one of the coordinate axes, the equations of the curves may be obtained from the previous equations by use of a translation of coordinates.

When the translation equations

$$x' = x - h \qquad \text{and} \qquad y' = y - h$$

are substituted in the previous equations of the parabola, ellipse, and hyperbola, the following **standard equations** of those curves are obtained.

EQUATIONS OF CONIC SECTIONS (STANDARD FORM)

A. Parabola with vertex at (h, k)

$(y - k)^2 = 4c(x - h)$	Axis parallel to x-axis
$(x - h)^2 = 4c(y - k)$	Axis parallel to y-axis

B. Ellipse with center at (h, k)

$\dfrac{(x - h)^2}{a^2} + \dfrac{(y - k)^2}{b^2} = 1$	Major axis parallel to x-axis

$\dfrac{(y-k)^2}{a^2}+\dfrac{(x-h)^2}{b^2}=1$	Major axis parallel to y-axis
C. Hyperbola with center at (h, k)	
$\dfrac{(x-h)^2}{a^2}-\dfrac{(y-k)^2}{b^2}=1$	Transverse axis parallel to x-axis
$\dfrac{(y-k)^2}{a^2}-\dfrac{(x-h)^2}{b^2}=1$	Transverse axis parallel to y-axis
D. Circle with center at (h, k)	
$(x-h)^2+(y-k)^2=r^2$	r is the radius

EXAMPLE 1 Find the equation of the parabola whose vertex is at $(2, -3)$ and whose focus is at $(2, -1)$.

SOLUTION By plotting the vertex $(2, -3)$ and focus $(2, -1)$ (see Figure 31), it can be seen that the axis of the parabola is parallel to

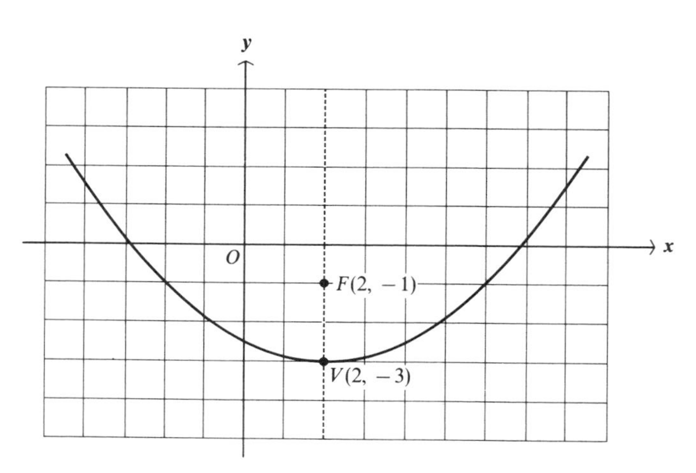

FIGURE 31

the y-axis; the equation is of the form $(x-h)^2 = 4c(y-k)$, with $h = 2$ and $k = -3$. Because the parabola opens upwards, c is positive. The value of c is its distance from the vertex, which is 2. Therefore, the required equation is

$$(x-h)^2 = 4c(y-k)$$

$$(x-2)^2 = 8(y+3) \qquad \text{or} \qquad x^2 - 4x - 8y - 20 = 0.$$

After expanding and rearranging terms, the conic equations take the general form

$$Ax^2 + Cy^2 + Dx + Ey + F = 0, \qquad \textbf{(1)}$$

where A and C are not both zero.

Any equation of the form

$$Ax^2 + Bxy + Cy^2 + Dx + Ey + F = 0, \qquad \textbf{(2)}$$

where A, B, and C are not all zero, is called an equation of the second degree and its graph, if it has one, is a conic section.

Any general second-degree equation can be reduced to a simple standard form of a conic by a suitable rotation and translation of axes. Rotation is used to eliminate the term Bxy. We will not consider rotation, so we shall only consider equations such as Equation 1. To transform Equation 1 to a standard form of a conic, translation is used.

GENERAL EQUATION OF CONIC SECTIONS (NOT HAVING A Bxy TERM)

Every conic with axis or axes parallel to or on a coordinate axis or axes may be represented by an equation of the form

$$Ax^2 + Cy^2 + Dx + Ey + F = 0, \qquad \textbf{(1)}$$

where A and C are not both zero.

If $A = C \neq 0$,	the conic is a circle.
If $AC = 0$,	the conic is a parabola.
If $AC > 0$ and $A \neq C$,	the conic is an ellipse.
If $AC < 0$,	the conic is a hyperbola.

EXAMPLE 2 Transform the equation $4x^2 - 9y^2 + 16x + 18y - 29 = 0$ to standard form, name the type of curve, draw both sets of axes, and plot the curve.

SOLUTION Since $A = 4$ and $C = -9$, $AC < 0$ and the curve is a hyperbola. We reduce the given equation to that of a hyperbola as follows.

$$4x^2 - 9y^2 + 16x + 18y - 29 = 0$$

$$4(x^2 + 4x + \quad) - 9(y^2 - 2y + \quad) = 29$$

$$4(x^2 + 4x + 4) - 9(y^2 - 2y + 1) = 29 + 16 - 9$$

$$4(x + 2)^2 - 9(y - 1)^2 = 36$$

$$\frac{(x + 2)^2}{9} - \frac{(y - 1)^2}{4} = 1$$

When we translate the origin to $(-2, 1)$, the transformed equation is

$$\frac{(x')^2}{9} - \frac{(y')^2}{4} = 1.$$

On the $x'y'$-axes, $a = 3$, $b = 2$. The center of the hyperbola is the origin of the translated axes. The equations of the asymptotes are $y' = \pm\frac{2}{3}x'$. To sketch the hyperbola, in addition to having its vertices and asymptotes, it is helpful to have $c = \sqrt{a^2 + b^2} = \sqrt{13} \doteq 3.6$ and two additional points such as the end points of its latus rectum. The coordinates of these points can be found by $y = b^2/a = \frac{4}{3}$. With this information, draw both sets of axes, the asymptotes, and sketch the curve (see Figure 32).

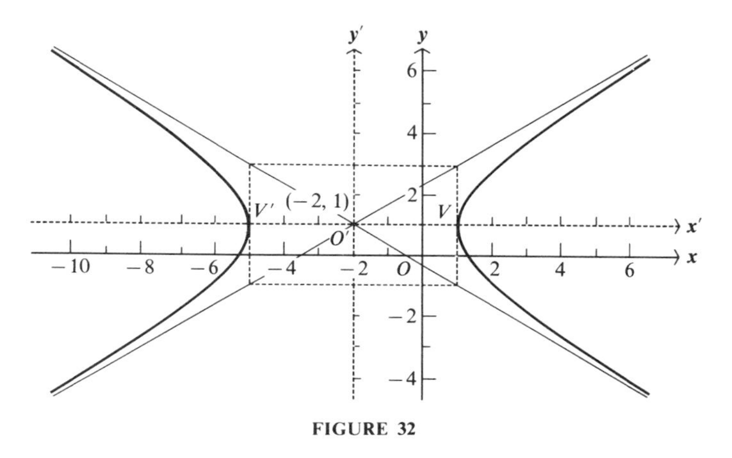

FIGURE 32

EXAMPLE 3 Transform the equation $x^2 + y^2 - 10x + 4y + 20 = 0$ to a standard conic form, name the type of curve, draw both sets of axes, and sketch its graph.

SOLUTION Since $A = C$, the curve is a circle. We reduce the given equation to that of a circle as follows.

$$x^2 + y^2 - 10x + 4y + 20 = 0$$
$$(x^2 - 10x + \quad\;) + (y^2 + 4y + \quad\;) = -20$$
$$(x^2 - 10x + 25) + (y^2 + 4y + 4) = -20 + 25 + 4$$
$$(x - 5)^2 + (y + 2)^2 = 9$$

Of course, we could graph the circle easily from the above equation, but if we translate the origin to $(5, -2)$, the transformed equation is

$$(x')^2 + (y')^2 = 9.$$

The center of the circle is at the translated origin $(5, -2)$ and its radius is $\sqrt{9} = 3$ (see Figure 33).

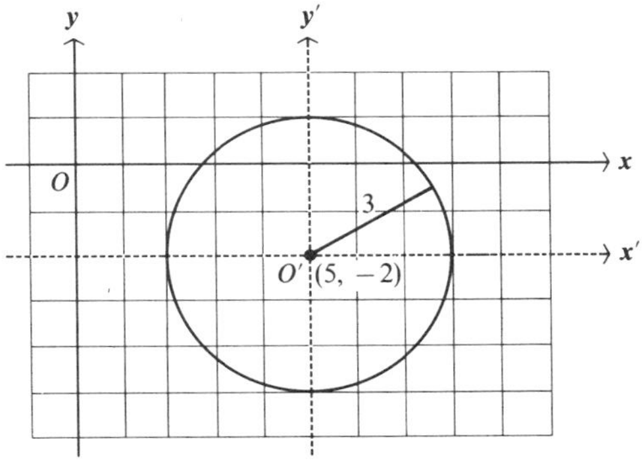

FIGURE 33

EXAMPLE 4 Find the equation of the ellipse with vertices at (4, 4) and (−2, 4) and length of minor axis 4.

SOLUTION By plotting the vertices (see Figure 34), the major axis is seen to be parallel to OX and the center of the ellipse is at (1, 4).

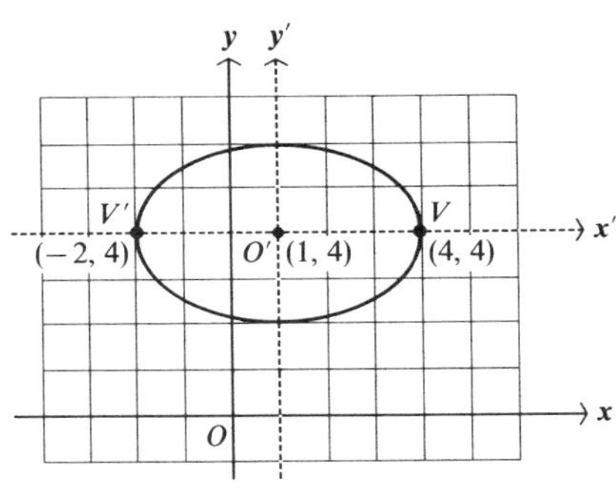

FIGURE 34

Therefore, $h = 1$, $k = 4$, and $a = 3$. Because the length of the minor axis is $2b = 4$, $b = 2$. Therefore, the required equation is

$$\frac{(x-1)^2}{9} + \frac{(y-4)^2}{4} = 1.$$

Exercises 10

A

1. Find an equation of the parabola with vertex (1, 3) and focus (5, 3).
2. Find an equation of the ellipse with vertices at (−4, 3) and (6, 3) and minor axis of length 8.
3. Find an equation of the hyperbola with foci at (−7, −3) and (3, −3) and one vertex at (2, −3).

In Exercises 4–6, simplify each equation by a translation of axes, name the type of curve defined by the equation, draw both sets of axes, and sketch the curve when it exists.

4. $x^2 + 4y^2 - 4x - 8y = 0$
5. $3y^2 + 6y + 6x - 9 = 0$
6. $9x^2 - 16y^2 - 18x + 96y - 279 = 0$

B

1. Find an equation of the parabola with vertex $(-2, -2)$ and focus $(-2, 2)$.
2. Find an equation of the ellipse with center at $(2, 1)$, semiaxes $a = 5$ and $b = 3$, and major axis parallel to the x-axis.
3. Find the equation of the hyperbola with vertices at $(4, 4)$ and $(-2, 4)$ and conjugate axis of length 8.

In Exercises 4–6, simplify each equation by a translation of axes, name the type of curve defined by the equation, draw both sets of axes, and sketch the curve when it exists.

4. $x^2 + 5y^2 + 4x - 10y + 4 = 0$
5. $y^2 - 4x - 6y + 5 = 0$
6. $3x^2 - 3y^2 - 12x + 18y + 12 = 0$

Summary

The **inclination of a straight line** not parallel to the x-axis is defined to be the least angle α measured counterclockwise from the positive x-axis to the line; the inclination of a line parallel to the x-axis is defined to be 0°. The inclination is a positive angle less than 180° or is 0°.

The **slope *m* of a line** is

$$m = \frac{y_2 - y_1}{x_2 - x_1}, \qquad x_2 \neq x_1$$

or $m = \tan \alpha$, where α is the angle of inclination.

The **general equation of a line** is $Ax + By + C = 0$, A and B not both zero. The **point-slope form of the equation of a line** is $y - y_1 = m(x - x_1)$. The **slope-intercept form of the equation of a line** is $y = mx + b$, where b is the y-intercept. The **intercept form of the equation of a line** is $x/a + y/b = 1$, where a is the x-intercept and b is the y-intercept.

Lines parallel to the *x*-axis are of the form $y = b$, where b is the y-intercept. **Lines parallel to the *y*-axis** are of the form $x = a$, where a is the x-intercept.

The **midpoint of a line segment** is the point $(\bar{x}, \bar{y})$ with

$$\bar{x} = \frac{x_2 + x_1}{2} \qquad \text{and} \qquad \bar{y} = \frac{y_2 + y_1}{2}.$$

The **distance between two points** (x_1, y_1) and (x_2, y_2) is

$$d = \sqrt{(x_2 - x_1)^2 + (y_2 - y_1)^2}.$$

Two lines with slopes m_1 and m_2 are **perpendicular** if $m_1 = -1/m_2$, $m_2 = -1/m_1$, or $m_1 m_2 = -1$, provided neither slope is zero. Also, if one slope is zero and the other is undefined, the lines are perpendicular.

The angle ϕ from a line l_1 with slope m_1 to a line l_2 with slope m_2 is given by the formula

$$\tan \phi = \frac{m_2 - m_1}{1 + m_2 m_1},$$

provided neither line is parallel to the y-axis and that the given lines are not perpendicular to each other.

Two points (x, y) and $(x, -y)$ are **symmetric with respect to the x-axis**. Points (x, y) and $(-x, y)$ are **symmetric with respect to the y-axis**. Points (x, y) and $(-x, -y)$ are **symmetric with respect to the origin**.

Conic sections are curves formed by planes intersecting right circular cones. These conic sections include the *circle, parabola, ellipse*, and *hyperbola*. A **circle** is a plane curve consisting of all points at a given distance from a fixed point in the plane. A **parabola** is the set of all points in a plane equidistant from a given fixed point and a given fixed line not containing the given point. An **ellipse** is the set of all points $P(x, y)$ in a plane such that the sum of the distances from P to a pair of fixed points called the foci is a constant. A **hyperbola** is the set of all points $P(x, y)$ in a plane such that the difference between the distances from P to a pair of fixed points called the foci is a constant.

In the following general equations of the conic sections, the point (h, k) represents the center of the circle, the ellipse, and the hyperbola, and the vertex of the parabola. The letter a represents the semimajor axis of an ellipse or the semitransverse axis of a hyperbola and the letter b represents the semiminor axis of an ellipse or the semiconjugate axis of a hyperbola.

Equations of conic sections (standard form)

Circle: $(x - h)^2 + (y - k)^2 = r^2$, where r is the radius.

Parabola: $(y - k)^2 = 4c(x - h)$, with axis parallel to the x-axis.

Parabola: $(x - h)^2 = 4c(y - k)$, with axis parallel to the y-axis.

Ellipse: $$\frac{(x - h)^2}{a^2} + \frac{(y - k)^2}{b^2} = 1,$$

with the major axis parallel to the x-axis.

Ellipse: $$\frac{(y - k)^2}{a^2} + \frac{(x - h)^2}{b^2} = 1,$$

with the major axis parallel to the y-axis.

Hyperbola: $$\frac{(x - h)^2}{a^2} - \frac{(y - k)^2}{b^2} = 1,$$

with the transverse axis parallel to the x-axis.

Hyperbola: $$\frac{(y - k)^2}{a^2} - \frac{(x - h)^2}{b^2} = 1,$$

with the transverse axis parallel to the y-axis.

Every conic with axis or axes parallel to or on a coordinate axis or axes may be represented by a general equation of the form

$$Ax^2 + Cy^2 + Dx + Ey + F = 0$$

where A and C are not both zero.

If $A = C \neq 0$,	the conic is a circle.
If $AC = 0$,	the conic is a parabola.
If $AC > 0$ and $A \neq C$,	the conic is an ellipse.
If $AC < 0$,	the conic is a hyperbola.

Review exercises

A

1. Given triangle ABC with vertices $A(1, -3)$, $B(7, 2)$, and $C(-4, 3)$.
   a. Write the equation of the line through B parallel to side AC.
   b. Find the midpoint of AB.
   c. Find the length of BC.
   d. Use the slope formula to show that angle A is a right angle.
   e. Find angle C.
2. Write the coordinates of the point symmetric to $(-2, 6)$, with respect to the x-axis.

In Exercises 3–6, find an equation of the curve satisfying the given conditions. Write each equation in both standard and general form.

3. Circle with center at $(3, -4)$ and radius 5
4. Parabola with vertex at $(-4, 2)$ and focus at $(-4, 0)$
5. Ellipse with center at $(2, -3)$, semimajor axis 3, semiminor axis 2, and major axis parallel to the y-axis.
6. Hyperbola with center at $(3, 1)$, one vertex at $(3, -2)$, and conjugate axis of length 4

In Exercises 7–10, use translation to transform each equation to a standard conic form, name the type of curve, draw both sets of axes, and sketch the curve.

7. $2x^2 + 2y^2 + 8x - 4y - 15 = 0$
8. $4x^2 - y^2 + 24x + 8y + 16 = 0$
9. $x^2 + 4x - 8y - 36 = 0$
10. $9x^2 + 16y^2 + 18x - 128y + 121 = 0$

B

1. Given triangle ABC with vertices $A(-3, 2)$, $B(3, 2)$, and $C(-1, 8)$.
   a. Write an equation of the line through C parallel to AB.
   b. Find the midpoint of BC.

c. Find the length of AC.

d. Find angle A.

2. Write the coordinates of the point symmetric to $(4, -3)$, with respect to the origin.

In Exercises 3–6, find an equation of the curve satisfying the given conditions. Write each equation in both standard and general form.

3. Circle with center at $(5, -2)$ and radius 3
4. Parabola with vertex at $(3, -1)$ and focus at $(3, 1)$
5. Ellipse with center at $(-4, 1)$ and tangent to both x- and y-axes
6. Hyperbola with center at $(0, 3)$, one vertex at $(4, 3)$, and one focus at $(5, 3)$

In Exercises 7–10, use translation to transform each equation to a standard conic form, name the type of curve, draw both sets of axes, and sketch the curve.

7. $3x^2 + 3y^2 - 12x + 6y + 3 = 0$
8. $9x^2 - 4y^2 + 54x + 24y + 9 = 0$
9. $x^2 + 6x - 8y + 41 = 0$
10. $4x^2 + y^2 - 24x + 8y + 48 = 0$

Diagnostic test

The purpose of this test is to see how well you understand the work in this chapter. Allow yourself approximately 60 minutes to do the test. Solutions to the problems, together with section references, are given in the Answers section at the end of this book. We suggest that you study the sections referred to for the problems you do incorrectly.

1. Given triangle ABC with vertices $A(-4, -2)$, $B(5, -3)$, and $C(1, 2)$.
   a. Write the equation of the line through C perpendicular to AB.
   b. Find the length of AB.
   c. Use the slope formula to show that angle C is a right angle.
   d. Find angle B.

2. Find the equation of each curve satisfying the given conditions. Write each equation in both standard and general form.
   a. Circle with center at $(4, -2)$ and passing through the origin
   b. Parabola with vertex at $(-2, -1)$ and focus at $(-2, 0)$
   c. Ellipse with center at $(3, -2)$ and tangent to both x- and y-coordinate axes

3. By translation, transform each of the following equations to standard form, name the type of curve, draw both sets of axes, and sketch the curve if it exists.

a. $x^2 - 4y^2 - 2x - 20y - 28 = 0$

b. $y^2 - 8x - 6y - 15 = 0$

c. $2x^2 + 2y^2 + 12x - 8y + 18 = 0$

10 Systems of linear and quadratic equations

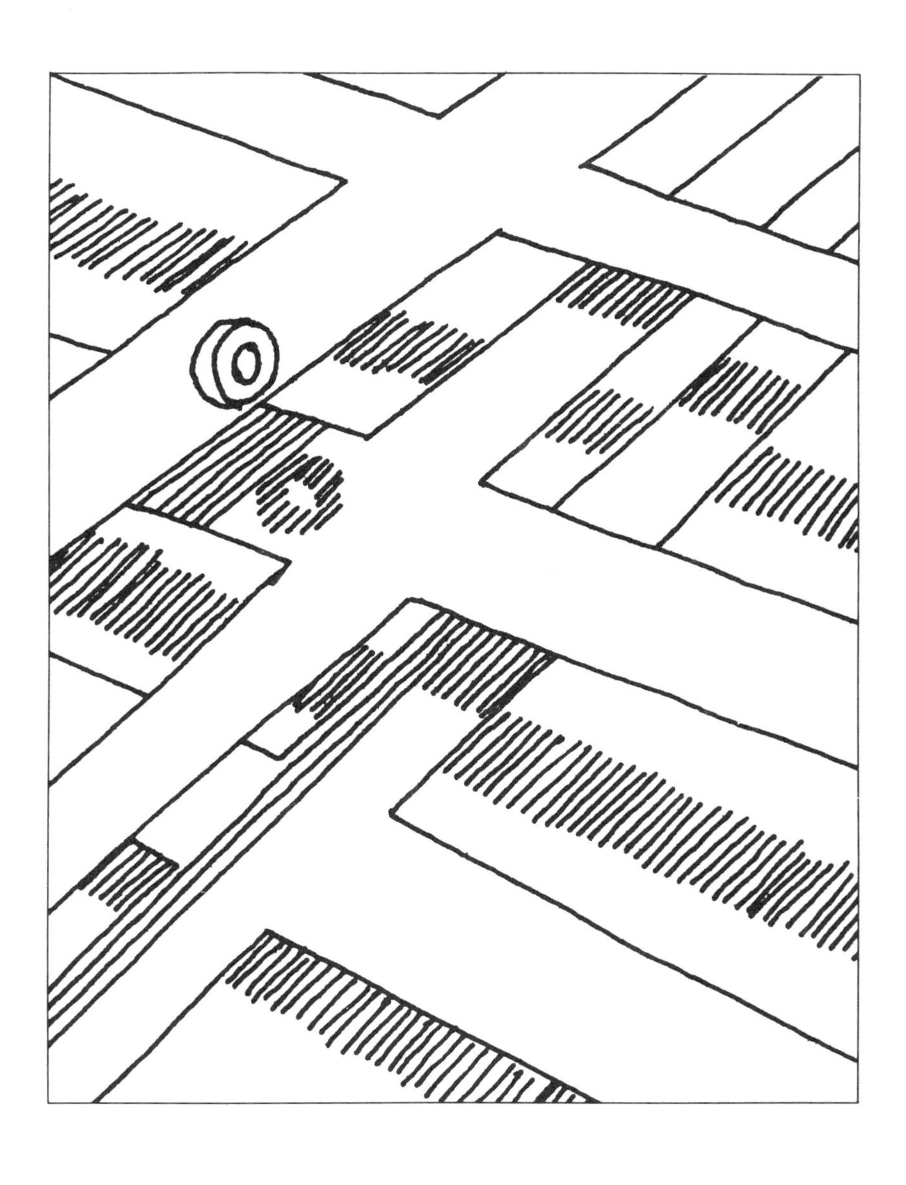

In mathematics and its applications, it is often necessary to work simultaneously with more than one equation or inequality in several variables.

In this chapter, we shall show certain types of systems of equations and inequalities and develop methods for finding their common solutions.

1 Definitions

An equation of the form

$$ax + by = c,$$

where a, b, and c are real number constants and a and b are not both zero is a **linear equation** in x and y.

Suppose we have two linear equations such as

$$a_1 x + b_1 y = c_1$$

and

$$a_2 x + b_2 y = c_2.$$

We call this a system of two equations in two variables. We seek an ordered pair (x, y) which, when substituted in these two equations, will make both equations true. Such a pair of values is a **solution** of the two equations.

2 Graphical method for solving a linear system

In Section 2 of Chapter 6, we graphed linear equations. To solve a system of two linear equations in two unknowns graphically, we first graph each equation. If the graphs intersect, the solution of the system is the point of intersection.

EXAMPLE 1 Solve the system

$$\boxed{\begin{aligned} 2x - 3y &= 2 \\ x + y &= 6 \end{aligned}}$$

graphically.

SOLUTION The graphs of the lines are constructed using their intercepts.

l_1: $2x - 3y = 2$

x	y
0	$-\frac{2}{3}$
1	0

l_2: $x + y = 6$

x	y
0	6
6	0

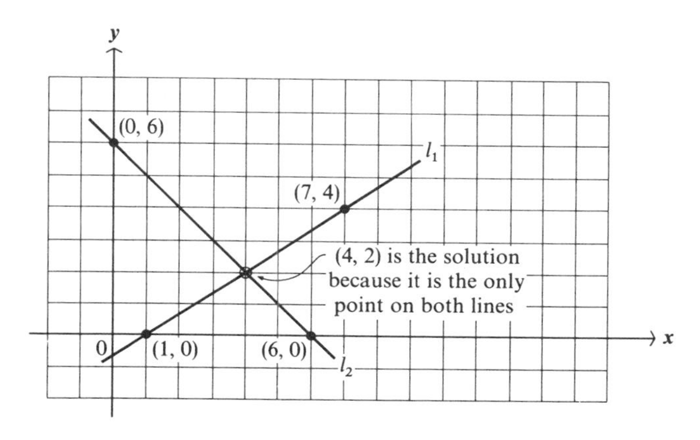

FIGURE 1

Identify the coordinates of their intersection and check them in the given equations.

$$\begin{aligned} 2x - 3y &= 2 \\ 2(4) - 3(2) &= 2 \\ 8 - 6 &= 2 \\ 2 &= 2 \end{aligned} \qquad \begin{aligned} x + y &= 6 \\ (4) + (2) &= 6 \\ 6 &= 6 \end{aligned}$$

Since the coordinates check in both equations, the solution set is $\{(4, 2)\}$.

Two linear equations are said to be consistent and independent if they have one and only one common solution. That is, their graphs are lines which intersect in exactly one point.

EXAMPLE 2 Solve

$$\begin{aligned} 3x + 2y &= 12 \\ 6x + 4y &= 9 \end{aligned}$$

graphically.

SOLUTION The graphs of the lines are constructed using their intercepts.

l_1: $3x + 2y = 12$

x	y
0	6
4	0

l_2: $6x + 4y = 8$

x	y
0	2
$1\frac{1}{3}$	0

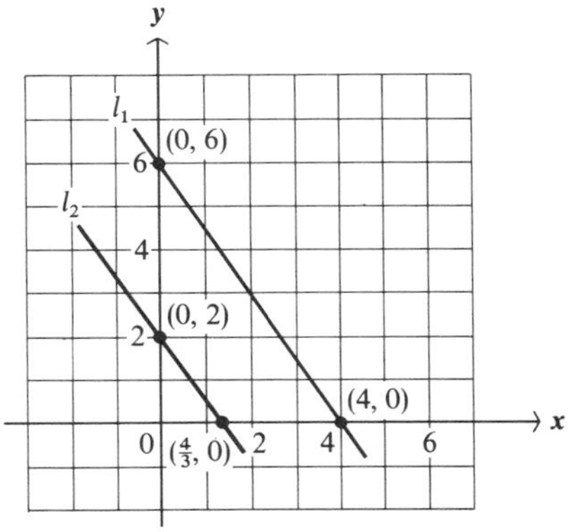

FIGURE 2

The lines appear to be nonintersecting (parallel). Hence, the system has no solution. The validity of this conclusion can be verified by reducing the given equations to slope-intercept form.

l_1: $3x + 2y = 12$ $\qquad$ l_2: $6x + 4y = 8$

$$2y = -3x + 12 \qquad\qquad 4y = -6x + 8$$

$$y = -\frac{3}{2}x + 6 \qquad\qquad y = -\frac{3}{2}x + 2$$

The lines have equal slopes ($m = -\frac{3}{2}$) but different y-intercepts (6 and 2).

Two linear equations are said to be **inconsistent** if they have equal slopes but different y-intercepts. That is, their graphs are nonintersecting lines.

EXAMPLE 3 Solve

$$2x - 4y = 4$$
$$3x - 6y = 6$$

graphically.

SOLUTION When we draw the graphs of these equations, we find that they coincide (see Figure 3). Therefore, every point that lies on one of the lines lies on the other as well. Thus every solution of one of the equations must also satisfy the other one.

When we multiply the first equation by $\frac{1}{2}$ and the second equation by $\frac{1}{3}$, we have:

$$\left.\begin{array}{l} \frac{1}{2}(2x - 4y = 4) = x - 2y = 2 \\ \\ \frac{1}{3}(3x - 6y = 6) = x - 2y = 2 \end{array}\right\} \boxed{\text{Same equation}}$$

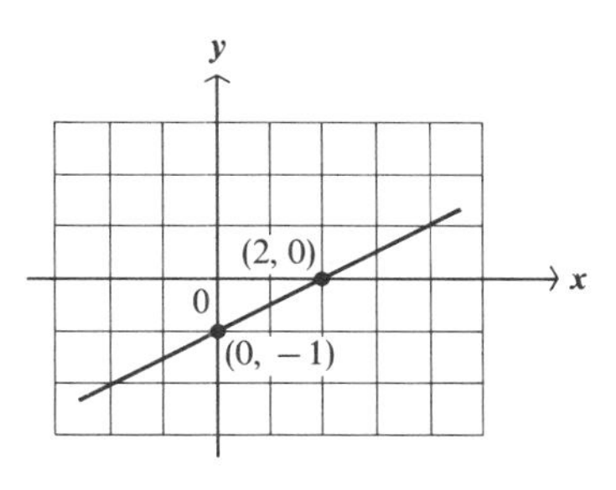

FIGURE 3

Two linear equations are said to be **consistent** and **dependent** if, by suitable algebraic operations, they can be reduced to the same equation; that is, their graphs are identical. When this is the case, every solution of one of the equations is also a solution of the other.

The accuracy obtained in solving a linear system graphically is limited. Plotting accurate points and estimating the coordinates of the solution is often difficult. More efficient methods are explained in the following sections of this chapter. However, the graphical method helps explain the different kinds of systems that exist.

SOLUTION OF A SYSTEM OF LINEAR EQUATIONS

Because the graph of each equation in a system of linear equations is a straight line, one of the three following possibilities must occur.

a. The lines intersect in exactly one point. The system has exactly one solution (see Figure 1).	A **consistent** system.
b. The lines are parallel and nonintersecting. The system has no solution (see Figure 2).	An **inconsistent** system.
c. The lines are identical. The system has an infinite number of solutions (see Figure 3).	A **consistent** and **dependent** system. The equations are **equivalent** equations.

Exercises 2

Solve each system graphically and determine whether the systems are consistent and independent, consistent and dependent, or inconsistent.

A

1. $x + 3y = 1$
$5x - 4y = 24$

2. $x - y = 1$
$2x + 3y = 8$

3. $3x - 2y = 1$
$3x + 2y = 5$

4. $3x - 2y = 6$
$3x + 4y = 6$

5. $2x + 4y = -2$
$5x + 10y = -5$

6. $2x + 3y = 6$
$4x + 6y = 4$

B

1. $x - y = -5$
$x + 2y = -2$

2. $8x - 3y = -2$
$3x - 2y = 1$

3. $2x + 7y = 13$
$5x + y = 16$

4. $3x - 7 = y$
$x + y = 5$

5. $2x - 6y = 6$
$3x - 9y = 9$

6. $-3x + 4y = 12$
$-6x + 8y = 8$

3 Substitution method for solving a linear system

When using the substitution method, we first solve one equation for one variable in terms of the other. We then substitute the resulting expression in the second equation, eliminating one of the variables. This method is illustrated with an example.

EXAMPLE 1 Solve for x and y.

$$2x + y = 8 \tag{1}$$

$$3x + 2y = 2 \tag{2}$$

SOLUTION We first solve Equation 1 for y in terms of x.

$$2x + y = 8$$

$$y = -2x + 8 \tag{3}$$

Next we substitute $-2x + 8$ for y in Equation 2, remove grouping symbols, and simplify.

$$3x + 2y = 2$$
$$3x + 2(-2x + 8) = 2$$
$$3x - 4x + 16 = 2$$
$$-x = -14$$
$$x = 14$$

If we substitute $x = 14$ in Equation 2, we get $y = -20$. Therefore, the solution set for the system is $\{(14, -20)\}$.

In general, solve one of the given equations for whichever unknown gives the simpler expression, avoiding fractions when possible.

When the given systems are inconsistent or dependent, the substitution method eliminates both unknowns and leaves us with statements such as $6 = 2$ or $8 = 8$. If the system is inconsistent, we are left with a *false* statement such as $6 = 2$. If the equations are dependent equations, we are left with a *true* statement such as $8 = 8$.

4 Addition-subtraction method for solving a linear system

In solving a linear system by the addition-subtraction method, we apply suitable multipliers to the equations which make the coefficients of at least one variable numerically equal. Then, by adding or subtracting the resulting equations, we obtain a new equation free of that variable. Before illustrating this method, we shall introduce two new symbols.

We shall use the symbol of implication $\Rightarrow$ in the following manner.

Read "implies"

$$x - 5 = 0 \Rightarrow x = 5$$

To indicate that an equation is multiplied by a specified number, such as 3, we introduce the symbol $3\big]$. For example,

$$3\big]\ x - 2y = 1 \Rightarrow 3x - 6y = 3.$$

EXAMPLE 1 Solve the system

$$\begin{aligned} x + y &= 5 \\ 2x + 5y &= 4. \end{aligned}$$

SOLUTION

$$\begin{aligned} 2\big]\ \ x + y = 5 &\Rightarrow 2x + 2y = 10 \\ -1\big]\ \ 2x + 5y = 4 &\Rightarrow -2x - 5y = -4 \qquad \text{Add equations.} \\ &\ -3y = 6 \\ &\ y = -2 \end{aligned}$$

We next substitute $y = -2$ in the first equation.

$$\begin{aligned} x + y &= 5 \\ x + (-2) &= 5 \\ x &= 7 \end{aligned}$$

Therefore, the solution for the system is the ordered pair $(7, -2)$.

EXAMPLE 2 Solve the system

$$\begin{aligned} 2x + 4y &= 9 \\ 5x - 3y &= 3. \end{aligned}$$

SOLUTION

$$\begin{aligned} 5\,]\quad 2x + 4y = 9 &\Rightarrow 10x + 20y = 45 \\ 2\,]\quad 5x - 3y = 3 &\Rightarrow 10x - 6y = 6 \end{aligned}$$ By subtraction.

$$26y = 39$$

This pair of numbers is found by interchanging the coefficients of x.

$$y = \frac{39}{26} = \frac{3}{2}$$

We now substitute $y = \frac{3}{2}$ in the first equation.

$$2x + 4y = 9$$

$$2x + 4\left(\frac{3}{2}\right) = 9$$

$$2x + 6 = 9$$

$$x = \frac{3}{2}$$

Therefore, the solution set for the system is $\{(\frac{3}{2}, \frac{3}{2})\}$. We could also have solved the system by first multiplying the top equation by 3 and the bottom equation by 4 to eliminate the y term.

EXAMPLE 3 Solve the system

$$13x - 9y = 17$$

$$11x + 6y = 28.$$

SOLUTION

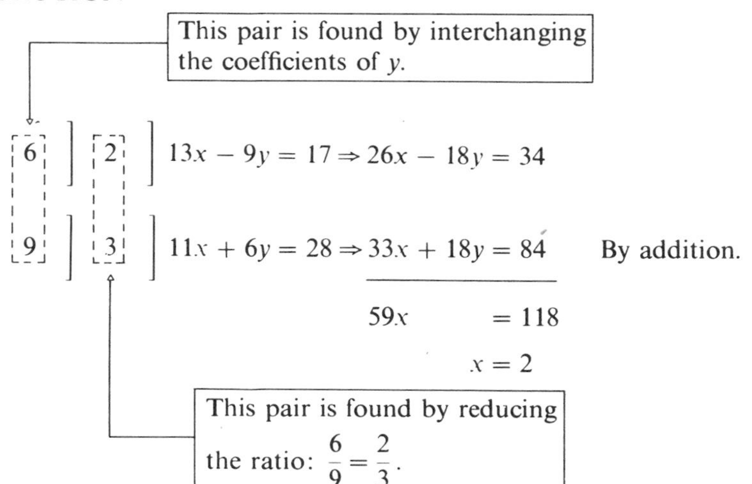

We now substitute $x = 2$ in the first equation.

$$\begin{aligned} 13x - 9y &= 17 \\ 13(2) - 9y &= 17 \\ 26 - 9y &= 17 \\ y &= 1 \end{aligned}$$

Therefore, the solution set for the system is $\{(2, 1)\}$.

EXAMPLE 4 Solve the system

$$\begin{aligned} 2x - 4y &= 2 \\ 3x - 6y &= 5. \end{aligned}$$

SOLUTION These equations have the same slope and are, therefore, equations of parallel lines.

$$\text{Top equation: } m = -\frac{a}{b} = -\frac{2}{-4} = \frac{1}{2}$$

$$\text{Bottom equation: } m = -\frac{a}{b} = -\frac{3}{-6} = \frac{1}{2}$$

The equations are not equivalent because they can not be reduced to the same equation.

If we did not notice that the lines were parallel and tried to solve the system by subtraction, the following false statement would result.

$$3 \left| \; 2x - 4y = 2 \Rightarrow 6x - 12y = 6 \right.$$

$$2 \left| \; 3x - 6y = 5 \Rightarrow 6x - 12y = 10 \right. \qquad \text{By subtraction.}$$

$$0 = -4$$

The system is an inconsistent system.

EXAMPLE 5 Solve the system

$$\begin{aligned} \frac{6}{x} + \frac{1}{y} &= 3 \\ \frac{7}{x} - \frac{3}{y} &= 1. \end{aligned}$$

SOLUTION We multiply the top equation by 3, then add the two equations.

$$3\left]\ \frac{6}{x}+\frac{1}{y}=3 \Rightarrow \frac{18}{x}+\frac{3}{y}=9\right.$$

$$1\left]\ \frac{7}{x}-\frac{3}{y}=1 \Rightarrow \frac{7}{x}-\frac{3}{y}=1\right. \qquad \text{By addition.}$$

$$\frac{25}{x} = 10$$

$$10x = 25$$

$$x = \frac{25}{10} = \frac{5}{2}$$

We now substitute $\frac{5}{2}$ for x in the first equation.

$$\frac{6}{\frac{5}{2}}+\frac{1}{y}=3$$

$$\frac{12}{5}+\frac{1}{y}=\frac{15}{5}$$

$$\frac{1}{y}=\frac{3}{5}$$

$$y=\frac{5}{3}$$

CHECK

Top equation	**Bottom equation**
$\frac{6}{\frac{5}{2}}+\frac{1}{\frac{5}{3}}=3$	$\frac{7}{\frac{5}{2}}-\frac{3}{\frac{5}{3}}=1$
$\frac{12}{5}+\frac{3}{5}=3$	$\frac{14}{5}-\frac{9}{5}=1$
$\frac{15}{5}=3$	$\frac{5}{5}=1$
$3=3$	$1=1$

Therefore, the solution set for the system is $\{(\frac{5}{2}, \frac{5}{3})\}$.

Exercises 4

A In Exercises 1–4, solve each system by the substitution method.

1. $x - 2y = -4$
$3x + y = -5$

2. $3x - 5y = 22$
$6x + y = 22$

3. $3x + 4y = 11$
$3x - 2y = -1$

4. $8a = 7b - 1$
$10a = 11b - 35$

In Exercises 5–10, solve each system by the addition-subtraction method. If the equations are inconsistent or dependent, show that their graphs are parallel lines or the same line.

5. $3x - y = 15$
$2x + y = 10$

6. $x + 3y = 8$
$3x + 9y = 12$

7. $2x - 5y = 11$
$7x + 2y = 6$

8. $2x + 5y = 3$
$3x + 6y = 3$

9. $\frac{1}{x} + \frac{1}{y} = \frac{5}{6}$
$\frac{2}{x} + \frac{3}{y} = 2$

10. $6x - 9y = 12$
$2x - 3y = 4$

B In Exercises 1–4, solve each system by the substitution method.

1. $x = 2y + 1$
$2x - 3y = 3$

2. $x = 2y + 7$
$2x - 3y = 13$

3. $11s - 5t + 4 = 0$
$4s - 3t + 5 = 0$

4. $4x + 3y = 8$
$3x + 2y = 7$

In Exercises 5–10, solve each system by the addition-subtraction method. If the equations are inconsistent or dependent, show that their graphs are parallel lines or the same line.

5. $4x - y = 5$
$x - 3y = -7$

6. $x - 2y = 7$
$2x - 3y = 13$

7. $7x + 3y = 14$
$3x + 2y = 4$

8. $3x - 2y + 1 = 0$
$5x - 3y - 9 = 0$

9. $\frac{1}{x} + \frac{2}{y} = 2$
$\frac{1}{x} + \frac{4}{y} = 5$

10. $2x + 3y = 6$
$4x + 6y = 5$

5 Systems of linear equations in three unknowns

To solve three linear equations in three unknowns, we first eliminate one of the unknowns in any of two pairs of the given equations. We then solve the resulting two equations for the two unknowns by any one of the methods of the preceding sections. When we have found the values of these two unknowns, we substitute their values in one of the given equations and solve for the third variable. As a check, we substitute the values found for the three variables in the remaining two equations to assure ourselves that these equations are also satisfied.

The solution set of three numbers, for a system of three equations in three unknowns, is called an **ordered triple**. *An ordered triple is a set with three members, for which one member is designated as the first, another as the second, and the remaining one as the third member.* For example, the number triple (2, 3, 7) in the solution of Example 1 indicates that $x = 2$, $y = 3$, and $z = 7$.

In special cases, the given equations may be **inconsistent** and have no solution, or they may be **dependent** and have an infinite number of solutions.

EXAMPLE 1 Solve the system

$$2x + y + z = 14 \quad \textbf{(1)}$$

$$3x + 2y + 6z = 54 \quad \textbf{(2)}$$

$$8x + y - z = 12. \quad \textbf{(3)}$$

SOLUTION By inspection, we see that y is an easy variable to eliminate.

First we multiply Equation 1 by 2 and subtract Equation 2. This gives us a new equation, Equation 4.

$$2 \,\Big|\; 2x + y + z = 14 \Rightarrow 4x + 2y + 2z = 28$$

$$3x + 2y + 6z = 54 \quad \textbf{(2)}$$

$$x \qquad -4z = -26 \quad \textbf{(4)}$$

We must now use Equation 3 and either Equation 1 or 2 to eliminate the same variable we eliminated before. We subtract Equation 1 from Equation 3.

$$8x + y - z = 12 \quad \textbf{(3)}$$

$$2x + y + z = 14 \quad \textbf{(1)}$$

$$6x \qquad - 2z = -2$$

We simplify this last equation by dividing it by 2. The resulting equation will be called Equation 5.

$$3x - z = -1 \quad \textbf{(5)}$$

We solve Equations 4 and 5 for x and z and find that $x = 2$ and $z = 7$. Substitute these values of x and z in Equation 1 to find y.

$$2x + y + z = 14 \quad \textbf{(1)}$$

$$2(2) + y + (7) = 14$$

$$y = 3$$

Therefore, the solution for the system is the ordered triple (2, 3, 7). As a check, we substitute these values in Equations 2 and 3.

CHECK

Equation 2	**Equation 3**
$3x + 2y + 6z = 54$	$8x + y - z = 12$
$3(2) + 2(3) + 6(7) = 54$	$8(2) + (3) - (7) = 12$
$6 + 6 + 42 = 54$	$16 + 3 - 7 = 12$
$54 = 54$	$12 = 12$

This same method can be extended to solve systems of any order. However, because of the amount of work involved in solving higher-order systems, solutions are usually carried out by computer.

ECHELON FORM

There are other methods of solving systems of equations. We show another method in the following example.

EXAMPLE 2 Solve the following system.

$$x + 2y - 3z = -7 \quad \textbf{(6)}$$

$$-x + 3y + 2z = -1 \quad \textbf{(7)}$$

$$2x - 5y + z = 16 \quad \textbf{(8)}$$

SOLUTION We shall transform the given system of equations to an **equivalent system** in a special form called **echelon form**. *An equivalent system of linear equations is one having the same solution as the given system.* The equations of the equivalent system in echelon form will consist of:

a. Equation 6 of the given system as its first equation.

b. A new equation with no x-term as its second equation.

c. A new equation with no x- or y-terms as its third equation.

To solve the system:

$$x + 2y - 3z = -7 \quad \textbf{(6)}$$

$$-x + 3y + 2z = -1 \quad \textbf{(7)}$$

$$2x - 5y + z = 16 \quad \textbf{(8)}$$

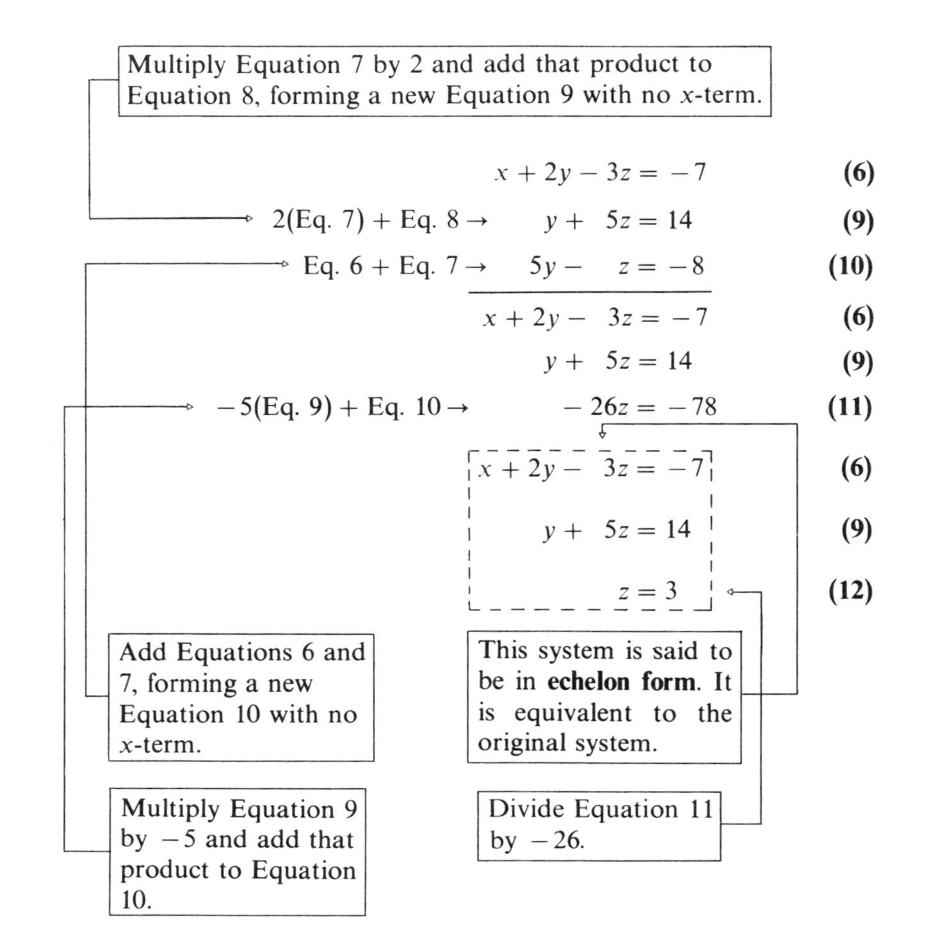

Then, substituting $z = 3$ in Equation 9, we find $y = -1$; and substituting $y = -1$ and $z = 3$ in Equation 6, we find $x = 4$. Therefore, the solution set for the system is $\{(4, -1, 3)\}$.

Once a system of equations has been reduced to echelon form, the solution is easy to obtain. However, to reduce a system to this form often involves insight and tedious work. The example we used was an exception to the general level of difficulty usually encountered. In the next section, we will show how this transformation of a system of equations can be done using **matrix methods**.

Exercises 5

Solve the following systems of equations.

A

1. $$\begin{aligned} x + y - z &= 6 \\ 2x - y - 4z &= 15 \\ 3x + 5y - 2z &= 17 \end{aligned}$$

2. $$\begin{aligned} 2x + 4y - 3z &= -9 \\ 3x + y - 2z &= 4 \\ 5x + 2y + 4z &= 28 \end{aligned}$$

3. $\begin{aligned} 3x + 2y - z &= 7 \\ 4x + 3y + 2z &= 1 \\ 5x - 3y - 6z &= -7 \end{aligned}$

4. $\begin{aligned} u + v + 4w + 3 &= 0 \\ 2u - 3v - w + 4 &= 0 \\ u + 2v - 3w - 1 &= 0 \end{aligned}$

5. $\begin{aligned} x - y &= 7 \\ y + z &= -2 \\ 2x - z &= 4 \end{aligned}$

B

1. $\begin{aligned} x - y + z &= 4 \\ 2x - y - z &= 9 \\ x - 3y + 2z &= 12 \end{aligned}$

2. $\begin{aligned} 2x + 2y + 5z &= 5 \\ 4x + 2y - 2z &= 1 \\ 5x + 4y - z &= 3 \end{aligned}$

3. $\begin{aligned} 2x + 3y - z &= 5 \\ 3x - 2y + 4z &= 2 \\ x + y + 3z &= -1 \end{aligned}$

4. $\begin{aligned} 3r + 4s + 2t &= 14 \\ 2r - 3s + 4t &= 3 \\ 5r + 2s + 3t &= 8 \end{aligned}$

5. $\begin{aligned} 3x + 2y &= 1 \\ 4y - 3z &= 23 \\ 5x - 2z &= -13 \end{aligned}$

6 Solving linear systems by matrices

To show the meaning of the terms used with matrices, define a matrix, and show how matrices are used in the solution of a linear system, we shall solve the following system.

$$x + 2y - 3z = -7 \quad \textbf{(1)}$$

$$-x + 3y + 2z = -1 \quad \textbf{(2)}$$

$$2x - 5y + z = 16 \quad \textbf{(3)}$$

We observe that the coefficients of the variables are important. The variables could be x, y, z, or r, s, t, or any group of three letters.

When synthetic division (Section 6, Chapter 2) was used, the main part of the calculation was done without using the variables. In solving this example, we shall use a method in which the main part of the calculation is done without using the variables. To do this, we introduce some new terminology.

The ordered array of coefficients of the variables in this example can be represented by a matrix.

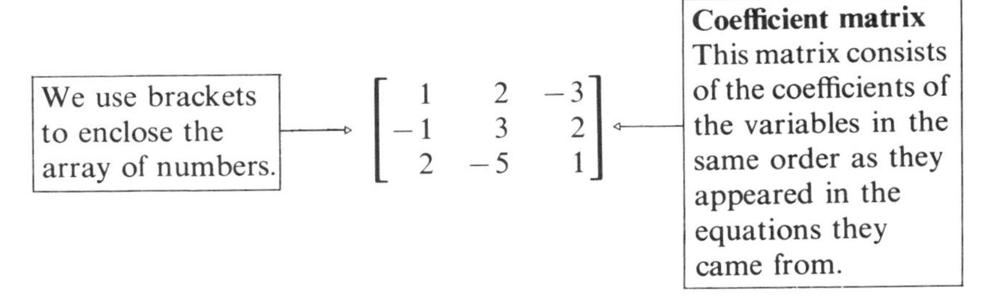

A ***matrix*** *is a rectangular ordered array of numbers.* The numbers are called the **elements** of the matrix. In this course, all numbers in the matrix array will be real.

The elements of a matrix are written in **rows** and **columns**.

An **augmented matrix** of a system of linear equations is one in which the column of constants of the system of linear equations is included to the right of the coefficients. The augmented matrix for our example is written as follows.

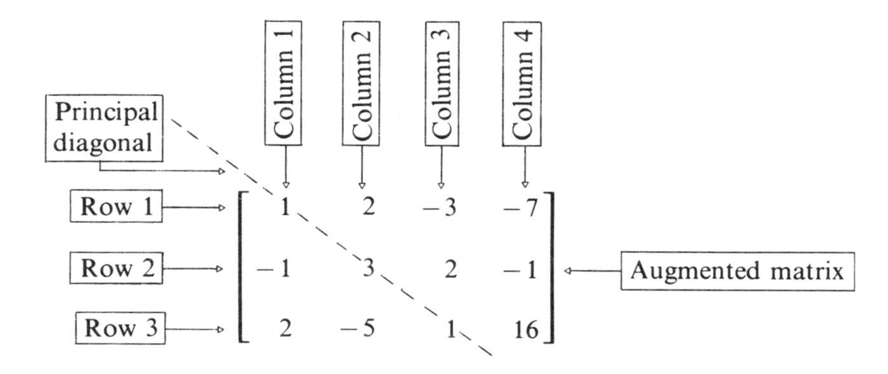

We work with the rows of the augmented matrix just as we worked with the equations of the system; however, we can simplify our previous work by making use of the following theorem.

THEOREM 1

TRANSFORMATIONS OF A MATRIX

Given k a nonzero real number and an augmented matrix of a linear system of equations, the following transformations give a matrix of an equivalent system of linear equations.

1. Any two rows may be interchanged.
2. The elements in any row may be multiplied by k.
3. We may add the elements in any row to k times the corresponding elements of another row.

TO REDUCE AN AUGMENTED MATRIX TO ECHELON FORM:

1. Arrange to have a 1 in the first position of the first row.
2. By multiplying the first row by an appropriate number and adding that product to the next row, get a 0 in the first position of the second row. Then use the same process to get a 0 in the first position of the third row, and so on if there are more rows.

> **3.** Divide the second row of the new matrix by the coefficient of its second term, if that coefficient is not 0 or 1, so that the number in the second position is a 1.
>
> **4.** By the method described in Step 2, get zero in the second position of the third row. Continue this process until only the right two numbers of the last row are either 1 or 0.

We now start with the augmented matrix of the system given in our example and, by using Theorem 1, we complete the solution. The objective is to get zeros below the principal diagonal line.

$$\begin{bmatrix} 1 & 2 & -3 & -7 \\ -1 & 3 & 2 & -1 \\ 2 & -5 & 1 & 16 \end{bmatrix} \Rightarrow \begin{bmatrix} 1 & 2 & -3 & -7 \\ 0 & 5 & -1 & -8 \\ 2 & -5 & 1 & 16 \end{bmatrix} \quad \text{Add one times the first row to the second row.}$$

$$\Rightarrow \begin{bmatrix} 1 & 2 & -3 & -7 \\ 0 & 5 & -1 & -8 \\ 0 & -9 & 7 & 30 \end{bmatrix} \quad \text{Add } -2 \text{ times the first row to the third row.}$$

$$\Rightarrow \begin{bmatrix} 1 & 2 & -3 & -7 \\ 0 & 1 & -\frac{1}{5} & -\frac{8}{5} \\ 0 & -9 & 7 & 30 \end{bmatrix} \quad \text{Multiply row 2 by } \frac{1}{5}.$$

$$\Rightarrow \begin{bmatrix} 1 & 2 & -3 & -7 \\ 0 & 1 & -\frac{1}{5} & -\frac{8}{5} \\ 0 & 0 & \frac{26}{5} & \frac{78}{5} \end{bmatrix} \quad \text{Add 9 times the second row to the third row}$$

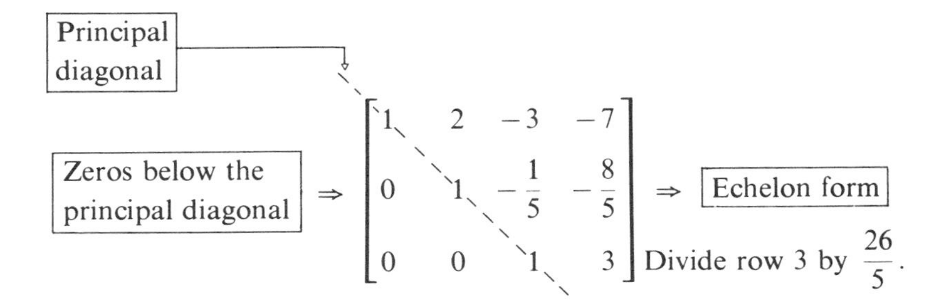

$$\Rightarrow \begin{bmatrix} 1 & 2 & -3 & -7 \\ 0 & 1 & -\frac{1}{5} & -\frac{8}{5} \\ 0 & 0 & 1 & 3 \end{bmatrix}$$

The last matrix is said to be in **echelon form** because it has zeros below the principal diagonal line and the farthest left nonzero element in each row is 1. Using the information contained in this last matrix, we write a linear system of equations that is equivalent to the given system.

$$x + 2y - 3z = -7 \quad (1)$$

$$y - \frac{1}{5}z = -\frac{8}{5} \quad (4)$$

$$z = 3 \quad (5)$$

From Equation 5, we have $z = 3$. Substituting $z = 3$ in Equation 4 gives $y = -1$. Substituting $y = -1$ and $z = 3$ in Equation 1 gives $x = 4$. Therefore, the solution of the system is $(4, -1, 3)$.

This method of solving linear systems can be applied to linear systems of any number of equations. The main objective is to transform the given system to an equivalent system in echelon form. Once this has been done, the solution is easily found.

EXAMPLE 1 Solve the following system by means of matrices.

$$\begin{aligned} 3x + y - 2z - w &= 3 \\ x - y + z - w &= 6 \\ 5x + 2y - z + w &= 7 \\ x + y + z + 3w &= 0 \end{aligned}$$

SOLUTION First write the augmented matrix; then use Theorem 1 to reduce the augmented matrix to echelon form. From the echelon form of the augmented matrix, write an equivalent system of linear equations. Then the solution of the given system is obtained by finding the solution of the equivalent system.

1. $\begin{bmatrix} 3 & 1 & -2 & -1 & 3 \\ 1 & -1 & 1 & -1 & 6 \\ 5 & 2 & -1 & 1 & 7 \\ 1 & 1 & 1 & 3 & 0 \end{bmatrix}$ Augmented matrix.

2. $\Rightarrow \begin{bmatrix} 1 & 1 & 1 & 3 & 0 \\ 1 & -1 & 1 & -1 & 6 \\ 3 & 1 & -2 & -1 & 3 \\ 5 & 2 & -1 & 1 & 7 \end{bmatrix}$ Interchange rows. It is helpful to start with a 1 in the upper left corner. A zero in that row also simplifies the calculation.

3. $\Rightarrow \begin{bmatrix} 1 & 1 & 1 & 3 & 0 \\ 0 & -2 & 0 & -4 & 6 \\ 0 & -2 & -5 & -10 & 3 \\ 0 & -3 & -6 & -14 & 7 \end{bmatrix}$ Add -1(row 1) to row 2. Add -3(row 1) to row 3. Add -5(row 1) to row 4.

4. $\Rightarrow \begin{bmatrix} 1 & 1 & 1 & 3 & 0 \\ 0 & 1 & 0 & 2 & -3 \\ 0 & -2 & -5 & -10 & 3 \\ 0 & -3 & -6 & -14 & 7 \end{bmatrix}$ Divide row 2 by -2.

5. $\Rightarrow \begin{bmatrix} 1 & 1 & 1 & 3 & 0 \\ 0 & 1 & 0 & 2 & -3 \\ 0 & 0 & -5 & -6 & -3 \\ 0 & 0 & -6 & -8 & -2 \end{bmatrix}$ Add 2(row 2) to row 3. Add 3(row 2) to row 4.

6. $\Rightarrow \begin{bmatrix} 1 & 1 & 1 & 3 & 0 \\ 0 & 1 & 0 & 2 & -3 \\ 0 & 0 & 1 & \frac{6}{5} & \frac{3}{5} \\ 0 & 0 & -6 & -8 & -2 \end{bmatrix}$ Divide row 3 by -5.

7. $\Rightarrow \begin{bmatrix} 1 & 1 & 1 & 3 & 0 \\ 0 & 1 & 0 & 2 & -3 \\ 0 & 0 & 1 & \frac{6}{5} & \frac{3}{5} \\ 0 & 0 & 0 & -\frac{4}{5} & \frac{8}{5} \end{bmatrix}$ Add 6(row 3) to row 4.

8. $\Rightarrow \begin{bmatrix} 1 & 1 & 1 & 3 & 0 \\ 0 & 1 & 0 & 2 & -3 \\ 0 & 0 & 1 & \frac{6}{5} & \frac{3}{5} \\ 0 & 0 & 0 & 1 & -2 \end{bmatrix}$ Divide row 4 by $-\frac{4}{5}$. Augmented matrix in echelon form.

From this last matrix, we write an equivalent system of linear equations.

$$\begin{aligned} x + y + z + 3w &= 0 \\ y \quad + 2w &= -3 \\ z + \frac{6}{5}w &= \frac{3}{5} \\ w &= -2 \end{aligned}$$

From the last equation, we have $w = -2$. Substituting $w = -2$ in the third equation gives $z = 3$. Substituting $w = -2$ in the second equation gives $y = 1$. Substituting $w = -2$, $z = 3$, and $y = 1$ in the top equation gives $x = 2$. Therefore, the solution set is $(2, 1, 3, -2)$.

We leave the check of this solution in the original equations for the student.

We will extend the operations on Step 8, the augmented matrix.

8. $\Rightarrow \begin{bmatrix} 1 & 1 & 1 & 3 & 0 \\ 0 & 1 & 0 & 2 & -3 \\ 0 & 0 & 1 & \frac{6}{5} & \frac{3}{5} \\ 0 & 0 & 0 & 1 & -2 \end{bmatrix}$

9. $\Rightarrow \begin{bmatrix} 1 & 1 & 1 & 0 & 6 \\ 0 & 1 & 0 & 0 & 1 \\ 0 & 0 & 1 & 0 & 3 \\ 0 & 0 & 0 & 1 & -2 \end{bmatrix}$ Add -3(row 4) to row 1. Add -2(row 4) to row 2. Add $-\frac{6}{5}$(row 4) to row 3.

10. $\Rightarrow \begin{bmatrix} 1 & 1 & 0 & 0 & 3 \\ 0 & 1 & 0 & 0 & 1 \\ 0 & 0 & 1 & 0 & 3 \\ 0 & 0 & 0 & 1 & -2 \end{bmatrix}$ Add -1(row 3) to row 1.

Principal diagonal

11. $\Rightarrow \begin{bmatrix} 1 & 0 & 0 & 0 & 2 \\ 0 & 1 & 0 & 0 & 1 \\ 0 & 0 & 1 & 0 & 3 \\ 0 & 0 & 0 & 1 & -2 \end{bmatrix}$ Add -1(row 2) to row 1.

Zeros above and below the principal diagonal.

Notice that by extending the operations on the augmented matrix through Step 11, we obtain a matrix with zeros both above and below the principal diagonal. In this latter form, the values of x, y, z, and w are the numbers in the right column reading from top to bottom. That is, $x = 2$, $y = 1$, $z = 3$, and $w = -2$.

The equivalent system thus obtained is

$$x = 2$$
$$y = 1$$
$$z = 3$$
$$w = -2.$$

The matrix method of solving linear systems of equations is easily adapted for use on a computer.

Exercises 6

A Use matrices to solve Exercises 1–5 of the A Exercises of Section 5.

B Use matrices to solve Exercises 1–5 of the B Exercises of Section 5.

7 Determinants

A **determinant** is a real number that is associated with a square ordered array of quantities, called **elements**, symbolizing the sum of certain products of these elements. The square array of elements is enclosed between two vertical bars. The diagonal, from the upper left corner to the lower right corner, is the **principal diagonal**. The diagonal from the lower left corner to the upper right corner is the **secondary diagonal**. A determinant of the second order is a square array of the following type.

$$\begin{vmatrix} a_1 & b_1 \\ a_2 & b_2 \end{vmatrix}$$

By definition,

$$\begin{vmatrix} a_1 & b_1 \\ a_2 & b_2 \end{vmatrix} = a_1 b_2 - a_2 b_1.$$

In words, the value of a second-order determinant is the product of the elements in its principal diagonal minus the product of the elements in its secondary diagonal.

A determinant of the third order is a square array of the type shown in Figure 4. By definition, the value of this third-order determinant is $a_1 b_2 c_3 + a_2 b_3 c_1 + a_3 b_1 c_2 - a_3 b_2 c_1 - a_2 b_1 c_3 - a_1 b_3 c_2$. (Methods for evaluating determinants will be given later in this section.)

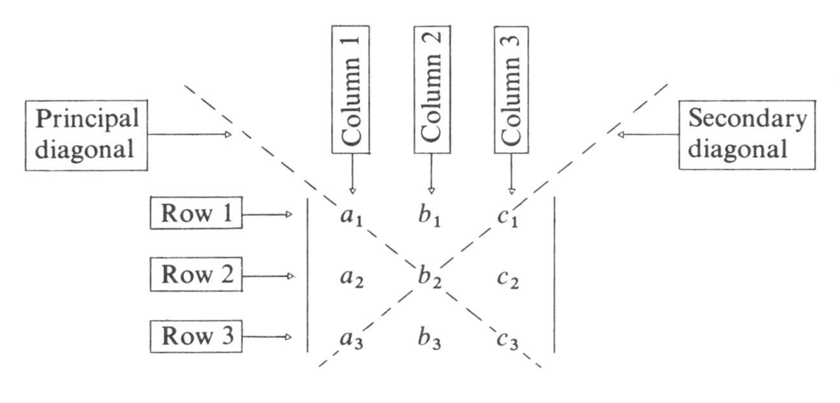

FIGURE 4

A determinant of the fourth order is a square array consisting of four rows and four columns of elements. Determinants of larger orders are defined in the same way.

The vertical lines that enclose the elements of a determinant are *not* absolute value symbols. The value of a determinant may be positive, zero, or negative, depending on the enclosed elements.

EXAMPLE 1 Evaluating second-order determinants.

a. $\begin{vmatrix} 5 & 2 \\ 3 & 4 \end{vmatrix} = (5)(4) - (3)(2) = 20 - 6 = 14$

b. $\begin{vmatrix} 4 & -3 \\ 2 & -5 \end{vmatrix} = (4)(-5) - (2)(-3) = -20 - (-6) = -20 + 6 = -14$

c. $\begin{vmatrix} x & b \\ a & y \end{vmatrix} = xy - ab$

Exercises 7A

Find the value of each of the following determinants.

A

1. $\begin{vmatrix} 8 & 5 \\ 3 & 2 \end{vmatrix}$ **2.** $\begin{vmatrix} 7 & 3 \\ 5 & 4 \end{vmatrix}$

3. $\begin{vmatrix} 4 & -3 \\ -2 & 5 \end{vmatrix}$ **4.** $\begin{vmatrix} -5 & -2 \\ 3 & -4 \end{vmatrix}$

5. $\begin{vmatrix} 5 & -2 \\ 0 & -3 \end{vmatrix}$ **6.** $\begin{vmatrix} x & 3 \\ 5 & 4 \end{vmatrix}$

B

1. $\begin{vmatrix} 7 & 3 \\ 4 & 5 \end{vmatrix}$ **2.** $\begin{vmatrix} 8 & 5 \\ 2 & 3 \end{vmatrix}$

3. $\begin{vmatrix} 3 & -2 \\ 5 & -4 \end{vmatrix}$ **4.** $\begin{vmatrix} -4 & -3 \\ -5 & 2 \end{vmatrix}$

5. $\begin{vmatrix} 0 & -6 \\ 2 & 8 \end{vmatrix}$ **6.** $\begin{vmatrix} a & 2 \\ -3 & 5 \end{vmatrix}$

MINOR OF AN ELEMENT

The minor of an element is the determinant that remains after striking out the row and column in which that element appears (see Example 2).

EXAMPLE 2 Given the determinant

$$\begin{vmatrix} 2 & 0 & -1 \\ 5 & -4 & 6 \\ -3 & 1 & 7 \end{vmatrix}.$$

a. The minor of the element 5 is found as follows.

$$\begin{vmatrix} 2 & 0 & -1 \\ (5) & -4 & 6 \\ -3 & 1 & 7 \end{vmatrix}$$

Strike out the row and column containing 5.

Therefore, the minor of 5 is $\begin{vmatrix} 0 & -1 \\ 1 & 7 \end{vmatrix}$.

b. The minor of 1 is $\begin{vmatrix} 2 & -1 \\ 5 & 6 \end{vmatrix}$.

$$\begin{vmatrix} 2 & 0 & -1 \\ 5 & -4 & 6 \\ -3 & \textcircled{1} & 7 \end{vmatrix} \Rightarrow \begin{vmatrix} 2 & -1 \\ 5 & 6 \end{vmatrix}$$

c. The minor of 7 is $\begin{vmatrix} 2 & 0 \\ 5 & -4 \end{vmatrix}$.

$$\begin{vmatrix} 2 & 0 & -1 \\ 5 & -4 & 6 \\ -3 & 1 & \textcircled{7} \end{vmatrix} \Rightarrow \begin{vmatrix} 2 & 0 \\ 5 & -4 \end{vmatrix}$$

A **determinant of *n*th order** is a determinant having n rows and n columns (where n is a whole number greater than 1). It is represented by

$$D = \begin{vmatrix} a_{11} & a_{12} & \cdots & a_{1j} & \cdots & a_{1n} \\ a_{21} & a_{22} & \cdots & a_{2j} & \cdots & a_{2n} \\ \cdots & \cdots & \cdots & \cdots & \cdots & \cdots \\ a_{i1} & a_{i2} & \cdots & a_{ij} & \cdots & a_{in} \\ \cdots & \cdots & \cdots & \cdots & \cdots & \cdots \\ a_{n1} & a_{n2} & \cdots & a_{nj} & \cdots & a_{nn} \end{vmatrix}.$$

THEOREM 2 The value of a determinant is the sum of the products obtained by multiplying each element of a column (or row) by its minor with a properly prefixed sign. The sign to be used with the element in the ith row and jth column is $(-1)^{i+j}$. This is called *expanding by minors.*

Since $(-1)^{i+j}$ is $+1$ or -1 according as $i+j$ is even or odd, we have the following pattern of signs corresponding to the position of the elements of the determinant.

$$\begin{vmatrix} + & - & + & - & \cdots \\ - & + & - & + & \cdots \\ + & - & + & - & \cdots \\ - & + & - & + & \cdots \\ \cdot & \cdot & \cdot & \cdot & \cdots \end{vmatrix}$$

Notice in the $+$, $-$, grid that the upper left position is $+$ and, when starting at that position and moving in a rectangular direction, the signs always alternate.

EXAMPLE 3 Expand

$$\begin{vmatrix} a_1 & b_1 & c_1 \\ a_2 & b_2 & c_2 \\ a_3 & b_3 & c_3 \end{vmatrix}$$

by minors according to the elements of the first column.

SOLUTION

$$\begin{aligned} \begin{vmatrix} a_1 & b_1 & c_1 \\ a_2 & b_2 & c_2 \\ a_3 & b_3 & c_3 \end{vmatrix} &= a_1 \begin{vmatrix} b_2 & c_2 \\ b_3 & c_3 \end{vmatrix} - a_2 \begin{vmatrix} b_1 & c_1 \\ b_3 & c_3 \end{vmatrix} + a_3 \begin{vmatrix} b_1 & c_1 \\ b_2 & c_2 \end{vmatrix} \\ &= a_1(b_2c_3 - b_3c_2) - a_2(b_1c_3 - b_3c_1) \\ &\quad + a_3(b_1c_2 - b_2c_1) \\ &= a_1b_2c_3 - a_1b_3c_2 - a_2b_1c_3 + a_2b_3c_1 \\ &\quad + a_3b_1c_2 - a_3b_2c_1 \end{aligned}$$

EXAMPLE 4 Expand

$$\begin{vmatrix} 2 & 3 & -5 \\ -1 & 2 & 3 \\ 0 & 5 & 6 \end{vmatrix}$$

by minors and find its value.

SOLUTION Because the first column contains a zero element, we expand by column 1. When zero is multiplied by its minor, that product is zero, thus shortening the calculation.

$$\begin{aligned} \begin{vmatrix} 2 & 3 & -5 \\ -1 & 2 & 3 \\ 0 & 5 & 6 \end{vmatrix} &= 2 \begin{vmatrix} 2 & 3 \\ 5 & 6 \end{vmatrix} - (-1) \begin{vmatrix} 3 & -5 \\ 5 & 6 \end{vmatrix} + 0 \begin{vmatrix} 3 & -5 \\ 2 & 3 \end{vmatrix} \\ &= 2(12 - 15) + 1(18 + 25) + 0 \\ &= 2(-3) + 43 \\ &= -6 + 43 \\ &= 37 \end{aligned}$$

EVALUATING DETERMINANTS

THEOREM 3 Transformations of a determinant. Let k be any nonzero real number.

1. If any two columns (or rows) of a determinant are interchanged, the sign of the determinant is changed.
2. If each element of any column (or row) is multiplied by a number k, the value of the determinant is multiplied by k.

Conversely, if a common factor k is factored from each element of any column (or row), the value of the determinant is divided by k.

3. The value of a determinant is not changed if, to each element of one column (or row), there is added k times the corresponding element of another column (or row).
4. A determinant with two identical or proportional columns (or rows) has value zero.

EXAMPLE 5 Evaluate

$$\begin{vmatrix} 3 & 5 & -2 & 3 \\ 1 & 4 & 2 & 1 \\ 2 & -6 & -7 & 2 \\ -4 & 5 & 8 & -4 \end{vmatrix}.$$

SOLUTION The first and fourth columns are identical. Therefore, the value of the determinant is zero (see Theorem 3).

EXAMPLE 6 Evaluate

$$\begin{vmatrix} 2 & 3 & -2 \\ 3 & 0 & 0 \\ -5 & 0 & 1 \end{vmatrix}.$$

SOLUTION Because the second row has two zeros, we expand by that row.

$$\begin{vmatrix} 2 & 3 & -2 \\ 3 & 0 & 0 \\ -5 & 0 & 1 \end{vmatrix} = -3\begin{vmatrix} 3 & -2 \\ 0 & 1 \end{vmatrix} = -3(3-0) = -9$$

EXAMPLE 7 Evaluate

$$\begin{vmatrix} 3 & 2 & 3 & -2 \\ 6 & 3 & 4 & 5 \\ -9 & -1 & -2 & 1 \\ 6 & 5 & 1 & 6 \end{vmatrix}.$$

SOLUTION Our objective is to transform the determinant into an equivalent determinant having three zeros in one column (or row) and replace as many other elements with zeros as is convenient.

$$\begin{vmatrix} 3 & 2 & 3 & -2 \\ 6 & 3 & 4 & 5 \\ -9 & -1 & -2 & 1 \\ 6 & 5 & 1 & 6 \end{vmatrix}$$

$$= 3\begin{vmatrix} 1 & 2 & 3 & -2 \\ 2 & 3 & 4 & 5 \\ -3 & -1 & -2 & 1 \\ 2 & 5 & 1 & 6 \end{vmatrix}$$ Factor 3 from each element in column 1.

$$= 3\begin{vmatrix} 1 & 0 & 0 & 0 \\ 2 & -1 & -2 & 9 \\ -3 & 5 & 7 & -5 \\ 2 & 1 & -5 & 10 \end{vmatrix}$$

Add -2 times column 1 to column 2.
Add -3 times column 1 to column 3.
Add 2 times column 1 to column 4.

The last expression shows the value of the determinant to be

$$(3)(1)\begin{vmatrix} -1 & -2 & 9 \\ 5 & 7 & -5 \\ 1 & -5 & 10 \end{vmatrix}$$

$$= (3)\left\{+(-1)\begin{vmatrix} 7 & -5 \\ -5 & 10 \end{vmatrix} - 5\begin{vmatrix} -2 & 9 \\ -5 & 10 \end{vmatrix} + 1\begin{vmatrix} -2 & 9 \\ 7 & -5 \end{vmatrix}\right\}$$

$$= 3\{-1(70-25) - 5(-20+45) + 1(10-63)\}$$

$$= 3\{(-45 - 125 - 53)\}$$

$$= 3(-223)$$

$$= -669$$

Exercises 7B

Find the value of each determinant.

A

1. $\begin{vmatrix} 1 & 0 & -5 \\ 3 & 4 & 2 \\ -1 & 2 & -2 \end{vmatrix}$

2. $\begin{vmatrix} 1 & 0 & -4 \\ 2 & 6 & 4 \\ -1 & 3 & -3 \end{vmatrix}$

3. $\begin{vmatrix} 0 & 3 & 0 \\ 2 & 4 & 8 \\ 5 & 3 & 1 \end{vmatrix}$

4. $\begin{vmatrix} 0 & -4 & 2 \\ 1 & 3 & 5 \\ 5 & 0 & 0 \end{vmatrix}$

5. $\begin{vmatrix} 2 & 1 & 3 & 4 \\ 5 & 2 & -1 & 2 \\ 4 & 3 & 5 & 1 \\ 1 & 2 & 2 & 1 \end{vmatrix}$

6. $\begin{vmatrix} 3 & 1 & 0 & 2 \\ 2 & 2 & 1 & -1 \\ 1 & 3 & 1 & 2 \\ 3 & 0 & 2 & 1 \end{vmatrix}$

7. The area of a triangle can be found by using the formula

$$A = \frac{1}{2}\begin{vmatrix} x_1 & y_1 & 1 \\ x_2 & y_2 & 1 \\ x_3 & y_3 & 1 \end{vmatrix}$$

where (x_1, y_1), (x_2, y_2), and (x_3, y_3) are the vertices of the triangle taken in order counterclockwise around the triangle. Use this formula to find the area of the triangle whose vertices are $A(-3, -4)$, $B(2, 1)$, and $C(-4, 4)$.

B

1. $\begin{vmatrix} 1 & 0 & -2 \\ 2 & -3 & 5 \\ -3 & 4 & 6 \end{vmatrix}$

2. $\begin{vmatrix} 2 & -5 & 1 \\ 3 & -1 & 2 \\ 3 & -1 & 2 \end{vmatrix}$

3. $\begin{vmatrix} 1 & 2 & 1 \\ 0 & 3 & 0 \\ 4 & 2 & 4 \end{vmatrix}$

4. $\begin{vmatrix} 0 & -5 & 0 \\ 2 & 3 & 4 \\ 3 & 6 & 9 \end{vmatrix}$

5. $\begin{vmatrix} 4 & 2 & 5 & 2 \\ 7 & 3 & 1 & 3 \\ 2 & 1 & 4 & 2 \\ 1 & 2 & 3 & 4 \end{vmatrix}$

6. $\begin{vmatrix} 2 & 1 & 3 & 5 \\ 0 & 2 & 3 & 4 \\ 1 & 3 & 5 & 1 \\ 2 & 3 & 5 & 4 \end{vmatrix}$

7. Use the formula given in Exercise 7 of Part A to find the area of the triangle whose vertices are $A(-2, -1)$, $B(3, 2)$, and $C(-1, 4)$.

---

8 Cramer's rule

Determinants have many applications, one of which is their use in solving systems of linear equations.

Consider the following system of two equations in two unknowns.

$$\begin{aligned} a_1 x + b_1 y &= c_1 \\ a_2 x + b_2 y &= c_2 \end{aligned} \tag{1}$$

We first eliminate y and then x by the addition-subtraction method and obtain the two equations

$$\begin{aligned} (a_1 b_2 - a_2 b_1)x &= c_1 b_2 - c_2 b_1 \\ (a_1 b_2 - a_2 b_1)y &= a_1 c_2 - a_2 c_1. \end{aligned} \tag{2}$$

If the number $a_1 b_2 - a_2 b_1 \neq 0$ and we divide each equation by it, we have

$$x = \frac{c_1 b_2 - c_2 b_1}{a_1 b_2 - a_2 b_1} \quad \text{and} \quad y = \frac{a_1 c_2 - a_2 c_1}{a_1 b_2 - a_2 b_1}. \tag{3}$$

Notice that the denominators in Equations 3 are the expansions of the determinant

$$\begin{vmatrix} a_1 & b_1 \\ a_2 & b_2 \end{vmatrix}.$$

The numerator for the value of x differs from the denominator only in that a_1 and a_2 are replaced by c_1 and c_2, respectively, and the numerator for the value of y differs from the denominator only in that b_1 and b_2 are replaced by c_1 and c_2. Therefore, we may write Equations 3 in the following form.

SOLUTION OF A SECOND-ORDER SYSTEM OF EQUATIONS

Given the system

$$\begin{aligned} a_1x + b_1y &= c_1 \\ a_2x + b_2y &= c_2 \end{aligned}, \qquad \text{where} \qquad \begin{vmatrix} a_1 & b_1 \\ a_2 & b_2 \end{vmatrix} \neq 0.$$

Then

$$x = \frac{\begin{vmatrix} c_1 & b_1 \\ c_2 & b_2 \end{vmatrix}}{\begin{vmatrix} a_1 & b_1 \\ a_2 & b_2 \end{vmatrix}} \qquad \text{and} \qquad y = \frac{\begin{vmatrix} a_1 & c_1 \\ a_2 & c_2 \end{vmatrix}}{\begin{vmatrix} a_1 & b_1 \\ a_2 & b_2 \end{vmatrix}}. \qquad \textbf{(4)}$$

We represent the determinants in 4 as follows.

$$\begin{vmatrix} a_1 & b_1 \\ a_2 & b_2 \end{vmatrix} = D, \qquad \begin{vmatrix} c_1 & b_1 \\ c_2 & b_2 \end{vmatrix} = D_x, \qquad \text{and} \qquad \begin{vmatrix} a_1 & c_1 \\ a_2 & c_2 \end{vmatrix} = D_y$$

Then Equations 4 take on a shorter form.

$$x = \frac{D_x}{D} \qquad \text{and} \qquad y = \frac{D_y}{D} \qquad \textbf{(5)}$$

Equations 4 constitute what is known as **Cramer's Rule** for the solution of a system of two linear equations in two variables.

Before applying Cramer's Rule to a system of equations, arrange the equations in **standard form**, $ax + by = c$.

EXAMPLE 1 Use Cramer's Rule to solve the system

$$\begin{aligned} 4x + 5y &= 13 \\ 2x - 3y &= -21. \end{aligned}$$

SOLUTION The system is already arranged in standard form. Use Cramer's Rule.

When solving for x, the constants c_1 and c_2 are placed in the first position.

$$x = \frac{\begin{vmatrix} 13 & 5 \\ -21 & -3 \end{vmatrix}}{\begin{vmatrix} 4 & 5 \\ 2 & -3 \end{vmatrix}} = \frac{-39 + 105}{-12 - 10} = \frac{66}{-22} = -3$$

When solving for y, the constants c_1 and c_2 are placed in the second position.

$$y = \frac{\begin{vmatrix} 4 & 13 \\ 2 & -21 \end{vmatrix}}{\begin{vmatrix} 4 & 5 \\ 2 & -3 \end{vmatrix}} = \frac{-84 - 26}{-12 - 10} = \frac{-110}{-22} = 5$$

Therefore, the solution of the system is $(-3, 5)$.

NATURE OF TWO LINEAR EQUATIONS

If	Graphs of lines that represent the equations	Nature of equations	Nature of solution
$D \neq 0$	Intersecting in one point	Consistent and independent equations	Unique
$D = 0$ and either D_x or $D_y \neq 0$	Parallel lines	Inconsistent equations	No solution
$D = 0$ and $D_x = D_y = 0$	Coincident lines	Consistent and dependent equations	Infinitely many solutions

Cramer's Rule can be used to solve systems of equations of higher order. In Example 2, we use Cramer's Rule to solve a third-order system of equations.

EXAMPLE 2 Solve the system

$$\begin{aligned} x - 3y + 2z &= 3 \\ 3x - 4y + 2z &= -2 \\ x + 5y - z &= -1. \end{aligned}$$

SOLUTION

$$\begin{cases} x - 3y + 2z = 3 \\ 3x - 4y + 2z = -2 \\ x + 5y - z = -1 \end{cases}$$

Column of constants replaces the column of constants of the unknown.

$$x = \frac{\begin{vmatrix} 3 & -3 & 2 \\ -2 & -4 & 2 \\ -1 & 5 & -1 \end{vmatrix}}{\begin{vmatrix} 1 & -3 & 2 \\ 3 & -4 & 2 \\ 1 & 5 & -1 \end{vmatrix}} = \frac{D_x}{D}, \quad y = \frac{\begin{vmatrix} 1 & 3 & 2 \\ 3 & -2 & 2 \\ 1 & -1 & -1 \end{vmatrix}}{\begin{vmatrix} 1 & -3 & 2 \\ 3 & -4 & 2 \\ 1 & 5 & -1 \end{vmatrix}} = \frac{D_y}{D},$$

$$z = \frac{\begin{vmatrix} 1 & -3 & 3 \\ 3 & -4 & -2 \\ 1 & 5 & -1 \end{vmatrix}}{\begin{vmatrix} 1 & -3 & 2 \\ 3 & -4 & 2 \\ 1 & 5 & -1 \end{vmatrix}} = \frac{D_z}{D}$$

$$D = \begin{vmatrix} 1 & -3 & 2 \\ 3 & -4 & 2 \\ 1 & 5 & -1 \end{vmatrix} = +(1)\begin{vmatrix} -4 & 2 \\ 5 & -1 \end{vmatrix} - (-3)\begin{vmatrix} 3 & 2 \\ 1 & -1 \end{vmatrix} + (2)\begin{vmatrix} 3 & -4 \\ 1 & 5 \end{vmatrix}$$
$$= 1(4 - 10) + 3(-3 - 2) + 2(15 + 4) = 17$$

$$D_x = \begin{vmatrix} 3 & -3 & 2 \\ -2 & -4 & 2 \\ -1 & 5 & -1 \end{vmatrix} = +(3)\begin{vmatrix} -4 & 2 \\ 5 & -1 \end{vmatrix} - (-2)\begin{vmatrix} -3 & 2 \\ 5 & -1 \end{vmatrix} + (-1)\begin{vmatrix} -3 & 2 \\ -4 & 2 \end{vmatrix}$$
$$= 3(4 - 10) + 2(3 - 10) - 1(-6 + 8)$$
$$= -34$$

$$D_y = \begin{vmatrix} 1 & 3 & 2 \\ 3 & -2 & 2 \\ 1 & -1 & -1 \end{vmatrix} = -(3)\begin{vmatrix} 3 & 2 \\ 1 & -1 \end{vmatrix} + (-2)\begin{vmatrix} 1 & 2 \\ 1 & -1 \end{vmatrix} - (-1)\begin{vmatrix} 1 & 2 \\ 3 & 2 \end{vmatrix}$$

$$= -3(-3-2) - 2(-1-2) + 1(2-6) = 17$$

$$D_z = \begin{vmatrix} 1 & -3 & 3 \\ 3 & -4 & -2 \\ 1 & 5 & -1 \end{vmatrix} = +(3)\begin{vmatrix} 3 & -4 \\ 1 & 5 \end{vmatrix} - (-2)\begin{vmatrix} 1 & -3 \\ 1 & 5 \end{vmatrix} + (-1)\begin{vmatrix} 1 & -3 \\ 3 & -4 \end{vmatrix}$$

$$= 3(15+4) + 2(5+3) - 1(-4+9) = 68$$

$$x = \frac{D_x}{D} = \frac{-34}{17} = -2, \qquad y = \frac{D_y}{D} = \frac{17}{17} = 1, \qquad z = \frac{D_z}{D} = \frac{68}{17} = 4$$

Therefore, the solution is $(-2, 1, 4)$.

Exercises 8

Use Cramer's Rule in solving each of the following systems of equations.

A

1. $3x + 2y = 3$, $3x - 2y = -9$

2. $2x - y + 5 = 0$, $x + 2y - 5 = 0$

3. $2x + y = 19$, $x + 2y = 8$

4. $x - 1 = 4y$, $2x - 7 = -y$

5. $-3x + 6y = 6$, $-4x + 8y = 4$

6. $x + y - z = 9$, $x - y + z = -3$, $x - y - z = 1$

7. $A + B + 3C = 4$, $3A - 2B + C = 10$, $-2A + 3B + 4C = -8$

8. $x - y = 5$, $y + z = 1$, $x - z = -2$

B

1. $x - y = -5$, $x + 2y = -2$

2. $8x - 3y = -2$, $3x - 2y = 1$

3. $2x + 7y = 13$, $5x + y = 16$

4. $3x - 7 = y$, $5y + 7 = 4x$

5. $-3x + 4y = 12$, $6x - 8y = -8$

6. $x - y + z = 4$, $2x - y - z = 9$, $x - 3y + 2z = 12$

7. $2x + 3y - z = 5$, $3x - 2y + 4z = 2$, $x + y + 3z = -1$

8. $3x + 2y = 1$, $4y - 3z = 23$, $5x - 2z = -13$

9 Homogeneous linear equations

A linear equation in which the constant term is zero is called a **homogeneous linear equation**. Any homogeneous linear equation has solution (0, 0, 0, ...). The solution of zeros for all the variables is called the **trivial solution**. Consider the equation

$$ax + by + cz + dw = 0,$$

where a, b, c, and d are constants and x, y, z, and w are variables. It is obvious that when each of the variables has a value of zero, the equation is a true statement.

The following can be proved in more advanced mathematics texts.

> A system of n homogeneous linear equations in n unknowns can have no solution except the trivial solution (0, 0, 0, ...) unless the determinant of its coefficients is zero. If the determinant of the coefficients is zero, the homogeneous system has an infinite number of nontrivial solutions.

EXAMPLE 1 Solve the system

$$\begin{aligned} 2x + 3y &= 0 \\ x - y &= 0. \end{aligned}$$

SOLUTION

$$D = \begin{vmatrix} 2 & 3 \\ 1 & -1 \end{vmatrix} = -2 - 3 = -5$$

Since $D \neq 0$, the *only* solution is (0, 0).

EXAMPLE 2 Solve the system

$$x + 2y + 3z = 0 \qquad (1)$$

$$5x - y - z = 0 \qquad (2)$$

$$3x - 5y - 7z = 0. \qquad (3)$$

SOLUTION Since

$$D = \begin{vmatrix} 1 & 2 & 3 \\ 5 & -1 & -1 \\ 3 & -5 & -7 \end{vmatrix} = 0,$$

the system has nontrivial solutions.

We now reduce the augmented matrix to echelon form.

$$\begin{bmatrix} 1 & 2 & 3 & 0 \\ 5 & -1 & -1 & 0 \\ 3 & -5 & -7 & 0 \end{bmatrix}$$

$$\Rightarrow \begin{bmatrix} 1 & 2 & 3 & 0 \\ 0 & -11 & -16 & 0 \\ 0 & -11 & -16 & 0 \end{bmatrix} \quad \begin{matrix} \text{Add } -5(\text{row } 1) \text{ to row } 2. \\ \text{Add } -3(\text{row } 1) \text{ to row } 3. \end{matrix}$$

$$\Rightarrow \begin{bmatrix} 1 & 2 & 3 & 0 \\ 0 & -11 & -16 & 0 \\ 0 & 0 & 0 & 0 \end{bmatrix} \quad \text{Add } -1(\text{row } 2) \text{ to row } 3.$$

$$\Rightarrow \begin{bmatrix} 1 & 2 & 3 & 0 \\ 0 & 1 & \frac{16}{11} & 0 \\ 0 & 0 & 0 & 0 \end{bmatrix} \quad \text{Echelon form.}$$

From the last matrix, we obtain the equations

$$x + 2y + 3z = 0$$

$$y + \frac{16}{11} z = 0$$

which is a system of equations equivalent to the original system. The row of zeros does not give an equation for the system. Therefore,

$$y = -\frac{16}{11} z.$$

Substituting $y = -\frac{16}{11}z$ for y in the first equation, we have

$$x + 2y + 3z = 0$$

$$x + 2\left(-\frac{16}{11} z\right) + 3z = 0$$

$$x - \frac{32}{11} z + \frac{33}{11} z = 0$$

$$x = -\frac{1}{11} z.$$

Therefore a general solution of the original equations is

$$x = -\frac{1}{11} z$$

$$y = -\frac{16}{11} z$$

$$z = z.$$

Any value of z in these three reduced equations gives a solution of the original system of equations. We write several specific solutions, then write a simplified general solution for the system.

	$z = 0$	$z = 1$	$z = -11$	$z = -11k$
$x = -\frac{1}{11}z$	0	$-\frac{1}{11}$	1	k
$y = -\frac{16}{11}z$	0	$-\frac{16}{11}$	16	$16k$
$z = z$	0	1	-11	$-11k$

The general solution is $(k, 16k, -11k)$, where k is any real number.

CHECK We will check the general solution in the original system of equations.

In Equation 1:
$$\begin{aligned} x + 2y + 3z &= 0 \\ (k) + 2(16k) + 3(-11k) &= 0 \\ k + 32k - 33k &= 0 \\ 0 &= 0 \end{aligned}$$

In Equation 2:
$$\begin{aligned} 5x - y - z &= 0 \\ 5(k) - (16k) - (-11k) &= 0 \\ 5k - 16k + 11k &= 0 \\ 0 &= 0 \end{aligned}$$

In Equation 3:
$$\begin{aligned} 3x - 5y - 7z &= 0 \\ 3(k) - 5(16k) - 7(-11k) &= 0 \\ 3k - 80k + 77k &= 0 \\ 0 &= 0 \end{aligned}$$

Exercises 9

For each of the following systems of equations, find the nontrivial solutions if there are any.

A

1. $\begin{aligned} 2x + y - 4z &= 0 \\ x - 3y + z &= 0 \end{aligned}$

2. $\begin{aligned} r + 5s - 2t &= 0 \\ 4r - s + 7t &= 0 \end{aligned}$

3. $\begin{aligned} 6u + 7v - 5w &= 0 \\ u + v + 3w &= 0 \\ 4u + 5v - 11w &= 0 \end{aligned}$

4. $\begin{aligned} 5A - 2B + 6C &= 0 \\ 3A + 4B + 6C &= 0 \\ 26B + 9C &= 0 \end{aligned}$

B

1. $6A + 4B + C = 0$
$5A + 2B + C = 0$

2. $3u - 2v + 4w = 0$
$2u + 5v - 3w = 0$

3. $2x + y - 3z = 0$
$x - 2y - z = 0$
$4x + 7y - 7z = 0$

4. $x - 3y + 4z = 0$
$2x - 5y + 2z = 0$
$5x + y - 3z = 0$

10 Systems of second-degree equations in two variables

Recall from Chapter 9 that conic sections can be represented by second-degree equations.

In solving a system of two equations in which both equations are second-degree equations or a system having one first-degree and one second-degree equation, zero, one, two, three, or four roots may be found. In Figures 5 and 6, some of the conditions that give these roots are shown.

GRAPHS OF A SECOND-DEGREE AND A FIRST-DEGREE EQUATION, BOTH IN TWO VARIABLES

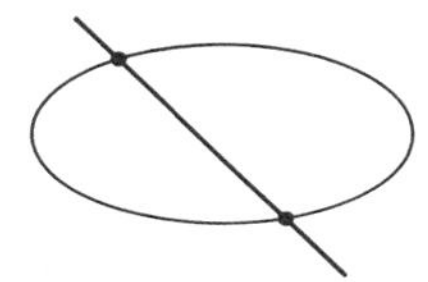

Intersecting in two points; 2 real roots

a.

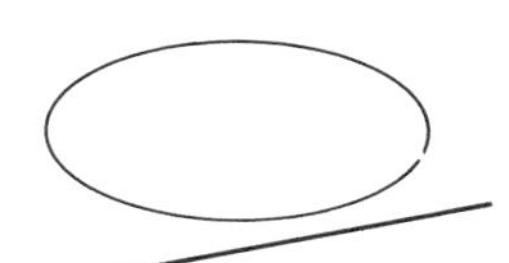

Nonintersecting; 2 complex roots, no real roots

b.

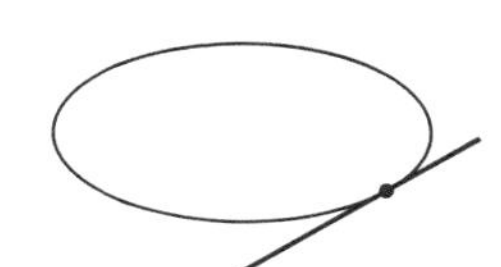

Tangent; intersecting in one point; 2 equal real roots

c.

FIGURE 5

Ellipse and straight line

We suggest that you make sketches comparable to those in Figure 5 showing the possible relations between a circle, parabola, hyperbola, and straight line.

GRAPHS OF TWO SECOND-DEGREE EQUATIONS, BOTH IN TWO VARIABLES

We have not shown all the possibilities of intersections and nonintersections, but it is hoped that the drawings shown in Figures 5 and 6 will give you an intuitive appreciation of what happens when we solve systems involving second-degree equations in two variables.

EXAMPLE 1 Solve the system

$$x^2 + y^2 = 25$$

$$3x - y = 5.$$

Then draw the graph for each equation and label the solution points.

SOLUTION We solve the system by substitution, solving the second equation for y in terms of x and then substituting that value for y in the

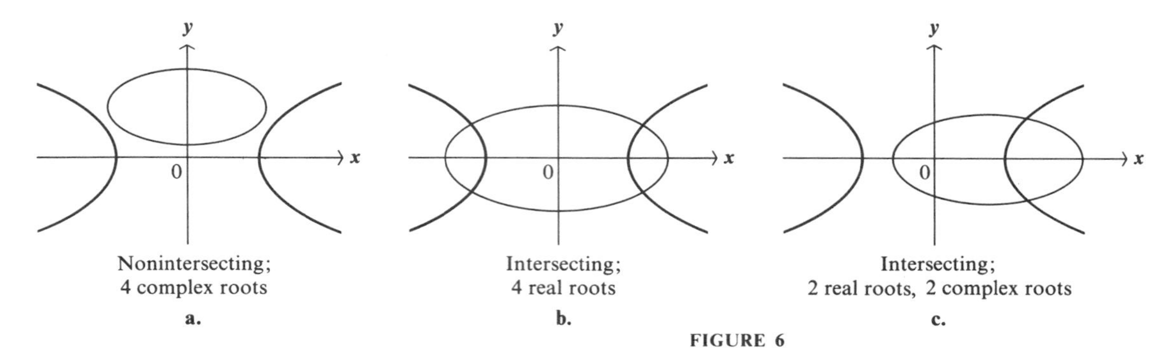

FIGURE 6

Ellipse and hyperbola

first equation. From $3x - y = 5$ we have $y = 3x - 5$.

$$
\begin{aligned}
x^2 + y^2 &= 25 \\
x^2 + (3x - 5)^2 &= 25 \\
x^2 + 9x^2 - 30x + 25 &= 25 \\
10x^2 - 30x &= 0 \\
10x(x - 3) &= 0 \\
x &= 0 \text{ and } 3
\end{aligned}
$$

We substitute these numbers in the linear equation. If they were substituted in the second-degree equation, we would get the extraneous roots $(0, 5)$ and $(3, -4)$.

$$\text{When } x = 0: \quad y = 3(0) - 5 = -5$$

$$\text{When } x = 3: \quad y = 3(3) - 5 = 4$$

Therefore, the solutions are $(0, -5)$ and $(3, 4)$ (see Figure 7).

EXAMPLE 2 Solve the system

$$
\begin{aligned}
4x^2 + 9y^2 &= 36 \\
x - y &= 5
\end{aligned}
$$

and draw the graph of each equation.

SOLUTION Solve the second equation for y in terms of x. Then substitute that value of y for y in the first equation. From the second equation, $y = x - 5$.

$$
\begin{aligned}
4x^2 + 9y^2 &= 36 \\
4x^2 + 9(x - 5)^2 &= 36 \\
4x^2 + 9x^2 - 90x + 225 &= 36 \\
13x^2 - 90x + 189 &= 0
\end{aligned}
$$

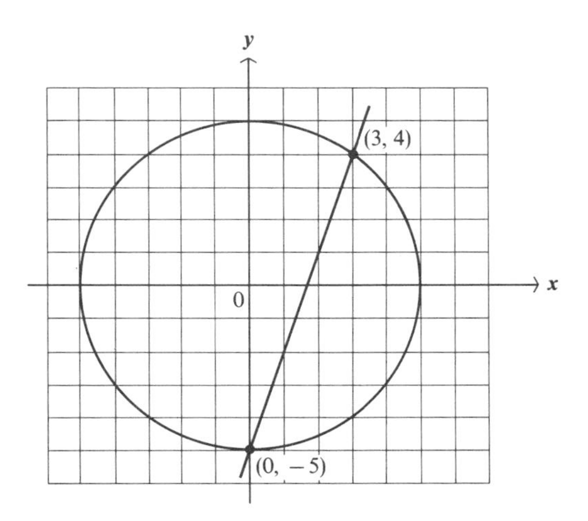

FIGURE 7

We use the quadratic formula $x = \dfrac{-b \pm \sqrt{b^2 - 4ac}}{2a}$.

$$x = \frac{-(-90) \pm \sqrt{(-90)^2 - 4(13)(189)}}{2(13)} = \frac{90 \pm \sqrt{-1{,}728}}{26}$$
$$= \frac{90 \pm \sqrt{576(-3)}}{26} = \frac{90 \pm 24i\sqrt{3}}{26}$$
$$= \frac{45}{13} \pm \frac{12}{13}i\sqrt{3}.$$

Using $y = x - 5$, we obtain

$$y = \frac{45}{13} \pm \frac{12}{13}i\sqrt{3} - 5$$
$$= -\frac{20}{13} \pm \frac{12}{13}i\sqrt{3}.$$

Therefore, the solutions of the system are

$$\left(\frac{45}{13} + \frac{12}{13}i\sqrt{3},\ -\frac{20}{13} + \frac{12}{13}i\sqrt{3}\right) \text{ and } \left(\frac{45}{13} - \frac{12}{13}i\sqrt{3},\ -\frac{20}{13} - \frac{12}{13}i\sqrt{3}\right).$$

See Figure 8.

EXAMPLE 3 Solve the system

$$3x^2 + y = 4$$
$$y = x^2$$

and draw the graph of each equation. Indicate the solution points.

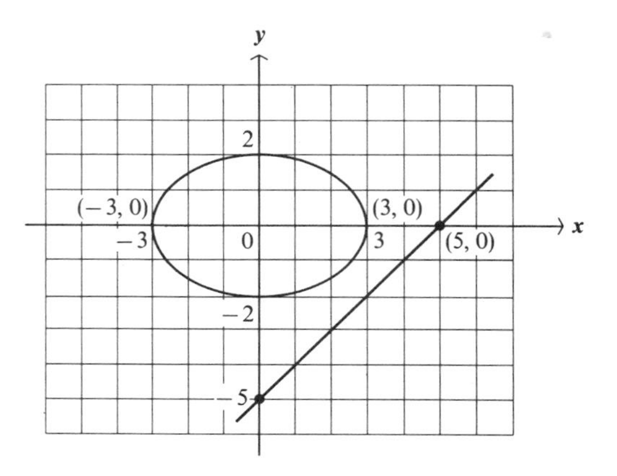

FIGURE 8

SOLUTION Substituting the value of y, in terms of x, in the first equation we have

$$\begin{aligned} 3x^2 + y \quad &= 4 \\ 3x^2 + (x^2) &= 4 \\ 4x^2 &= 4 \\ x &= \pm 1. \end{aligned}$$

When we substitute $x = \pm 1$ in the second equation, $y = 1$. Therefore, the two solutions are (1, 1) and (−1, 1).

By using the methods of Section 9, Chapter 9, we translate the axes to (0, 4), and, in so doing, reduce $y - 4 = -3x^2$ to standard form.

$$\begin{aligned} 3x^2 + y &= 4 \\ 3x^2 \quad &= -y + 4 \\ x^2 \quad &= -\frac{1}{3}(y - 4) \\ x^2 \quad &= 4\left(-\frac{1}{12}\right)(y - 4) \end{aligned}$$

Translating the origin to (0, 4) gives

$$(x')^2 = 4\left(-\frac{1}{12}\right)y'.$$

This is a parabola with vertex at (0, 4) and opening down. The second equation is already in standard form and is a parabola. We show the graphs and the points of intersection in Figure 9.

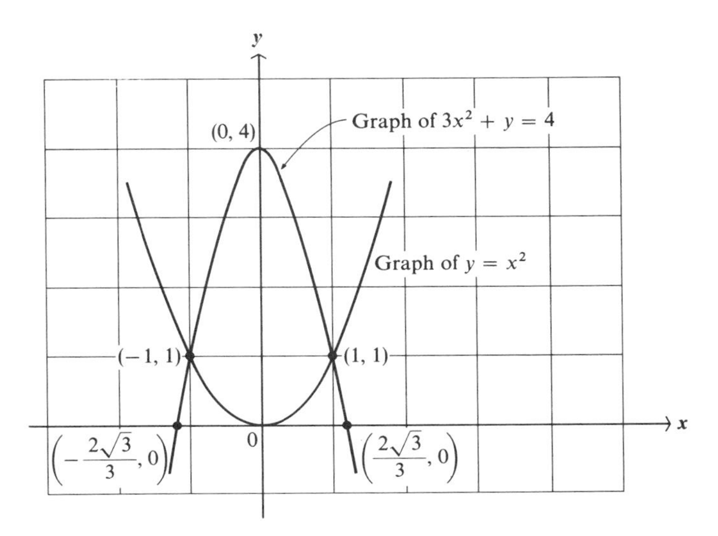

FIGURE 9

EXAMPLE 4 Solve the system

$$x^2 + y^2 = 20$$
$$xy = 8.$$

SOLUTION We first solve the second equation for y in terms of x and then substitute that value for y in the first equation.

$$xy = 8$$
$$y = \frac{8}{x}$$
$$x^2 + y^2 = 20$$
$$x^2 + \left(\frac{8}{x}\right)^2 = 20$$
$$x^4 + 64 = 20x^2$$
$$x^4 - 20x^2 + 64 = 0$$
$$(x^2 - 16)(x^2 - 4) = 0$$
$$(x + 4)(x - 4)(x + 2)(x - 2) = 0$$

Thus $x = \pm 4$ and ± 2. Then using these values of x in $y = 8/x$ we get the following points of intersection.

x	y
4	2
−4	−2
2	4
−2	−4

Solutions

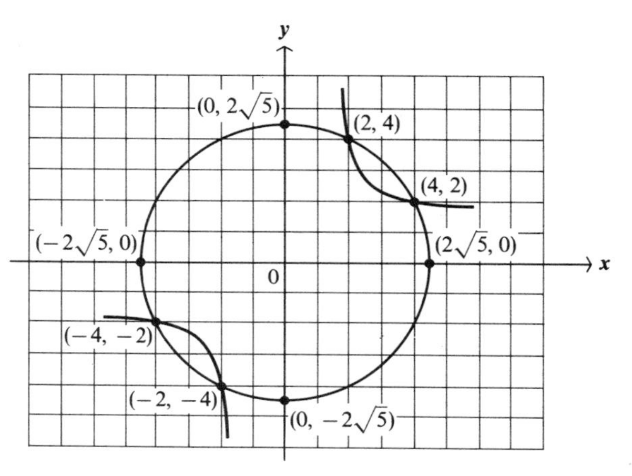

FIGURE 10

EXAMPLE 5 Solve the system of equations

$$x^2 + y^2 = 5$$
$$xy + 4y = 6.$$

SOLUTION Solving the second equation for x, we obtain

$$x = \frac{6 - 4y}{y}.$$

When we substitute this expression for x in the first equation and simplify, we have

$$y^4 + 11y^2 - 48y + 36 = 0.$$

We use synthetic division to factor the left member of this equation and thus find its roots. Possible rational roots are

$$\pm(1, 2, 3, 4, 6, 9, 12, 18, 36).$$

$$y^4 + 11y^2 - 48y + 36 = 0$$

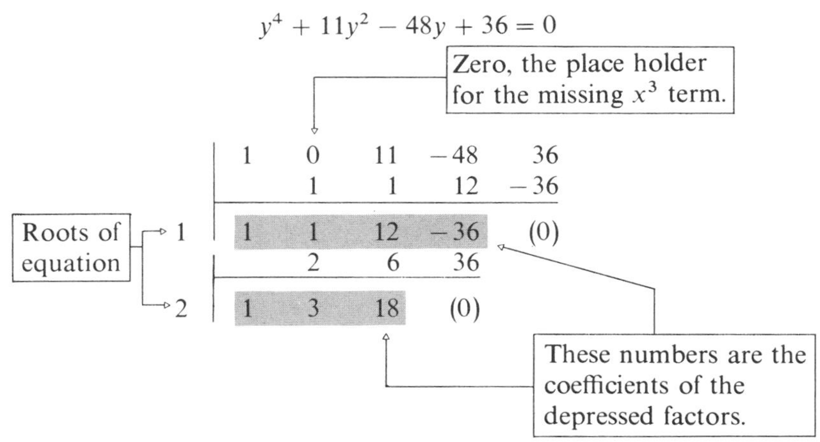

Thus $y^4 + 11y^2 - 48y + 36 = 0$ becomes

$$(y - 1)(y - 2)(y^2 + 3y + 18) = 0.$$

We use the quadratic equation to solve $y^2 + 3y + 18 = 0$.

$y - 1 = 0$	$y - 2 = 0$	$y^2 + 3y + 18 = 0$
$y = 1$	$y = 2$	$y = \dfrac{-3 \pm i3\sqrt{7}}{2}$

By substituting these values of y in the equation $x = (6 - 4y)/y$, we find the solutions

$$(2, 1), \quad (-1, 2),$$

$$\left(-\frac{9}{2} - i\frac{1}{2}\sqrt{7}, -\frac{3}{2} + i\frac{3}{2}\sqrt{7}\right), \quad \left(-\frac{9}{2} + i\frac{1}{2}\sqrt{7}, -\frac{3}{2} - i\frac{3}{2}\sqrt{7}\right).$$

HOMOGENEOUS QUADRATIC EQUATIONS

Equations whose terms are all of the second degree are called **homogeneous quadratics**. Thus, $2x^2 + xy - 3y^2 = 0$ is homogeneous, but $2x^2 + xy - 3y^2 = 1$ is not homogeneous. The following method of solution is applicable to any quadratic system of two equations in two unknowns which contains or can be made to contain homogeneous quadratic equations.

TO SOLVE A QUADRATIC SYSTEM OF HOMOGENEOUS EQUATIONS IN TWO UNKNOWNS

1. By the addition-subtraction method, eliminate the constant terms between the two equations, obtaining a single equation of the form

$$Ax^2 + Bxy + Cy^2 = 0.$$

This is a homogeneous equation.

2. Solve this homogeneous quadratic equation for one of the variables as a linear function of the other variable.
3. Substitute this linear expression into one of the two given equations; solve the resulting equation for the remaining unknown. Then obtain the corresponding values of the other unknown by substitution.

EXAMPLE 6 Solve the system

$$x^2 - 2xy + 2y^2 = 5 \qquad \textbf{(1)}$$

$$x^2 + xy = 15. \qquad \textbf{(2)}$$

SOLUTION By multiplying Equation 1 by 3 and subtracting Equation 2, we eliminate the constant term. We then factor and solve for x in terms of y.

$$3\,\big]\; x^2 - 2xy + 2y^2 = 5 \Rightarrow 3x^2 - 6xy + 6y^2 = 15$$

$$\begin{aligned} x^2 + \ \ xy \qquad\quad &= 15 \\ \hline 2x^2 - 7xy + 6y^2 &= 0 \\ (2x - 3y)(x - 2y) &= 0 \end{aligned}$$

$$\begin{array}{r|r} 2x - 3y = 0 & x - 2y = 0 \\ x = \dfrac{3}{2}y & x = 2y \end{array}$$

By substituting these linear expressions in either of the original equations of the system, we can find the solutions. We substitute $x = 2y$ in Equation 2.

$$\begin{aligned} x^2 + \quad xy &= 15 \\ (2y)^2 + (2y)y &= 15 \\ 6y^2 &= 15 \\ y &= \pm\frac{1}{2}\sqrt{10} \end{aligned}$$

Substituting $y = \pm\frac{1}{2}\sqrt{10}$ in $x = 2y$, we get the two points $(\sqrt{10}, \frac{1}{2}\sqrt{10})$ and $(-\sqrt{10}, -\frac{1}{2}\sqrt{10})$. In like manner, by substituting $x = 3y/2$ in either Equation 1 or 2, the solutions $(3, 2)$ and $(-3, -2)$ can be found. Therefore, the four solutions of the system are $(3, 2)$, $(-3, -2)$, $(\sqrt{10}, \frac{1}{2}\sqrt{10})$, and $(-\sqrt{10}, -\frac{1}{2}\sqrt{10})$.

Exercises 10

A In Exercises 1–7, solve each system of equations.

1. $2x - y - 3 = 0$
$y = x^2 - 3x + 1$

2. $5x - 3y - 12 = 0$
$x^2 - y^2 = 8$

3. $x + y = 3$
$3x^2 - 4y^2 = 12$

4. $4x^2 + 9y^2 = 25$
$x^2 + y^2 = 5$

5. $x^2 - y^2 = 5$
$5x^2 = 57 - 3y^2$

6. $x^2 - 4y^2 = 21$
$2x^2 - 5y^2 = 117$

7. $3x^2 + xy - y^2 = 4$
$3x^2 - xy = 4$

In Exercises 8–10, use a system of equations to solve the exercises.

8. The sum of two angles is 90° and their difference is 40°. Find the angles.

9. A rectangular field containing 126 square rods is enclosed by a fence and is divided into three plots by fences parallel to its shorter sides. The total length of the fence is 64 rods. Find the dimensions of the field.

10. A reservoir can be filled and then drained in a total time of 10 hours. If the inlet and outlet pumps work at the same time, the reservoir will be filled in 12 hours. Find the time required to fill the reservoir and the time required to empty it when each pump operates alone.

B In Exercises 1–7, solve each system of equations.

1. $2x + y = 20$
$x^2 + y^2 = 85$

2. $x + 3y - 12 = 0$
$x^2 + 9y^2 = 90$

3. $y = x - 6$
$x^2 + 2y^2 = 21$

4. $x^2 + 4y^2 = 12$
$2x^2 + 3y^2 = 19$

5. $x^2 - y^2 = 9$
$2x^2 - 3y^2 = 2$

6. $y^2 = 16 + x^2$
$9x^2 = 156 - 16y^2$

7. $2x^2 + xy = 12$
$y^2 - 2xy = 16$

In Exercises 8–10, use a system of equations to solve the exercises.

8. The sum of two numbers is 50 and their difference is 22. What are the numbers?

9. The area of a rectangle is 161 square inches. Its length exceeds 3 times its width by 2 inches. Find its dimensions.

10. A man has part of \$30,000 invested at 8% interest and the rest at 10% interest. If the annual income from the money invested at 8% is \$240 more than that invested at 10%, what is the amount of each investment?

Summary

An equation of the form $ax + by = c$, where a, b, and c are real number constants and a and b are not both zero, is called a **linear equation** in x and y.

The **slope m of a linear equation** $ax + by = c$ is $-a/b$, where $b \neq 0$.

A **system of two linear equations in two variables** is of the form:

$$a_1x + b_1y = c_1$$

$$a_2x + b_2y = c_2.$$

A **system of equations** may consist of more than two equations, each containing one or more variables of any degree. Because the graph of each equation in a linear system of equations is a straight line, one of the three following possibilities must occur.

1. The lines intersect in one point: a **consistent** and **independent** system.
2. The lines are parallel: an **inconsistent** system.
3. The lines are identical: a **consistent** and **dependent** system. Every solution of one equation is also a solution of the other. The equations are also called **equivalent equations**.

An **ordered triple** is a set with three members in which the order makes a difference. For example, the number triple (2, 3, 1), is a solution of the equation $2x - y + z = 2$.

A **matrix** is a rectangular ordered array of numbers. The numbers are called the **elements** of the matrix. The **coefficient matrix** of a linear system is the matrix consisting of the coefficients of the variables in the correct order. An **augmented matrix** is one in which the column of constants of a system of linear equations is included with the coefficient constants.

Given an augmented matrix of a linear system of equations and k a nonzero real number, the following **transformations of a matrix** give an augmented matrix of an equivalent system of linear equations.

1. Any two rows may be interchanged.
2. The elements in any row may be multiplied by k.
3. We may add k times the elements of any row to the corresponding elements in any other row.

A **determinant** is a square array of quantities, called **elements**, symbolizing the sum of certain products of these elements. The square array of elements is enclosed between two vertical bars. The diagonal, from the upper left corner to the lower right corner, is the **principal diagonal**. The diagonal from the lower left corner to the upper right corner is the **secondary diagonal**. An nth-order determinant is an n by n square array of quantities. The **minor of an element** in a determinant is the determinant that remains after striking out the row and column in which that element appears. The **value of a determinant** is the sum of the products obtained by multiplying each element of a column (or row) by its minor with a properly prefixed sign. The sign to be used with the element in the ith row and jth column is $(-1)^{i+j}$.

The following **transformations of a determinant** can be made, where k is any nonzero real number.

1. If any two columns (or rows) of a determinant are interchanged, the sign of the determinant is changed.
2. If each element of any column (or row) is multiplied by a number k, the value of the determinant is multiplied by k. Conversely, if a common factor k is factored from each element of any column (or row), the value of the determinant is divided by k.
3. The value of the determinant is not changed if to each element of one column (or row) is added k times the corresponding element of another column (or row).
4. A determinant with two identical columns (or rows) has the value zero.

Cramer's Rule can be used to solve a system of linear equations.

$$\begin{aligned} x - 3y + 2z &= 3 \\ 3x - 4y + 2z &= -2 \\ x + 5y - z &= -1 \end{aligned}$$

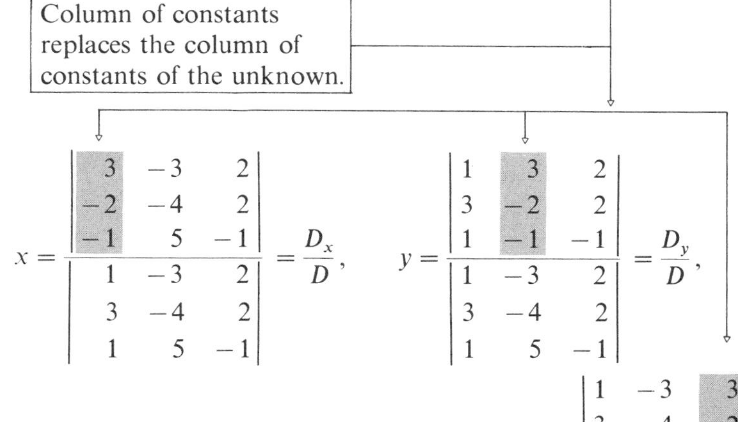

$$x = \frac{\begin{vmatrix} 3 & -3 & 2 \\ -2 & -4 & 2 \\ -1 & 5 & -1 \end{vmatrix}}{\begin{vmatrix} 1 & -3 & 2 \\ 3 & -4 & 2 \\ 1 & 5 & -1 \end{vmatrix}} = \frac{D_x}{D}, \qquad y = \frac{\begin{vmatrix} 1 & 3 & 2 \\ 3 & -2 & 2 \\ 1 & -1 & -1 \end{vmatrix}}{\begin{vmatrix} 1 & -3 & 2 \\ 3 & -4 & 2 \\ 1 & 5 & -1 \end{vmatrix}} = \frac{D_y}{D},$$

$$z = \frac{\begin{vmatrix} 1 & -3 & 3 \\ 3 & -4 & -2 \\ 1 & 5 & -1 \end{vmatrix}}{\begin{vmatrix} 1 & -3 & 2 \\ 3 & -4 & 2 \\ 1 & 5 & -1 \end{vmatrix}} = \frac{D_z}{D}$$

$$x = \frac{D_x}{D} = \frac{-34}{17} = -2, \qquad y = \frac{D_y}{D} = \frac{17}{17} = 1, \qquad z = \frac{D_z}{D} = \frac{68}{17} = 4$$

Therefore, the solution is $(-2, 1, 4)$.

A linear equation in which the constant term is zero is called a **homogeneous linear equation**. A system of n homogeneous linear equations in n unknowns can have no solution except the trivial solution $(0, 0, 0, \ldots)$ unless the determinant of its coefficients is zero. If the determinant of the coefficients is zero, the homogeneous system has an infinite number of nontrivial solutions.

Equations whose terms are all of the second degree are called **homogeneous quadratics**. We use the following steps to solve a system of two quadratic equations in two unknowns that contain or can be made to contain homogeneous quadratic equations.

1. Eliminate the constant terms between the two equations, obtaining a single equation of the form

$$Ax^2 + Bxy + Cy^2 = 0.$$

2. Solve this latter equation for one of the variables as a linear function of the other variable.
3. Substitute the linear expression in one of the two given equations, solve the resulting equations for the remaining unknown, and obtain the corresponding values of the other unknown by substitution.

Review exercises

A In Exercises 1 and 2, evaluate each determinant.

1. $\begin{vmatrix} 3 & -2 \\ -1 & 5 \end{vmatrix}$ **2.** $\begin{vmatrix} 4 & 6 & 4 & 3 \\ 3 & 1 & -7 & 8 \\ 2 & 3 & -1 & 4 \\ 4 & 6 & -2 & 8 \end{vmatrix}$

In Exercises 3–11, solve each system of equations.

3. $\begin{aligned} 2x - 3y &= 36 \\ 3x + 2y &= 2 \end{aligned}$

4. $\begin{aligned} x + 3y &= 8 \\ 3x + 9y &= 12 \end{aligned}$

5. $\begin{aligned} 3x + 2y - z &= 7 \\ 4x + 3y + 2z &= 1 \\ 5x - 3y - 6z &= -7 \end{aligned}$

6. $\begin{aligned} 2x^2 + 3y^2 &= 5 \\ x + 2y &= 3 \end{aligned}$

7. $\begin{aligned} 4x^2 + 9y^2 &= 145 \\ 3x^2 - 4y^2 &= 12 \end{aligned}$

8. $\begin{aligned} xy + 6y^2 &= 10 \\ 2x^2 + 3xy - 4y^2 &= 40 \end{aligned}$

9. $\begin{aligned} x^2 + 2xy &= y^2 + 7 \\ x^2 + y^2 + 5 &= 3xy \end{aligned}$

10. $\begin{aligned} 3x^2 + y^2 - 2y - 27 &= 0 \\ x^2 - y^2 + 2y + 23 &= 0 \end{aligned}$

11. $\begin{aligned} \frac{1}{x^2} - \frac{1}{xy} + \frac{1}{y^2} &= 3 \\ \frac{1}{x^2} + \frac{2}{xy} \qquad &= 5 \end{aligned}$

$\left(\textit{Hint}: \text{First find the value of } \frac{1}{x} \text{ and } \frac{1}{y}.\right)$

12. When the digits of a two-digit number are reversed, the value of the number is reduced by 54. If the sum of the digits is 8, find the original number.

13. A pilot takes $2\frac{1}{2}$ hours to fly 1,200 miles against the wind, and only 2 hours to return with the wind. Find the average speed of the plane in still air and the average speed of the wind.

14. **a.** Find the values of A, C, D, and E in the equation $Ax^2 + Cy^2 + Dx + Ey + 1 = 0$ so that the graph of the equation passes through the points $(0, 1)$, $(-2, 4)$, $(-2, -2)$, and $(-4, 1)$.

b. Write this new equation, reduce it to the standard form of a conic, and sketch its graph.

B In Exercises 1 and 2, evaluate each determinant.

1. $\begin{vmatrix} 3 & 1 \\ -2 & -4 \end{vmatrix}$

2. $\begin{vmatrix} 2 & 5 & 3 & 0 \\ 1 & 4 & 1 & 3 \\ -4 & 2 & -1 & 3 \\ 3 & 2 & -2 & 1 \end{vmatrix}$

In Exercises 3–11, solve each system of equations.

3. $5u + 2v = 11$
$3u + 4v = 1$

4. $3x - 2y = 6$
$6x - 4y = 5$

5. $2x + 3y - z = 5$
$3x - 2y + 4z = 2$
$x + y + 3z = -1$

6. $x^2 + y^2 = 25$
$x = y + 1$

7. $x^2 + y^2 = 13$
$3x^2 + 2y^2 = 30$

8. $8x^2 - 3y^2 = 5$
$7x^2 - 3xy = 10$

9. $xy + 24 = 0$
$x^2 - 2y^2 + 4 = 0$

10. $2x^2 + y^2 - y - 8 = 0$
$4x^2 + 3y^2 - 6y - 13 = 0$

11. $\dfrac{1}{x} + \dfrac{2}{y} = 4$

$\dfrac{1}{x^2} + \dfrac{4}{y^2} = 10$

$\left(\textit{Hint:} \text{ First find the value of } \dfrac{1}{x} \text{ and } \dfrac{1}{y}.\right)$

12. When the digits of a two-digit number are reversed the value of the number is reduced by 36. If the sum of the digits is 10, find the original number.

13. A plane travels 1,050 miles in 2 hours 30 minutes with a tail wind and returns against the wind in 3 hours 30 minutes. Find the speed of the plane in still air and the speed of the wind.

14. **a.** Find the values of A, C, D, and E in the equation $Ax^2 + Cy^2 + Dx + Ey + 1 = 0$ so that the graph of the equation passes through the points (1, 2), (1, 0), (5, 1), and (−3, 1).

b. Write this new equation, reduce it to the standard form of a conic, and sketch its graph.

Diagnostic test

The purpose of this test is to see how well you understand the work in this chapter. Allow yourself approximately 60 minutes to do the test. Solutions to all the problems, together with section references, are given in the Answers section at the end of this book. We suggest that you study the sections referred to for the problems you do incorrectly.

1. Evaluate the determinant

$$\begin{vmatrix} 1 & -2 & 0 & 3 \\ -1 & 2 & 1 & 2 \\ 2 & 0 & 0 & 4 \\ -2 & -1 & 0 & 5 \end{vmatrix}.$$

2. **a.** Transform the following system of linear equations to an equivalent system in echelon form.

$$\begin{aligned} x - 2y - z &= 2 \\ 2x + 3y + 2z &= 2 \\ 3x + y - z &= -2 \end{aligned}$$

b. Find the solution of the system.

3. Explain why the following system of linear equations does not have a solution.

$$\begin{aligned} 2x - 4y &= 2 \\ -3x + 6y &= 5 \end{aligned}$$

4. Solve the system

$$\begin{aligned} u^2 + v^2 &= 20 \\ u + v &= 6\ . \end{aligned}$$

5. Solve the following system of equations.

$$\begin{aligned} 17xy - 12x^2 &= 56 \\ 3y^2 - 5xy &= 28 \end{aligned}$$

6. Many people give the same incorrect answer when solving the following problem. Show a solution for this problem by using a system of linear equations. Problem: A tie and a pin cost \$1.10. The tie costs \$1.00 more than the pin. What is the cost of each?

11 Partial fractions

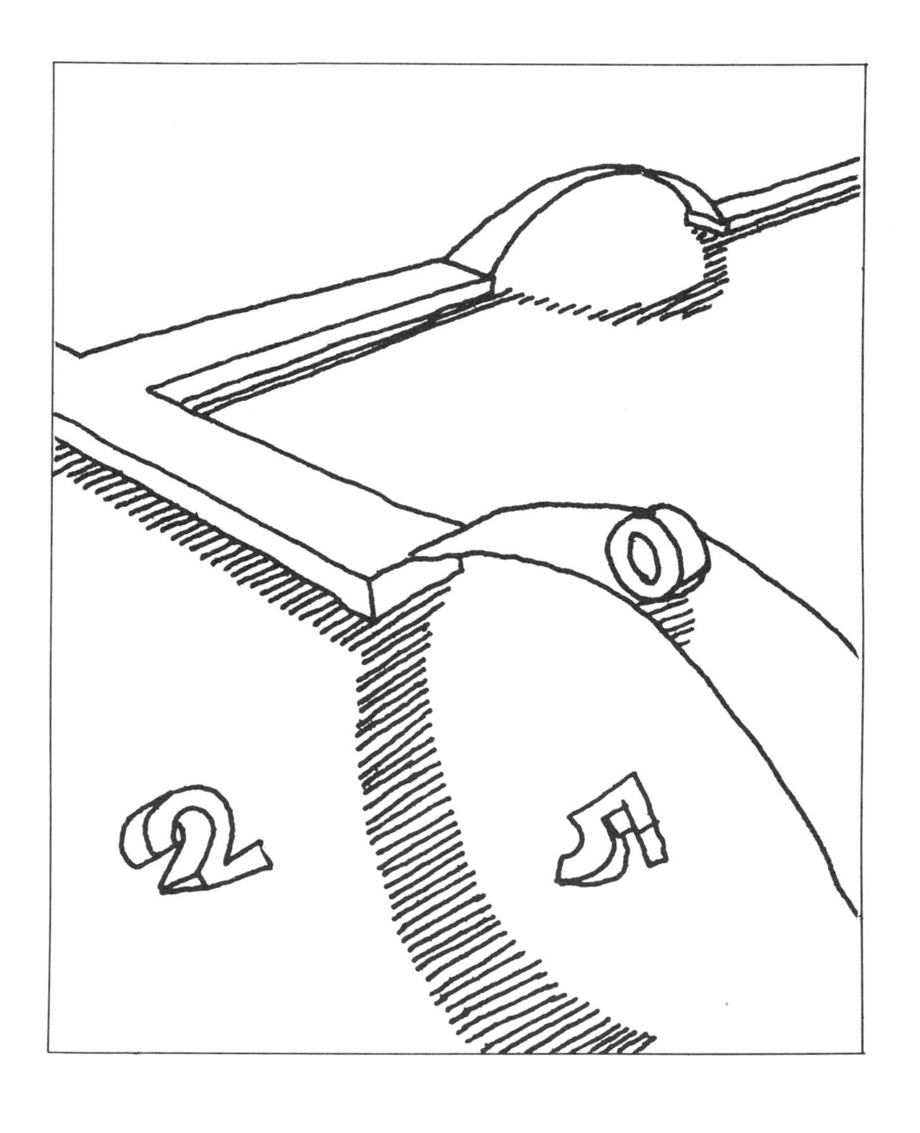

In Section 3 of Chapter 3, we discussed combining a group of simple algebraic fractions joined by plus and minus signs into a single fraction. In this chapter, we discuss the methods of performing the reverse operation.

1 Partial fractions

In many problems in calculus and other areas, it is necessary to perform the operation of resolving a given fraction into the sum of two or more simpler fractions. The simpler fractions that add up to a single fraction are called **partial fractions**. Consider the following example.

EXAMPLE 1 $\frac{3}{x-1} + \frac{2}{x+1} = \frac{5x+1}{x^2-1}$

The two fractions $3/(x-1)$ and $2/(x+1)$ are the partial fractions of the fraction $(5x+1)/(x^2-1)$.

We shall consider only fractions of the form where $P(x)$ and $Q(x)$ are polynomials with real coefficients. A fraction of this form is a proper fraction if the numerator, $P(x)$, is of lower degree than the denominator, $Q(x)$. If the fraction we wish to resolve into partial fractions is an improper fraction, we must first reduce the fraction to a mixed expression by dividing the numerator by the denominator until a remainder is obtained which is of lower degree than the denominator. Thus, if we wish to resolve the improper fraction

$$\frac{2x^3 + 7x^2 - 20x - 38}{x^2 + 2x - 15}$$

into partial fractions, we must first divide the numerator by the denominator and write the fraction as a mixed expression in the form

$$\frac{2x^3 + 7x^2 - 20x - 38}{x^2 + 2x - 15} = 2x + 3 + \frac{4x + 7}{x^2 + 2x - 15}.$$

The fractional part,

$$\frac{4x + 7}{x^2 + 2x - 15},$$

can then be resolved by the methods which will be explained in the following sections.

After reducing the fraction to be resolved to a proper fraction, we must next factor the denominator. We shall deal only with fractions in which the coefficients are real numbers and whose denominators can be factored into a product of linear and quadratic factors with real coefficients. Complex factors are not discussed. After factoring the denominator of the fraction that is to be resolved, we set up an identity between the fraction and the sum of its partial fractions. This is done according to the rules of Theorem 1.

THEOREM 1 A proper fraction in lowest terms can be expressed as a sum of partial fractions, as follows.

A. *Nonrepeating linear factors:* To a linear factor $ax + b$, that occurs exactly once as a factor of the denominator, there corresponds a partial fraction of the form

$$\frac{A}{ax + b},$$

where A is a constant, the value of which is to be determined.

B. *Repeating linear factors:* To each linear factor $(ax + b)^k$ that occurs k times as a factor of the denominator, there corresponds the sum of k partial fractions

$$\frac{A_1}{ax + b} + \frac{A_2}{(ax + b)^2} + \cdots + \frac{A_k}{(ax + b)^k},$$

where $A_1, A_2, \ldots, A_k$ are constants, the values of which are to be determined.

C. *Nonrepeating quadratic factors:* To each quadratic factor $ax^2 + bx + c$, that occurs just once as a factor of the denominator, there corresponds a partial fraction of the form

$$\frac{Ax + B}{ax^2 + bx + c},$$

where A and B are constants to be determined.

D. *Repeating quadratic factors:* To each quadratic factor $(ax^2 + bx + c)^k$ that occurs k times as a factor of the denominator, there corresponds the sum of k partial fractions

$$\frac{A_1x + B_1}{ax^2 + bx + c} + \frac{A_2x + B_2}{(ax^2 + bx + c)^2} + \cdots + \frac{A_kx + B_k}{(ax^2 + bx + c)^k},$$

where $A_1, B_1, A_2, B_2, \ldots, A_k, B_k$ are constants to be determined.

Only proper fractions whose denominators can be factored can be resolved into partial fractions.

PROCEDURE FOR RESOLVING A PROPER FRACTION INTO PARTIAL FRACTIONS

1. Factor the denominator.
2. Write an identity that equates the fraction to be resolved to the sum of its partial fractions, the partial fractions being written using the letters A, B, C, ..., as explained in Theorem 1.
3. Clear fractions in the identity by multiplying by the LCD, and combine like terms on both sides of the identity.
4. Form a system of equations in $A, B, C, \ldots$ by equating the coefficients of corresponding powers of the variables and the constants found on the two sides of the identity.
5. Solve the system of equations (see Chapter 10).
6. Use the constants found in Step 5 to write the partial fractions.
7. Check the solution to see if the sum of the obtained partial fractions is the original single fraction.

In the balance of this chapter, we will use examples to show how to find the numerical constants that appear in the numerators of the partial fractions.

2 Denominator containing nonrepeating linear and quadratic factors

EXAMPLE 1 Resolve into partial fractions.

$$\frac{5x+1}{x^2-1}$$

SOLUTION The denominator $x^2 - 1$ factors into $(x+1)(x-1)$.

A different letter is used for the numerator of each fraction.

Thus,

$$\frac{5x+1}{x^2-1} = \frac{A}{x+1} + \frac{B}{x-1}$$

$$5x + 1 = A(x-1) + B(x+1) \qquad \text{Clearing fractions.}$$

$$= Ax - A + Bx + B$$

$$5x + 1 = (A+B)x + (-A+B)1 \qquad \text{Combining like terms.} \qquad \textbf{(1)}$$

Because Equation 1 is an identity, the coefficient of x on the right side of the identity must equal the coefficient of x on the left side and the constant on the right side must equal the constant on the left side. This leads to the following system of equations, which we solve by addition.

$$A + B = 5 \qquad \text{Coefficients of } x.$$

$$-A + B = 1 \qquad \text{Constants.}$$

$$2B = 6 \qquad \text{Solve by addition.}$$

$$B = 3$$

Then, if $B = 3$ and $A + B = 5$, $A + 3 = 5$ and $A = 2$. Therefore,

$$\frac{5x + 1}{x^2 - 1} = \frac{2}{x + 1} + \frac{3}{x - 1}.$$

A similar method was used in Section 4 of Chapter 4 in the solution of complex equations.

CHECK To check the solution, add the partial fractions and see if their sum is the same as the fraction being resolved.

$$\frac{2}{x+1} + \frac{3}{x-1} = \boxed{\frac{x-1}{x-1}}\,\frac{2}{x+1} + \boxed{\frac{x+1}{x+1}}\,\frac{3}{x-1}$$

$$= \frac{2x - 2}{x^2 - 1} + \frac{3x + 3}{x^2 - 1}$$

$$= \frac{2x - 2 + 3x + 3}{x^2 - 1}$$

$$= \frac{5x + 1}{x^2 - 1} \qquad \text{The original fraction.}$$

EXAMPLE 2 Resolve into partial fractions.

$$\frac{x^4 + 5x^3 + 2x^2 + 2}{x^3 + 2x^2 - 2x}$$

SOLUTION Because this is an improper fraction, we first divide the numerator by the denominator.

$$\frac{x^4 + 5x^3 + 2x^2 + 2}{x^3 + 2x^2 - 2x} = x + 3 + \frac{-2x^2 + 6x + 2}{x^3 + 2x^2 - 2x}$$

We next factor the denominator of the fractional part and set up an identity for the fractional part.

$$\frac{-2x^2 + 6x + 2}{x^3 + 2x^2 - 2x} = \frac{-2x^2 + 6x + 2}{x(x^2 + 2x - 2)} = \frac{A}{x} + \frac{Bx + C}{x^2 + 2x - 2}$$

Clearing fractions and collecting like terms, we have

$$\begin{aligned} -2x^2 + 6x + 2 &= A(x^2 + 2x - 2) + (Bx + C)x \\ &= Ax^2 + 2Ax - 2A + Bx^2 + Cx \qquad (2) \\ &= (A + B)x^2 + (2A + C)x - 2A. \end{aligned}$$

Since the equation is an identity, the coefficients of like powers of x on either side must be equal. Equating these coefficients gives the system

$$\begin{aligned} A + B \qquad &= -2 \\ 2A \qquad + C &= \ \ 6 \\ -2A \qquad &= \ \ 2. \end{aligned}$$

The solution is $A = -1$, $B = -1$, $C = 8$. Therefore,

$$\frac{-2x^2 + 6x + 2}{x^3 + 2x^2 - 2x} = -\frac{1}{x} + \frac{-x + 8}{x^2 + 2x - 2}.$$

Finally,

$$\frac{x^4 + 5x^3 + 2x^2 + 2}{x^3 + 2x^2 - 2x} = x + 3 - \frac{1}{x} - \frac{x - 8}{x^2 + 2x - 2}.$$

Exercises 2

Resolve the following fractions into partial fractions.

A

1. $\dfrac{3x + 5}{x^2 + x - 12}$ **2.** $\dfrac{4x + 1}{x^2 + 5x - 14}$

3. $\dfrac{2x^2 - 13x - 6}{x^3 - x^2 - 2x}$ **4.** $\dfrac{-4x^2 + 16x - 24}{x^3 - 8x^2 + 12x}$

5. $\dfrac{6x^3 - 17x^2 + 5}{6x^2 - 5x - 6}$

B

1. $\dfrac{4x + 7}{x^2 + x - 6}$ **2.** $\dfrac{4x + 1}{x^2 - 3x - 10}$

3. $\dfrac{4x^2 - 9x - 4}{x^3 + x^2 - 2x}$ **4.** $\dfrac{-6x^2 + 17x + 75}{x^3 + 2x^2 - 15x}$

5. $\dfrac{6x^3 + 11x^2 - 2x - 14}{6x^2 + 5x - 6}$

---

3 Denominator containing repeating linear factors

When the denominator contains linear factors, some of which appear to powers higher than the first degree, care must be taken to write all of the partial fractions called for by such a repeated factor (see Part B of Theorem 1).

EXAMPLE 1 Resolve into partial fractions.

$$\frac{-2x^3 + 13x^2 - 13x + 4}{(x-1)^3(x+1)}$$

SOLUTION Applying Theorem 1, we have

$$\frac{-2x^3 + 13x^2 - 13x + 4}{(x-1)^3(x+1)} = \frac{A}{x-1} + \frac{B}{(x-1)^2} + \frac{C}{(x-1)^3} + \frac{D}{x+1}$$

We next clear fractions and collect like terms.

$$\begin{aligned} -2x^3 + 13x^2 - 13x + 4 &= A(x-1)^2(x+1) + B(x-1)(x+1) \\ &\quad + C(x+1) + D(x-1)^3 \\ &= A(x^3 - x^2 - x + 1) + B(x^2 - 1) \\ &\quad + C(x+1) + D(x^3 - 3x^2 + 3x - 1) \\ &= (A + D)x^3 + (-A + B - 3D)x^2 \\ &\quad + (-A + C + 3D)x + (A - B + C - D). \end{aligned}$$

Finally we form a system of equations in A, B, C, and D by equating the respective coefficients of x^3, x^2, and x and the constants of the two sides of the identity.

$$\begin{aligned} A \qquad\qquad + D &= -2 && \text{Coefficients of } x^3. \\ -A + B \qquad - 3D &= 13 && \text{Coefficients of } x^2. \\ -A \qquad + C + 3D &= -13 && \text{Coefficients of } x. \\ A - B + C - D &= 4 && \text{Constants.} \end{aligned}$$

The solution of the above system is $(2, 3, 1, -4)$. Therefore,

$$\frac{-2x^3 + 13x^2 - 13x + 4}{(x-1)^3(x+1)} = \frac{2}{x-1} + \frac{3}{(x-1)^2} + \frac{1}{(x-1)^3} - \frac{4}{x+1}.$$

Exercises 3

Resolve the following fractions into partial fractions.

A

1. $\dfrac{x^3 + 1}{x(x-1)^3}$
2. $\dfrac{4}{x^2(x-2)^2}$
3. $\dfrac{2x^3 + 19x^2 - 4x - 52}{(x^2 - 4)^2}$
4. $\dfrac{14x^2 + 2x - 1}{(x^2 + x)^3}$

B

1. $\dfrac{3x^2 + 13x - 10}{x^2(x-2)}$
2. $\dfrac{11x^2 - 13x - 10}{x^3(3x+2)}$
3. $\dfrac{9x^3 - 8x^2 - 4x + 48}{(x^2 - 4)^2}$
4. $\dfrac{w^2 - 13w + 16}{(w+3)(w-5)^2}$

4 Denominator containing quadratic and linear factors, some repeating

EXAMPLE 1 Resolve into partial fractions.

$$\frac{9x - 7}{(x^2 + 1)^2(x + 2)}$$

SOLUTION Applying Theorem 1, we have

$$\frac{9x - 7}{(x^2 + 1)^2(x + 2)} = \frac{Ax + B}{x^2 + 1} + \frac{Cx + D}{(x^2 + 1)^2} + \frac{E}{x + 2}$$

We next clear fractions and collect like terms.

$$\begin{aligned} 9x - 7 &= (Ax + B)(x^2 + 1)(x + 2) \\ &\quad + (Cx + D)(x + 2) + E(x^2 + 1)^2 \\ &= (Ax + B)(x^3 + 2x^2 + x + 2) \\ &\quad + (Cx + D)(x + 2) + E(x^4 + 2x^2 + 1). \end{aligned}$$

$$\begin{aligned} 9x - 7 &= (A + E)x^4 + (2A + B)x^3 + (A + 2B + C + 2E)x^2 \\ &\quad + (2A + B + 2C + D)x + (2B + 2D + E) \end{aligned}$$

Finally, we form a system of equations in A, B, C, D, and E by equating the respective coefficients of x^4, x^3, x^2, and x, and the constants of the two sides of the identity. Notice that the coefficients of the x^4, x^3, and x^2 terms in the above identity are zero. These zeros make our system of equations in A, B, C, D, and E simpler.

$A + E = 0$	Coefficients of x^4.	**(1)**
$2A + B = 0$	Coefficients of x^3.	**(2)**
$A + 2B + C + 2E = 0$	Coefficients of x^2.	**(3)**
$2A + B + 2C + D = 9$	Coefficients of x.	**(4)**
$2B + 2D + E = -7$	Constants.	**(5)**

We solve this system of linear equations in the following manner. From Equation 1, $A + E = 0 \Rightarrow E = -A$. From Equation 2, $2A + B = 0 \Rightarrow B = -2A$. Substituting $-A$ for E and $-2A$ for B in Equations 3, 4, and 5, we have:

$$\begin{aligned} A + 2B + C + 2E &= 0 \qquad \textbf{(3)} \\ A + 2(-2A) + C + 2(-A) &= 0 \\ A - 4A + C - 2A &= 0 \\ -5A + C &= 0 \Rightarrow C = 5A \end{aligned}$$

$$\begin{aligned} 2A + B + 2C + D &= 9 \qquad \textbf{(4)} \\ 2A + (-2A) + 2(5A) + D &= 9 \\ 10A + D &= 9 \end{aligned}$$

$$\begin{aligned} 2B + 2D + \quad E &= -7 \\ 2(-2A) + 2D + (-A) &= -7 \\ -5A + \quad 2D &= -7 \end{aligned} \qquad \textbf{(5)}$$

Next we solve

$$\begin{array}{l|l} 1 & 10A + D = 9 \Rightarrow \quad 10A + D = 9 \\ 2 & -5A + 2D = -7 \Rightarrow -10A + 4D = -14 \end{array}$$

$$\begin{aligned} 5D &= -5 \\ D &= -1 \\ A &= 1 \end{aligned}$$

Then $E = -1$, $B = -2$, and $C = 5$. The solution of the system is $(1, -2, 5, -1, -1)$. Therefore,

$$\frac{9x - 7}{(x^2 + 1)^2(x + 2)} = \frac{x - 2}{x^2 + 1} + \frac{5x - 1}{(x^2 + 1)^2} - \frac{1}{x + 2}.$$

Exercises 4

Resolve the following fractions into partial fractions.

A

1. $\dfrac{3x^2 - 9x + 8}{(x^2 + 3)(x + 1)}$
2. $\dfrac{9y^2 + 29}{(y - 2)(y^2 + 2y + 5)}$
3. $\dfrac{x^4 - x^2 + 9}{x(x^2 + 3x + 3)^2}$
4. $\dfrac{2x^2 + 8x + 41}{(x^2 + 2x + 10)^2}$

B

1. $\dfrac{11x^2 + 11x + 8}{(2x^2 + 3)(x + 4)}$
2. $\dfrac{3x^2 - x + 4}{x(x^2 + x + 1)}$
3. $\dfrac{x^2 + 4x + 126}{(x^2 - x + 1)^2(x - 3)}$
4. $\dfrac{2x^3 + 5x^2 + x - 7}{(x^2 + 4x + 5)^2}$

Summary

In this chapter, we considered only fractions of the form $P(x)/Q(x)$, where $P(x)$ and $Q(x)$ were polynomials with real coefficients. The fraction $P(x)/Q(x)$ is a **proper fraction** if the numerator $P(x)$ is of lower degree than the denominator $Q(x)$.

Partial fractions are the simple algebraic fractions which, when added, give a single fraction. If the fraction to be resolved into partial fractions is an improper fraction, we first reduce it to a mixed expression by dividing the numerator by the denominator until a remainder is obtained which is of lower degree than the denominator. Only proper fractions whose denominators can be factored can be resolved into partial fractions. Complex factors are not included in our discussion.

The following process is used to resolve a proper fraction into partial fractions.

1. Factor the denominator.
2. Write an identity that equates the fraction to be resolved to the sum of its partial fractions, the partial fractions being written using the letters A, B, C, ..., as explained in Theorem 1.
3. Clear fractions in the identity by multiplying by the LCD, and combine like terms on both sides of the identity.
4. Form a system of equations in A, B, C, ... by equating the coefficients of corresponding powers of the variables and the constants found on the two sides of the identity.
5. Solve the system of equations.
6. Substitute the values of A, B, C, ... found in Step 5 into their respective places in the fractions on the right side of the identity in Step 2. These fractions are the required partial fractions.
7. Check the solution to see if the sum of the obtained partial fractions is the original single fraction.

Review exercises

Resolve the following fractions into partial fractions.

A

1. $\dfrac{2y-2}{(y+5)(y+2)}$
2. $\dfrac{2z^2-z+8}{z(z-2)^2}$
3. $\dfrac{2x^3-3}{x^3+x}$ (Show your check.)
4. $\dfrac{4x^4+4x^3+2x^2+1}{x^2(2x^2+x+1)^2}$

B

1. $\dfrac{3x-5}{(x-1)(x-3)}$
2. $\dfrac{9w^2+2w-14}{w^2(w-7)}$
3. $\dfrac{x^3+8x^2+9x-1}{(x+1)(x^2+2x+2)}$ (Show your check.)
4. $\dfrac{2x^4-4x^3+10x^2-5x+8}{x(x^2-x+2)^2}$

Diagnostic test

The purpose of this test is to see how well you understand the work in this chapter. Allow yourself approximately 60 minutes to do the test. Solutions to the problems, together with section references, are given in the Answers section at the end of this book. We suggest that you study the sections referred to for the problems you do incorrectly.

Resolve each of the following fractions into partial fractions. Show the check for the solution of Problem 3.

1. $\dfrac{7x + 4}{3x^2 - 17x + 10}$

2. $\dfrac{x^3 + 3x^2 - 3x - 2}{x^2(x + 1)^2}$

3. $\dfrac{2x^3 + 4x^2 - x + 1}{x^3 + x}$

4. $\dfrac{2x^4 + 13x^2 - x + 18}{x(x^2 + 3)^2}$

5. $\dfrac{7x^3 - 4x^2 + 2x - 1}{x^2(2x^2 + 1)}$

12 The binomial theorem and mathematical induction

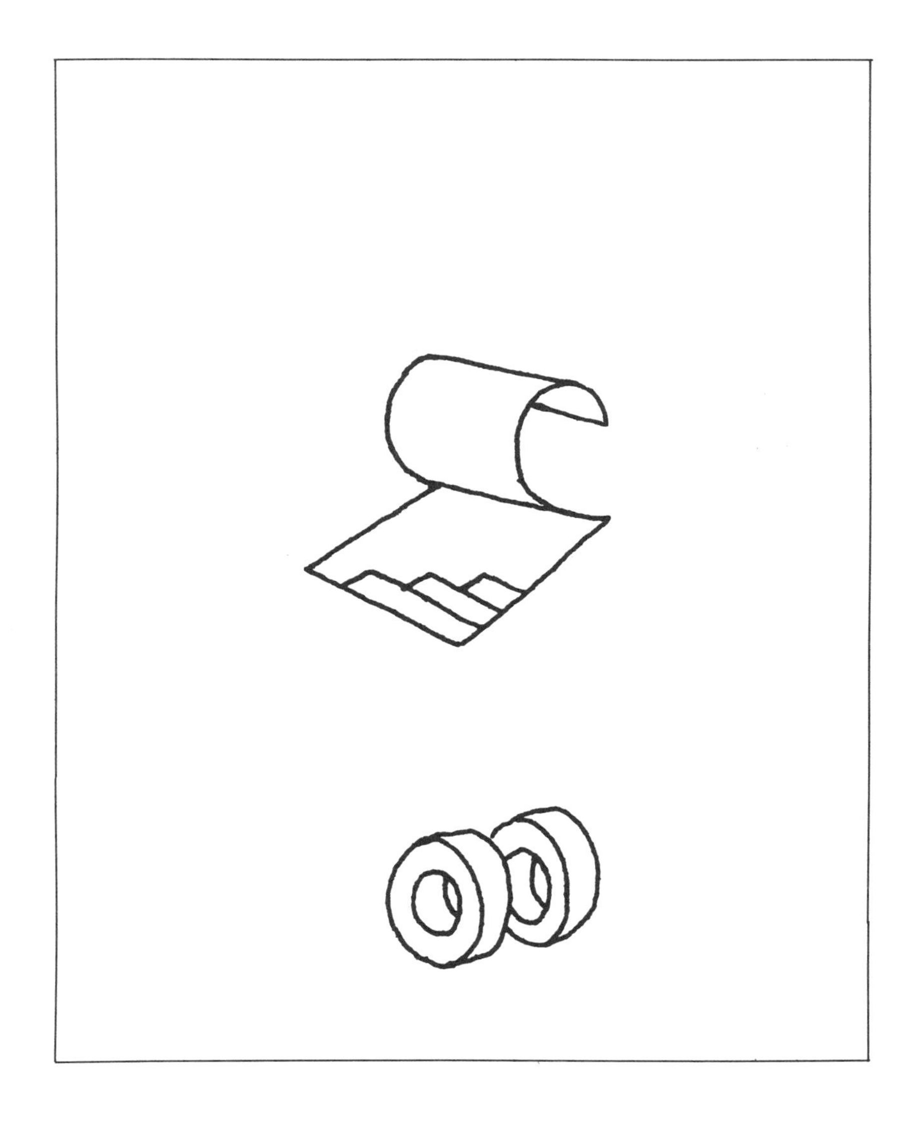

1 Factorial notation

In mathematical calculations, the product of all the positive integers from 1 to n occurs so frequently that a special notation has been devised to represent it. The symbol is written $n!$ and is read "n factorial."

n FACTORIAL

$$n! = 1 \cdot 2 \cdot 3 \cdot \cdots \cdot n, \qquad n \text{ a positive integer} \tag{1}$$

By definition,

$$0! = 1. \tag{2}$$

From Definition 1, it follows that

$$(n+1)! = n!(n+1) \qquad \text{or} \qquad n! = \frac{(n+1)!}{n+1} \tag{3}$$

and

$$\binom{n}{r} = \frac{n!}{(n-r)!\,r!} \qquad r \in N \tag{4}$$

Factorial numbers can be broken down in many different ways. We will illustrate with examples.

EXAMPLE 1

a. $6! = 6 \cdot 5! = 6 \cdot 5 \cdot 4! = 6 \cdot 5 \cdot 4 \cdot 3!$
$\quad = 6 \cdot 5 \cdot 4 \cdot 3 \cdot 2! = 6 \cdot 5 \cdot 4 \cdot 3 \cdot 2 \cdot 1$

b. $(n+2)! = (n+2)(n+1)! = (n+2)(n+1)(n)!$
$\quad = (n+2)(n+1)(n)(n-1)!$ and so on.

EXAMPLE 2

a. $3! = 1 \cdot 2 \cdot 3 = 6$

b. $5! = 1 \cdot 2 \cdot 3 \cdot 4 \cdot 5 = 120$

c. $10! = 1 \cdot 2 \cdot 3 \cdot 4 \cdot 5 \cdot 6 \cdot 7 \cdot 8 \cdot 9 \cdot 10 = 3{,}628{,}800$

d. $\dfrac{8!}{4!} = \dfrac{\cancel{1 \cdot 2 \cdot 3 \cdot 4} \cdot 5 \cdot 6 \cdot 7 \cdot 8}{\cancel{1 \cdot 2 \cdot 3 \cdot 4}} = 1{,}680$

e. $\dfrac{6!}{2!\,3!} = \dfrac{\cancel{1 \cdot 2} \cdot 3 \cdot 4 \cdot 5 \cdot \cancel{6}}{\cancel{1 \cdot 2} \cdot 1 \cdot \cancel{2 \cdot 3}} = 60$

f. $69! = 1.711224524 \times 10^{98}$

EXAMPLE 3

a. $5!\,6 = 6!$ because $6! = 6 \cdot \boxed{5 \cdot 4 \cdot 3 \cdot 2 \cdot 1}$

$$= 6 \cdot \boxed{5!}$$

b. $(n+1)!\,(n+2) = (n+2)!$

because $(n+2)! = (n+2)\ \boxed{(n+1)(n)(n-1) \cdot \cdots \cdot 1}$

$$= (n+2)\ \boxed{(n+1)!}$$

c. $(n-r-1)!\,(n-r) = (n-r)!$

because $(n-r)!$

$$= (n-r)\ \boxed{(n-r-1)(n-r-2)(n-r-3) \cdot \cdots \cdot 1}$$

$$= (n-r)\ \boxed{(n-r-1)!}$$

d. $$\frac{n!}{(n-3)!3!} = \frac{n(n-1)(n-2)[\cancel{(n-3)}!]}{\cancel{(n-3)}!\,3!} = \frac{n(n-1)(n-2)}{6}$$

Exercises 1

A In Exercises 1–5, find the value of each expression.

1. $7!$ **2.** $\frac{8!}{5!}$ **3.** $\frac{9!}{4!\,5!}$ **4.** $\frac{(r+1)!}{r!}$ **5.** $\frac{0!\,5!}{3!}$

In Exercises 6–10, simplify each expression.

6. $6!\,7$ **7.** $n!\,(n+1)$ **8.** $\frac{(n+2)!}{n!}$

9. $\frac{6!\,(n+2)!}{5!\,(n-1)!\,n}$ **10.** $\frac{(n+1)!\,2!}{(n-1)!\,n}$

B In Exercises 1–5, find the value of each expression.

1. $6!$ **2.** $\frac{7!}{6!}$ **3.** $\frac{7!}{2!\,3!}$ **4.** $\frac{(n+1)!}{n!}$ **5.** $\frac{0!\,4!}{3!}$

In Exercises 6–10, simplify each expression.

6. $4!\,5$ **7.** $(n-1)!\,n$ **8.** $\frac{(n+1)!}{(n-1)!}$

9. $\dfrac{7!\,(r+2)!}{6!\,(r-1)!\,r}$ 10. $\dfrac{n!}{(n-2)!\,2!}$

2 The binomial theorem for positive integral exponents

It is often necessary to find the expansion of an expression of the type $(a + b)^n$, where a and b are mathematical expressions and n is a large integer. Since $(a + b)$ is a binomial, the theorem for the expansion of $(a + b)^n$ is called the **Binomial Theorem**.

We show the expansion of $(a + b)^n$ for relatively small positive integral values of n and give rules for this expansion. A formal proof for the expansion of $(a + b)^n$ is given in Section 5 of this chapter.

The following expansions are obtained by multiplication.

$$(a + b)^1 = a + b$$
$$(a + b)^2 = a^2 + 2ab + b^2$$
$$(a + b)^3 = a^3 + 3a^2b + 3ab^2 + b^3$$
$$(a + b)^4 = a^4 + 4a^3b + 6a^2b^2 + 4ab^3 + b^4$$
$$(a + b)^5 = a^5 + 5a^4b + 10a^3b^2 + 10a^2b^3 + 5ab^4 + b^5$$

These five identities are special cases of a general formula for the expansion of $(a + b)^n$. By close inspection of the pattern of terms in the above expansions, the rule for the expansion of $(a + b)^n$ can be discovered.

Notice that:

1. The first term of the expansion is a^n and the last term is b^n.
2. The exponent of a decreases by 1 with each succeeding term and the exponent of b increases by one in each succeeding term. The first term does not contain b. The degree of each term in a and b is n.
3. The coefficient of the second term is always n.

EXAMPLE 1 Write the letters in each term of the expansion of $(a + b)^6$.

SOLUTION The exponent of a decreases by 1, while the exponent of b increases by 1.

$$(a + b)^6 = a^6 + \_\_\_a^5b^1 + \_\_\_a^4b^2 + \_\_\_a^3b^3 + \_\_\_a^2b^4 + \_\_\_ab^5 + \_\_\_b^6$$

Notice that the degree of each term is 6.

EXAMPLE 2 Write the first two terms of the expansion of $(a + b)^n$ for $n = 3$, 5, 6, and 10.

SOLUTION

a. $(a + b)^3 = a^3 + 3a^2b + \cdots$

b. $(a + b)^5 = a^5 + 5a^4b + \cdots$

c. $(a + b)^6 = a^6 + 6a^5b + \cdots$

d. $(a + b)^{10} = a^{10} + 10a^9b + \cdots$

4. We further observe that, after having found the second term of the expansion of $(a + b)^n$, we can find the third term by multiplying the exponent of a by its coefficient and dividing the product by one more than the exponent of b.

EXAMPLE 3 Write the first three terms of the expansion of $(a + b)^n$ for $n = 4$, 6, 7, and 10.

$$\frac{(\text{Coefficient of second term})\ (\text{Exponent of } a)}{(\text{Exponent of } b) + 1}$$

$$= \text{Coefficient of third term}$$

SOLUTION

a. $(a + b)^4 = a^4 + 4a^3b^1 + 6a^2b^2 + \cdots$

$$\frac{4 \cdot 3}{2} = 6$$

The exponent of b is 1, so we divide by 2.

b. $(a + b)^6 = a^6 + 6a^5b + 15a^4b^2 + \cdots$

$$\frac{6 \cdot 5}{2} = 15$$

c. $(a + b)^7 = a^7 + 7a^6b + 21a^5b^2 + \cdots$

$$\frac{7 \cdot 6}{2} = 21$$

d. $(a + b)^{10} = a^{10} + 10a^9b + 45a^8b^2 + \cdots$

$$\frac{10 \cdot 9}{2} = 45$$

5. If we know any term, we can get the following term by the same method used in Example 3.

EXAMPLE 4 Expand $(a + b)^5$.

SOLUTION We make use of the information given in Statements 1–4.

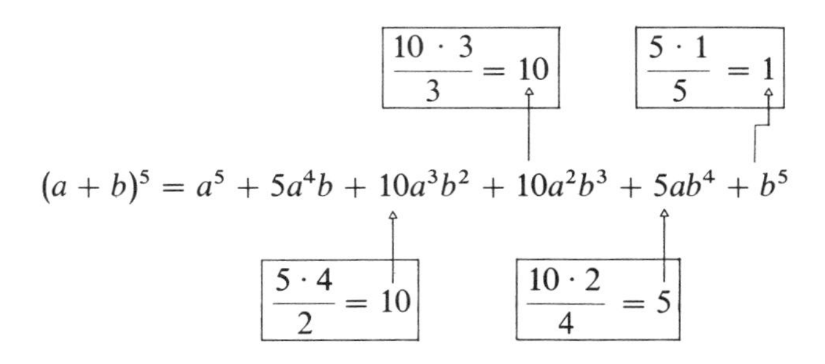

EXAMPLE 5 Expand $(a + 2)^7$.

SOLUTION We apply the information given in Statements 1–4.

$$\begin{aligned}(a + 2)^7 &= a^7 + 7a^6 2 + 21a^5 2^2 + 35a^4 2^3 + 35a^3 2^4 \\ &\quad + 21a^2 2^5 + 7a2^6 + 2^7 \\ &= a^7 + 14a^6 + 84a^5 + 280a^4 + 560a^3 \\ &\quad + 672a^2 + 448a + 128\end{aligned}$$

In addition to the Statements 1–5, we may also make the following observations.

6. The expansion of $(a + b)^n$ has $n + 1$ terms.
7. The coefficients are symmetric, that is, the same coefficients can be observed beginning at the left or at the right.

Notice the symmetrical pattern of the coefficients in Examples 4 and 5. In Example 4, the coefficients are

$$1 \qquad 5 \qquad 10 \qquad 10 \qquad 5 \qquad 1.$$

In the first step of Example 5, the coefficients are

$$1 \qquad 7 \qquad 21 \qquad 35 \qquad 35 \qquad 21 \qquad 7 \qquad 1.$$

Also, observe these symmetrical patterns in the identities given at the beginning of this section. Because of this symmetry, we only have to work out the left-hand half of the coefficients.

EXAMPLE 6 Expand $\left(2 + \frac{x^2}{2}\right)^6$.

SOLUTION Let $a = 2$ and $b = \frac{x^2}{2}$.

$$\left(2 + \frac{x^2}{2}\right)^6 = (2)^6 + 6(2)^5\left(\frac{x^2}{2}\right)^1 + 15(2)^4\left(\frac{x^2}{2}\right)^2 + 20(2)^3\left(\frac{x^2}{2}\right)^3$$
$$+ 15(2)^2\left(\frac{x^2}{2}\right)^4 + 6(2)^1\left(\frac{x^2}{2}\right)^5 + \left(\frac{x^2}{2}\right)^6$$
$$= 64 + 96x^2 + 60x^4 + 20x^6 + \frac{15x^8}{4} + \frac{3x^{10}}{8} + \frac{x^{12}}{64}$$

So far, we have considered only binomials with plus signs in the expansion of $(a + b)^n$. We now examine the expansion of $(a - b)^n$.

We let $(a - b)^n = [a + (-b)]^n$, and then determine the plus or minus signs of the terms in the expansion by disregarding the coefficients of the terms and writing only the letters of the respective terms of the expansion.

$$(a - b)^n = [a + (-b)]^n$$
$$= a^n + \underline{\qquad} a^{n-1}(-b)^1 + \underline{\qquad} a^{n-2}(-b)^2$$
$$+ \underline{\qquad} a^{n-2}(-b)^3 + \cdots + (-b)^n$$

Note that terms having $-b$ raised to an odd power are negative and terms having $-b$ raised to an even power are positive. Therefore, the signs of the terms in the expansion $(a - b)^n$ alternate, with the first term always positive.

EXAMPLE 7 Expand $(2x - y)^5$.

SOLUTION

$$(2x - y)^5 = (2x)^5 - 5(2x)^4y + 10(2x)^3y^2 - 10(2x)^2y^3$$
$$+ 5(2x)^1y^4 - y^5$$
$$= 32x^5 - 80x^4y + 80x^3y^2 - 40x^2y^3$$
$$+ 10xy^4 - y^5$$

Assuming that the characteristics given in Statements 1–7 are maintained for all positive integral values of n, we have the following theorem.

THE BINOMIAL THEOREM

$$(a+b)^n = a^n + \frac{n}{1!}a^{n-1}b + \frac{n(n-1)}{2!}a^{n-2}b^2 + \frac{n(n-1)(n-2)}{3!}a^{n-3}b^3 + \cdots + \frac{n(n-1)\cdots(n-r+1)}{r!}a^{n-r}b^r + \cdots + b^n,$$

where a and b are mathematical expressions and n and r are positive integers.

So far we have verified the Binomial Theorem only for small values of n. The general proof is given in Section 5 of this chapter.

DEFINITIONS ASSOCIATED WITH THE SYMBOL $\binom{n}{r}$

$$\binom{n}{r} = \frac{n!}{r!(n-r)!} = \frac{n(n-1)\cdots(n-r+1)(n-r)!}{r!(n-r)!} = \frac{n(n-1)\cdots(n-r+1)}{r!}$$

$$\binom{n}{0} = 1, \qquad \binom{n}{n} = 1,$$

where n is a positive integer and r is a whole number.

The numbers $\binom{n}{r}$ are called **binomial coefficients**. We illustrate the meaning of the symbol $\binom{n}{r}$ in the following example.

EXAMPLE 8

$$(a+b)^4 = \frac{1}{0!}a^4 + \frac{4}{1!}a^3b^1 + \frac{4\cdot 3}{2!}a^2b^2 + \frac{4\cdot 3\cdot 2}{3!}ab^3 + \frac{4\cdot 3\cdot 2\cdot 1}{4!}b^4$$

$$= \binom{4}{0}a^4 + \binom{4}{1}a^3b^1 + \binom{4}{2}a^2b^2 + \binom{4}{3}ab^3 + \binom{4}{4}b^4$$

That is,

$$\binom{4}{0} = 1 \quad \leftarrow \text{By definition}$$

$$\binom{4}{1} = \frac{4}{1!} = 4 \quad \leftarrow \text{One factor, } 1!$$

$$\binom{4}{2} = \frac{4 \cdot 3}{2!} = 6 \quad \leftarrow \text{Two factors, } 2!$$

$$\binom{4}{3} = \frac{4 \cdot 3 \cdot 2}{3!} = \frac{4 \cdot 3 \cdot 2}{3 \cdot 2 \cdot 1} = 4 \quad \leftarrow \text{Three factors, } 3!$$

$$\binom{4}{4} = \frac{4 \cdot 3 \cdot 2 \cdot 1}{4!} = 1 \quad \leftarrow \text{Four factors, } 4!$$

We now rewrite the binomial theorem using the $\binom{n}{r}$ notation.

THE BINOMIAL THEOREM WRITTEN WITH $\binom{n}{r}$ NOTATION

$$(a + b)^n = \binom{n}{0}a^n b^0 + \binom{n}{1}a^{n-1}b + \binom{n}{2}a^{n-2}b^2 + \cdots$$

$$+ \binom{n}{r}a^{n-r}b^r + \cdots \quad \leftarrow (r + 1)\text{st term}$$

$$+ \binom{n}{n-1}a^{n-(n-1)}b^{n-1} + \binom{n}{n}a^0 b^n$$

The expansion of $(a + b)^n$ may be obtained either by using properties in statements 1–7 or by using the Binomial Theorem.

You should familiarize yourself with both methods for evaluating $(a + b)^n$ and use one method as a check on the other.

Exercises 2

A In Exercises 1–6, expand and simplify. Leave the exponent of a in Exercise 6 negative.

1. $(a + b)^7$ **2.** $(x - y)^6$

3. $\left(2x + \frac{y}{2}\right)^5$ **4.** $\left(\frac{x}{3} - \frac{3}{y}\right)^4$

5. $(\sqrt{a} + \sqrt[4]{b})^6$ **6.** $\left(a^{-2} - \frac{b^2}{\sqrt{2}}\right)^6$

In Exercises 7–10, write the first three terms of the expansion.

7. $(2x + y)^{10}$ **8.** $\left(\frac{a^2}{2} - 2b^3\right)^{14}$

9. $(a^{1/2} - b^{1/2})^8$ **10.** $\left(\frac{1}{x} + \frac{1}{y}\right)^{12}$

B In Exercises 1–6, expand and simplify. Leave the exponent of h in Exercise 6 negative.

1. $(x + y)^6$ **2.** $(a - b)^7$

3. $\left(3a + \frac{b}{3}\right)^5$ **4.** $\left(\frac{x}{2} - \frac{2}{y}\right)^4$

5. $(\sqrt{x} + \sqrt[4]{y})^5$ **6.** $\left(h^{-2} - \frac{k^2}{\sqrt{3}}\right)^6$

In Exercises 7–10, write the first three terms of the expansion.

7. $(a + 2b)^{10}$ **8.** $\left(x^3 - \frac{y^2}{2}\right)^{14}$

9. $(x^{1/2} - y^{-1/2})^{12}$ **10.** $\left(\frac{1}{a} + \frac{1}{b}\right)^8$

3 The general term of the binomial expansion

It is sometimes necessary to write a specified term of the expansion of $(a + b)^n$ without finding the preceding ones. To work problems of this type, we need a formula for the rth term of the expansion.

In the second form of the Binomial Theorem (see Section 2), the $(r + 1)$st term was shown to be $\binom{n}{r}a^{n-r}b^r$. To obtain a formula for the rth term, we substitute $(r - 1)$ for r in the expression $\binom{n}{r}a^{n-r}b^r$.

rth TERM OF THE EXPANSION $(a + b)^n$

$$\binom{n}{r-1}a^{n-(r-1)}b^{r-1} = \binom{n}{r-1}a^{n-r+1}b^{r-1}$$

EXAMPLE 1 Find the fifth term in the expansion $(a + x)^{11}$ without finding the preceding terms.

SOLUTION We use the formula for the rth term.

$$r\text{th term} = \binom{n}{r-1} a^{n-r+1} x^{r-1}, \qquad n = 11 \text{ and } r = 5$$

$$5\text{th term} = \binom{11}{5-1} a^{11-5+1} x^{5-1}$$

$$= \binom{11}{4} a^7 x^4$$

$$= \frac{11 \cdot 10 \cdot \overset{3}{\cancel{9}} \cdot \cancel{8}}{1 \cdot \cancel{2} \cdot \cancel{3} \cdot \cancel{4}} a^7 x^4 = 330 a^7 x^4$$

EXAMPLE 2 Find the eighth term in the expansion

$$\left(\frac{\sqrt{x}}{2} - \frac{4}{x}\right)^{15}$$

without finding the preceding terms.

SOLUTION

$$r\text{th term} = \binom{n}{r-1} a^{n-r+1} b^{r-1}$$

In this example, $r = 8$, $n = 15$, $a = \sqrt{x}/2$, and $b = -4/x$.

$$8\text{th term} = \binom{15}{8-1} a^{15-8+1} b^{8-1}$$

$$= \binom{15}{7} \left(\frac{\sqrt{x}}{2}\right)^8 \left(-\frac{4}{x}\right)^7$$

$$= \frac{\overset{3}{\cancel{15}} \cdot \overset{2}{\cancel{14}} \cdot 13 \cdot \cancel{12} \cdot 11 \cdot \overset{5}{\cancel{10}} \cdot 9}{1 \cdot \cancel{2} \cdot \cancel{3} \cdot \underset{2}{\cancel{4}} \cdot \cancel{5} \cdot \cancel{6} \cdot \cancel{7}} \left(\frac{x^4}{\cancel{2^8}}\right)\left(-\frac{\overset{2^6}{\cancel{4^7}}}{x^7}\right)$$

$$= -\frac{411{,}840}{x^3}$$

Notice that when we have 7 in the lower part of $\binom{15}{7}$, there are 7 factors in the numerator of the coefficients.

We recommend that you use your calculator in computing the coefficients of the terms in expansions of this type.

Exercises 3

Find the indicated term in each expansion without writing the preceding terms.

A

1. $(a - x)^{15}$; eighth term
2. $(x + y)^{10}$; seventh term
3. $\left(7y + \dfrac{1}{7z^2}\right)^{12}$; sixth term
4. $\left(x - \dfrac{1}{2y}\right)^{10}$; fourth term
5. $(x^{-2} - \frac{2}{3}y^{-3})^{18}$; fourth term

B

1. $(y - 2)^{13}$; seventh term
2. $(u^3 + v^4)^{12}$; eighth term
3. $\left(\dfrac{x}{y} + \dfrac{z^2}{c}\right)^{14}$; ninth term
4. $\left(t^5 - \dfrac{5}{t^2}\right)^6$; fourth term
5. $(e^{1/2} + e^{-1/2})^{14}$; ninth term

4 Mathematical induction

Mathematical induction is a form of reasoning that enables us to prove that certain formulas or theorems are true for all integral values of n that are greater than some fixed integer.

Suppose we wish to prove the statement "the sum of the first n consecutive odd natural numbers is n^2." That is, we wish to prove the formula $1 + 3 + 5 + \cdots + (2n - 1) = n^2$. We can easily establish the truth of the formula for the first few natural numbers.

One natural number $\quad 1 = 1^2$

$$1 = 1$$

Two natural numbers $\quad 1 + 3 = 2^2$

$$4 = 4$$

Three natural numbers $\quad 1 + 3 + 5 = 3^2$

$$9 = 9$$

Four natural numbers $\quad 1 + 3 + 5 + 7 = 4^2$

$$16 = 16$$

While the evidence given above makes the formula quite plausible, nothing has actually been proved beyond the fact that the formula is correct for $n = 1, 2, 3,$ and 4. There is always the possibility, for example, that a proposed formula may be valid for $n = 1, 2, 3, \ldots, 1{,}000$ and then fail for $n = 1{,}001$. The formal proof of this formula may be carried out by **mathematical induction**.

FUNDAMENTAL PRINCIPLE OF MATHEMATICAL INDUCTION

The proof by mathematical induction of a formula or theorem, where the formula or theorem is written in terms of a positive integer n, consists of the following two parts.

1. A verification that the formula or theorem is true for some one integral value of n, usually $n = 1$.
2. A proof that, if k is any integral value of n for which the formula or theorem is true, then $k + 1$ is also a value for which it is true.

If both parts have been proved, then the formula or theorem is true for all integral values of n greater than, or equal to, the value used in Part 1.

EXAMPLE 1 Prove by mathematical induction that the sum of the first n even integers is $n(n + 1)$, that is,

$$2 + 4 + 6 + \cdots + 2n = n(n + 1). \tag{1}$$

STEP 1 We verify that the formula is true for $n = 1$.

$$2(1) = 1(1 + 1)$$
$$2 = 2$$

STEP 2 We now prove that whenever Equation 1 holds true for some value of n, say $n = k$, it holds true for the next greater integer, $n = k + 1$. We assume $n = k$ in Equation 1.

$$2 + 4 + 6 + \cdots + 2k = k(k + 1) \tag{2}$$

Next we add the even number $2k + 2$, the even integer following $2k$, to both sides of Equation 2.

$$\begin{aligned} 2 + 4 + 6 + \cdots + 2k + \boxed{2k + 2} &= k(k + 1) + \boxed{2k + 2} \\ &= k(k + 1) + 2(k + 1) \\ &= (k + 1)(k + 2) \end{aligned}$$

or $$2 + 4 + 6 + \cdots + 2k + 2(k + 1) = (k + 1)(k + 2) \tag{3}$$

This latter formula is the original Equation 1 with $n = k + 1$. This completes the proof of Step 2. Notice that the left side of Equation 3 is equivalent to the left side of Equation 1 if $n = k + 1$. Also, the right side of Equation 3 is equivalent to the right side of Equation 1 if $n = k + 1$. Therefore, $2 + 4 + 6 + \cdots + 2n = n(n + 1)$ is true for all positive integral values of n.

Mathematical induction states that if a formula is true for some integral value k and can be proved for the integral value $k + 1$, then it can be proved for the integral values $k + 2$, $k + 3$, and so on, and therefore is true for all positive integral values of n.

EXAMPLE 2 Prove by mathematical induction that

$$3 + 3^2 + 3^3 + \cdots + 3^n = \frac{3}{2}(3^n - 1). \qquad \textbf{(4)}$$

STEP 1 Verify the formula for $n = 1$.

$$\begin{aligned} 3^1 &= \frac{3}{2}(3^1 - 1) \\ &= \frac{3}{2}(2) \\ 3 &= 3 \end{aligned}$$

STEP 2 We assume Equation 4 is true for $n = k$.

$$3 + 3^2 + 3^3 + \cdots + 3^k = \frac{3}{2}(3^k - 1) \qquad \textbf{(5)}$$

To obtain the left side of Equation 4 for $n = k + 1$, we add 3^{k+1} to both sides of Equation 5. This gives

$$\begin{aligned} 3 + 3^2 + 3^3 + \cdots + 3^k + \boxed{3^{k+1}} &= \frac{3}{2}(3^k - 1) + \boxed{3^{k+1}} \qquad \textbf{(6)} \\ &= \frac{3}{2} \cdot 3^k - \frac{3}{2} + 3^{k+1} \\ &= \frac{1}{2} \cdot 3^{k+1} - \frac{3}{2} + 3^{k+1} \\ &= \frac{3}{2} \cdot 3^{k+1} - \frac{3}{2} \\ &= \frac{3}{2}(3^{k+1} - 1) \end{aligned}$$

But this is just Equation 4 with $n = k + 1$. Hence, by the principle of mathematical induction Equation 4 is valid for every positive integer.

The general term of Example 1 is $2n$. That is,

$$\begin{aligned} \text{First term} &= 2(1) = 2 \\ \text{Second term} &= 2(2) = 4 \\ \text{Fifth term} &= 2(5) = 10 \\ (k + 1)\text{st term} &= 2(k + 1) \end{aligned}$$

If we want to write the first four terms, we have

$$2(1) + 2(2) + 2(3) + 2(4) = 2 + 4 + 6 + 8.$$

The general term of Example 2 is 3^n. That is,

$$\text{First term} = 3^1 = 3$$

$$\text{Second term} = 3^2 = 9$$

$$(k+1)\text{st term} = 3^{k+1} \quad \text{or} \quad 3^k \cdot 3$$

Students often wonder what to add to both sides of the assumed equation. In the problems we will be doing add the $(k + 1)$st term of the assumed equation to both sides of the assumed equation. Then reduce this new equation to the form the equation to be proved would have if $n = k + 1$ in that equation.

In Example 1, the $(k + 1)$st term was $2k + 2$, which was added to both sides of the assumed equation. In Example 2, the $(k + 1)$st term was 3^{k+1}, which was added to both sides of the assumed equation.

Exercises 4

Prove the following theorems by mathematical induction for all positive integral values of n.

A

1. $1 + 3 + 5 + \cdots + (2n - 1) = n^2$
2. $1 + 5 + 9 + \cdots + (4n - 3) = n(2n - 1)$
3. $4 + 4^2 + 4^3 + \cdots + 4^n = \dfrac{4(4^n - 1)}{3}$
4. $2 \cdot 4 + 4 \cdot 6 + 6 \cdot 8 + \cdots + 2n(2n + 2) = \dfrac{4n}{3}(n + 1)(n + 2)$
5. $1 \cdot 2 \cdot 3 + 2 \cdot 3 \cdot 4 + 3 \cdot 4 \cdot 5 + \cdots + n(n + 1)(n + 2)$
   $= \dfrac{1}{4} n(n + 1)(n + 2)(n + 3)$
6. $\dfrac{1}{1 \cdot 3} + \dfrac{1}{3 \cdot 5} + \dfrac{1}{5 \cdot 7} + \cdots + \dfrac{1}{(2n - 1)(2n + 1)} = \dfrac{n}{2n + 1}$
7. $1^2 + 3^2 + 5^2 + \cdots + (2n - 1)^2 = \dfrac{4n^3 - n}{3}$

B

1. $1 + 2 + 3 + \cdots + n = \dfrac{1}{2} n(n + 1)$
2. $4 + 8 + 12 + \cdots + 4n = 2n(n + 1)$
3. $1^2 + 2^2 + 3^2 + \cdots + n^2 = \dfrac{1}{6} n(n + 1)(2n + 1)$
4. $1 \cdot 6 + 2 \cdot 9 + 3 \cdot 12 + \cdots + n(3n + 3) = n(n + 1)(n + 2)$
5. $1 \cdot 2 + 2 \cdot 3 + 3 \cdot 4 + \cdots + n(n + 1) = \dfrac{n(n + 1)(n + 2)}{3}$
6. $\dfrac{1}{1 \cdot 2} + \dfrac{1}{2 \cdot 3} + \dfrac{1}{3 \cdot 4} + \cdots + \dfrac{1}{n(n + 1)} = \dfrac{n}{n + 1}$
7. $2 + 2^2 + 2^3 + \cdots + 2^n = 2(2^n - 1)$

5 Proof of the binomial theorem

We use the method of mathematical induction to prove that the Binomial Theorem

$$(a+b)^n = \binom{n}{0}a^n + \binom{n}{1}a^{n-1}b + \binom{n}{2}a^{n-2}b^2 + \cdots + \binom{n}{r-1}a^{n-r+1}b^{r-1} + \binom{n}{r}a^{n-r}b^r + \cdots + \binom{n}{n}b^n \quad (1)$$

is true for all positive integral values of n. The letter r designates the $(r+1)$st term. In Equation 1, we have included both the rth and the $(r+1)$st term.

STEP 1 We verify the formula for $n = 1$.

$$(a+b)^1 = a+b$$

The verification was also done for $n = 1, 2, 3, 4$, and 5 at the beginning of Section 2.

STEP 2 Let k be any positive integral value of n for which Equation 1 is true. Then, by hypothesis, we have Equation 2.

$$(a+b)^k = \binom{k}{0}a^k + \binom{k}{1}a^{k-1}b + \binom{k}{2}a^{k-2}b^2 + \cdots + \binom{k}{r-1}a^{k-r+1}b^{r-1} + \binom{k}{r}a^{k-r}b^r + \cdots + \binom{k}{k}b^k \quad (2)$$

We next multiply both sides of Equation 2 by $a+b$ to generate the $(k+1)$st term. The product of the left side by $a+b$ gives

$$(a+b)^k(a+b) = (a+b)^{k+1}.$$

This is equivalent to the entire left side of Equation 1 if $n = k+1$.

To multiply the right side by $a+b$, we must multiply each of its terms by a and by b and add the results. Multiplying the right side of Equation 2 by a, we obtain

$$= \binom{k}{0}a^{k+1} + \binom{k}{1}a^k b + \binom{k}{2}a^{k-1}b^2 + \cdots + \binom{k}{r-1}a^{k-r+2}b^{r-1} + \binom{k}{r}a^{k-r+1}b^r + \cdots + \binom{k}{k}ab^k \quad (3)$$

Multiplying the right side of Equation 2 by b, we obtain

$$= \binom{k}{0}a^k b + \binom{k}{1}a^{k-1}b^2 + \binom{k}{2}a^{k-2}b^3 + \cdots + \binom{k}{r-1}a^{k-r+1}b^r + \binom{k}{r}a^{k-r}b^{r+1} + \cdots + \binom{k}{k}b^{k+1} \quad (4)$$

By adding Expressions 3 and 4 we get the right side of Equation 2 after it has been multiplied by $(a + b)$.

$$(a+b)^{k+1} = \binom{k}{0}a^{k+1} + \left[\binom{k}{0} + \binom{k}{1}\right]a^k b + \left[\binom{k}{1} + \binom{k}{2}\right]a^{k-1}b^2 + \cdots + \left[\binom{k}{r-1} + \binom{k}{r}\right]a^{k-r+1}b^r + \cdots + \binom{k}{k}b^{k+1} \quad \textbf{(5)}$$

This equals

$$\binom{k+1}{0}a^{k+1} + \binom{k+1}{1}a^k b + \binom{k+1}{2}a^{k-1}b^2 + \cdots + \binom{k+1}{k+1}b^{k+1}$$

which is equal to the entire right side of Equation 1. Equating Expressions 2 and 5, we see that if Equation 1 is true for $n = k$, it is also true for $n = k + 1$. Thus, if the binomial formula is correct for $n = k$, we must have

$$(a+b)^{k+1} = \binom{k+1}{0}a^{k+1} + \binom{k+1}{1}a^k b + \binom{k+1}{2}a^{k-1}b^2 + \cdots + \binom{k+1}{r}a^{k-r+1}b^r + \cdots + \binom{k+1}{k+1}b^{k+1},$$

which is the binomial formula for $n = k + 1$. Therefore Step 2 has been completed. This completes the proof of the Binomial Theorem for positive integral exponents.

Summary

We write ***n* factorial** as $n!$, where n is a nonnegative integer. By definition, $n! = n(n-1)(n-2)\cdots 1$ and $0! = 1$.

The binomial coefficients are represented by the symbol

$$\binom{n}{r} = \frac{n(n-1)(n-2)\cdots(n-r+1)}{r!}.$$

By definition

$$\binom{n}{0} = 1, \quad \text{and} \quad \binom{n}{n} = \frac{n!}{n!} = 1.$$

The **Binomial Theorem** is as follows.

$$(a+b)^n = \binom{n}{0}a^n b^0 + \binom{n}{1}a^{n-1}b + \binom{n}{2}a^{n-2}b^2 + \cdots + \underbrace{\binom{n}{r}a^{n-r}b^r}_{(r+1)\text{st term}} + \cdots + \binom{n}{n}a^0 b^n$$

The **rth term** of the expansion of $(a + b)^n$ is

$$\binom{n}{r-1} a^{n-r+1} b^{r-1}.$$

Mathematical induction is a method of proving a formula or theorem. The proof of a formula or theorem by mathematical induction, where the formula or theorem is written in terms of a positive integer n, always consists of two parts.

1. A verification that the formula or theorem is true for some one integral value of n.
2. A proof that, if k is any integral value of n for which the formula or theorem is true, then $k + 1$ is also a value for which it is true.

If both parts have been proved, then the formula or theorem is true for all integral values of n greater than or equal to the value used in Step 1.

Review exercises

A In Exercises 1–4, find the value of each expression.

1. $6!$ **2.** $\dfrac{10!}{8!}$ **3.** $\dbinom{7}{5}$ **4.** $\dbinom{5}{0}$

In Exercises 5–8, simplify each expression.

5. $7!8$ **6.** $\dfrac{(2)(5!)}{3!}$ **7.** $\dfrac{n!}{(n-2)!}$ **8.** $\dbinom{n}{n-1}$

In Exercises 9–12, expand and simplify.

9. $(x - 3y)^4$ **10.** $\left(\dfrac{x}{2} + 2y\right)^5$

11. $(\sqrt[3]{a} + \sqrt[3]{b})^6$ **12.** $\left(\dfrac{x^3}{2} - 2y^2\right)^7$

13. Write the first three terms of the expansion $(x - y)^{20}$.

14. Write the tenth term of the expansion $(e^{1/2} - e^{-1/2})^{12}$.

15. Prove the following theorem by mathematical induction for all positive integral values of n.

$$\frac{1}{2} + \frac{1}{2^2} + \frac{1}{2^3} + \cdots + \frac{1}{2^n} = \frac{2^n - 1}{2^n}$$

B In Exercises 1–4, find the value of each expression.

1. $5!$ **2.** $\dfrac{8!}{6!}$ **3.** $\dbinom{8}{7}$ **4.** $\dbinom{10}{0}$

In Exercises 5–8, simplify each expression.

5. $6!7$ **6.** $\dfrac{10!}{8!}$ **7.** $\dfrac{(n+2)!}{n!}$ **8.** $\dbinom{n}{n-2}$

In Exercises 9–12, expand and simplify.

9. $(2x - y)^5$ **10.** $\left(\dfrac{x}{3} - \dfrac{3}{y}\right)^4$

11. $(\sqrt{x} - \sqrt{y})^6$ **12.** $\left(2x^2 + \dfrac{y^3}{2}\right)^7$

13. Write the first three terms of the expansion $(a - b)^{100}$.

14. Write the eighth term of the expansion $(e^{1/2} + e^{-1/2})^{10}$.

15. Prove the following theorem by mathematical induction for all positive integral values of n.

$$3 + 3^2 + \cdots + 3^n = \frac{3}{2}(3^n - 1)$$

Diagnostic test

The purposes of this test is to see how well you understand the work in this chapter. Allow yourself approximately 60 minutes to finish the test. Solutions to the problems, together with section references, are given in the Answers section at the end of this book. We suggest that you study the sections referred to for the problems you do incorrectly.

1. Find the value of each of the following expressions.

a. $5!$ **b.** $\dfrac{7!}{4!}$ **c.** $\dbinom{6}{2}$ **d.** $\dbinom{8}{0}$

2. Simplify each of the following expressions.

a. $9!10$ **b.** $\dfrac{9!}{8!}$ **c.** $\dfrac{n(n+1)[(n-1)!]}{n!}$

3. Write the first three terms of the expansion $(a - 2b)^{50}$.

4. Write the ninth term of the expansion $(\sqrt{a} - b)^{12}$ without writing the preceding terms.

5. Prove the following theorem by mathematical induction for all positive integral values of n.

$$2 + 5 + 12 + \cdots + (5n - 3) = \frac{1}{2}n(5n - 1)$$

In Problems 6 and 7, expand and simplify.

6. $(a - 2b)^5$ **7.** $(\sqrt{x} + 2)^6$

8. Write the first four terms of the expansion $\left(\dfrac{x^2}{2} - 2y^3\right)^7$.

13 Topics from the theory of equations

The concept of a polynomial was introduced in Chapter 2. In that chapter, our main emphasis was on the basic operations and factoring of polynomials. In this chapter, we shall be concerned mainly with solving polynomial equations and graphing polynomial functions.

1 Polynomial equations

A **polynomial function** of degree n in the single variable x can be written in the form

$$f(x) = a_0x^n + a_1x^{n-1} + a_2x^{n-2} + \cdots + a_{n-1}x + a_n, \tag{1}$$

where n is a positive integer or zero, the coefficients $a_0, a_1, \ldots, a_n$ are real constants and $a_0 \neq 0$. With the same conditions on the coefficients, the corresponding equation

$$a_0x^n + a_1x^{n-1} + \cdots + a_{n-1}x + a_n = 0 \tag{2}$$

represents a **polynomial equation** of degree n in x.

In this chapter, functional symbols such as $f(x)$, $q(x)$, and $P(x)$, will be understood to represent a polynomial in x, and the word **equation** will mean a polynomial equation of degree at least equal to one in one unknown. The words *linear*, *quadratic*, *cubic*, and *quartic* are applied to polynomials and equations which are of degrees 1, 2, 3, and 4, respectively. A nonzero constant may be considered as a polynomial of degree zero.

A polynomial $f(x)$ will be said to be in **standard form** if it is written in the above form, where the terms are arranged in descending order of powers of x from a_0x^n to a_n, and in which no term is missing. A zero coefficient is supplied for any term when necessary. An **equation is in standard form** if it is written in the form $f(x) = 0$, where $f(x)$ is in standard form, and the leading coefficient a_0 if it is real, is taken to be positive.

EXAMPLE 1 Write the equation $3x^2 - 2x = 3 - 5x^4$ in standard form.

SOLUTION $5x^4 + 0x^3 + 3x^2 - 2x - 3 = 0$.

DEFINITION A number r is said to be a **root** of the equation $f(x) = 0$ and a zero of the function $f(x)$ if and only if $f(r) = 0$.

FUNDAMENTAL THEOREM OF ALGEBRA

Every polynomial equation of degree greater than zero has at least one root.

We shall assume the Fundamental Theorem to be true. The proof of this theorem is left for courses beyond the level of this book. Notice that the theorem does not say that an equation necessarily has a *real* root. A polynomial equation may have complex roots, or it may have a combination of real and complex roots.

2 The remainder theorem

Before stating the Remainder Theorem, we shall find the value of a polynomial by direct substitution.

EXAMPLE 1 Find the value of $f(x) = x^3 - 3x^2 + 6x - 5$ when $x = 2$.

SOLUTION Substituting 2 for x, we obtain the following value.

$$\begin{aligned} f(x) &= x^3 - 3x^2 + 6x - 5 \\ f(2) &= (2)^3 - 3(2^2) + 6(2) - 5 \\ &= 8 - 12 + 12 - 5 \\ &= \boxed{3} \end{aligned}$$

Dividing $f(x)$ by $x - 2$ using synthetic division, we obtain the quotient $x^2 - x + 4$ with a remainder of 3.

$$\begin{array}{r|rrrr} & 1 & -3 & 6 & -5 \\ & & 2 & -2 & 8 \\ \hline 2 & 1 & -1 & 4 & \boxed{3} \end{array}$$

$x^2 \quad -x \quad +4$ Quotient

In this case, we observe that the value of the function when x equals 2—that is, $f(2)$—is exactly equal to the remainder obtained by dividing the function by $x - 2$.

It is not difficult to show that the preceding statement must be true. Suppose we had checked the division.

$$\text{Dividend} = \text{Divisor} \times \text{Quotient} + \text{Remainder}$$

In this case, we would have

$$f(x) = x^3 - 3x^2 + 6x - 5 = (x - 2)(x^2 - x + 4) + 3.$$

This equation, which is true for all values of x, is true, in particular, for $x = 2$. Substituting $x = 2$, we have the following.

$$\begin{aligned} f(2) &= (2)^3 - 3(2^2) + 6(2) - 5 = [(2) - 2][(2)^2 - (2) + 4] + 3 \\ &= 8 - 12 + 12 - 5 = [2 - 2][4 - 2 + 4] + 3 \\ &= 3 = [0][6] + 3 \\ &= 3 = 0 + 3 \end{aligned}$$

That is, $f(2) = 3$.

The Remainder Theorem affirms that this equality between the value of the functions and remainder always exists. We state the theorem as follows.

REMAINDER THEOREM

If $f(x)$ is divided by $x - r$ until a remainder independent of x is obtained so that $f(x) = (x - r)q(x) + R$, then this remainder equals $f(r)$. That is,

$$f(r) = R.$$

Let

$$f(x) = a_0 x^n + a_1 x^{n-1} + a_2 x^{n-2} + \cdots + a_n.$$

Then

$$f(r) = a_0 r^n + a_1 r^{n-1} + a_2 r^{n-2} + \cdots + a_n.$$

PROOF OF THE REMAINDER THEOREM Divide $f(x)$ by $x - r$; let the quotient be $q(x)$ and the remainder be R. Then from the complete check formula, we have

$$f(x) = (x - r)[q(x)] + R.$$

Substituting $x = r$,

$$\begin{aligned} f(r) &= (r - r)[q(r)] + R \\ &= (0)[q(r)] + R \\ &= R. \end{aligned}$$

The use of the Remainder Theorem simplifies considerably the calculation in the evaluation of a polynomial function of higher degree.

EXAMPLE 2 Find the value of $f(x) = x^5 - 3x^4 + 5x^3 - 14x^2 - 3x + 10$ when $x = 3$.

SOLUTION We use synthetic division to find the remainder by dividing $f(x)$ by $x - 3$.

$$f(x) = x^5 - 3x^4 + 5x^3 - 14x^2 - 3x + 10$$

	1	−3	5	−14	−3	10
		3	0	15	3	0
3	1	0	5	1	0	10

Therefore, $f(3) = 10$.

EXAMPLE 3 Find $f(-2)$ by dividing $f(x) = x^7 + 3x^6 + 5x - 2$ by $x + 2$.

SOLUTION We write the function in standard form, use synthetic division to divide the function by $x + 2$, and thus find $f(-2)$.

$$f(x) = x^7 + 3x^6 + 0x^5 + 0x^4 + 0x^3 + 0x^2 + 5x - 2$$

	1	3	0	0	0	0	5	−2
		−2	−2	4	−8	16	−32	54
−2	1	1	−2	4	−8	16	−27	52

Therefore, the remainder is $f(-2) = 52$.

EXAMPLE 4 Given $f(x) = x^4 + 3x^3 - 8x^2 - 12x + 22$, use synthetic division to find the quotient and $f(-4)$ when $f(x)$ is divided by $x + 4$.

SOLUTION To divide $f(x)$ by $x + 4$, we use -4 in the synthetic division calculation. This gives us the quotient, the remainder, and $f(-4)$.

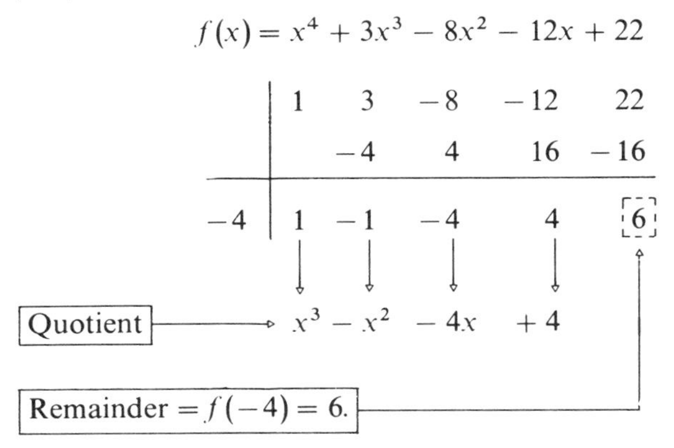

Exercises 2

A

1. Given $f(x) = x^5 - 16x^3 - 2x^2 + 10x - 6$, find $f(4)$.
2. Given $f(x) = x^4 - 4x^3 + 5x^2 - 4x + 10$, find $f(3)$.

In Exercises 3–5, use synthetic division to find the quotient and remainder.

3. $(x^3 + 4x^2 - 3x + 7) \div (x + 2)$
4. $(x^8 - 5x^3 + 3x^2 + 5x - 6) \div (x + 1)$
5. $(2x^3 + 5x^2 - 6x + 2) \div \left(x - \frac{3}{2}\right)$

B

1. Given $f(x) = x^4 - 6x^3 + 3x^2 + 12x - 4$. find $f(5)$.
2. Given $f(x) = x^4 - 7x^3 + 8x^2 - 10x - 8$, find $f(6)$.

In Exercises 3–5, use synthetic division to find the quotient and remainder.

3. $(x^4 - x^3 - x^2 + x + 3) \div (x - 3)$
4. $(2x^8 + 5x^3 - x - 6) \div (x + 1)$
5. $(3x^4 + 4x^3 - x^2 - 8x - 4) \div \left(x - \frac{2}{3}\right)$

3 The factor theorem

We made use of the Factor Theorem in Section 6 of Chapter 2, but at that time we did not call it the Factor Theorem.

> **THE FACTOR THEOREM**
>
> If r is a root of the equation $f(x) = 0$, then $x - r$ is a factor of the polynomial $f(x)$; conversely, if $x - r$ is a factor of $f(x)$, then r is a root of the equation $f(x) = 0$.

PROOF From the complete check formula for division and the Remainder Theorem, we have

$$f(x) = (x - r)[q(x)] + R, \qquad \text{where } R = f(r).$$

If r is a root of $f(x) = 0$, then $f(r) = 0$. Therefore $R = 0$ and

$$f(x) = (x - r)q(x); \tag{1}$$

that is, $(x - r)$ is a factor of $f(x)$. On the other hand, if $x - r$ is a factor of $f(x)$, the remainder R is equal to zero and $f(r) = 0$; then r is a root of $f(x) = 0$.

Equation 1 shows that when $R = 0$, both $x - r$ and the quotient $q(x)$ are factors of $f(x)$.

EXAMPLE 1 Show that $x + 3$ is a factor of $f(x) = x^4 + 5x^3 + 2x^2 - 20x - 24$.

SOLUTION We use synthetic division to show that $f(x)$ is exactly divisible by $x + 3$. That is, the remainder for $f(-3)$ is 0.

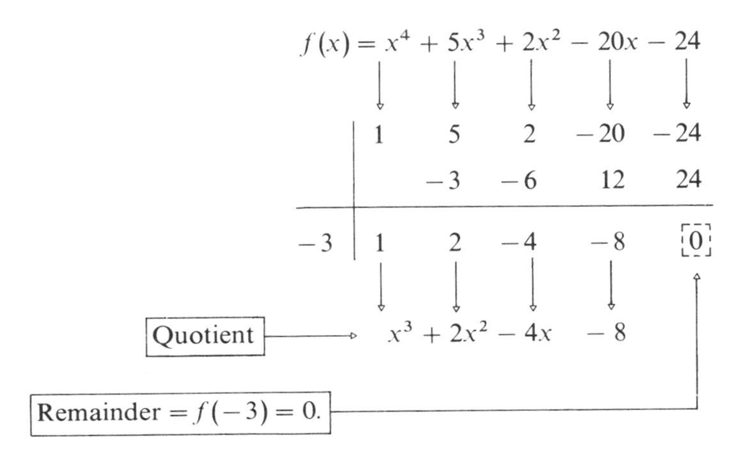

Also, note that both $x + 3$ and $x^3 + 2x^2 - 4x - 8$ are factors of $f(x)$. Of the two factors, $x + 3$ and $x^3 + 2x^2 - 4x - 8$, the second factor is called the **depressed factor**.

EXAMPLE 2 Use the factor theorem to show that $x^7 - a^7$ is exactly divisible by $x - a$; then find the depressed factor of $x^7 - a^7$.

SOLUTION Let $f(x) = x^7 - a^7$. Then $f(a) = a^7 - a^7 = 0$. Because the remainder $f(a) = 0$, $x - a$ is a factor of $f(x)$. We use synthetic division to find the depressed factor.

$$f(x) = x^7 + 0x^6 + 0x^5 + 0x^4 + 0x^3 + 0x^2 + 0x - a^7$$

$$\begin{array}{c|cccccccc} & 1 & 0 & 0 & 0 & 0 & 0 & 0 & -a^7 \\ & & a & a^2 & a^3 & a^4 & a^5 & a^6 & +a^7 \\ \hline a & 1 & a & a^2 & a^3 & a^4 & a^5 & a^6 & \boxed{0} \end{array}$$

The depressed factor. → $x^6 + ax^5 + a^2x^4 + a^3x^3 + a^4x^2 + a^5x + a^6$

Remainder $= f(a) = 0$.

Exercises 3

Use the Factor Theorem and synthetic division to determine the answer for each of the following questions.

A

1. Is $x - 7$ a factor of $x^3 - 5x^2 - 17x + 21$?
2. Is $x + 3$ a factor of $x^4 - 2x^3 - 12x^2 + 11x + 6$?
3. Is $x - 7$ a factor of $x^4 - 6x^3 - 5x^2 - 12x - 15$?
4. Is $x + a$ a factor of $x^6 + a^6$?

B

1. Is $x - 3$ a factor of $x^3 - 6x^2 + 11x - 6$?
2. Is $x - 2$ a factor of $x^4 - 2x^3 - 12x^2 + 11x + 6$?
3. Is $x - 8$ a factor of $x^4 - 7x^3 - 6x^2 - 15x - 8$?
4. Is $x + a$ a factor of $x^4 + a^4$?

4 Number and nature of the roots of a polynomial equation

We will give several theorems without proof. The proof for some is beyond the level of this book, while for the others, we do not wish to take the space required to go through the proof.

THEOREM 1 Any polynomial of degree n can be expressed as the product of exactly n linear factors. That is,

$$f(x) = a_0(x - r_1)(x - r_2)(x - r_3) \cdots (x - r_n).$$

THEOREM 2 An equation of degree n, where $n > 0$, has at most n distinct roots.

The roots $r_1, r_2, \ldots, r_n$ are not necessarily all distinct roots. If r represents one of the roots $r_1, r_2, \ldots, r_n$, and if r occurs exactly k times among these roots, that is, if $(x - r)^k$ is the highest power of $x - r$ that is a factor of $f(x)$, then r is said to be a root of **multiplicity k**. If $k = 2$, r is a **double root**, if $k = 3$, r is a **triple root**, etc.

THEOREM 3 If a root of multiplicity k is counted as k roots, every equation of degree n has exactly n roots.

EXAMPLE 1 For the equation $4(x - 2)^3(x + 1)^2(x - 5) = 0$, the root 5 is a simple root, -1 is a double root and 2 is a triple root. This equation is a sixth-degree equation having the six roots 5, -1, -1, 2, 2, and 2.

EXAMPLE 2 Find the three roots of the cubic equation $x^3 = 8$.

SOLUTION We express this equation in standard form, factor, equate each factor to zero, and solve.

$$x^3 - 8 = 0$$

$$(x - 2)(x^2 + 2x + 4) = 0 \qquad \text{Factor the difference of two cubes.}$$

$$x - 2 = 0 \quad \Big| \quad x^2 + 2x + 4 = 0$$

$$x = 2 \quad \Big| \quad x = \frac{-2 \pm \sqrt{4 - 16}}{2} \qquad \text{Apply the quadratic formula.}$$

$$= \frac{-2 \pm i2\sqrt{3}}{2}$$

$$= -1 \pm i\sqrt{3}$$

Therefore, the solution set is

$$\{2, -1 + i\sqrt{3}, -1 - i\sqrt{3}\}$$

5 Formation of an equation with given roots

Let $r_1, r_2, \ldots, r_n$ be any n given numbers. We wish to write a polynomial equation of degree n having these n numbers as roots.

The equation

$$f(x) = a_0(x - r_1)(x - r_2) \cdots (x - r_n) = 0,$$

where a_0 is any constant different from zero, clearly has the given numbers as roots since, if x is replaced by any of the numbers, one factor of the product is equal to zero, and hence the entire product is equal to zero.

If we multiply together the factors in this product, we obtain a polynomial equation in which the term of highest degree in x is $a_0 x^n$. It follows that this equation is the required polynomial equation.

EXAMPLE 1 Write an equation in standard form of degree 4 with 2, -2, $\frac{2}{3}$, and $\frac{1}{6}$ as its roots.

SOLUTION We form the equation

$$f(x) = a_0(x - 2)(x + 2)\left(x - \frac{2}{3}\right)\left(x - \frac{1}{6}\right) = 0,$$

which is satisfied by any one of the given numbers. We let $a_0 = 18$ to avoid a fractional equation and multiply the factors to obtain the required equation.

$$f(x) = 18x^4 - 15x^3 - 70x^2 + 60x - 8 = 0$$

CHECK We use synthetic division to show that this equation has the required roots.

$$f(x) = 18x^4 - 15x^3 - 70x^2 + 60x - 8 = 0$$

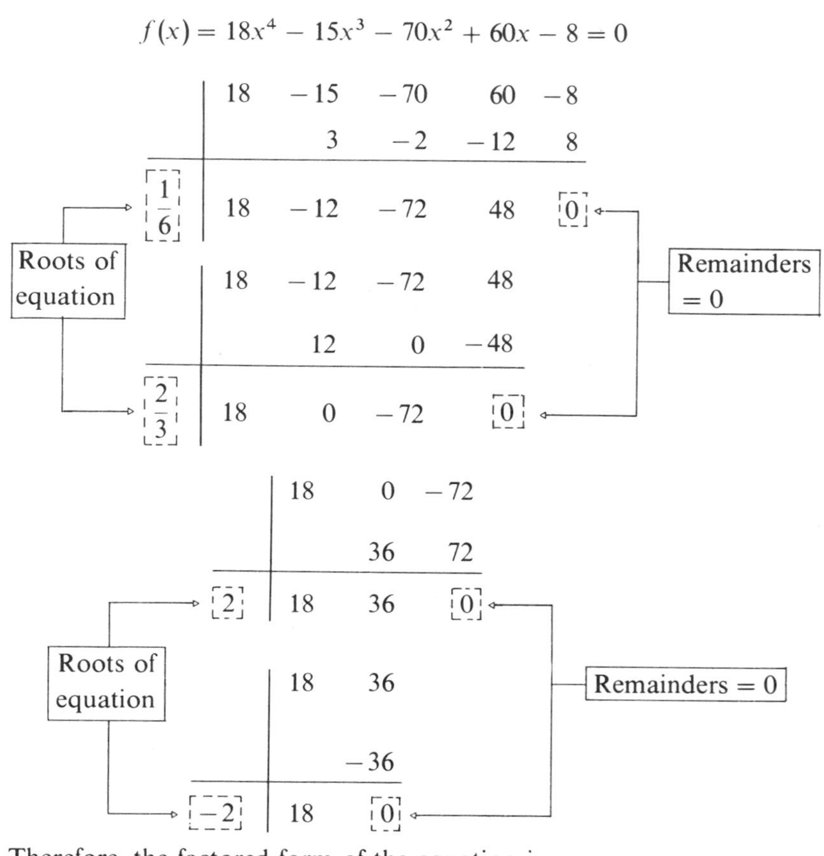

Therefore, the factored form of the equation is

$$(x - 2)(x + 2)\left(x - \frac{2}{3}\right)\left(x - \frac{1}{6}\right) = 0, \qquad \text{with } a_0 = 1.$$

When using synthetic division to find or check the roots of an equation, always use the remaining depressed factor in the test for additional roots after one root has been determined.

EXAMPLE 2 Write an equation of degree 5 having roots $\sqrt{2}$, $-\sqrt{2}$, $3i$, $-3i$, and 2 as its roots.

SOLUTION In order to have these roots, the equation must have the factors $(x - \sqrt{2})$, $(x + \sqrt{2})$, $(x - 3i)$, $(x + 3i)$, and $(x - 2)$. In factored form, the equation is

$$a_0(x - \sqrt{2})(x + \sqrt{2})(x - 3i)(x + 3i)(x - 2) = 0.$$

Since there are no fractions, we choose $a_0 = 1$ and multiply the factors.

$$x^5 - 2x^4 + 7x^3 - 14x^2 - 18x + 36 = 0$$

CHECK We use synthetic division to see if the given roots satisfy this equation.

$$x^5 - 2x^4 + 7x^3 - 14x^2 - 18x + 36 = 0$$

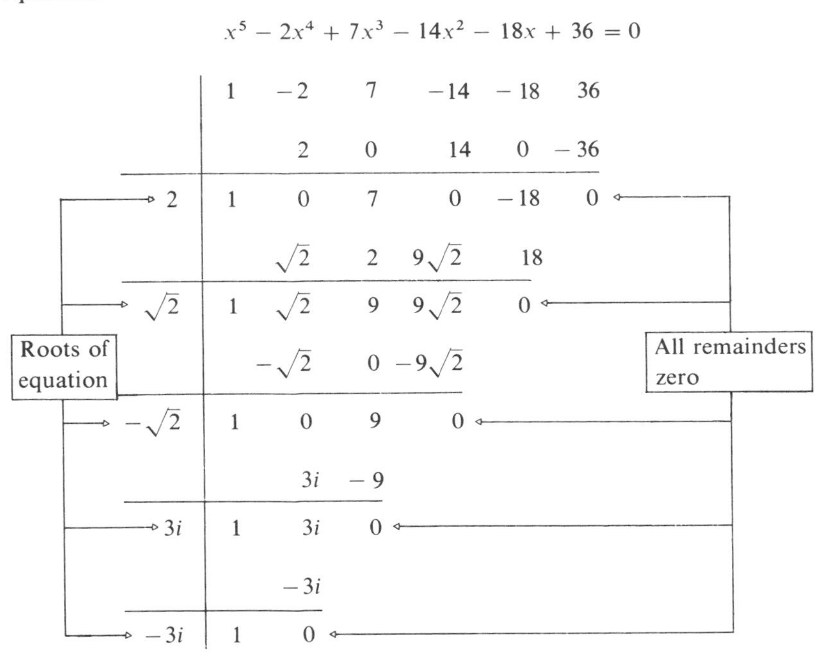

Exercises 5

A

1. Find all the roots of the equation $(3x + 2)(5x - 3)(2x - 7) = 0$.
2. Write a third-degree equation with roots -1, 3, and 4.
3. Write a fourth-degree equation with roots 1, -1, i, and $-i$.
4. Write a fourth-degree equation with roots -3, -3, $\frac{5}{2}$, and $\frac{5}{2}$.
5. Write $f(x)$ in factored form, given that the roots of the equation $f(x) = 0$ are -3, 1, and 7.
6. Show that 2 is a double root of the equation $x^3 - 9x^2 + 24x - 20 = 0$. Find the other root.
7. Form an equation with integral coefficients if $-2 + \sqrt{3}$ and $-2 - \sqrt{3}$ are single roots and 1 is a double root.

B

1. Find all the roots of the equation $(x - 2 + 3i)(x - 2 - 3i) \times (x - 5)^2 = 0$.
2. Write a third-degree equation with roots 2, $1 + \sqrt{2}$, and $1 - \sqrt{2}$.
3. Write a third-degree equation with roots 2, $-1 + i\sqrt{3}$, and $-1 - i\sqrt{3}$.

4. Write a fourth-degree equation with roots $\frac{5}{3}$, $-\frac{1}{4}$, $2+\sqrt{5}$, and $2-\sqrt{5}$.
5. Write $f(x)$ in factored form, given that the roots of the equation $f(x)=0$ are -3, 1, i, and $-i$.
6. Show that 3 is a double root of the equation $x^3-2x^2-15x+36=0$. Find the other root.
7. Form an equation with integral coefficients if $2+i$ and $2-i$ are single roots and 2 is a double root.

6 Rational roots of an equation

We have shown that an nth degree polynomial equation $f(x)=0$ has n roots and that $f(r_i)=0$, $i=1, 2, 3, \ldots, n$.

We shall now give some theorems that are helpful in finding rational roots of polynomial equations having integral coefficients. Before stating the theorems, let us consider an example.

EXAMPLE 1 Form a cubic equation having roots 3, $\frac{2}{3}$, and $-\frac{1}{3}$.

SOLUTION If the equation is to have these roots, then it must have factors of $(x-3)$, $(x-\frac{2}{3})$, and $(x+\frac{1}{3})$. The equation (in factored form) is therefore

$$f(x)=a_0(x-3)\left(x-\frac{2}{3}\right)\left(x+\frac{1}{3}\right)=0. \tag{1}$$

To clear the equation of fractions, we let $a_0=9$.

$$f(x)=(x-3)(3x-2)(3x+1)=0$$

$$f(x)=9x^3-30x^2+7x+6=0 \tag{2}$$

Note that if the factors in Equation 1 were multiplied, then the last term in the product would be $(-3)(-\frac{2}{3})(\frac{1}{3})=\frac{6}{9}$. This is the same fraction as that formed by dividing the last term of Equation 2 by the numerical coefficient of its first term.

Let $p/q=6/9$ and list the factors of p and q.

$$p=6: \pm 1, \pm 2, \pm 3, \pm 6$$

$$q=9: \pm 1, \pm 3, \pm 9$$

The distinct fractions of the form p/q are

$$\pm\left[1, 2, \boxed{3}, 6, \boxed{\frac{1}{3}}, \frac{1}{9}, \boxed{\frac{2}{3}}, \frac{2}{9}\right].$$

Note that this set of numbers includes the roots of Equation 1. Knowing the set of possible roots of an equation is helpful in limiting the number of trials needed to find the roots of that equation. The relation that appears to be true in Example 1 is stated in the following theorem.

THEOREM 4 Let p/q, a real rational fraction in lowest terms, be a root of the equation

$$f(x) = a_0 x^n + a_1 x^{n-1} + \cdots + a_{n-1} x + a_n = 0, \qquad a_0 \neq 0$$

where $a_0, a_1, \ldots, a_{n-1}, a_n$ are integers. Then p is a factor of a_n, the constant term, and q is a factor of a_0, the coefficient of x^n.

As a corollary to this theorem, if the first coefficient of $f(x)$ is 1 and all the other coefficients are either integers or zeros, then every rational root of $f(x) = 0$ is a factor of a_n and all rational roots are integers.

EXAMPLE 2 Find a rational root of $6x^3 + 11x^2 - 3x - 2 = 0$. Then equate the depressed second-degree factor to zero to find the remaining roots.

SOLUTION In this equation p is -2 and q is 6. We form the possible roots p/q by factoring p and q then writing all combinations of the factors of p over the factors of q.

Factors of p: $\pm 1, \pm 2$

Factors of q: $\pm 1, \pm 2, \pm 3, \pm 6$

$$\text{Possible roots } \frac{p}{q}: \quad \pm \left| 1, 2, \frac{1}{2}, \frac{1}{3}, \frac{1}{6}, \frac{2}{3} \right|$$

By direct substitution, it is easy to show that neither $+1$ nor -1 satisfy the equation. We use synthetic division to see if 2 is a root.

$$6x^3 + 11x^2 - 3x - 2 = 0$$

	6	11	−3	−2	
		12	46	86	
2	6	23	43	84	$R \neq 0$; therefore 2 is not a root.

Next we try -2.

	6	11	−3	−2	
		−12	2	2	
−2	6	−1	−1	0	$R = 0$; therefore -2 is a root.

Rather than make more tries using synthetic division, we equate the depressed second-degree factor to zero and solve it by factoring.

$$6x^2 - x - 1 = 0$$

$$(2x - 1)(3x + 1) = 0$$

$$2x - 1 = 0 \quad \Big| \quad 3x + 1 = 0$$

$$x = \frac{1}{2} \quad \Big| \quad x = -\frac{1}{3}$$

Therefore, the solution set is $\{-2, \frac{1}{2}, -\frac{1}{3}\}$.

EXAMPLE 3 Find the solution set of the equation $x^5 - 3x^4 - 3x^3 + 9x^2 + 2x - 6 = 0$.

SOLUTION Because the coefficient of x^5 is 1 and all other coefficients are integers, all rational roots of this equation are factors of -6. Thus, the possible roots are $\pm[1, 2, 3, 6]$. We use synthetic division as an aid in determining the roots.

$$x^5 - 3x^4 - 3x^3 + 9x^2 + 2x - 6 = 0$$

	1	−3	−3	9	2	−6	
		1	−2	−5	4	6	
1	1	−2	−5	4	6	[0]	$R = 0$; therefore 1 is a root.
		−1	3	2	−6		
−1	1	−3	−2	6	[0]		$R = 0$; therefore −1 is a root.
		2	−2	−8			
2	1	−1	−4	[−2]			$R \neq 0$; therefore 2 is not a root.
	1	−3	−2	6			
		3	0	−6			
3	1	0	−2	[0]			$R = 0$; therefore 3 is a root.

We equate the last depressed factor to zero and solve.

$$x^2 - 2 = 0$$

$$x = \pm\sqrt{2}$$

Therefore, the solution set is $\{1, -1, 3, \sqrt{2}, -\sqrt{2}\}$.

IMAGINARY ROOTS

> **THEOREM 5** If a polynomial equation $f(x) = 0$ with real coefficients has an imaginary root $a + bi$ $(b \neq 0)$, then it also has a root $a - bi$; that is, complex roots occur in conjugate pairs.

EXAMPLE 4 Find the solution set of the equation $x^3 - 3x^2 + 7x - 5 = 0$.

SOLUTION Because the leading coefficient is 1 and all other coefficients are integers, all rational roots of this equation are factors of -5. The possible rational roots are ± 1, and ± 5. Using synthetic division, we find

$$x^3 - 3x^2 + 7x - 5 = 0$$

$$\begin{array}{r|rrrr} & 1 & -3 & 7 & -5 \\ & & 1 & -2 & 5 \\ \hline 1 & 1 & -2 & 5 & \boxed{0} \end{array} \qquad R = 0\text{; therefore 1 is a root.}$$

Equating the depressed factor to zero, we have

$$x^2 - 2x + 5 = 0$$

$$x = \frac{-(-2) \pm \sqrt{(-2)^2 - 4(1)(5)}}{2(1)}$$

$$= \frac{2 \pm \sqrt{4 - 20}}{2}$$

$$= \frac{2 \pm 4i}{2} = 1 \pm 2i$$

Therefore, the solution set is $\{1, 1 + 2i, 1 - 2i\}$.

THE EQUATIONS $f(x) = 0$ AND $f(-x) = 0$

> **THEOREM 6** If, in an equation $f(x) = 0$, x is replaced by $-x$, the resulting equation $f(-x)$ has roots which are the negatives of the roots of the given equation.

PROOF Suppose that the equation $f(x) = 0$ has the roots a, b, c, $\ldots$, k. That is, the equation is satisfied if we substitute $x = a$, $x = b$, $x = c$, $\ldots$, $x = k$. If we replace x by $-x$ in the original equation, we obtain a new equation $f(-x) = 0$ in which $-x$ plays the same role that

x does in the original equation. The equation $f(-x) = 0$ is therefore satisfied if and only if

$$-x = a, \quad -x = b, \quad -x = c, \ldots, \quad -x = k,$$

or if

$$x = -a, \quad x = -b, \quad x = -c, \ldots, \quad x = -k,$$

which is what we wished to show. Therefore, to obtain an equation whose roots are the negatives of the roots of a given polynomial equation, reverse the signs of the terms of odd degree.

EXAMPLE 5 Write an equation whose roots are the negatives of the roots of the equation

$$2x^4 - 7x^3 + x^2 + 7x - 3 = 0. \tag{3}$$

SOLUTION Changing the signs of the terms of odd degree, we have

$$2x^4 + 7x^3 + x^2 - 7x - 3 = 0. \tag{4}$$

The solution set for Equation 3 is $\{3, \frac{1}{2}, 1, -1\}$, while the solution set for Equation 4 is $\{-3, -\frac{1}{2}, -1, 1\}$. We show the check of these roots for Equations 3 and 4 as follows.

CHECK Equation 3:

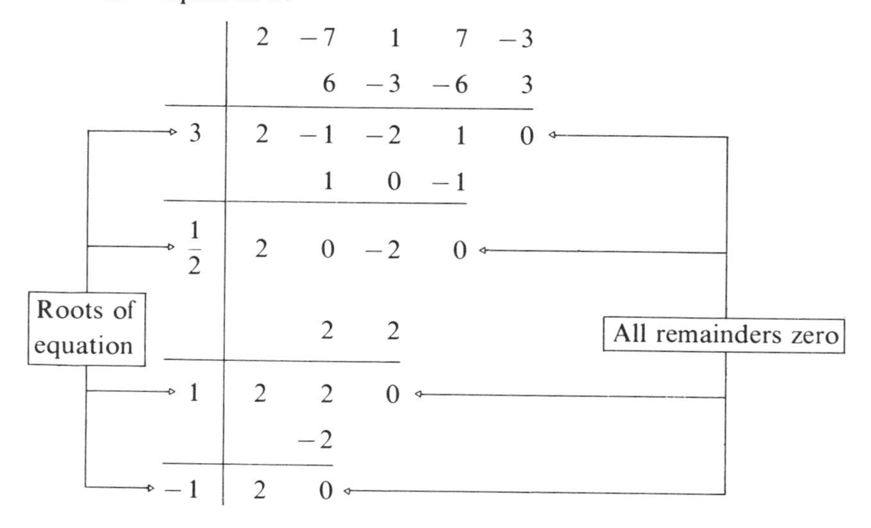

CHECK Equation 4:

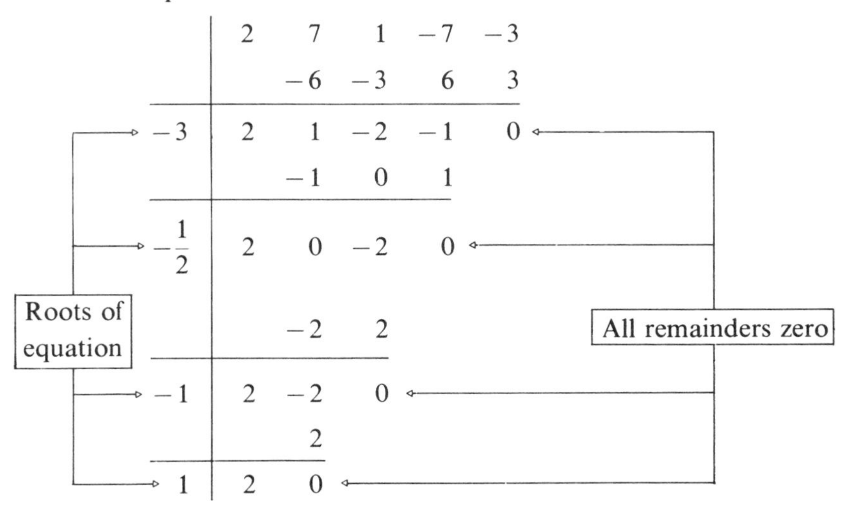

VARIATIONS OF SIGN

If a polynomial is written in standard form, a **variation of sign** is said to occur whenever the coefficients of two successive terms have opposite signs. In counting the variations of sign, one may neglect any term whose coefficient is zero, since such a term is assumed to have the same sign as the preceding term.

EXAMPLE 6

$x^5 - 3x^4 - 2x^3 + 4x^2 + x - 7$ has three variations of sign.

$x^9 - x^7 + 5x^6 - 3x^5 - x^3 + 2x - 8$ has five variations of sign.

Once the possible rational roots have been determined, it is helpful to have a method for eliminating some of the possibilities without having to test all of them by using synthetic division. Descartes' Rule of Signs gives some help in this respect.

> **DESCARTES' RULE OF SIGNS**
>
> The number of positive real roots of a polynomial equation $f(x) = 0$ with real coefficients is either equal to the number of variations in sign of $f(x)$ or is less than that number by an even integer.

In particular, if there are no variations of sign in $f(x)$, there are no positive real roots of $f(x) = 0$ and, if there is just one variation, then

$f(x) = 0$ has just one positive root. In the remaining cases, the rule does not determine the exact number of positive roots but gives a number that cannot be exceeded by the number of positive roots.

A corresponding limitation on the number of negative roots of $f(x) = 0$ can be obtained by transforming $f(x) = 0$ into an equation whose roots are opposite in sign to those of $f(x) = 0$. Since every negative root of $f(x) = 0$ is transformed into a positive root of $f(-x) = 0$, and conversely, it follows that the number of negative roots of $f(x) = 0$ is either equal to the number of variations in sign in $f(-x) = 0$ or is less than that number by an even integer.

EXAMPLE 7 Discuss the number of positive, negative, zero, and complex roots of $f(x) = x^7 - 3x^6 - 2x^5 + 4x^4 + x^3 - 7x^2 = 0$.

SOLUTION Since $f(x) = 0$ has three variations in sign, $f(x) = 0$ has either three positive roots, or one positive root. The transformed equation

$$f(-x) = -x^7 - 3x^6 + 2x^5 + 4x^4 - x^3 - 7x^2 = 0$$

has two variations in sign. Therefore, $f(-x)$ has either two or no positive roots, which means that $f(x)$ has either two or no negative roots.

Since x^2 is a factor of $f(x)$, zero is a double root of $f(x) = 0$.

Because $f(x) = 0$ is a seventh degree equation, it has seven roots. We make a chart showing the possible number of zero, positive, negative, and complex roots. Each row of the chart shows a possible collection of roots.

Zero roots	Positive roots	Negative roots	Complex roots (must come in conjugate pairs)	Total number of roots
2	3	2	0	7
2	3	0	2	7
2	1	2	2	7
2	1	0	4	7

LIMITS FOR THE MAGNITUDES OF THE REAL ROOTS

Descartes' Rule gives a limit for the number of positive roots of $f(x) = 0$. The following theorems tell how to find upper and lower limits of an interval that includes the roots of a polynomial equation.

THEOREM 7 If L is a positive number such that, when $f(x)$ is divided by $x - L$ using synthetic division every number in the third line of the synthetic division is positive or zero, then L is an upper limit of the roots of the equation $f(x) = 0$.

THEOREM 8 If a is an upper limit of the roots of $f(-x) = 0$, then $l = -a$ is a lower limit of the roots of $f(x) = 0$.

Thus, in searching for roots of an equation, we need not look for any roots outside the interval $[l, L]$.

EXAMPLE 8 Determine an interval that contains all the roots of the equation $3x^4 + 5x^3 - 14x^2 - 20x + 8 = 0$.

SOLUTION We show the test for an upper positive limit of 3.

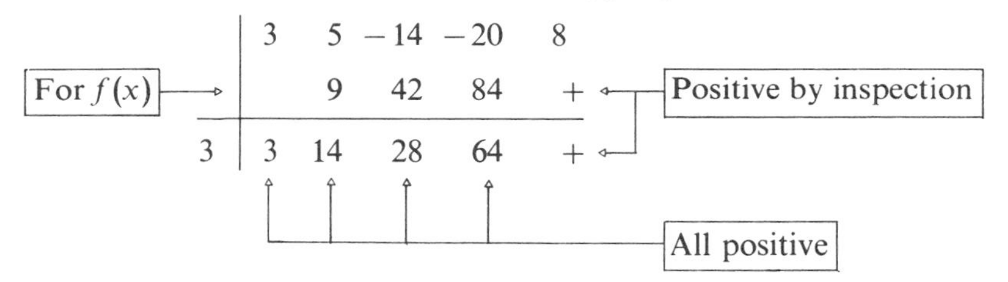

For $f(-x)$:

	3	−5	−14	20	8
		12	28	+	+
4	3	7	14	+	+

Positive by inspection

Since this last test was on $f(-x)$, there are no roots less than -4. Therefore, all roots of $f(x) = 0$ are in the interval $[-4, 3]$.

IRRATIONAL ROOTS

Most of the equations dealt with so far have had either rational or complex roots. In applied problems, the polynomial graph may cross the x-axis at a point where x is an irrational number. In this case the solution of the equation $f(x) = 0$ will have irrational roots.

EXAMPLE 9 Find the positive root of $f(x) = x^3 + 2x^2 - 5x - 5 = 0$ correct to two decimal places.

SOLUTION By the usual tests, we find that the equation has no rational roots. There is just one positive root, since there is only one variation of sign and $f(0) < 0$. We also find that 2 is an upper limit of the roots. By synthetic division, we find the coordinates of enough points to draw the graph of $f(x)$, as shown in Figure 1.

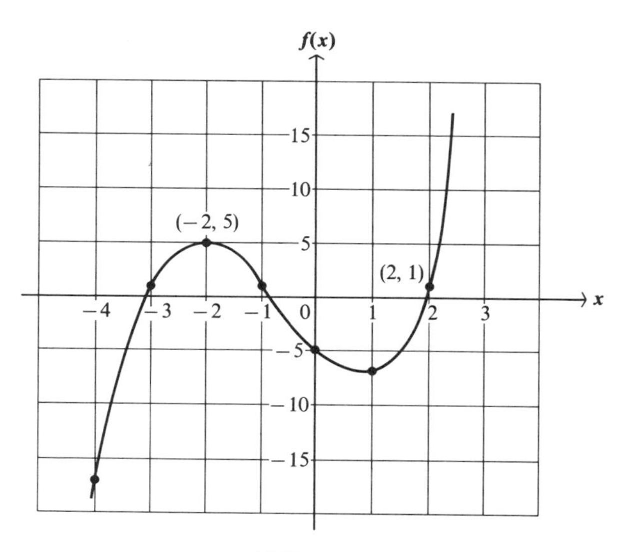

FIGURE 1

From the graph, to the nearest tenth, we estimate the positive root to be 1.9.

By synthetic division and the remainder theorem, we find that $f(1.9) = -0.421$ and $f(2) = 1$. Since $f(x)$ changes sign between 1.9 and 2.0, the root we seek must lie between 1.9 and 2.

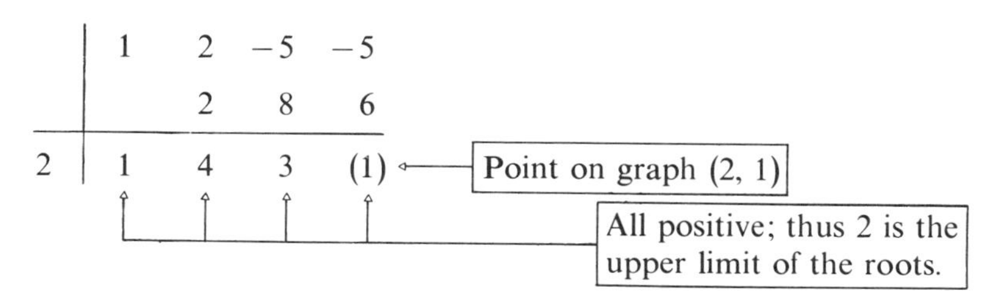

	1	2	−5	−5
		1.9	7.41	4.579
1.9	1	3.9	2.41	(−0.421)

Point on graph (1.9, −0.421)

Although interpolation does not give exact results, we assume that for small variations it gives a close approximation of the value of x for which $f(x) = 0$.

We interpolate between the two points where $x = 1.9$ and 2.

$$0.1\left\{\begin{array}{ll} 2 & 1 \\ d\left\{\begin{array}{l} x \\ 1.9 \end{array}\right. & \left.\begin{array}{r} 0 \\ -.421 \end{array}\right\} 0.421 \end{array}\right\} 1.421$$

$$\frac{d}{0.1} = \frac{0.421}{1.421}$$

$$d = 0.029627 \doteq 0.03$$

Thus, the value of x is approximately $1.9 + 0.03 = 1.93$. The positive root of $f(x) = x^3 + 2x^2 - 5x - 5 = 0$ is 1.93, correct to two decimal places.

Exercises 6

A In Exercises 1 and 2, how many variations in sign does each polynomial have?

1. $5x^4 + 2x^3 - x^2 - x + 6$ **2.** $x^6 - 3x^5 + x^4 - x^2 + 5$

In Exercises 3–7, find the solution set for each equation.

3. $x^4 + 4x^2 - 5 = 0$ **4.** $x^4 - 7x^2 - 6x = 0$

5. $9x^5 + 36x^4 - 10x^3 - 40x^2 + x + 4 = 0$

6. $x^4 + 11x^3 + 44x^2 + 74x + 40 = 0$

7. $9x^4 - 30x^3 + 16x^2 + 30x - 25 = 0$

8. Find the real root of $x^3 - 3x^2 + 3x - 7 = 0$ accurate to two decimal places.

B In Exercises 1 and 2, how many variations in sign does each polynomial have?

1. $3x^5 - 2x^3 - x + 4$ **2.** $x^6 + 3x^5 - x^3 + x - 6$

In Exercises 3–7, find the solution set for each equation.

3. $x^4 - 3x^2 - 4 = 0$ **4.** $x^4 - 2x^3 - 5x^2 + 6x = 0$

5. $4x^5 - 16x^4 - 5x^3 + 20x^2 + x - 4 = 0$

6. $4x^3 - 24x^2 + 61x - 75 = 0$

7. $36x^4 - 66x^3 + 42x^2 - 11x + 1 = 0$

8. Find the positive root of $x^3 + 3x^2 - 6x - 2 = 0$ accurate to two decimal places.

7 The graph of a polynomial

The graph of a polynomial may be sketched by plotting points on the coordinate axes and drawing a smooth curve through these points. To find the value of $f(x)$ corresponding to any value of x, use synthetic division and the Remainder Theorem. For $x = 0$, $x = 1$, or $x = -1$, direct substitution is easier.

Some typical forms that the graph of a polynomial function may assume are shown in Figures 2*a*, 2*b*, 2*c* and 2*d*. Since these are functions, any vertical line meets the graph in just one point. A horizontal line meets it in at most n points, where n is the degree of $f(x)$. In particular, the x values of the points where the graph crosses the x-axis (x-intercepts) are the real roots of $f(x) = 0$, since they are the values of x which, when substituted in the given equation, make $f(x) = 0$.

A **relative maximum** point on the graph is a point, such as H (see Figure 2), that is higher than any nearby point on the graph. A **relative minimum** point is a point such as M (see Figure 2), that is lower than

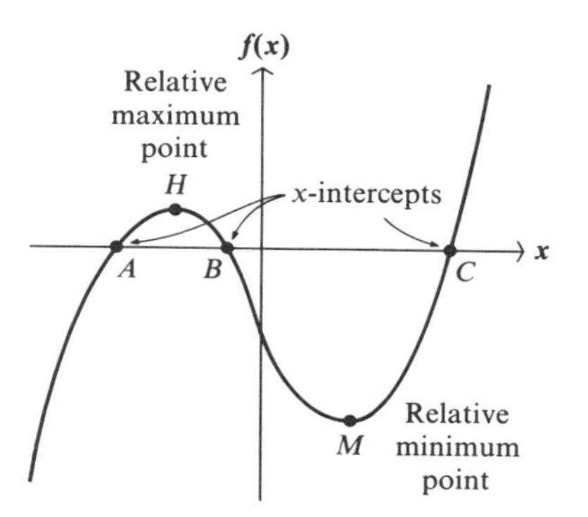

$f(x) = 0$ has 3 real roots

a.

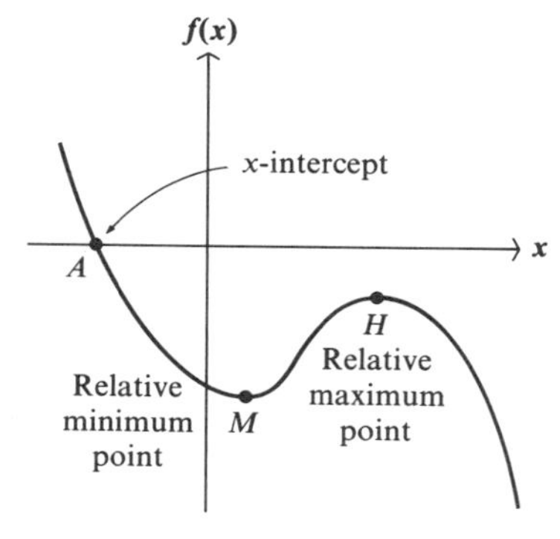

$f(x) = 0$ has 1 real and 2 complex roots

b.

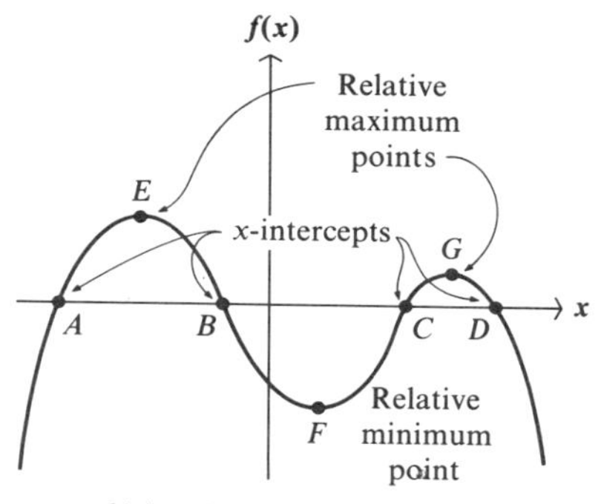

$f(x) = 0$ has 4 real roots

c.

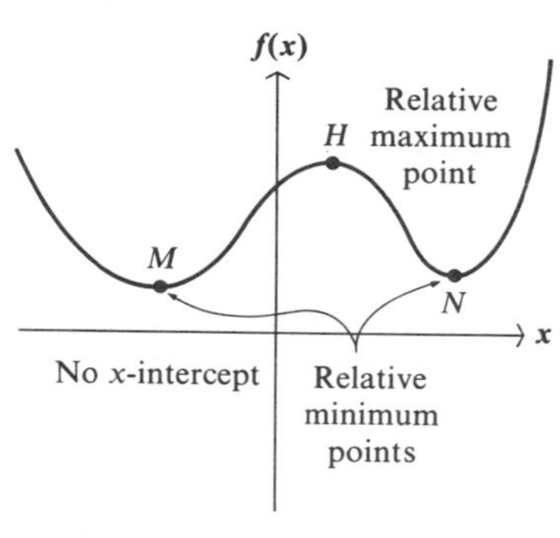

$f(x) = 0$ has 4 complex roots

d.

FIGURE 2

any nearby point. These maximum and minimum points on a graph have important meaning in applied mathematics. To locate them accurately by our methods of graphing is a tedious process. We will, in most cases, only approximate their position on the graph. An important part of calculus is concerned with techniques for finding maximum and minimum points on a graph.

EXAMPLE 1 Graph $f(x) = x^3 - 5x^2 + 2x + 8$.

SOLUTION We use synthetic division and the Remainder Theorem to find points on the graph. Cubic curves are usually S-shaped curves.

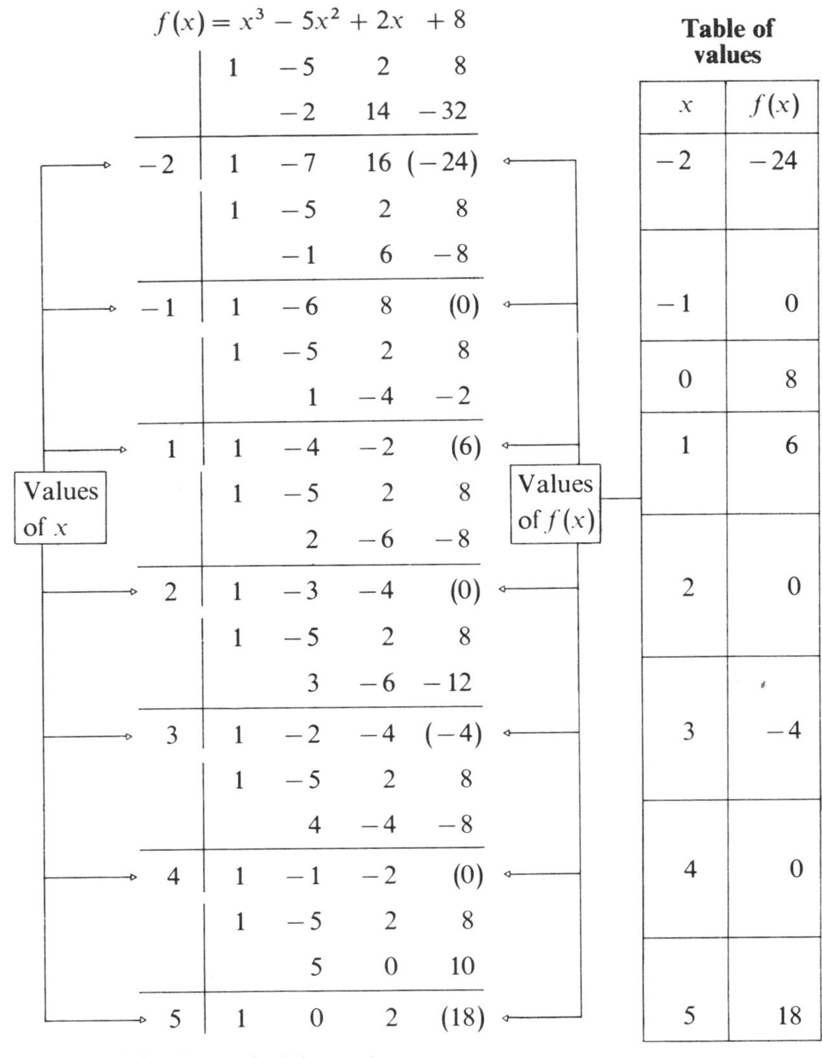

Table of values

x	$f(x)$
-2	-24
-1	0
0	8
1	6
2	0
3	-4
4	0
5	18

The graph is shown in Figure 3.

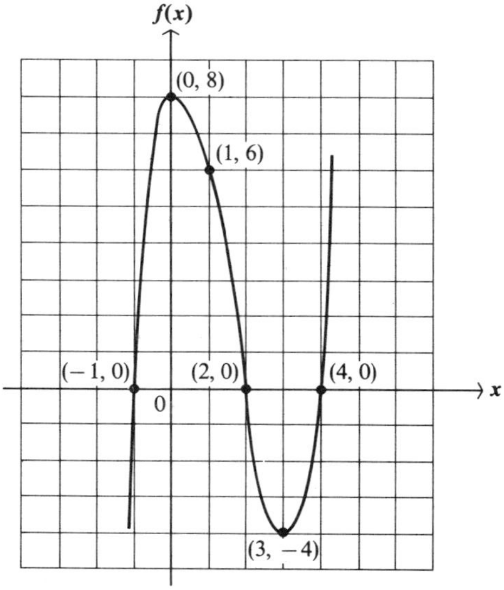

FIGURE 3

Exercises 7

Graph each of the following functions.

A

1. $f(x) = x^3 - 4x^2 + 2x + 3$
2. $f(x) = 2x^3 - x^2 - x + 5$
3. $f(x) = x^4 + x^2 - 3x - 5$
4. $f(x) = x^3$
5. $f(x) = x^4$

B

1. $f(x) = x^3 - 3x^2 + 6$
2. $f(x) = 2x^3 - 9x^2 + 12x - 3$
3. $f(x) = x^3 - 3x^2 - 4x + 10$
4. $f(x) = x^3 - 12x$
5. $f(x) = x^4 - x^3 + 2$

Summary

A **polynomial function** of degree n in the single variable x can be written in the form

$$f(x) = a_0 x^n + a_1 x^{n-1} + a_2 x^{n-2} + \cdots + a_{n-1} x + a_n,$$

where n is a positive integer or zero and the coefficients $a_0, a_1, \ldots, a_n$ are constants and $a_0 \neq 0$. The corresponding equation,

$$a_0 x^n + a_1 x^{n-1} + \cdots + a_{n-1} x + a_n = 0,$$

represents the general **polynomial equation** of degree n in x. A number r is said to be a **root** of the equation $f(x) = 0$, if and only if $f(r) = 0$. Every polynomial equation has at least one root.

REMAINDER THEOREM If $f(x)$ is divided by $x - r$ until a constant remainder is obtained, then this remainder equals $f(r)$. That is $f(r) = R$.

THE FACTOR THEOREM If r is a root of the equation $f(x) = 0$, then $x - r$ is a factor of the polynomial $f(x)$; conversely, if $x - r$ is a factor of $f(x)$, then r is a root of the equation $f(x) = 0$

THEOREM 1 Any polynomial of degree n can be expressed as the product of exactly n linear factors.

THEOREM 2 An equation of degree n, where $n > 0$, has at most n distinct roots.

THEOREM 3 If a root of multiplicity k is counted as k roots, every equation of degree n has exactly n roots.

THEOREM 4 Let p/q, a real rational fraction in lowest terms, be a root of the equation

$$f(x) = a_0 x^n + a_1 x^{n-1} + \cdots + a_{n-1} x + a_n = 0, \qquad a_0 \neq 0$$

where $a_0, a_1, \ldots, a_{n-1}, a_n$ are integers. Then p is a factor of a_n, the constant term, and q is a factor of a_0, the coefficient of x^n.

THEOREM 5 If a polynomial equation $f(x) = 0$ with real coefficients has an imaginary root $a + bi$ ($b \neq 0$), then it also has a root $a - bi$; that is, complex roots occur in conjugate pairs.

THEOREM 6 If, in an equation $f(x) = 0$, x is replaced by $-x$, the resulting equation $f(-x) = 0$ has roots which are the negatives of the roots of the given equation.

If a polynomial is written in standard form, a **variation of sign** is said to occur whenever the coefficients of two successive terms have opposite signs.

DESCARTES' RULE OF SIGNS The number of positive roots of a polynomial equation $f(x) = 0$ with real coefficients is either equal to the number of variations in sign of $f(x)$ or is less than that number by an even integer. If there are no variations of sign in $f(x)$, there are no positive roots of $f(x) = 0$ and, if there is just one variation, then $f(x) = 0$ has just one positive root.

THEOREM 7 If L is a positive number such that, when $f(x)$ is divided by $x - L$ using synthetic division every number in the third line of the synthetic division is positive or zero, then L is an upper limit of the roots of the equation $f(x) = 0$.

THEOREM 8 If a is an upper limit of the roots of $f(-x) = 0$, then $l = -a$ is a lower limit of the roots of $f(x) = 0$.

A **relative maximum point** on a graph is a point that is higher than any nearby point on the graph. A **relative minimum point** on a graph is a point that is lower than any nearby point on the graph.

Review exercises

A

1. Given $f(x) = x^5 - 4x^4 + x^2 - 6x + 14$, find $f(4)$.

In Exercises 2 and 3, use synthetic division to find the quotient and remainder.

2. $(x^7 + 4x^6 - 9x^4 - 2x^2 + x + 6) \div (x + 3)$
3. $(x^6 + 1) \div (x - 1)$
4. Is $x + 6$ a factor of $x^5 + 6x^4 + 2x^2 + 14x + 12$? Show proof for your answer.
5. Write a fifth-degree equation with 1, -1, 2, $(1 + i\sqrt{3})/2$, and $(1 - i\sqrt{3})/2$ as its roots.
6. Find the solution set of the equation $2x^4 - 9x^3 + 6x^2 + 11x - 6 = 0$.
7. Let the graph shown in Figure 4 be that of a fourth-degree polynomial function $f(x)$.

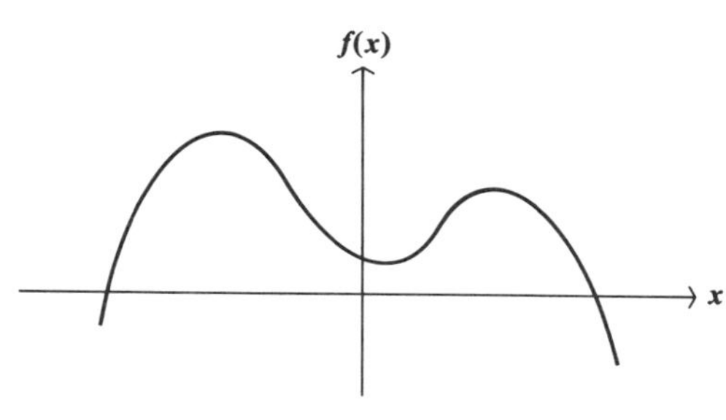

FIGURE 4

   a. How many real roots does the equation $f(x) = 0$ have?
   b. How many imaginary roots does the equation $f(x) = 0$ have?
8. Graph the function $f(x) = x^3 - 4x$.

9. Find the positive root of $x^3 + 2x^2 - 13x - 19 = 0$ to two decimal places.

B

1. Given $f(x) = 3x^6 - 30x^4 + 9x^3 + x + 8$, find $f(3)$.

In Exercises 2 and 3, use synthetic division to find the quotient and remainder.

2. $(x^7 - 4x^5 + x^2 - 4x + 3) \div (x + 2)$
3. $(x^5 + 1) \div (x - 1)$
4. Is $x - 5$ a factor of $x^6 - 6x^5 + 5x^4 + x^2 - 25$? Show proof for your answer.
5. Write a fourth-degree equation with 2, -2, $(1 + i\sqrt{2})/3$ and $(1 - i\sqrt{2})/3$ as its roots.
6. Find the solution set of the equation $3x^4 - 7x^3 - x^2 + 7x - 2 = 0$.
7. Let the graph shown in Figure 5 be that of a third-degree polynomial function $f(x)$.

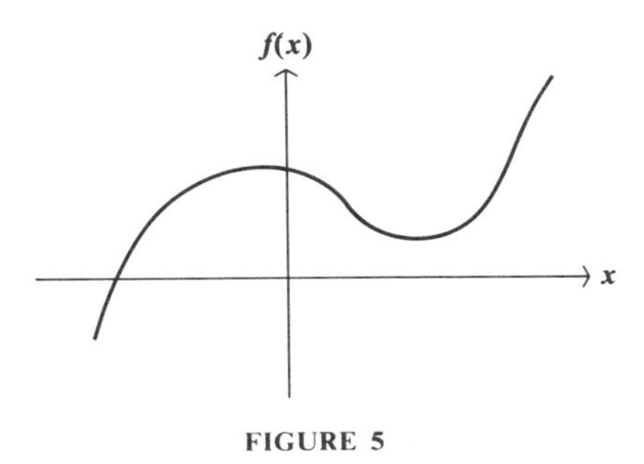

FIGURE 5

   a. How many real roots does the equation $f(x) = 0$ have?
   b. How many imaginary roots does the equation $f(x) = 0$ have?
8. Graph the function $f(x) = x^3 - 6x + 4$.

---

9. Find the positive root of $x^3 + x^2 - 2x - 1 = 0$ to two decimal places.

Diagnostic test

The purpose of this test is to see how well you understand the work in this chapter. Allow yourself approximately 60 minutes to finish this test. Solutions to the problems, together with section references, are given in the Answers section at the end of this book. We suggest that you study the sections referred to for the problems you do incorrectly.

1. Given $f(x) = x^7 - 16x^5 + x^3 + x^2 - 10x - 20$, find $f(4)$.
2. Use synthetic division to find the quotient and remainder when $2x^6 + 4x^5 + x^3 + 2x^2 + 3x + 10$ is divided by $x + 2$.

3. Is $x - 4$ a factor of $x^5 - 14x^3 - 9x^2 + 4x + 6$? Show proof for your answer.
4. Write a fourth-degree equation with 1, -2, $(1 + 5i)/2$, and $(1 - 5i)/2$ as its roots.
5. Find the solution set of the equation $x^4 + 2x^3 - 8x - 16 = 0$.
6. Let the graph shown in Figure 6 be that of a fourth-degree polynomial function $f(x)$.

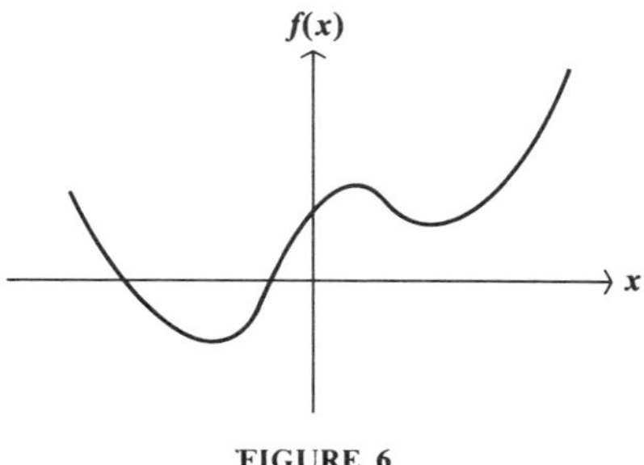

FIGURE 6

   a. How many real roots does the equation $f(x) = 0$ have?
   b. How many complex roots does the equation $f(x) = 0$ have?
7. Graph the function $f(x) = x^3 - 3x + 5$.

8. The equation $3x^3 - 3x^2 + x - 7 = 0$ has a real root between 1 and 2. Express the root accurate to two decimal places.

14 Sequences and series

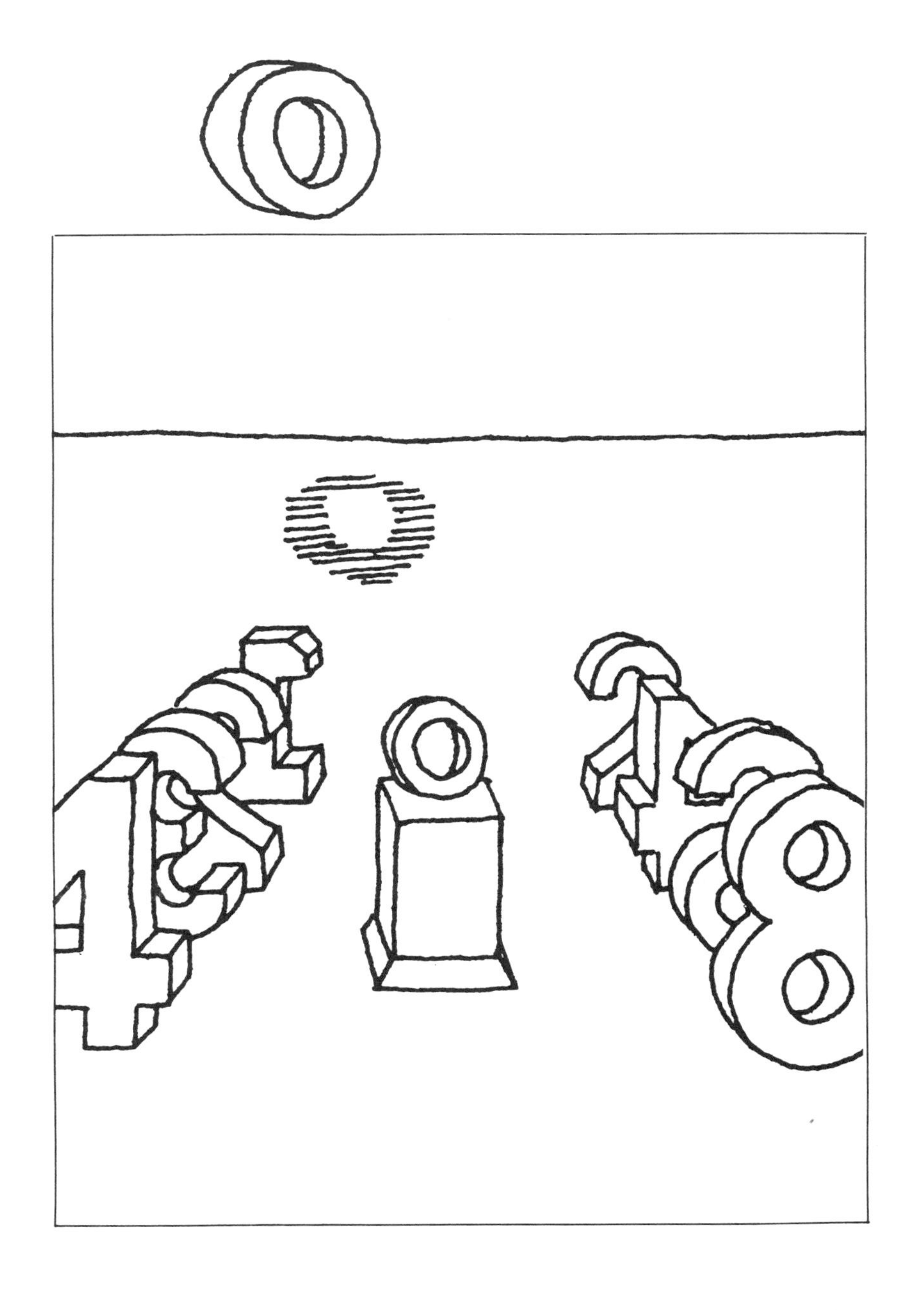

In this chapter, we shall discuss sequences and series. There are many applications of sequences and series in mathematics and the sciences. For example, the numbers in the square root, logarithm, and trigonometric tables were calculated using series. Many formulas, such as those used to calculate compound interest, annuities, and mortgage loans, are derived using series.

1 Basic definitions

SEQUENCES

A set of numbers arranged in a definite order is called a **sequence of numbers**. The numbers themselves are called the **terms** of the sequence and are spoken of as the first term, the second term, and so on, according to their positions in the sequence. A sequence is usually written

$$a_1, a_2, a_3, \ldots, a_n, \ldots, \qquad n \text{ a natural number,}$$

where

a_1 is the first term

a_2 is the second term

a_3 is the third term

$\vdots$

a_n is the nth term (also called the general term of the sequence)

$\vdots$

The subscript of each term represents the number of the term. The dots "..." are read "and so on." A **finite sequence** is one that has a last term. The last term of a finite sequence is represented by the symbol a_n or l.

EXAMPLE 1 The digits, 0, 1, 2, 3, 4, 5, 6, 7, 8, 9 represent a finite sequence.

An **infinite sequence** is nonterminating. That is, any term is followed by another term. An infinite sequence is written

$$\{a_1, a_2, a_3, \ldots, a_n, \ldots\}, \qquad \text{or} \qquad \text{just } \{a_n\}.$$

EXAMPLE 2 An example of an infinite sequence is a function whose domain is the set of positive integers.

Often each term of a sequence is a function of n, where n is the number of the term and n is a natural number.

$$a_n = f(n)$$

EXAMPLE 3

a. If $a_n = f(n) = \frac{n}{3}$,

then $a_1 = f(1) = \frac{1}{3}$,
$a_2 = f(2) = \frac{2}{3}$,
$a_3 = f(3) = \frac{3}{3} = 1$,

and so on. Then

$$\{a_n\} = \left\{\frac{1}{3}, \frac{2}{3}, \ldots, \frac{n}{3}, \ldots\right\}.$$

b. If $a_n = f(n) = 2n - 1$,

then $a_1 = f(1) = 2(1) - 1 = 1$,
$a_2 = f(2) = 2(2) - 1 = 3$,
$a_3 = f(3) = 2(3) - 1 = 5$,

and so on. Then

$$\{a_n\} = \{1, 3, 5, \ldots, 2n - 1, \ldots\}.$$

SERIES

A **series** is the indicated sum of a finite or infinite sequence of terms. It is a finite or an infinite series, depending on whether the number of terms is finite or infinite. An infinite series can be written in the form

$$a_1 + a_2 + a_3 + \cdots + a_n + \cdots$$

where a_n is the general term or the nth term. The expression *infinite series* is usually shortened to *series*. That is, when we say series, we usually mean infinite series.

A **partial sum** of a series is the sum of a finite number of consecutive terms of the series, beginning with the first term.

$S_1 = a_1$	The first partial sum.
$S_2 = a_1 + a_2$	The second partial sum.
$S_3 = a_1 + a_2 + a_3$	The third partial sum.
$\vdots$	
$S_n = a_1 + a_2 + a_3 + \cdots + a_n$	The nth partial sum.
$\vdots$	

EXAMPLE 4 Given the infinite sequence $f(n) = 2n$.

The *sequence* is: 2, 4, 6, ...

The *series* for this sequence is: $2 + 4 + 6 + \cdots$

The first three *partial sums* are: $S_1 = 2$

$S_2 = 2 + 4 = 6$

$S_3 = 2 + 4 + 6 = 12$

Exercises 1

A In Exercises 1–5, write each of the sequences and find its sum.

1. The odd integers from 3 to 11
2. The even integers from 4 to 12
3. The squares of the odd integers from 1 to 7
4. The first three terms of the sequence whose general term is $(2n - 1)/n$
5. The first four terms of the sequence whose general term is $(n - 1)^2$

In Exercises 6–10, find the indicated partial sum of the series with the given nth term.

6. $S_3, \quad a_n = \dfrac{n-1}{n+1}$
7. $S_4, \quad a_n = 1 + (0.1)^n$
8. $S_5, \quad a_n = (n + 1)^2$
9. $S_{11}, \quad a_n = (-1)^{2n}$
10. $S_4, \quad a_n = (\frac{1}{2})^n$

B In Exercises 1–5, write each of the sequences and find its sum.

1. The even integers from 2 to 10
2. The odd integers from 5 to 13
3. The reciprocals of the integers from 2 to 5.
4. The first three terms of the sequence whose general term is $(n + 1)/n$
5. The first three terms of the sequence whose general term is n^2

In Exercises 6–10, find the indicated partial sum of the series with the given nth term.

6. $S_3, \quad a_n = \dfrac{n}{n+1}$
7. $S_4, \quad a_n = n^2$
8. $S_3, \quad a_n = (n - 1)(n - 2)$
9. $S_7, \quad a_n = (-1)^n$
10. $S_5, \quad a_n = 10 - 2n$

2 The summation symbol

If the general term of a sequence is $a_n = (2n - 1)^2$, then the sum of the first three terms of the sequence can be indicated by the symbol

$$\sum_{i=1}^{3} (2i - 1)^2$$

where i is replaced by each integer in turn from 1 to 3 to form each of the first three terms. The symbol $\sum$ is used to indicate that the given expression is a sum. The form of each term is $(2i - 1)^2$, where the i represents an integer. The subscript $(i = 1)$ and superscript (3) on $\sum$ indicate the values of i to be used for the first and last terms, respectively, of the sum. The above expression is read "the sum from $i = 1$ to 3 of $2i - 1$ squared."

We write $\sum_{i=1}^{3} (2i - 1)^2$ in expanded form.

$$\begin{aligned}\sum_{i=1}^{3} (2i - 1)^2 &= [2(1) - 1]^2 + [2(2) - 1]^2 + [2(3) - 1]^2 \\ &= 1 + 9 + 25 = 35\end{aligned}$$

Any other symbol can be used in place of i with the same result. In many cases, the letter below the sigma symbol is omitted, as is shown in the following example.

EXAMPLE 1 Use of the $\sum$ symbol.

a. $\sum_{1}^{3} 2^n = 2^1 + 2^2 + 2^3 = 2 + 4 + 8 = 14$

b. $$\begin{aligned}\sum_{0}^{3} (n + 1) &= [(0) + 1] + [(1) + 1] + [(2) + 1] + [(3) + 1] \\ &= 1 + 2 + 3 + 4 = 10\end{aligned}$$

If a_i is the same for all positive integers i, say $a_i = c$, where c is a real number, then

$$\begin{aligned}\sum_{i=1}^{n} a_i &= a_1 + a_2 + a_3 + \cdots + a_n \\ &= c + c + c + \cdots + c \\ &= nc.\end{aligned}$$

With this in mind we state the following rule.

$$\boxed{\sum_{i=1}^{n} c = nc}$$

EXAMPLE 2 Use of the symbol $\sum_{i=a}^{n}$, where a, i, and n are natural numbers.

a. $$\begin{aligned}\sum_{n=2}^{5} 2n &= 2(2) + 2(3) + 2(4) + 2(5) \\ &= 4 + 6 + 8 + 10 = 28\end{aligned}$$

b. $$\sum_{n=1}^{3}\left(\frac{1}{n}\right)^2 = \left(\frac{1}{1}\right)^2 + \left(\frac{1}{2}\right)^2 + \left(\frac{1}{3}\right)^2$$

$$= 1 + \frac{1}{4} + \frac{1}{9} = \frac{49}{36}$$

Exercises 2

Evaluate each expression.

A

1. $\sum_{n=1}^{4} 4n$
2. $\sum_{n=2}^{6} \frac{1}{2}(n-1)$
3. $\sum_{n=1}^{3} \frac{n(n-1)(n+2)}{3}$
4. $\sum_{n=0}^{3} \frac{n}{2n+1}$
5. $\sum_{k=1}^{100} \pi$

B

1. $\sum_{n=1}^{4} 3n$
2. $\sum_{n=1}^{5} \frac{1}{2}n$
3. $\sum_{n=1}^{3} n(3n+3)$
4. $\sum_{n=0}^{4} 2(2^n - 1)$
5. $\sum_{k=1}^{50} 4$

3 Arithmetic progressions

An **arithmetic progression** (abbreviated **A.P.**) is a sequence in which each term after the first is formed by adding the same fixed number to the preceding term. The number which is added in this manner is called the **common difference**. For example, 5, 7, 9, 11, is an A.P. with common difference 2.

To determine whether a given sequence of numbers is an A.P., subtract each member of the sequence, except the last, from the following member. If the differences are equal, the numbers constitute an A.P., while if any two of the differences are unequal, the sequence is not an A.P.

Denoting the first term of an A.P. by a_1 and the common difference by d, the progression becomes Expression 1.

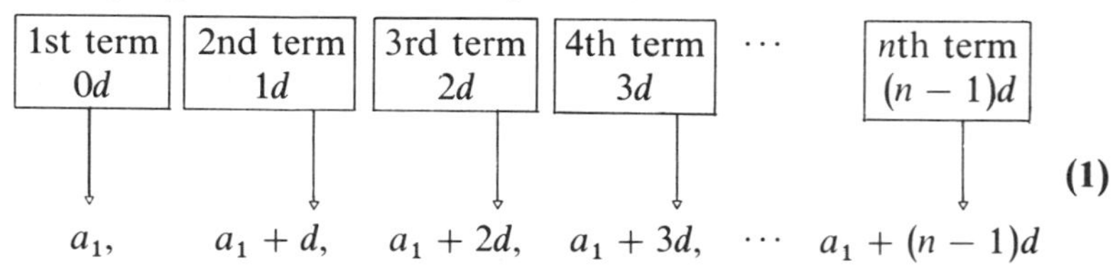

(1)

From Expression 1, notice that the coefficient of d in the second term is 1, in the third term is 2, and, in general, *in any term the coefficient of* d *is one less than the number of the term.* For the nth term, which we have denoted by a_n, we have the following.

$$a_n = a_1 + (n - 1)d \tag{2}$$

Associated with these first n terms of the A.P., there are five numbers of special importance which are called its elements.

THE ELEMENTS OF AN ARITHMETIC PROGRESSION

- a_1, the first term
- d, the common difference
- n, the number of terms
- a_n, the nth term
- S_n, the sum of the first n terms

EXAMPLE 1 The sequence 3, 6, 9, 12, ... is an arithmetic progression in which each term after the first is formed by adding 3 to the preceding term: $a_1 = 3$, $d = 3$.

EXAMPLE 2 The sequence 15, 13, 11, 9, ... is an arithmetic progression in which each term after the first is formed by adding -2 to the preceding term: $a_1 = 15$, $d = -2$.

EXAMPLE 3 Find the thirty-first term of the arithmetic progression 3, 5, 7, 9, ... without writing the preceding terms.

SOLUTION By inspection, we can see that $d = 2$. We were asked for the thirty-first term, so $n = 31$. The first term, a_1, in the A.P. is 3.

Then

$$\begin{aligned} a_n &= a_1 + (n - 1)d \\ a_{31} &= 3 + (31 - 1)(2) \\ &= 3 + (30)(2) \\ &= 63. \end{aligned}$$

An **arithmetic series** is the sum of the terms of an arithmetic progression. An infinite arithmetic series can be written in the form

$$a_1 + (a_1 + d) + (a_1 + 2d) + \cdots + [a_1 + (n - 1)d] + \cdots. \tag{3}$$

For a finite arithmetic series of n terms, the sum, S_n, of those terms is

$$S_n = a_1 + (a_1 + d) + (a_1 + 2d) + \cdots + (a_n - 2d) + (a_n - d) + a_n.$$

Subtracting d from the last term gives the preceding term.

Another formula for S_n can be found by adding the reverse of S_n to itself.

$$\begin{aligned} S_n &= a_1 + (a_1 + d) + (a_1 + 2d) + \cdots + (a_n - 2d) + (a_n - d) + a_n \\ S_n &= a_n + (a_n - d) + (a_n - 2d) + \cdots + (a_1 + 2d) + (a_1 + d) + a_1 \\ \hline 2S_n &= (a_1 + a_n) + (a_1 + a_n) + \cdots + (a_1 + a_n) + (a_1 + a_n) \end{aligned}$$

The right side of the last equation has n terms of $a_1 + a_n$, so that

$$2S_n = n(a_1 + a_n).$$

Therefore, we have the following value for S_n.

$$S_n = \frac{n(a_1 + a_n)}{2} \tag{4}$$

EXAMPLE 4 Find the sum of the first 100 natural numbers.

SOLUTION

$$\left.\begin{aligned} a_1 &= 1 \\ a_n &= 100 \\ n &= 100 \end{aligned}\right\} \qquad \begin{aligned} S_n &= \frac{n(a_1 + a_n)}{2} \\ S_{100} &= \frac{100(1 + 100)}{2} = 50(101) = 5050 \end{aligned}$$

The story is told that the famous German mathematician, Carl Friedrich Gauss, very quickly solved a problem like this at the age of ten when it was presented in his first arithmetic class.

EXAMPLE 5 A falling object, starting from rest, falls 16 feet during the first second, 48 feet during the second second, 80 feet during the third second, and so on. Find the distance it falls during the twelfth second and the total distance it falls during the first 12 seconds.

SOLUTION This is an A.P. because $48 - 16 = 32$ and $80 - 48 = 32$. That is, when we subtract the first term from the second term and the second term from the third term we get the same number, 32. Thus $d = 32$ and $a_1 = 16$. Since we are asked to solve for the twelfth term and the sum of the first 12 terms (the total distance), we use $n = 12$ and solve for a_{12} and S_{12}.

1st term	2nd term	3rd term		
16	+ 48	+ 80	+ ⋯ +	a_n
= 16	+ 16 + 32	+ 16 + 2(32)	+ ⋯ +	16 + (n − 1)32

$$a_n = a_1 + (n-1)d \qquad S_n = \frac{n(a_1 + a_n)}{2}$$

$$a_{12} = 16 + (12 - 1)32$$

$$= 16 + (11)(32) \qquad S_{12} = \frac{12(16 + 368)}{2}$$

$$= 368 \qquad = 2{,}304$$

The object falls 368 feet during the twelfth second and it falls 2,304 feet during the first 12 seconds.

Exercises 3

A In Exercises 1–4, determine if each sequence is an arithmetic progression. If it is an A.P.:

a. Find the common difference.

b. Determine the a_n term.

c. Write the twenty-first term.

d. Find the sum of the first 11 terms.

1. 3, 7, 11, ... **2.** 6, 4, 2, ...

3. 67, 54, 41, 28, ... **4.** $4a + 6 \cdot 2,\ 3a + 3 \cdot 2,\ 2a, \ldots$

5. Find the eleventh and seventeenth terms of the arithmetic progression $a + 24e,\ 4a + 20e,\ 7a + 16e, \ldots$.

6. Use the value of log 3 = 0.477 and find the value of the fifth term of the arithmetic progression log 3, log 9, log 27, ...

7. Find the tenth term of the arithmetic sequence whose first two terms are $2 + \sqrt{3}$ and $\sqrt{3}$.

8. How many terms are in the arithmetic sequence with first term 15, common difference $-\frac{3}{2}$, and sum $28\frac{1}{2}$?

In Exercises 9–11, find the remaining two elements.

9. $a_1 = 17;\quad d = -3;\quad a_n = -13$

10. $d = -5;\quad a_n = -32;\quad n = 18$

11. $a_1 = -5;\quad d = 4;\quad S_n = 30$

12. Evaluate the sum $\sum_{k=4}^{9} (5k + 6)$.

B In Exercises 1–4, determine if each sequence is an arithmetic progression. If it is an A.P.:

a. Find the common difference.
b. Determine the a_n term.
c. Write the twenty-first term.
d. Find the sum of the first 11 terms.

1. 8, 10, 12, ... **2.** 11, 8, 5, ...
3. 7, 12, 17, 22, ... **4.** $1 + 7 \cdot 5, 6 \cdot 5, -1 + 5 \cdot 5, \ldots$
5. Find the fifteenth and thirty-third terms of the arithmetic progression 19, $17\frac{1}{4}$, $15\frac{1}{2}$, $13\frac{3}{4}$, ...
6. Use the value of $\log 2 = 0.301$ and find the value of the eighth term of the arithmetic progression log 2, log 4, log 8, log 16, ...
7. Find the eleventh term of the arithmetic sequence whose first two terms are $1 + \sqrt{2}$ and 3.
8. How many terms are in the finite arithmetic sequence with first term -2, common difference $\frac{1}{4}$, and sum 25?

In Exercises 9–11, find the remaining two elements.

9. $a_1 = 7; \quad a_n = 31; \quad S_n = 931$
10. $a_n = -5; \quad n = 17; \quad S_n = -170$
11. $a_1 = 2; \quad d = \frac{4}{5}; \quad n = 11$
12. Evaluate the sum $\sum_{k=1}^{6} (3k + 2)$.

4 Geometric progressions

A sequence of numbers is a **geometric progression**, abbreviated **G.P.**, if each term after the first is obtained by multiplying the preceding term by a fixed number called the **common ratio** r. For example, 1, 2, 4, 8, 16, ..., with $a_1 = 1$ and $r = 2$.

Denoting the first term of a G.P. by a_1 and the common ratio by r, the progression becomes Expression 1.

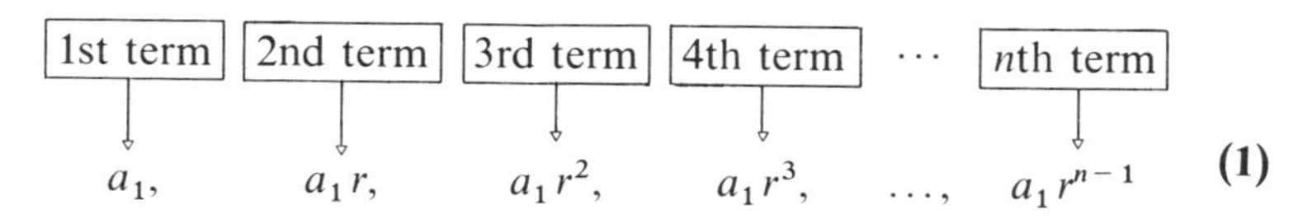

$$a_1, \quad a_1 r, \quad a_1 r^2, \quad a_1 r^3, \quad \ldots, \quad a_1 r^{n-1} \tag{1}$$

Notice in Expression 1 that the exponent of r is 1 in the second term, 2 in the third term, and, in general, is one less than the number of the term. For the nth term, which we shall denote by a_n, we have the following definition.

$$a_n = a_1 r^{n-1} \qquad \textbf{(2)}$$

To determine whether a given sequence is a G.P., divide each number after the first by the preceding number. If the quotients are equal, the numbers constitute a G.P.; if any two of the quotients are unequal, the sequence is not a G.P.

THE ELEMENTS OF A GEOMETRIC PROGRESSION

- a_1, the first term
- r, the common ratio
- n, the number of terms
- a_n, the nth term
- S_n, the sum of the first n terms

EXAMPLE 1 Examples of geometric progressions.

a. $2,\ 2(2),\ 2(2^2),\ 2(2^3),\ \ldots$
$= 2,\ 4,\ 8,\ 16,\ \ldots;\ a_1 = 2,\ r = 2$

b. $4,\ 4(\frac{1}{2}),\ 4(\frac{1}{2})^2,\ 4(\frac{1}{2})^3,\ \ldots$
$= 4,\ 2,\ 1,\ \frac{1}{2},\ \ldots;\ a_1 = 4,\ r = \frac{1}{2}$

c. $1,\ 1(3),\ 1(3^2),\ 1(3^3),\ \ldots$
$= 1,\ 3,\ 9,\ 27,\ \ldots;\ a_1 = 1,\ r = 3$

EXAMPLE 2 Write the general term for the geometric progression $4, 2, 1, \frac{1}{2}, \ldots$.

SOLUTION Dividing each term by its preceding term, we find

$$r = \frac{a_2}{a_1} = \frac{2}{4} = \frac{1}{2}.$$

Since $a_1 = 4$, from the formula $a_n = a_1 r^{n-1}$, we have

$$a_n = 4\left(\frac{1}{2}\right)^{n-1} = 2^2 \cdot \frac{1}{2^{n-1}} = \frac{1}{2^{n-3}} \qquad \text{or} \qquad 2^{3-n}.$$

EXAMPLE 3 Find the tenth term of the geometric progression 32, 48, 72, ...

SOLUTION

$$\frac{a_2}{a_1} = \frac{48}{32} = \frac{3}{2}, \quad \frac{a_3}{a_2} = \frac{72}{48} = \frac{3}{2}$$

Thus $r = \frac{3}{2}$ and $a_1 = 32$. We use the formula $a_n = a_1 r^{n-1}$ to find the tenth term.

$$\begin{aligned} a_{10} &= 32\left(\frac{3}{2}\right)^9 \\ &= \frac{2^5(3^9)}{2^9} \\ &= \frac{3^9}{2^4} \\ &= \frac{19{,}683}{16} \\ &= 1{,}230.1875 \end{aligned}$$

A **geometric series** is the sum of the terms of a geometric progression. An infinite geometric series can be written

$$a_1 + a_1 r^1 + a_1 r^2 + \cdots + a_1 r^{n-1} + \cdots = \sum_{n=1}^{\infty} a_1 r^{n-1}.$$

For a finite geometric series of n terms, the sum of those terms, S_n, is

$$S_n = a_1 + a_1 r^1 + a_1 r^2 + \cdots + a_1 r^{n-1} = \sum_{i=1}^{n} a_1 r^{i-1}. \tag{3}$$

Another formula for S_n can be found as follows: Multiply both sides of Equation 3 by r, then subtract the resulting equation from Equation 3.

$$\begin{aligned} S_n &= a_1 + a_1 r^1 + a_1 r^2 + \cdots + a_1 r^{n-1} \\ rS_n &= \phantom{a_1 + {}} a_1 r^1 + a_1 r^2 + \cdots + a_1 r^{n-1} + a_1 r^n \\ \hline S_n - rS_n &= a_1 \qquad\qquad\qquad\qquad\qquad\qquad - a_1 r^n \\ (1 - r)S_n &= a_1(1 - r^n) \end{aligned}$$

By factoring both sides and then dividing by $(1 - r)$ we have the following.

$$S_n = \frac{a_1(1 - r^n)}{1 - r}, \qquad r \neq 1 \tag{4}$$

EXAMPLE 4 Find the sum of the first ten terms of the geometric progression $1 + 2 + 4 + 8 + \cdots$ using Equation 4.

SOLUTION

$$\frac{a_2}{a_1} = \frac{2}{1}$$

Thus $r = 2$ and $a_1 = 1$. Then using Formula 4, we have

$$S_n = \frac{a_1(1 - r^n)}{1 - r}$$

$$S_{10} = \frac{1(1 - 2^{10})}{1 - 2}$$

$$= \frac{1 - 2^{10}}{-1}$$

$$= 2^{10} - 1$$

$$= 1{,}024 - 1 = 1{,}023. \quad \text{The sum of the first 10 terms.}$$

EXAMPLE 5 The number of bacteria in a culture increased from 425,000 to 3,575,000 in 5 days. Find the daily rate of increase if this rate is assumed to be constant.

SOLUTION In this problem $a_1 = 425{,}000$, $a_5 = 3{,}575{,}000$, and $n = 5$.

$$a_n = a_1 r^{n-1}$$

$$3{,}575{,}000 = 425{,}000 r^{5-1}$$

$$r^4 = \frac{3{,}575{,}000}{425{,}000}$$

$$r = \sqrt[4]{\frac{3{,}575{,}000}{425{,}000}}$$

Key operations	Display
[C] 3 5 7 5 [÷] 4 2 5 [y^x] [.] 2 5 [=]	1.703027963

(*Note:* We also could use logarithms to solve this problem.)

The number of bacteria increases by about 1.7 times each day.

EXAMPLE 6 Write the first four terms of the series $\sum_{n=1}^{\infty} 1/3^n$.

SOLUTION The symbol $\sum_{n=1}^{\infty}$ indicates that this is an infinite series. That is, it is a nonterminating series. We use the formula

$$a_n = a_1 r^{n-1}.$$

Then

$$a_1 = \frac{1}{3^1} = \frac{1}{3}$$

$$a_2 = \frac{1}{3^2} = \frac{1}{9}$$

$$a_3 = \frac{1}{3^3} = \frac{1}{27}$$

$$a_4 = \frac{1}{3^4} = \frac{1}{81}.$$

Therefore, the first four terms of the series are

$$\frac{1}{3} + \frac{1}{9} + \frac{1}{27} + \frac{1}{81} + \cdots.$$

An **ordinary annuity** is a geometric series of equal payments at the end of regular intervals. The value of the annuity at the end of n interest periods is found by means of the following formula.

SUM OF AN ORDINARY ANNUITY

$$S = R\left[\frac{(1+i)^n - 1}{i}\right] \quad \textbf{(5)}$$

R = Amount of each payment
n = Number of interest periods
i = Interest rate per period expressed as a decimal
S = Value of the annuity at the end of n interest periods

EXAMPLE 7 $100 is deposited at the end of each month into an account that pays 8% per year, compounded monthly. Find the value of the annuity at the end of 5 years.

SOLUTION

$$S = R\left[\frac{(1+i)^n - 1}{i}\right]$$

$$R = 100, n = 5 \times 12 = 60$$

$$1 + i = 1 + \frac{0.08}{12} = 1.00666\cdots$$

Interest rate per month

Therefore,

$$S = 100\left[\frac{(1.00666666)^{60} - 1}{0.006666666}\right]$$

Key operations	Display
[C] [·] 0 8 [÷] 1 2 [=] [STO] [1] [RCL] [1] [+] 1 [=] [y^x]	
6 0 [=] [−] 1 [=] [÷] [RCL] [1] [X] 1 0 0 [=]	[7347.685624]

Therefore, $S \doteq \$7{,}347.69$

Exercises 4

A In Exercises 1–4, determine if each sequence is a geometric progression. If it is a G.P., find the common ratio.

1. 3, 6, 9, 12, ... **2.** 4, 8, 16, 32, ...
3. $-2, 4, -8, 16, \ldots$ **4.** $5, 5(1 - x), 5(1 - x)^2, \ldots$

In Exercises 5–6, write the next three terms of each G.P.

5. 80, 120, 180, ... **6.** $1, (1 + x), (1 + x)^2, \ldots$
7. Find the seventh term of the G.P. $-2, 4, -8, 16, \ldots$.

In Exercises 8–10, find the indicated elements of the given G.P.

8. $r = \frac{1}{2}$; $n = 10$; $a_1 = \frac{16}{3}$; a_n; S_{10}
9. $a_1 = 32$; $r = \frac{3}{2}$; $S_n = 422$; n; a_n
10. $a_1 = 81$; $a_n = 1$; $r = \frac{1}{3}$; n; S_n
11. Write the first four terms of a geometric progression in which the fifth term is $\frac{1}{7}$ and the seventh term is $\frac{4}{343}$.

12. The population of a city in a recent census year was 25,000. If the population increases 20% every ten years, what will the population be 40 years after the census?

13. Suppose you agree to work for $\frac{1}{2}$¢ the first day, 1¢ the second day, 2¢ the third day, and so on in this manner for a month.
- **a.** What is your pay for the tenth day?
- **b.** What is your pay for the thirtieth day?
- **c.** What is the total pay for a month of 30 days?

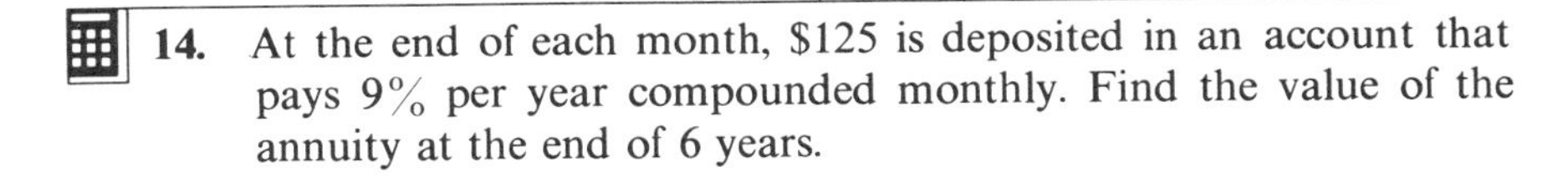
14. At the end of each month, $125 is deposited in an account that pays 9% per year compounded monthly. Find the value of the annuity at the end of 6 years.

B In Exercises 1–4, determine if each sequence is a geometric progression. If it is a G.P., find the common ratio.

1. 4, 8, 12, 16, ... **2.** 3, 6, 12, 24, ...
3. $-5, 10, -20, 40, \ldots$ **4.** $2, 2(1 + x), 2(1 + x)^2, \ldots$

In Exercises 5–6, write the next three terms of each G.P.

5. $\frac{1}{6}, \frac{1}{2}, \frac{3}{2}, \ldots$ **6.** $x - 2, 3x - 6, 9x - 18, \ldots$
7. Find the seventh term of the G.P. $-3, 6, -12, 24, \ldots$.

In Exercises 8–10, find the indicated elements of the given G.P.

8. $a_n = 80$; $r = \frac{2}{3}$; $n = 5$; a_1; S_5
9. $a_1 = 5$; $S_4 = 200$; a_4; r
10. $a_1 = 250$; $r = \frac{3}{5}$; $a_n = 32\frac{2}{5}$; n; S_n
11. Write the first three terms of a geometric progression in which the fourth term is 2 and the seventh term is 54.

12. If the population of a certain country increases at the rate of 5 percent per year and the present population is 3,900,000, what will the population be 5 years from now?

13. Suppose you agree to work for 1¢ the first day, 2¢ the second day, 4¢ the third day, and so on in this manner throughout the month.
a. What is your pay for the tenth day?
b. What is your pay for the thirtieth day?
c. What is the total pay for the month of 30 days?

14. At the end of each month, $150 is deposited in an account that pays 9% per year compounded monthly. Find the value of the annuity at the end of 4 years.

5 Infinite geometric series

Consider the formula

$$S_n = \frac{a_1(1 - r^n)}{1 - r}$$

which gives the sum of n terms of a geometric series. This formula can be written

$$S_n = \frac{a_1}{1 - r} - \frac{a_1 r^n}{1 - r}. \qquad (1)$$

Let us see what seems to be true when $|r| < 1$ and n becomes larger and larger.

Suppose $r = \frac{1}{2}$. Then

$$\left(\frac{1}{2}\right)^2 = 0.25$$

$$\left(\frac{1}{2}\right)^3 = 0.125$$

$$\left(\frac{1}{2}\right)^4 = 0.625$$

$$\left(\frac{1}{2}\right)^{10} \doteq 0.00098$$

$$\left(\frac{1}{2}\right)^{20} \doteq 0.00000095$$

These examples indicate that as n gets large, $(\frac{1}{2})^n$ approaches zero. This fact can be proved with calculus for all $|r| < 1$. Thus if r^n approaches zero when $|r| < 1$ and n grows large without limit, the second part of Formula 1, namely

$$\frac{a_1 r^n}{1 - r},$$

approaches zero. This leads to the formula for the sum of an infinite geometric series.

$$\sum_{k=1}^{\infty} ar^{k-1} = \frac{a_1}{1 - r}, \qquad |r| < 1 \tag{2}$$

This is read "the sum of terms ar^{k-1} as k goes from 1 to infinity is equal to $a_1/(1 - r)$" and means that by taking k sufficiently large, we can make the difference between the sum and $a_1/(1 - r)$ less than any previously designated positive number.

On the other hand, if $|r| > 1$, the absolute value of r^n increases when n increases. In fact, by taking n sufficiently large, we can make the absolute value of S_n as large as we wish. Consequently S_n does not approach a limit when n becomes infinite.

DEFINITION A series is said to be convergent if $\lim_{n \to \infty} S_n$ exists and to be divergent if such a limit does not exist.

THEOREM An infinite geometric series in which $a_1 \neq 0$ converges if $|r| < 1$ and diverges if $|r| > 1$.

If the series diverges, we do not give it a value. If the series converges we use Formula 2 to find its convergent value. The value of the series is the limit of the sum of the first n terms of the series when n becomes infinite.

The symbol S_∞ or $\sum_{k=1}^{\infty}$ is often used to represent S_n when n becomes infinitely large.

EXAMPLE 1 Evaluate $1 + \frac{1}{2} + \frac{1}{4} + \cdots$.

SOLUTION

$$1, \frac{1}{2}, \frac{1}{4}, \ldots$$

$$\frac{1}{2} \div 1 = \frac{1}{2} = r \qquad \text{Since } |r| < 1, \text{ use Formula 2.}$$

$$S_\infty = \frac{a_1}{1-r} = \frac{1}{1-\frac{1}{2}} = \frac{1}{\frac{1}{2}} = 2$$

EXAMPLE 2 Evaluate $\sum_{k=1}^{\infty} 6\left(\frac{2}{3}\right)^k$.

SOLUTION

$$\sum_{k=1}^{\infty} 6\left(\frac{2}{3}\right)^k = 6\left(\frac{2}{3}\right)^1 + 6\left(\frac{2}{3}\right)^2 + 6\left(\frac{2}{3}\right)^3 + \cdots$$

$$\frac{a_2}{a_1} = \frac{6(\frac{2}{3})^2}{6(\frac{2}{3})} = \frac{2}{3}, \qquad \frac{a_3}{a_2} = \frac{6(\frac{2}{3})^3}{6(\frac{2}{3})^2} = \frac{2}{3}$$

Thus $r = \frac{2}{3}$, which is less than 1, and $a_1 = 6(\frac{2}{3})$.

$$S_\infty = \frac{a_1}{1-r} = \frac{6(\frac{2}{3})}{1-(\frac{2}{3})} = \frac{4}{\frac{1}{3}} = 12$$

REPEATING DECIMALS

Any repeating decimal is an infinite geometric series and can be written as a fraction.

EXAMPLE 3 Write the repeating decimal 4.212121 ... as a fraction.

SOLUTION

$$4.212121\ldots = 4 + \boxed{0.21 + 0.0021 + 0.000021 + \ldots}$$

$$a_1 + \quad a_2 + \quad a_3 + \ldots$$

We will first change this sum to a fraction.

$$\frac{a_2}{a_1} = \frac{0.0021}{0.21} = 0.01, \qquad \frac{a_3}{a_2} = \frac{0.000021}{0.0021} = 0.01$$

Thus $r = 0.01$ and $a_1 = 0.21$.

$$S_\infty = \frac{a_1}{1-r} = \frac{0.21}{1-0.01} = \frac{0.21}{0.99} = \frac{21}{99} = \frac{7}{33}$$

Therefore, $4.212121\ldots = 4\frac{7}{33}$, or $\frac{139}{33}$.

Exercises 5

A In Exercises 1–5, determine whether the series converges or diverges. If it converges, find its sum.

1. $12 + 9 + \frac{27}{4} + \cdots$ **2.** $64 + 48 + 36 + \cdots$
3. $16 + 20 + 25 + \cdots$ **4.** $15 + 6 + \frac{12}{5} + \cdots$
5. $5 + 3 + 1.8 + \cdots$

In Exercises 6–9, evaluate each sum.

6. $\sum_{n=0}^{\infty} (\frac{5}{6})^n$ **7.** $\sum_{n=0}^{\infty} (\frac{9}{10})^n$
8. $\sum_{k=1}^{\infty} (0.3)^k$ **9.** $\sum_{k=1}^{\infty} (0.2)^k$

10. The first swing of a pendulum is 9 inches. Each succeeding swing is five-sixths as long as the preceding one. Find the total distance traveled by the pendulum before it comes to rest.

11. A rubber ball is dropped from a height of 10 feet. Each time it strikes the floor, it rebounds to a height which is five-sixths the height from which it last fell. Find the total distance the ball travels before coming to rest.

12. Express the repeating decimal 0.545454 ... as a fraction.

B In Exercises 1–5, determine whether the series converges or diverges. If it converges, find its sum.

1. $18 + 12 + 8 + \cdots$ **2.** $36 + 24 + 16 + \cdots$
3. $4 + 6 + 9 + \cdots$ **4.** $\frac{3}{4} + \frac{3}{5} + \frac{12}{25} + \cdots$
5. $10 + 9 + 8.1 + \cdots$

In Exercises 6–9, evaluate each sum.

6. $\sum_{k=0}^{\infty} (\frac{3}{4})^k$ **7.** $\sum_{k=0}^{\infty} (\frac{2}{3})^k$
8. $\sum_{n=1}^{\infty} (0.1)^n$ **9.** $\sum_{n=1}^{\infty} (0.4)^n$

10. The first swing of a pendulum is 15 inches. Each succeeding swing is fourth-fifths as long as the preceding one. Find the total distance traveled by the pendulum before it comes to rest.

11. A rubber ball is dropped from a height of 12 feet. Each time it strikes the floor, it rebounds to a height which is two-thirds the height from which it last fell. Find the total distance the ball travels before coming to rest.

12. Express the repeating decimal 0.351351351 ... as a fraction.

Summary

A **sequence of numbers** is a set of numbers arranged in a definite order. The numbers themselves are the **terms** of the sequence and are spoken of as the first term, the second term, and so on, according to their positions in the sequence. A **finite sequence** is one that has a last term. An **infinite sequence** does not terminate.

A **series** is the indicated sum of a finite or infinite sequence of terms.

An **arithmetic progression** (abbreviated **A.P.**) is a sequence in which each term after the first is formed by adding the same fixed number to the preceding term. The number which is added is called the **common difference**. An **arithmetic series** is the sum of the terms of an arithmetic progression.

A **geometric progression** (abbreviated **G.P.**) is a sequence of numbers in which each term after the first is formed by multiplying the same fixed number, called the **common ratio** r, times each preceding term.

The following terms are used in arithmetic and geometric progressions.

a_1 = First term; used in both A.P. and G.P.
d = Common difference; used in A.P.
r = Common ratio; used in G.P.
n = nth term; used in both A.P. and G.P.
a_n = nth or general term; used in both A.P. and G.P.
S_n = Sum of n terms; used in both A.P. and G.P.

The following formulas are used with arithmetic progressions.

$$a_n = a_1 + (n - 1)d \qquad n\text{th term.}$$

$$S_n = \frac{n(a_1 + a_n)}{2} \qquad \text{Sum of } n \text{ terms.}$$

The following formulas are used with geometric progressions.

$$a_n = a_1 r^{n-1} \qquad n\text{th term.}$$

$$S_n = \frac{a_1(1 - r^n)}{1 - r}, \quad \text{when } r \neq 1 \qquad \text{Sum of } n \text{ terms.}$$

$$S_\infty = \frac{a_1}{1 - r}, \quad \text{when } |r| < 1 \qquad \text{Sum of infinitely many terms.}$$

The Greek letter $\sum$ is used to indicate that the given expression is a sum. For example,

$$\sum_{n=1}^{k} a_n = a_1 + a_2 + \cdots + a_k.$$

DEFINITION A series is said to **converge** if $\lim_{n \to \infty} S_n$ exists and to **diverge** if such a limit does not exist.

THEOREM An infinite geometric series in which $a_1 \neq 0$ converges if $|r| < 1$ and diverges if $|r| \geq 1$.

We have discussed only two of the many kinds of series. After studying calculus you will be better equipped to extend your knowledge to other kinds of series.

Review exercises

A In Exercises 1–5, find the sum of each series if it exists.

1. $\sum_{n=0}^{5} n^2$ **2.** $\sum_{k=-1}^{4} (3k + 1)$

3. $\sum_{n=1}^{4} n(n + 1)$ **4.** $\sum_{n=0}^{\infty} (0.3)^n$

5. $\sum_{n=1}^{\infty} (\frac{1}{4})^n$

6. **a.** Find the forty-first term of the arithmetic progression 1, 3, 5, ...

b. Find the sum of the first thirty-one terms of this A.P.

7. **a.** Find the tenth term of the geometric progression 18, 6, 2, $\frac{2}{3}$, . . . without writing the preceding terms.

b. Find the sum of the first ten terms of this G.P.

8. Express the repeating decimal 4.515151 . . . as a fraction.

9. The number of bacteria in a culture increased from 562,500 to 4,500,000 in 4 days. Find the daily rate of increase, if this rate is assumed to be constant.

10. When money is being compounded for a period of time, the accumulation at any time is called the **compound amount** and is found by the formula $S = P(1 + i)^n$, where P = principal invested, i = interest rate per period expressed as a decimal, n = number of interest periods, and S = value of the accumulation at the end of n interest periods. Find S at the end of 4 years when \$2,500 is deposited in a savings account paying $7\frac{1}{2}\%$ interest.

B In Exercises 1–5, find the sum of each series if it exists.

1. $\sum_{n=0}^{4} 2n$ **2.** $\sum_{k=-1}^{5} (2k-1)$

3. $\sum_{n=0}^{3} 4^n$ **4.** $\sum_{n=0}^{\infty} (0.2)^n$

5. $\sum_{n=0}^{\infty} (\frac{2}{5})^n$

6. **a.** Find the forty-first term of the arithmetic progression 2, 4, 6, 8, ...

b. Find the sum of the first thirty-one terms of this A.P.

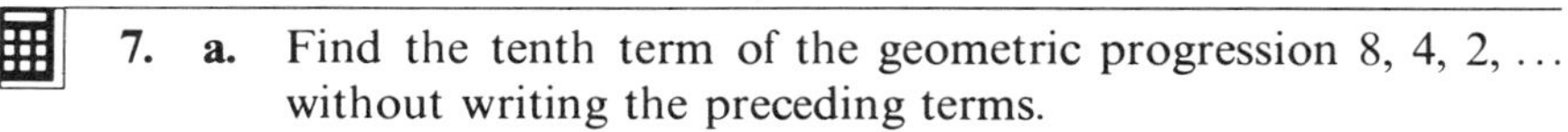

7. **a.** Find the tenth term of the geometric progression 8, 4, 2, ... without writing the preceding terms.

b. Find the sum of the first ten terms of this G.P.

8. Express the repeating decimal 2.313131 ... as a fraction.

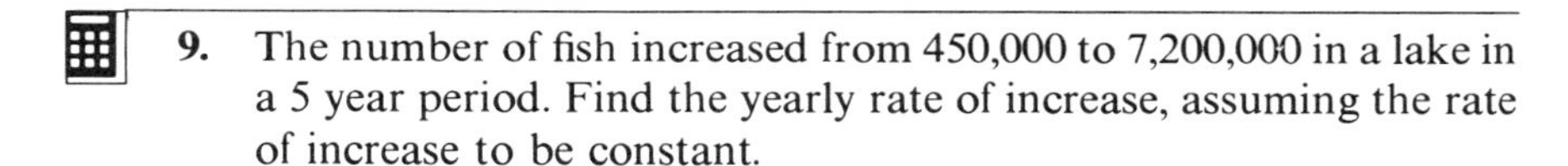

9. The number of fish increased from 450,000 to 7,200,000 in a lake in a 5 year period. Find the yearly rate of increase, assuming the rate of increase to be constant.

10. Find S at the end of $2\frac{1}{2}$ years when \$1,500 is deposited in a savings account paying $6\frac{3}{4}\%$ interest. Interest is compounded daily on a 360-day year. Use the formula in Exercise 10 of Part A.

Diagnostic test

The purpose of this test is to see how well you understand the work in this chapter. Allow yourself approximately 50 minutes to do the test. Solutions to the problems, together with section references, are given in the Answers section at the end of this book. We suggest that you study the sections referred to for the problems you do incorrectly.

In Problems 1–3, find the sum of each series.

1. $\sum_{n=0}^{3} (n+1)(n+2)$ **2.** $\sum_{n=0}^{3} 2^n$

3. $\sum_{n=1}^{\infty} (\frac{1}{3})^n$

4. Express 2.121212 ... as a fraction.

5. Find the sixty-first term of the arithmetic progression 2, $\frac{5}{2}$, 3, $\frac{7}{2}$, ...

6. Given the sequence 27, 9, 3, 1, ...

a. Find the seventh term.

b. Find the sum of the first 6 terms.

7. The fourth term of an arithmetic progression is 3 and the seventh term is 9. Write the first 5 terms.
8. A merchant's income during his first year in business was $12,000. If his income increased 20% every year, what was his total income for the first 4 years?
9. Does the series $\frac{1}{5} + \frac{1}{5} + \frac{1}{5} + \cdots$ diverge or converge? Explain.
10. The number of bacteria in a culture was 1,296,000 on the first day, 1,080,000 on the second day, and 900,000 on the third day. Assuming this same rate of decline continues, find the predicted count on the fifth day.

Appendix

TABLE 1 SQUARES AND SQUARE ROOTS

n	n^2	$\sqrt{n}$	$\sqrt{10n}$	n	n^2	$\sqrt{n}$	$\sqrt{10n}$
1	1	1.000	3.162	51	2601	7.141	22.583
2	4	1.414	4.472	52	2704	7.211	22.804
3	9	1.732	5.477	53	2809	7.280	23.022
4	16	2.000	6.325	54	2916	7.348	23.238
5	25	2.236	7.071	55	3025	7.416	23.452
6	36	2.449	7.746	56	3136	7.483	23.664
7	49	2.646	8.367	57	3249	7.550	23.875
8	64	2.828	8.944	58	3364	7.616	24.083
9	81	3.000	9.487	59	3481	7.681	24.290
10	100	3.162	10.000	60	3600	7.746	24.495
11	121	3.317	10.488	61	3721	7.810	24.698
12	144	3.464	10.954	62	3844	7.874	24.900
13	169	3.606	11.402	63	3969	7.937	25.100
14	196	3.742	11.832	64	4096	8.000	25.298
15	225	3.873	12.247	65	4225	8.062	25.495
16	256	4.000	12.649	66	4356	8.124	25.690
17	289	4.123	13.038	67	4489	8.185	25.884
18	324	4.243	13.416	68	4624	8.246	26.077
19	361	4.359	13.784	69	4761	8.307	26.268
20	400	4.472	14.142	70	4900	8.367	26.458
21	441	4.583	14.491	71	5041	8.426	26.646
22	484	4.690	14.832	72	5184	8.485	26.833
23	529	4.796	15.166	73	5329	8.544	27.019
24	576	4.899	15.492	74	5476	8.602	27.203
25	625	5.000	15.811	75	5625	8.660	27.386
26	676	5.099	16.125	76	5776	8.718	27.568
27	729	5.196	16.432	77	5929	8.775	27.749
28	784	5.292	16.733	78	6084	8.832	27.928
29	841	5.385	17.029	79	6241	8.888	28.107
30	900	5.477	17.321	80	6400	8.944	28.284
31	961	5.568	17.607	81	6561	9.000	28.460
32	1024	5.657	17.889	82	6724	9.055	28.636
33	1089	5.745	18.166	83	6889	9.110	28.810
34	1156	5.831	18.439	84	7056	9.165	28.983
35	1225	5.916	18.708	85	7225	9.220	29.155
36	1296	6.000	18.974	86	7396	9.274	29.326
37	1369	6.083	19.235	87	7569	9.327	29.496
38	1444	6.164	19.494	88	7744	9.381	29.665
39	1521	6.245	19.748	89	7921	9.434	29.833
40	1600	6.325	20.000	90	8100	9.487	30.000
41	1681	6.403	20.248	91	8281	9.539	30.166
42	1764	6.481	20.494	92	8464	9.592	30.332
43	1849	6.557	20.736	93	8649	9.644	30.496
44	1936	6.633	20.976	94	8836	9.695	30.659
45	2025	6.708	21.213	95	9025	9.747	30.822
46	2116	6.782	21.448	96	9216	9.798	30.984
47	2209	6.856	21.679	97	9409	9.849	31.145
48	2304	6.928	21.909	98	9604	9.899	31.305
49	2401	7.000	22.136	99	9801	9.950	31.464
50	2500	7.071	22.361	100	10000	10.000	31.623

TABLE 2 COMMON LOGARITHMS

n	0	1	2	3	4	5	6	7	8	9
1.0	.0000	.0043	.0086	.0128	.0170	.0212	.0253	.0294	.0334	.0374
1.1	.0414	.0453	.0492	.0531	.0569	.0607	.0645	.0682	.0719	.0755
1.2	.0792	.0828	.0864	.0899	.0934	.0969	.1004	.1038	.1072	.1106
1.3	.1139	.1173	.1206	.1239	.1271	.1303	.1335	.1367	.1399	.1430
1.4	.1461	.1492	.1523	.1553	.1584	.1614	.1644	.1673	.1703	.1732
1.5	.1761	.1790	.1818	.1847	.1875	.1903	.1931	.1959	.1987	.2014
1.6	.2041	.2068	.2095	.2122	.2148	.2175	.2201	.2227	.2253	.2279
1.7	.2304	.2330	.2355	.2380	.2405	.2430	.2455	.2480	.2504	.2529
1.8	.2553	.2577	.2601	.2625	.2648	.2672	.2695	.2718	.2742	.2765
1.9	.2788	.2810	.2833	.2856	.2878	.2900	.2923	.2945	.2967	.2989
2.0	.3010	.3032	.3054	.3075	.3096	.3118	.3139	.3160	.3181	.3201
2.1	.3222	.3243	.3263	.3284	.3304	.3324	.3345	.3365	.3385	.3404
2.2	.3424	.3444	.3464	.3483	.3502	.3522	.3541	.3560	.3579	.3598
2.3	.3617	.3636	.3655	.3674	.3692	.3711	.3729	.3747	.3766	.3784
2.4	.3802	.3820	.3838	.3856	.3874	.3892	.3909	.3927	.3945	.3962
2.5	.3979	.3997	.4014	.4031	.4048	.4065	.4082	.4099	.4116	.4133
2.6	.4150	.4166	.4183	.4200	.4216	.4232	.4249	.4265	.4281	.4298
2.7	.4314	.4330	.4346	.4362	.4378	.4393	.4409	.4425	.4440	.4456
2.8	.4472	.4487	.4502	.4518	.4533	.4548	.4564	.4579	.4594	.4609
2.9	.4624	.4639	.4654	.4669	.4683	.4698	.4713	.4728	.4742	.4757
3.0	.4771	.4786	.4800	.4814	.4829	.4843	.4857	.4871	.4886	.4900
3.1	.4914	.4928	.4942	.4955	.4969	.4983	.4997	.5011	.5024	.5038
3.2	.5051	.5065	.5079	.5092	.5105	.5119	.5132	.5145	.5159	.5172
3.3	.5185	.5198	.5211	.5224	.5237	.5250	.5263	.5276	.5289	.5302
3.4	.5315	.5328	.5340	.5353	.5366	.5378	.5391	.5403	.5416	.5428
3.5	.5441	.5453	.5465	.5478	.5490	.5502	.5514	.5527	.5539	.5551
3.6	.5563	.5575	.5587	.5599	.5611	.5623	.5635	.5647	.5658	.5670
3.7	.5682	.5694	.5705	.5717	.5729	.5740	.5752	.5763	.5775	.5786
3.8	.5798	.5809	.5821	.5832	.5843	.5855	.5866	.5877	.5888	.5899
3.9	.5911	.5922	.5933	.5944	.5955	.5966	.5977	.5988	.5999	.6010
4.0	.6021	.6031	.6042	.6053	.6064	.6075	.6085	.6096	.6107	.6117
4.1	.6128	.6138	.6149	.6160	.6170	.6180	.6191	.6201	.6212	.6222
4.2	.6232	.6243	.6253	.6263	.6274	.6284	.6294	.6304	.6314	.6325
4.3	.6335	.6345	.6355	.6365	.6375	.6385	.6395	.6405	.6415	.6425
4.4	.6435	.6444	.6454	.6464	.6474	.6484	.6493	.6503	.6513	.6522
4.5	.6532	.6542	.6551	.6561	.6571	.6580	.6590	.6599	.6609	.6618
4.6	.6628	.6637	.6646	.6656	.6665	.6675	.6684	.6693	.6702	.6712
4.7	.6721	.6730	.6739	.6749	.6758	.6767	.6776	.6785	.6794	.6803
4.8	.6812	.6821	.6830	.6839	.6848	.6857	.6866	.6875	.6884	.6893
4.9	.6902	.6911	.6920	.6928	.6937	.6946	.6955	.6964	.6972	.6981
5.0	.6990	.6998	.7007	.7016	.7024	.7033	.7042	.7050	.7059	.7067
5.1	.7076	.7084	.7093	.7101	.7110	.7118	.7126	.7135	.7143	.7152
5.2	.7160	.7168	.7177	.7185	.7193	.7202	.7210	.7218	.7226	.7235
5.3	.7243	.7251	.7259	.7267	.7275	.7284	.7292	.7300	.7308	.7316
5.4	.7324	.7332	.7340	.7348	.7356	.7364	.7372	.7380	.7388	.7396
5.5	.7404	.7412	.7419	.7427	.7435	.7443	.7451	.7459	.7466	.7474
5.6	.7482	.7490	.7497	.7505	.7513	.7520	.7528	.7536	.7543	.7551
5.7	.7559	.7566	.7574	.7582	.7589	.7597	.7604	.7612	.7619	.7627
5.8	.7634	.7642	.7649	.7657	.7664	.7672	.7679	.7686	.7694	.7701
5.9	.7709	.7716	.7723	.7731	.7738	.7745	.7752	.7760	.7767	.7774

(*continued*)

TABLE 2—(*continued*)

n	0	1	2	3	4	5	6	7	8	9
6.0	.7782	.7789	.7796	.7803	.7810	.7818	.7825	.7832	.7839	.7846
6.1	.7853	.7860	.7868	.7875	.7882	.7889	.7896	.7903	.7910	.7917
6.2	.7924	.7931	.7938	.7945	.7952	.7959	.7966	.7973	.7980	.7987
6.3	.7993	.8000	.8007	.8014	.8021	.8028	.8035	.8041	.8048	.8055
6.4	.8062	.8069	.8075	.8082	.8089	.8096	.8102	.8109	.8116	.8122
6.5	.8129	.8136	.8142	.8149	.8156	.8162	.8169	.8176	.8182	.8189
6.6	.8195	.8202	.8209	.8215	.8222	.8228	.8235	.8241	.8248	.8254
6.7	.8261	.8267	.8274	.8280	.8287	.8293	.8299	.8306	.8312	.8319
6.8	.8325	.8331	.8338	.8344	.8351	.8357	.8363	.8370	.8376	.8382
6.9	.8388	.8395	.8401	.8407	.8414	.8420	.8426	.8432	.8439	.8445
7.0	.8451	.8457	.8463	.8470	.8476	.8482	.8488	.8494	.8500	.8506
7.1	.8513	.8519	.8525	.8531	.8537	.8543	.8549	.8555	.8561	.8567
7.2	.8573	.8579	.8585	.8591	.8597	.8603	.8609	.8615	.8621	.8627
7.3	.8633	.8639	.8645	.8651	.8657	.8663	.8669	.8675	.8681	.8686
7.4	.8692	.8698	.8704	.8710	.8716	.8722	.8727	.8733	.8739	.8745
7.5	.8751	.8756	.8762	.8768	.8774	.8779	.8785	.8791	.8797	.8802
7.6	.8808	.8814	.8820	.8825	.8831	.8837	.8842	.8848	.8854	.8859
7.7	.8865	.8871	.8876	.8882	.8887	.8893	.8899	.8904	.8910	.8915
7.8	.8921	.8927	.8932	.8938	.8943	.8949	.8954	.8960	.8965	.8971
7.9	.8976	.8982	.8987	.8993	.8998	.9004	.9009	.9015	.9020	.9025
8.0	.9031	.9036	.9042	.9047	.9053	.9058	.9063	.9069	.9074	.9079
8.1	.9085	.9090	.9096	.9101	.9106	.9112	.9117	.9122	.9128	.9133
8.2	.9138	.9143	.9149	.9154	.9159	.9165	.9170	.9175	.9180	.9186
8.3	.9191	.9196	.9201	.9206	.9212	.9217	.9222	.9227	.9232	.9238
8.4	.9243	.9248	.9253	.9258	.9263	.9269	.9274	.9279	.9284	.9289
8.5	.9294	.9299	.9304	.9309	.9315	.9320	.9325	.9330	.9335	.9340
8.6	.9345	.9350	.9355	.9360	.9365	.9370	.9375	.9380	.9385	.9390
8.7	.9395	.9400	.9405	.9410	.9415	.9420	.9425	.9430	.9435	.9440
8.8	.9445	.9450	.9455	.9460	.9465	.9469	.9474	.9479	.9484	.9489
8.9	.9494	.9499	.9504	.9509	.9513	.9518	.9523	.9528	.9533	.9538
9.0	.9542	.9547	.9552	.9557	.9562	.9566	.9571	.9576	.9581	.9586
9.1	.9590	.9595	.9600	.9605	.9609	.9614	.9619	.9624	.9628	.9633
9.2	.9638	.9643	.9647	.9652	.9657	.9661	.9666	.9671	.9675	.9680
9.3	.9685	.9689	.9694	.9699	.9703	.9708	.9713	.9717	.9722	.9727
9.4	.9731	.9736	.9741	.9745	.9750	.9754	.9759	.9763	.9768	.9773
9.5	.9777	.9782	.9786	.9791	.9795	.9800	.9805	.9809	.9814	.9818
9.6	.9823	.9827	.9832	.9836	.9841	.9845	.9850	.9854	.9859	.9863
9.7	.9868	.9872	.9877	.9881	.9886	.9890	.9894	.9899	.9903	.9908
9.8	.9912	.9917	.9921	.9926	.9930	.9934	.9939	.9943	.9948	.9952
9.9	.9956	.9961	.9965	.9969	.9974	.9978	.9983	.9987	.9991	.9996

TABLE 3 FOUR-PLACE TRIGONOMETRIC FUNCTIONS

θ degrees	θ radians	sin θ	tan θ	cot θ	cos θ		
0° 00′	.0000	.0000	.0000	—	1.0000	1.5708	90° 00′
10	.0029	.0029	.0029	343.77	1.0000	1.5679	50
20	.0058	.0058	.0058	171.89	1.0000	1.5650	40
30	.0087	.0087	.0087	114.59	1.0000	1.5621	30
40	.0116	.0116	.0116	85.940	.9999	1.5592	20
50	.0145	.0145	.0145	68.750	.9999	1.5563	10
1° 00′	.0175	.0175	.0175	57.290	.9998	1.5533	89° 00′
10	.0204	.0204	.0204	49.104	.9998	1.5504	50
20	.0233	.0233	.0233	42.964	.9997	1.5475	40
30	.0262	.0262	.0262	38.188	.9997	1.5446	30
40	.0291	.0291	.0291	34.368	.9996	1.5417	20
50	.0320	.0320	.0320	31.242	.9995	1.5388	10
2° 00′	.0349	.0349	.0349	28.636	.9994	1.5359	88° 00′
10	.0378	.0378	.0378	26.432	.9993	1.5330	50
20	.0407	.0407	.0407	24.542	.9992	1.5301	40
30	.0436	.0436	.0437	22.904	.9990	1.5272	30
40	.0465	.0465	.0466	21.470	.9989	1.5243	20
50	.0495	.0494	.0495	20.206	.9988	1.5213	10
3° 00′	.0524	.0523	.0524	19.081	.9986	1.5184	87° 00′
10	.0553	.0552	.0553	18.075	.9985	1.5155	50
20	.0582	.0581	.0582	17.169	.9983	1.5126	40
30	.0611	.0610	.0612	16.350	.9981	1.5097	30
40	.0640	.0640	.0641	15.605	.9980	1.5068	20
50	.0669	.0669	.0670	14.924	.9978	1.5039	10
4° 00′	.0698	.0698	.0699	14.301	.9976	1.5010	86° 00′
10	.0727	.0727	.0729	13.727	.9974	1.4981	50
20	.0756	.0756	.0758	13.197	.9971	1.4952	40
30	.0785	.0785	.0787	12.706	.9969	1.4923	30
40	.0814	.0814	.0816	12.251	.9967	1.4893	20
50	.0844	.0843	.0846	11.826	.9964	1.4864	10
5° 00′	.0873	.0872	.0875	11.430	.9962	1.4835	85° 00′
10	.0902	.0901	.0904	11.059	.9959	1.4806	50
20	.0931	.0929	.0934	10.712	.9957	1.4777	40
30	.0960	.0958	.0963	10.385	.9954	1.4748	30
40	.0989	.0987	.0992	10.078	.9951	1.4719	20
50	.1018	.1016	.1022	9.7882	.9948	1.4690	10
6° 00′	.1047	.1045	.1051	9.5144	.9945	1.4661	84° 00′
10	.1076	.1074	.1080	9.2553	.9942	1.4632	50
20	.1105	.1103	.1110	9.0098	.9939	1.4603	40
30	.1134	.1132	.1139	8.7769	.9936	1.4573	30
40	.1164	.1161	.1169	8.5555	.9932	1.4544	20
50	.1193	.1190	.1198	8.3450	.9929	1.4515	10
		cos θ	cot θ	tan θ	sin θ	θ radians	θ degrees

(*continued*)

TABLE 3—(*continued*)

θ degrees	θ radians	sin θ	tan θ	cot θ	cos θ		
7° 00′	.1222	.1219	.1228	8.1443	.9925	1.4486	83° 00′
10	.1251	.1248	.1257	7.9530	.9922	1.4457	50
20	.1280	.1276	.1287	7.7704	.9918	1.4428	40
30	.1309	.1305	.1317	7.5958	.9914	1.4399	30
40	.1338	.1334	.1346	7.4287	.9911	1.4370	20
50	.1367	.1363	.1376	7.2687	.9907	1.4341	10
8° 00′	.1396	.1392	.1405	7.1154	.9903	1.4312	82° 00′
10	.1425	.1421	.1435	6.9682	.9899	1.4283	50
20	.1454	.1449	.1465	6.8269	.9894	1.4254	40
30	.1484	.1478	.1495	6.6912	.9890	1.4224	30
40	.1513	.1507	.1524	6.5606	.9886	1.4195	20
50	.1542	.1536	.1554	6.4348	.9881	1.4166	10
9° 00′	.1571	.1564	.1584	6.3138	.9877	1.4137	81° 00′
10	.1600	.1593	.1614	6.1970	.9872	1.4108	50
20	.1629	.1622	.1644	6.0844	.9868	1.4079	40
30	.1658	.1650	.1673	5.9758	.9863	1.4050	30
40	.1687	.1679	.1703	5.8708	.9858	1.4021	20
50	.1716	.1708	.1733	5.7694	.9853	1.3992	10
10° 00′	.1745	.1736	.1763	5.6713	.9848	1.3963	80° 00′
10	.1774	.1765	.1793	5.5764	.9843	1.3934	50
20	.1804	.1794	.1823	5.4845	.9838	1.3904	40
30	.1833	.1822	.1853	5.3955	.9833	1.3875	30
40	.1862	.1851	.1883	5.3093	.9827	1.3846	20
50	.1891	.1880	.1914	5.2257	.9822	1.3817	10
11° 00′	.1920	.1908	.1944	5.1446	.9816	1.3788	79° 00′
10	.1949	.1937	.1974	5.0658	.9811	1.3759	50
20	.1978	.1965	.2004	4.9894	.9805	1.3730	40
30	.2007	.1994	.2035	4.9152	.9799	1.3701	30
40	.2036	.2022	.2065	4.8430	.9793	1.3672	20
50	.2065	.2051	.2095	4.7729	.9787	1.3643	10
12° 00′	.2094	.2079	.2126	4.7046	.9781	1.3614	78° 00′
10	.2123	.2108	.2156	4.6382	.9775	1.3584	50
20	.2153	.2136	.2186	4.5736	.9769	1.3555	40
30	.2182	.2164	.2217	4.5107	.9763	1.3526	30
40	.2211	.2193	.2247	4.4494	.9757	1.3497	20
50	.2240	.2221	.2278	4.3897	.9750	1.3468	10
13° 00′	.2269	.2250	.2309	4.3315	.9744	1.3439	77° 00′
10	.2298	.2278	.2339	4.2747	.9737	1.3410	50
20	.2327	.2306	.2370	4.2193	.9730	1.3381	40
30	.2356	.2334	.2401	4.1653	.9724	1.3352	30
40	.2385	.2363	.2432	4.1126	.9717	1.3323	20
50	.2414	.2391	.2462	4.0611	.9710	1.3294	10
		cos θ	cot θ	tan θ	sin θ	θ radians	θ degrees

TABLE 3—(*continued*)

θ degrees	θ radians	sin θ	tan θ	cot θ	cos θ		
14° 00′	.2443	.2419	.2493	4.0108	.9703	1.3265	76° 00′
10	.2473	.2447	.2524	3.9617	.9696	1.3235	50
20	.2502	.2476	.2555	3.9136	.9689	1.3206	40
30	.2531	.2504	.2586	3.8667	.9681	1.3177	30
40	.2560	.2532	.2617	3.8208	.9674	1.3148	20
50	.2589	.2560	.2648	3.7760	.9667	1.3119	10
15° 00′	.2618	.2588	.2679	3.7321	.9659	1.3090	75° 00′
10	.2647	.2616	.2711	3.6891	.9652	1.3061	50
20	.2676	.2644	.2742	3.6470	.9644	1.3032	40
30	.2705	.2672	.2773	3.6059	.9636	1.3003	30
40	.2734	.2700	.2805	3.5656	.9628	1.2974	20
50	.2763	.2728	.2836	3.5261	.9621	1.2945	10
16° 00′	.2793	.2756	.2867	3.4874	.9613	1.2915	74° 00′
10	.2822	.2784	.2899	3.4495	.9605	1.2886	50
20	.2851	.2812	.2931	3.4124	.9596	1.2857	40
30	.2880	.2840	.2962	3.3759	.9588	1.2828	30
40	.2909	.2868	.2994	3.3402	.9580	1.2799	20
50	.2938	.2896	.3026	3.3052	.9572	1.2770	10
17° 00′	.2967	.2924	.3057	3.2709	.9563	1.2741	73° 00′
10	.2996	.2952	.3089	3.2371	.9555	1.2712	50
20	.3025	.2979	.3121	3.2041	.9546	1.2683	40
30	.3054	.3007	.3153	3.1716	.9537	1.2654	30
40	.3083	.3035	.3185	3.1397	.9528	1.2625	20
50	.3113	.3062	.3217	3.1084	.9520	1.2595	10
18° 00′	.3142	.3090	.3249	3.0777	.9511	1.2566	72° 00′
10	.3171	.3118	.3281	3.0475	.9502	1.2537	50
20	.3200	.3145	.3314	3.0178	.9492	1.2508	40
30	.3229	.3173	.3346	2.9887	.9483	1.2479	30
40	.3258	.3201	.3378	2.9600	.9474	1.2450	20
50	.3287	.3228	.3411	2.9319	.9465	1.2421	10
19° 00′	.3316	.3256	.3443	2.9042	.9455	1.2392	71° 00′
10	.3345	.3283	.3476	2.8770	.9446	1.2363	50
20	.3374	.3311	.3508	2.8502	.9436	1.2334	40
30	.3403	.3338	.3541	2.8239	.9426	1.2305	30
40	.3432	.3365	.3574	2.7980	.9417	1.2275	20
50	.3462	.3393	.3607	2.7725	.9407	1.2246	10
20° 00′	.3491	.3420	.3640	2.7475	.9397	1.2217	70° 00′
10	.3520	.3448	.3673	2.7228	.9387	1.2188	50
20	.3549	.3475	.3706	2.6985	.9377	1.2159	40
30	.3578	.3502	.3739	2.6746	.9367	1.2130	30
40	.3607	.3529	.3772	2.6511	.9356	1.2101	20
50	.3636	.3557	.3805	2.6279	.9346	1.2072	10
		cos θ	cot θ	tan θ	sin θ	θ radians	θ degrees

(*continued*)

TABLE 3—(*continued*)

θ degrees	θ radians	sin θ	tan θ	cot θ	cos θ		
21° 00′	.3665	.3584	.3839	2.6051	.9336	1.2043	69° 00′
10	.3694	.3611	.3872	2.5826	.9325	1.2014	50
20	.3723	.3638	.3906	2.5605	.9315	1.1985	40
30	.3752	.3665	.3939	2.5386	.9304	1.1956	30
40	.3782	.3692	.3973	2.5172	.9293	1.1926	20
50	.3811	.3719	.4006	2.4960	.9283	1.1897	10
22° 00′	.3840	.3746	.4040	2.4751	.9272	1.1868	68° 00′
10	.3869	.3773	.4074	2.4545	.9261	1.1839	50
20	.3898	.3800	.4108	2.4342	.9250	1.1810	40
30	.3927	.3827	.4142	2.4142	.9239	1.1781	30
40	.3956	.3854	.4176	2.3945	.9228	1.1752	20
50	.3985	.3881	.4210	2.3750	.9216	1.1723	10
23° 00′	.4014	.3907	.4245	2.3559	.9205	1.1694	67° 00′
10	.4043	.3934	.4279	2.3369	.9194	1.1665	50
20	.4072	.3961	.4314	2.3183	.9182	1.1636	40
30	.4102	.3987	.4348	2.2998	.9171	1.1606	30
40	.4131	.4014	.4383	2.2817	.9159	1.1577	20
50	.4160	.4041	.4417	2.2637	.9147	1.1548	10
24° 00′	.4189	.4067	.4452	2.2460	.9135	1.1519	66° 00′
10	.4218	.4094	.4487	2.2286	.9124	1.1490	50
20	.4247	.4120	.4522	2.2113	.9112	1.1461	40
30	.4276	.4147	.4557	2.1943	.9100	1.1432	30
40	.4305	.4173	.4592	2.1775	.9088	1.1403	20
50	.4334	.4200	.4628	2.1609	.9075	1.1374	10
25° 00′	.4363	.4226	.4663	2.1445	.9063	1.1345	65° 00′
10	.4392	.4253	.4699	2.1283	.9051	1.1316	50
20	.4422	.4279	.4734	2.1123	.9038	1.1286	40
30	.4451	.4305	.4770	2.0965	.9026	1.1257	30
40	.4480	.4331	.4806	2.0809	.9013	1.1228	20
50	.4509	.4358	.4841	2.0655	.9001	1.1199	10
26° 00′	.4538	.4384	.4877	2.0503	.8988	1.1170	64° 00′
10	.4567	.4410	.4913	2.0353	.8975	1.1141	50
20	.4596	.4436	.4950	2.0204	.8962	1.1112	40
30	.4625	.4462	.4986	2.0057	.8949	1.1083	30
40	.4654	.4488	.5022	1.9912	.8936	1.1054	20
50	.4683	.4514	.5059	1.9768	.8923	1.1025	10
27° 00′	.4712	.4540	.5095	1.9626	.8910	1.0996	63° 00′
10	.4741	.4566	.5132	1.9486	.8897	1.0966	50
20	.4771	.4592	.5169	1.9347	.8884	1.0937	40
30	.4800	.4617	.5206	1.9210	.8870	1.0908	30
40	.4829	.4643	.5243	1.9074	.8857	1.0879	20
50	.4858	.4669	.5280	1.8940	.8843	1.0850	10
		cos θ	cot θ	tan θ	sin θ	θ radians	θ degrees

TABLE 3—(*continued*)

θ degrees	θ radians	sin θ	tan θ	cot θ	cos θ		
28° 00′	.4887	.4695	.5317	1.8807	.8829	1.0821	62° 00′
10	.4916	.4720	.5354	1.8676	.8816	1.0792	50
20	.4945	.4746	.5392	1.8546	.8802	1.0763	40
30	.4974	.4772	.5430	1.8418	.8788	1.0734	30
40	.5003	.4797	.5467	1.8291	.8774	1.0705	20
50	.5032	.4823	.5505	1.8165	.8760	1.0676	10
29° 00′	.5061	.4848	.5543	1.8040	.8746	1.0647	61° 00′
10	.5091	.4874	.5581	1.7917	.8732	1.0617	50
20	.5120	.4899	.5619	1.7796	.8718	1.0588	40
30	.5149	.4924	.5658	1.7675	.8704	1.0559	30
40	.5178	.4950	.5696	1.7556	.8689	1.0530	20
50	.5207	.4975	.5735	1.7437	.8675	1.0501	10
30° 00′	.5236	.5000	.5774	1.7321	.8660	1.0472	60° 00′
10	.5265	.5025	.5812	1.7205	.8646	1.0443	50
20	.5294	.5050	.5851	1.7090	.8631	1.0414	40
30	.5323	.5075	.5890	1.6977	.8616	1.0385	30
40	.5352	.5100	.5930	1.6864	.8601	1.0356	20
50	.5381	.5125	.5969	1.6753	.8587	1.0327	10
31° 00′	.5411	.5150	.6009	1.6643	.8572	1.0297	59° 00′
10	.5440	.5175	.6048	1.6534	.8557	1.0268	50
20	.5469	.5200	.6088	1.6426	.8542	1.0239	40
30	.5498	.5225	.6128	1.6319	.8526	1.0210	30
40	.5527	.5250	.6168	1.6212	.8511	1.0181	20
50	.5556	.5275	.6208	1.6107	.8496	1.0152	10
32° 00′	.5585	.5299	.6249	1.6003	.8480	1.0123	58° 00′
10	.5614	.5324	.6289	1.5900	.8465	1.0094	50
20	.5643	.5348	.6330	1.5798	.8450	1.0065	40
30	.5672	.5373	.6371	1.5697	.8434	1.0036	30
40	.5701	.5398	.6412	1.5597	.8418	1.0007	20
50	.5730	.5422	.6453	1.5497	.8403	.9977	10
33° 00′	.5760	.5446	.6494	1.5399	.8387	.9948	57° 00′
10	.5789	.5471	.6536	1.5301	.8371	.9919	50
20	.5818	.5495	.6577	1.5204	.8355	.9890	40
30	.5847	.5519	.6619	1.5108	.8339	.9861	30
40	.5876	.5544	.6661	1.5013	.8323	.9832	20
50	.5905	.5568	.6703	1.4919	.8307	.9803	10
34° 00′	.5934	.5592	.6745	1.4826	.8290	.9774	56° 00′
10	.5963	.5616	.6787	1.4733	.8274	.9745	50
20	.5992	.5640	.6830	1.4641	.8258	.9716	40
30	.6021	.5664	.6873	1.4550	.8241	.9687	30
40	.6050	.5688	.6916	1.4460	.8225	.9657	20
50	.6080	.5712	.6959	1.4370	.8208	.9628	10
		cos θ	cot θ	tan θ	sin θ	θ radians	θ degrees

(*continued*)

TABLE 3—(*continued*)

θ degrees	θ radians	sin θ	tan θ	cot θ	cos θ		
35° 00′	.6109	.5736	.7002	1.4281	.8192	.9599	55° 00′
10	.6138	.5760	.7046	1.4193	.8175	.9570	50
20	.6167	.5783	.7089	1.4106	.8158	.9541	40
30	.6196	.5807	.7133	1.4019	.8141	.9512	30
40	.6225	.5831	.7177	1.3934	.8124	.9483	20
50	.6254	.5854	.7221	1.3848	.8107	.9454	10
36° 00′	.6283	.5878	.7265	1.3764	.8090	.9425	54° 00′
10	.6312	.5901	.7310	1.3680	.8073	.9396	50
20	.6341	.5925	.7355	1.3597	.8056	.9367	40
30	.6370	.5948	.7400	1.3514	.8039	.9338	30
40	.6400	.5972	.7445	1.3432	.8021	.9308	20
50	.6429	.5995	.7490	1.3351	.8004	.9279	10
37° 00′	.6458	.6018	.7536	1.3270	.7986	.9250	53° 00′
10	.6487	.6041	.7581	1.3190	.7969	.9221	50
20	.6516	.6065	.7627	1.3111	.7951	.9192	40
30	.6545	.6088	.7673	1.3032	.7934	.9163	30
40	.6574	.6111	.7720	1.2954	.7916	.9134	20
50	.6603	.6134	.7766	1.2876	.7898	.9105	10
38° 00′	.6632	.6157	.7813	1.2799	.7880	.9076	52° 00′
10	.6661	.6180	.7860	1.2723	.7862	.9047	50
20	.6690	.6202	.7907	1.2647	.7844	.9018	40
30	.6720	.6225	.7954	1.2572	.7826	.8988	30
40	.6749	.6248	.8002	1.2497	.7808	.8959	20
50	.6778	.6271	.8050	1.2423	.7790	.8930	10
39° 00′	.6807	.6293	.8098	1.2349	.7771	.8901	51° 00′
10	.6836	.6316	.8146	1.2276	.7753	.8872	50
20	.6865	.6338	.8195	1.2203	.7735	.8843	40
30	.6894	.6361	.8243	1.2131	.7716	.8814	30
40	.6923	.6383	.8292	1.2059	.7698	.8785	20
50	.6952	.6406	.8342	1.1988	.7679	.8756	10
40° 00′	.6981	.6428	.8391	1.1918	.7660	.8727	50° 00′
10	.7010	.6450	.8441	1.1847	.7642	.8698	50
20	.7039	.6472	.8491	1.1778	.7623	.8668	40
30	.7069	.6494	.8541	1.1708	.7604	.8639	30
40	.7098	.6517	.8591	1.1640	.7585	.8610	20
50	.7127	.6539	.8642	1.1571	.7566	.8581	10
41° 00′	.7156	.6561	.8693	1.1504	.7547	.8552	49° 00′
10	.7185	.6583	.8744	1.1436	.7528	.8523	50
20	.7214	.6604	.8796	1.1369	.7509	.8494	40
30	.7243	.6626	.8847	1.1303	.7490	.8465	30
40	.7272	.6648	.8899	1.1237	.7470	.8436	20
50	.7301	.6670	.8952	1.1171	.7451	.8407	10
		cos θ	cot θ	tan θ	sin θ	θ radians	θ degrees

TABLE 3—(*continued*)

θ degrees	θ radians	sin θ	tan θ	cot θ	cos θ		
42° 00′	.7330	.6691	.9004	1.1106	.7431	.8378	48° 00′
10	.7359	.6713	.9057	1.1041	.7412	.8348	50
20	.7389	.6734	.9110	1.0977	.7392	.8319	40
30	.7418	.6756	.9163	1.0913	.7373	.8290	30
40	.7447	.6777	.9217	1.0850	.7353	.8261	20
50	.7476	.6799	.9271	1.0786	.7333	.8232	10
43° 00′	.7505	.6820	.9325	1.0724	.7314	.8203	47° 00′
10	.7534	.6841	.9380	1.0661	.7294	.8174	50
20	.7563	.6862	.9435	1.0599	.7274	.8145	40
30	.7592	.6884	.9490	1.0538	.7254	.8116	30
40	.7621	.6905	.9545	1.0477	.7234	.8087	20
50	.7650	.6926	.9601	1.0416	.7214	.8058	10
44° 00′	.7679	.6947	.9657	1.0355	.7193	.8029	46° 00′
10	.7709	.6967	.9713	1.0295	.7173	.7999	50
20	.7738	.6988	.9770	1.0235	.7153	.7970	40
30	.7767	.7009	.9827	1.0176	.7133	.7941	30
40	.7796	.7030	.9884	1.0117	.7112	.7912	20
50	.7825	.7050	.9942	1.0058	.7092	.7883	10
45° 00′	.7854	.7071	1.0000	1.0000	.7071	.7854	45° 00′
		cos θ	cot θ	tan θ	sin θ	θ radians	θ degrees

Answers

CHAPTER 1

Exercises 1 (page 5) A Answers

1. Yes **2.** 1, b, 3, □ **3. a.** {3, 4, 5, 7, 8} **b.** {3} **4.** {4}, {e}, {4, e}, ∅
5. Yes **6.** E and F, F and G **7.** Yes. The null set is considered a set of every set.

(*Note:* Not every section has an exercise set. Since Sections 2 and 3 do not have exercise sets, the next exercise set is Exercises 4.)

Exercises 4 (page 9) A Answers

1. a. 2, 0 **b.** $\frac{3}{5}$, 2, 0 **c.** 2, 0 **d.** $-\sqrt{3}$, $-\pi$ **e.** $\frac{3}{5}$, 2, 0, $-\sqrt{3}$, $-\pi$ **f.** 0, 2
2. a. $8 > 4$ **b.** $-2 > -3$ **c.** $-5 < 0$ **d.** $3 < |-6|$ **3. a.** 7 **b.** 3
c. $5 - \sqrt{5}$ **d.** $\sqrt{7} - 2$ **4.** {7, 8, 9} **5.** {0, 1, 2}

Exercises 5 (page 13) A Answers

1. 4 **2.** -8 **3.** -3 **4.** -20 **5.** 14 **6.** -48 **7.** -3 **8.** 42
9. 243 **10.** 60 **11.** 70 **12.** $16 - 8b$ **13.** 30 **14.** 0 **15.** Not defined

Exercises 6 (page 15) A Answers

1. $2^7 = 128$ **2.** $3^6 = 729$ **3.** $10^3 = 1{,}000$ **4.** $8x^3$ **5.** $\frac{27}{64}$ **6.** 1 **7.** $\frac{1}{9}$
8. -27 **9.** 1 **10.** -1 **11.** $\frac{1}{9x^2}$ **12.** $\frac{5}{3}$ **13.** $-27y^6$ **14.** 81
15. $8a^6b^9c^{12}$ **16. a.** 3.2 **b.** 20.0 **c.** 1.79

Exercises 7 (page 17) A Answers

1. 8 **2.** -6 **3.** -2 **4.** -2 **5.** 7 **6.** -1 **7.** 0.3 **8.** 0.2
9. -0.2 **10.** 8 **11.** 8 and -8; 8 is the principal square root **12.** -3 **13.** 9
14. a. 2.20 **b.** 2.31 **c.** -2.49

Exercises 8 (page 22) A Answers

1. 5 **2.** $2 + 15a$ **3.** $12m - 2mn$ **4.** $10 - 4a$ **5.** $10ab$ **6.** $2x^6y^7z^{10}$
7. $31 - 12x$ **8.** $-x^2 + y^2$ **9.** $3a^2$ **10.** $5x^2 - y$

Exercises 9 (page 24) A Answers

1. 27; Division is done first. **2.** 8; Multiplication is done first. **3.** 75; Powers are done first. **4.** -9; Powers are done first. **5.** 1 **6.** 64 **7.** -15 **8.** 37 **9.** -28 **10.** 18 **11.** 1.800

Exercises 10 (page 28) A Answers

1. 5.7×10^4 **2.** 4.1×10^{-3} **3.** 0.00028 **4.** 970,000 **5.** 4.07×10^7 **6.** 0.014 **7.** 0.0068 **8.** 12.146 **9.** 0.56 **10.** 0.4694 **11.** 1.97×10^8 square miles **12.** 2.53×10^{13} miles

Exercises 11 (page 31) A Answers

1. 94.9 **2.** -23.3 **3.** 0.737 **4.** 6.28 **5.** 1,210 **6.** 104 **7.** 7,230 **8.** 201 **9.** $305.79

Review exercises (page 34) A Answers

1. a. $\{3, 5, 7, 8, 9\}$ **b.** $\varnothing$ **2. a.** $-2, 0, \frac{3}{5}$ **b.** 0 **c.** $-2, 0, \frac{3}{5}, \sqrt{2}$
3. a. 5 **b.** $\sqrt{7} - 1$ **c.** $1 - \sqrt{7}$ **4.** $\{-2, -1, 0\}$ **5.** 7.8×10^{-4}
6. a. -17 **b.** -22 **c.** 29 **7. a.** 23 **b.** $-x^3y^4z^3$ **8. a.** 0.0482 **b.** 929,000 **c.** 0.006 **9. a.** 3.45×10^7 **b.** 5.2×10^{-4} **c.** 5.006×10^4
10. 7.40×10^7 **11.** 0.000227

Diagnostic test (page 36)

(*Note:* Following each problem number in parentheses is the section number in which that kind of problem is discussed.)

1. (1) **a.** $\{2, 4, 5, 6, 7\}$ **b.** $\{5\}$ **c.** $\varnothing$ **2.** (2) **a.** Infinite **b.** $\{-2, -1, 0, 1, 2, 3, 4, 5, 6, 7, 8, 9\}$ is finite **c.** Infinite

3. (4, 5) **a.** $5|-2| - 5^2 = 5(2) - (25) = 10 - 25 = -15$
b. $30 \div (-2)(5) - 4\sqrt{36} = (-15)(5) - 4(6) = -75 - 24 = -99$
c. $0(15)^2 + (-1)^{25} + \frac{0}{6} = 0 - 1 + 0 = -1$ **4.** (6) **a.** $+4$ and -4 **b.** 4

5. (6) -2 **6.** (7) **a.** 1 **b.** $\frac{2^{-4}}{2^{-6}} = 2^{-4-(-6)} = 2^2 = 4$ **c.** $10^6 \cdot 10^{-4} = 10^{6-4} = 10^2 = 100$

d. $(2^{-3})^{-1} = 2^3 = 8$ **e.** $0^5 = 0 \cdot 0 \cdot 0 \cdot 0 \cdot 0 = 0$ **f.** $\frac{4}{0}$; not defined **g.** $\sqrt[3]{-64} = -4$

h. $\sqrt{49} = 7$ **i.** $\sqrt{(-6)^2} = 6$ **7.** (1) **a.** $\sqrt{3}, \pi$ **b.** $0, -5$ **c.** 0

d. $-5, 0, 2.4, \frac{2}{7}, \sqrt{3}, \pi$ **8.** (1, 2, 3) **a.** True **b.** False **c.** True

9. (10) **a.** 0.1564 **b.** 186,380 **c.** 7.2
10. (8) **a.** $12 - \{2x - 3[x - 2(4 - x)] - 24\} - 4x = 12 - \{2x - 3[x - 8 + 2x] - 24\} - 4x =$
$12 - \{2x - 3x + 24 - 6x - 24\} - 4x = 12 - 2x + 3x - 24 + 6x + 24 - 4x = 3x + 12$
b. $3x(4x^2y - 5y^2) - xy(12x^2 - 5y) = 12x^3y - 15xy^2 - 12x^3y + 5xy^2 = -10xy^2$

11. (10) 2.300×10^6 **12.** (11) $S = \dfrac{a(1-r^n)}{1-r} = \dfrac{-16\left[1-\left(\frac{1}{2}\right)^5\right]}{1-\frac{1}{2}} = \dfrac{-16\left(\frac{31}{32}\right)}{\frac{1}{2}} = -31$

CHAPTER 2

Exercises 1 (page 42) A Answers

1. Binomial; degree 7 **2.** Trinomial; degree 2 **3.** Monomial, degree 6
4. No; $\sqrt{x+y} = (x+y)^{\frac{1}{2}}$ with nonwhole number exponent. **5. a.** 34 **b.** -5
c. $12a^4 - 8a^2 - 5$ **6. a.** 0 **b.** -1 **c.** 0 **7. a.** 1 **b.** -27 **c.** -27

Exercises 2 (page 46) A Answers

1. $a^2 + 6a - 27$ **2.** $20x^5y - 12x^2y^3$ **3.** $x^2 - 6x - 27$ **4.** $\dfrac{x^2}{4} - \dfrac{4y^4}{9}$
5. $6a^2 - ab - 15b^2$ **6.** $25x^2 + 30xy + 9y^2$ **7.** $\dfrac{4x^2}{9} - 4x + 9$
8. $64x^3 - 48x^2 + 12x - 1$ **9.** $m^2 + \dfrac{2mr}{3} + \dfrac{r^2}{9}$ **10.** $25b^2 - 15bc + \dfrac{9c^2}{4}$
11. $3x^3 - 15x^2 + 2x - 10$ **12.** $27x^3 - 1$ **13.** $a^3 + b^3$ **14.** $-16x$
15. $24a^2b + 2b^3$ **16.** $x^6 - x^4 + 4x^2 - 4$

Exercises 3 (page 54) A Answers

1. $2x^2 + 1$ **2.** $2a^3b - 3ab^3 + 1$ **3.** $5x - 1$ **4.** $2x + 1$ **5.** $x^2 - 2x - 5$
6. $9x^2 - 3x + 1$ **7.** Quotient: $x^2 - 3x + 1$; remainder: $-6x + 3$
8. Quotient: $x^2 + x - 2$; remainder: 2 **9.** Quotient: $2x + 3$; remainder: 0
10. Quotient: $3x - 2$; remainder: 0 **11.** Quotient: $5x - 7$; remainder: 0
12. Quotient: $2x^3 - x^2 + 2x + 3$; remainder: 7
13. Quotient: $27x^3 + 9x^2 - 3x - 1$; remainder: -1
14. Quotient: $x^5 - x^4 + x^3 - x^2 + x - 1$; remainder: 2
15. Quotient: $x^4 + 2x^3 + 4x^2 + 8x + 16$; remainder: 0

Exercises 4 (page 61) A Answers

1. $4(3y - 1)$ **2.** $2m^2n^2(4n - 2m + 1)$ **3.** $(x + 6)(x - 4)$ **4.** $2(2a + 3)(3a - 5)$
5. $(4xy + 3)(4xy - 3)$ **6.** $(5r - 1)(5r - 1)$ **7.** $2(2 - n)(4 + 2n + n^2)$
8. $(2a - 3b)(3c - 4d)$ **9.** $(x - 1)(y - 1)$ **10.** $(4a + 5bc)(4a - 5bc)$
11. $(3x + 5y)^2$ **12.** $(3x + 3y - 4)(2x + 2y + 3)$ **13.** $(3u + 1)(3u - 1)$
14. $(3a - b + 1)(3a - b - 1)$ **15.** $(-2x)(x^2 + 48)$

Exercises 5 (page 63) A Answers

1. $(x^2 + x + 5)(x^2 - x + 5)$ **2.** $(3x^2 + x + 1)(3x^2 - x + 1)$ **3.** $(m^2 + 3m + 5)(m^2 - 3m + 5)$
4. $(3r^2 + 3r + 1)(3r^2 - 3r + 1)$ **5.** $(y^2 + 4yz + 8z^2)(y^2 - 4yz + 8z^2)$
6. $(4w^2 + 5wz + z^2)(4w^2 - 5wz + z^2)$ **7.** $2n(2m^2 + 2mn + n^2)(2m^2 - 2mn + n^2)$
8. $2y(5x^2 + 2xy + 2y^2)(5x^2 - 2xy + 2y^2)$ **9.** $4(x^2 + 3xy + 4y^2)(x^2 - 3xy + 4y^2)$

Exercises 6 (page 66) A Answers

1. $(x-1)(x+2)(x-3)$ **2.** $(x-1)(x-2)(x-4)$ **3.** $(x+2)(x+1)(x-3)$
4. $(x+2)(x-2)(x+1)$ **5.** $(x+1)(x-1)(x+2)(x-4)$ **6.** $(x+3)(x-1)(x+3)(x-1)$
7. $(3x)(x-2)(x+1)(x-3)$ **8.** $(x+1)(x-1)(x+2)(x-2)(x+3)$

Review exercises (page 67) A Answers

1. a. -9 **b.** -9 **c.** -5 **d.** $a^3 - 6a^2 + 9a - 9$ **2. a.** 8 **b.** 3
3. $8x^2 - 6x - 35$ **4.** $16x^4 - 9y^4$ **5.** $3a - 2b - 1$ **6.** $27a^3 - 64b^3$ **7.** $2x + 5$
8. $4a^2b + 12ab^2 + 10a - 1$ **9.** Quotient: $x^2 + 2x - 1$; remainder: 14
10. $(x^2+4)(x+2)(x-2)$ **11.** $(1-3a)(1+3a+9a^2)$ **12.** $(2x-1)(3x+5)$
13. $(2m+n)(2m+n-1)$ **14.** $(3a^2+2ab+b^2)(3a^2-2ab+b^2)$ **15.** $(x+1)(x-1)(x-3)(x+2)$

Diagnostic test (page 68)

(*Note:* Following each problem number in parentheses is the section number in which that kind of problem is discussed.)

1. (1) $3x^2 + 2x$, $5x^3y - 2xy^2 + y^3$ **2.** (1) **a.** $f(-2) = (-2)^2 - (-2) + 1 = 4 + 2 + 1 = 7$
b. $f(x+h) = (x+h)^2 - (x+h) + 1 = x^2 + 2hx + h^2 - x - h + 1$
3. (1) Yes, because this expression contains variables and real numbers, involved with powers, addition, and division **4.** (2) $9x^4 - \frac{1}{4}$ **5.** (2) $35x^2 + 11xy - 6y^2$ **6.** (2) $27a^3 - 64b^3$

7. (3) $\dfrac{4-12x}{4} = \dfrac{4}{4} + \dfrac{-12x}{4} = 1 - 3x$

8. (3) $\dfrac{12x^3y^2 + 8x^2y^3 - 4xy}{4x^2y^2} = \dfrac{12x^3y^2}{4x^2y^2} + \dfrac{8x^2y^3}{4x^2y^2} + \dfrac{-4xy}{4x^2y^2} = 3x + 2y - \dfrac{1}{xy}$

9. (2) $(10uv^2 + 8u^2v - 3) - (3u^2v - 5uv^2 + 7) = 15uv^2 + 5u^2v - 10$

10. (3)

$$\begin{array}{r|l} & -\ 5x + 1 \\ -2x^3 + 5x + 4 & 10x^4 - 2x^3 - 25x^2 - 15x + 4 \\ & 10x^4 \qquad\quad - 25x^2 - 20x \\ \hline & \quad - 2x^3 \qquad\qquad + 5x + 4 \\ & \quad - 2x^3 \qquad\qquad + 5x + 4 \\ \hline \end{array}$$

11. (4) $3x^3 - 12x = 3x(x^2 - 4) = 3x(x+2)(x-2)$ **12.** (4) $15x^2 - 4x - 4 = (5x+2)(3x-2)$
13. (4) $2ax - 2bx + ay - by = a(2x+y) - b(2x+y) = (2x+y)(a-b)$
14. (4)

$$\begin{array}{r|rrrrrl} & x^4 + & 5x^3 + & 0x^2 & -20x & -16 & \\ & 1 & 5 & 0 & -20 & -16 & \\ & & 2 & 14 & 28 & 16 & \\ \hline 2 & 1 & 7 & 14 & 8 & 0 & R = 0 \\ & & -2 & -10 & -8 & & \\ \hline -2 & 1 & 5 & 4 & 0 & & R = 0 \\ & & -1 & -4 & & & \\ \hline -1 & 1 & 4 & 0 & & & R = 0 \end{array}$$

$x^4 + 5x^3 - 20x - 16 = (x-2)(x+2)(x+1)(x+4)$

15. (4) $8a^3 + 27b^3 = (2a)^3 + (3b)^3 = [2a + 3b][4a^2 - 6ab + 9b^2]$
16. (5) $4x^4 + 11x^2y^2 + 9y^4 + x^2y^2 - x^2y^2 = 4x^4 + 12x^2y^2 + 9y^4 - x^2y^2 = (2x^2 + 3y^2)^2 - x^2y^2$
$= (2x^2 + 3y^2 + xy)(2x^2 + 3y^2 - xy) = (2x^2 + xy + 3y^2)(2x^2 - xy + 3y^2)$

CHAPTER 3

Exercises 1 (page 75) A Answers

1. **a.** Improper **b.** Improper **c.** Improper **d.** Improper **e.** Proper **f.** Improper **g.** Proper **h.** Proper **2.** 60 **3.** $3x^2(x-1)(x-3)(x+3)$ **4.** $2(x+1)(x-1)^2$ **5.** $\frac{1}{3}$ **6.** -1 **7.** $\frac{x+2}{x^2+2x+4}$ **8.** $\frac{3x-1}{2x+3}$ **9.** $\frac{x+2}{x-3}$ **10.** $\frac{x-3}{x+2}$

Exercises 2 (page 78) A Answers

1. 6 **2.** $2x^2-x$ **3.** $\frac{6ac}{b}$ **4.** $6x+3$ **5.** $\frac{9(x-5y)}{4(x+2y)}$ **6.** $\frac{3(x-y)}{2}$ **7.** $\frac{2}{5}$ **8.** $(x+1)(x+5)$ **9.** $3b^2$ **10.** $\frac{1}{x+1}$

Exercises 3 (page 83) A Answers

1. $\frac{34}{35}$ **2.** $\frac{13}{20}$ **3.** $\frac{7}{10}$ **4.** $\frac{m^2}{5(m+5)}$ **5.** $\frac{-a+7b}{12}$ **6.** $\frac{4ab}{a^2-b^2}$, or $\frac{4ab}{b^2-a^2}$ **7.** $\frac{15x+7}{60}$ **8.** $\frac{3-4z}{z^2}$ **9.** $\frac{z^2+7z-3}{(z+4)(z-1)}$ **10.** $-\frac{3H+8}{H+2}$ **11.** $\frac{1}{(x+1)(x+2)(x-3)}$ **12.** $-\frac{8}{x+4}$ **13.** $\frac{-3x^2-6x+1}{(x+1)^2(x-1)}$ **14.** $\frac{4}{x+2}$ **15.** $\frac{1}{(b-c)(a-b)}$ **16.** $\frac{-x^3+x^2-9x+6}{3x^2(x-1)}$

Exercises 4 (page 89) A Answers

1. $\frac{11}{17}$ **2.** $\frac{a^2-2}{a^2+1}$ **3.** $\frac{c^2}{3-2c}$ **4.** $\frac{n-m}{2}$ **5.** $c-3$ **6.** $\frac{a+4}{a-6}$ **7.** $x-y$ **8.** $\frac{15}{7}$ **9.** 1 **10.** $\frac{4y-3x}{xy}$

Review exercises (page 91) A Answers

1. **a.** $\frac{3}{5}, \frac{x+1}{x^2-1}, \frac{x+100}{x^2}$ **b.** $\frac{x^3-1}{x^2}, \frac{x+6}{x}, \frac{12}{12}$ **2.** $10x^2(x+2)^2$ **3.** -1 **4.** $x+2$ **5.** $\frac{x+2}{3x+1}$ **6.** $\frac{x^3-1}{x^3}$ **7.** $\frac{-3x-2}{(x-5)(x+2)(x+1)}$ **8.** $\frac{x+y}{y-x}$ **9.** $\frac{xy}{2y-3x}$ **10.** $(a+b)^2$

Diagnostic test (page 92)

(*Note:* Following each problem number in parentheses is the section number in which that kind of problem is discussed.)

1. (1) **a.** $\frac{x}{x^2+1}, \frac{x^2}{x^3+8}$ **b.** $\frac{x+10}{x}, \frac{7}{4}$

2. (3) $x^2 - 4 = (x + 2)(x - 2)$; $x^2 - x - 6 = (x - 3)(x + 2)$; $x^2 - 5x + 6 = (x - 3)(x - 2)$;
LCM $= (x + 2)(x - 2)(x - 3)$ **3.** (1) $\dfrac{2x - 1}{x^2 + 1}$, $x^2 - 4$

4. (3) $\dfrac{a}{a + b} - \dfrac{a - b}{a} = \dfrac{a^2 - (a + b)(a - b)}{(a + b)a} = \dfrac{a^2 - a^2 + b^2}{a^2 + ab} = \dfrac{b^2}{a^2 + ab}$

5. (4) $(x^{-2} + y^{-2})x^2y^2 = \left(\dfrac{1}{x^2} + \dfrac{1}{y^2}\right)x^2y^2 = \dfrac{y^2 + x^2}{\cancel{x^2y^2}} \cdot \dfrac{\cancel{x^2y^2}}{1} = x^2 + y^2$

6. (2) $\left(x - \dfrac{4}{x}\right)\left(\dfrac{x}{4 - x^2}\right) = \dfrac{\cancel{x^2 - 4}}{\cancel{x}} \cdot \dfrac{\cancel{x}}{(-1)\cancel{(x^2 - 4)}} = -1$

7. (3) $\dfrac{x + 1}{1} - \dfrac{x^3}{x^2 - x + 1} = \dfrac{(x + 1)(x^2 - x + 1) - 1(x^3)}{x^2 - x + 1} = \dfrac{x^3 - x^2 + x + x^2 - x + 1 - x^3}{x^2 - x + 1} = \dfrac{1}{x^2 - x + 1}$

8. (2) $\left(\dfrac{x}{y} + \dfrac{x + y}{x}\right)\left(\dfrac{x}{x^3 - y^3}\right) = \dfrac{\cancel{x^2 + xy + y^2}}{\cancel{x}y} \cdot \dfrac{x}{(x - y)\cancel{(x^2 + xy + y^2)}} = \dfrac{1}{y(x - y)}$

9. (4) $\dfrac{x + 2 - \dfrac{x}{x - 1}}{x - 3 + \dfrac{x}{x - 1}} \cdot \dfrac{\dfrac{x - 1}{1}}{\dfrac{x - 1}{1}} = \dfrac{(x + 2)(x - 1) - x}{(x - 3)(x - 1) + x} = \dfrac{x^2 - 2}{x^2 - 3x + 3}$

10. (4) $\dfrac{\dfrac{(x + 1)(x - 1)}{1}}{\dfrac{(x + 1)(x - 1)}{1}} \cdot \dfrac{\dfrac{x - 1}{x + 1} + \dfrac{x + 1}{x - 1}}{\dfrac{x + 1}{x - 1} - \dfrac{x - 1}{x + 1}} = \dfrac{(x - 1)^2 + (x + 1)^2}{(x + 1)^2 - (x - 1)^2} = \dfrac{x^2 - 2x + 1 + x^2 + 2x + 1}{x^2 + 2x + 1 - x^2 + 2x - 1}$

$= \dfrac{\cancel{2}(x^2 + 1)}{2(2x)} = \dfrac{x^2 + 1}{2x}$

11. (3) LCD $= (3x - 1)(x + 2)$; $\dfrac{x - 1}{x + 2} - \dfrac{3x^2}{(3x - 1)(x + 2)} + \dfrac{2}{3x - 1} =$

$\dfrac{3x - 1}{3x - 1} \cdot \dfrac{x - 1}{x + 2} + \dfrac{-3x^2}{(3x - 1)(x + 2)} + \dfrac{x + 2}{x + 2} \cdot \dfrac{2}{3x - 1} = \dfrac{(3x - 1)(x - 1) - 3x^2 + (x + 2)2}{(3x - 1)(x + 2)}$

$= \dfrac{3x^2 - 4x + 1 - 3x^2 + 2x + 4}{(3x - 1)(x + 2)} = \dfrac{-2x + 5}{(3x - 1)(x + 2)}$

12. (4) $x^{-1}y(x + y)^{-1}(x^2 - y^2)(y^2 - xy)^{-1} = \dfrac{1}{x} \cdot \cancel{y}\left(\dfrac{1}{\cancel{x + y}}\right) \cdot \dfrac{\cancel{(x + y)}(x - y)}{1} \cdot \dfrac{1}{\cancel{y}(y - x)}$

$= \dfrac{1}{x}\left(\dfrac{\cancel{x - y}}{1}\right) \cdot \dfrac{1}{(-1)\cancel{(x - y)}} = -\dfrac{1}{x}$

CHAPTER 4

Exercises 1 (page 99) A Answers

1. 6 **2.** 27 **3.** $\dfrac{1}{8}$ **4.** $\dfrac{5}{4}$ **5.** $\dfrac{13}{3}$ **6.** 15 **7.** $\dfrac{x^3}{y^5}$ **8.** $\dfrac{c^4}{a^3b^2}$

9. $\dfrac{2y^3 + 3x}{xy^3}$ **10.** $\dfrac{8}{3m}$ **11.** $\dfrac{3}{2}e^x$ **12.** $\dfrac{u(v - 1)}{v(u + 1)}$ **13.** $5u^{-3}v^{-1}$ **14.** $3xy^2w^{-5}z^{-3}$

15. $xy^{-1/2}$ **16.** $2a^{6/7}$ **17.** $x + x^{9/10}$ **18.** $a - 6a^{1/2}b^{1/2} + 9b$ **19.** $9a^4$ **20.** m^{x-2}

21. $x^{-4} - y^{-4} = \dfrac{1}{x^4} - \dfrac{1}{y^4} = \dfrac{y^4 - x^4}{x^4y^4}$ **22.** $x + y$

Exercises 2 (page 104) A Answers

1. $2y$ **2.** b **3.** $2\sqrt[4]{3}$ **4.** $3a^2\sqrt{2a}$ **5.** $-3x\sqrt[3]{x}$ **6.** $3x + 2$ **7.** $m\sqrt{m}$

8. u^3v^2 **9.** 2 **10.** $x\sqrt[12]{x}$ **11.** $\frac{2}{3}$ **12.** 5 **13.** $40\sqrt{3x} + 12x$ **14.** $\sqrt[10]{x^7y^9}$
15. $a\sqrt[8]{ab^6}$ **16.** $3\sqrt{3}$ **17.** $\frac{\sqrt{2m}}{m}$ **18.** $\frac{b\sqrt{b+1}}{b+1}$ **19.** $2(\sqrt{5} + 2\sqrt{3})$
20. $\frac{2a + b + 2\sqrt{a^2 + ab}}{b}$ **21. a.** 2.428 **b.** 2.426 **c.** 0.896 **d.** 0.894

(*Note:* There are no exercises for Section 3.)

Exercises 4 (page 110) A Answers

1. a. $7 - 2i$ **b.** -3 **c.** $5i$ **d.** $8 + i$ **2.** -3 **3.** $-4\sqrt{3}$ **4.** $9i\sqrt{3}$
5. $5i\sqrt{2}$ **6.** -10 **7.** $9 + 2i$ **8.** $10 + 8i$ **9.** 25 **10.** $6 + 15i$ **11.** $15 - 8i$
12. $-3\sqrt{3} + 4i\sqrt{3}$ **13.** $18 - 26i$ **14.** i **15.** 1 **16.** $8 - 2i$ **17.** i
18. $\frac{21}{29} - \frac{20}{29}i$

Review exercises (page 113) A Answers

1. 7 **2.** $\frac{8}{27}$ **3.** $\frac{2}{3}$ **4.** -5 **5.** 1 **6.** $32i$ **7.** i **8.** $\frac{7}{2}$ **9.** $3\sqrt{2}$
10. $5i$ **11.** $-3\sqrt[3]{2}$ **12.** $2\sqrt[4]{2}$ **13.** $\frac{x^2\sqrt[5]{y^3}}{2y}$ **14.** a^4b^3 **15.** 6 **16.** $3x^2$
17. $8x^3i$ **18.** $\sqrt[4]{x}$ **19.** $(x + y)\sqrt[4]{x + y}$ **20.** $\sqrt[10]{x^2y^9}$ **21.** 30 **22.** $-2\sqrt{6}$
23. $7 - 24i$ **24.** $-6 + 3\sqrt{3}$ **25.** $64x + y$ **26.** $\frac{a^2 - ab + b^2}{a^2b^2}$ **27.** $3\sqrt{3}$ **28.** $-5i$
29. $4 - 2i\sqrt{3}$ **30.** $\frac{a^2}{b}$ **31.** $\frac{x - 2\sqrt{xy} + y}{x - y}$ **32.** $\frac{\sqrt[4]{80x^6y^4} + 3}{2xy}$ **33.** $2\sqrt{6}$
34. $4a\sqrt[3]{4a}$ **35.** $6 + 4i$

Diagnostic test (page 115)

(*Note:* Following each problem number in parentheses is the section number in which that kind of problem is discussed.)

1. (1) **a.** $81^{-1/2} = \left(\frac{1}{81}\right)^{1/2} = \frac{1}{\sqrt{81}} = \frac{1}{9}$ **b.** $\left(\frac{27}{8}\right)^{2/3} = \frac{(\sqrt[3]{27})^2}{(\sqrt[3]{8})^2} = \frac{3^2}{2^2} = \frac{9}{4}$ **c.** $\sqrt{(-3)^2} = 3$

d. (3) $i^6 = (i^2)^3 = (-1)^3 = -1$ **e.** $\left(\frac{\frac{1}{6} - \frac{1}{8}}{\frac{1}{6} + \frac{1}{8}}\right)^{-1} = \left(\frac{\frac{2}{48}}{\frac{14}{48}}\right)^{-1} = \left(\frac{2}{14}\right)^{-1} = \left(\frac{1}{7}\right)^{-1} = \left(\frac{7}{1}\right)^{1} = 7$

2. (2) **a.** $\sqrt{98} = \sqrt{49 \cdot 2} = \sqrt{49}\sqrt{2} = 7\sqrt{2}$ **b.** $\sqrt[3]{-32x^5y^7} = \sqrt[3]{-8 \cdot 4x^3x^2y^6y} = -2xy^2\sqrt[3]{4x^2y}$

c. $\frac{4}{\sqrt[3]{2}} = \frac{2^2}{2^{1/3}} = 2^{2-1/3} = 2^{5/3} = \sqrt[3]{2^3 \cdot 2^2} = 2\sqrt[3]{4}$ **d.** $\sqrt[n]{a^{3n}b^n} = a^{3n/n}b^{n/n} = a^3b$

3. (2) **a.** $(2 + \sqrt{3})(4 - 3\sqrt{3}) = 8 + 4\sqrt{3} - 6\sqrt{3} - 9 = -1 - 2\sqrt{3}$

b. $\frac{6}{2 - \sqrt{5}} \cdot \frac{2 + \sqrt{5}}{2 + \sqrt{5}} = \frac{6(2 + \sqrt{5})}{4 - 5} = -12 - 6\sqrt{5}$

c. $\dfrac{3+\sqrt{2}}{3+\sqrt{8}} \cdot \dfrac{3-\sqrt{8}}{3-\sqrt{8}} = \dfrac{9+3\sqrt{2}-3\sqrt{8}-\sqrt{16}}{9-8} = \dfrac{9+3\sqrt{2}-3\sqrt{4}\sqrt{2}-4}{1} = 5+3\sqrt{2}-6\sqrt{2} = 5-3\sqrt{2}$

d. $2\sqrt[4]{\dfrac{y}{81x^2}} - \dfrac{1}{3}\sqrt[4]{\dfrac{y^2}{x^2y}} = 2\sqrt[4]{\dfrac{y}{81x^2} \cdot \dfrac{x^2}{x^2}} - \dfrac{1}{3}\sqrt[4]{\dfrac{y}{x^2} \cdot \dfrac{x^2}{x^2}} = 2\sqrt[4]{\dfrac{x^2y}{81x^4}} - \dfrac{1}{3}\sqrt[4]{\dfrac{x^2y}{x^4}} = \dfrac{2}{3x}\sqrt[4]{x^2y} - \dfrac{1}{3x}\sqrt[4]{x^2y}$

$= \left(\dfrac{2}{3x} - \dfrac{1}{3x}\right)\sqrt[4]{x^2y} = \dfrac{1}{3x}\sqrt[4]{x^2y}$

4. (2) $\sqrt{25-x^2} = \sqrt{25-9} = \sqrt{16} = 4$

5. (2) $\dfrac{(x^n)^{n+1}}{x^{n^2}} = \dfrac{x^{n^2+n}}{x^{n^2}} = x^{n^2+n-n^2} = x^n$

6. (2) $\dfrac{\sqrt{x+y}}{\sqrt{x+y}-\sqrt{y}} \cdot \dfrac{\sqrt{x+y}+\sqrt{y}}{\sqrt{x+y}+\sqrt{y}} = \dfrac{\sqrt{x+y}(\sqrt{x+y}+\sqrt{y})}{x+y-y} = \dfrac{x+y+\sqrt{xy+y^2}}{x}$

7. (3) $(2-3i)^2 = (2)^2 - 2(2)3i + (3i)^2 = 4 - 12i - 9 = -5 - 12i$

8. (3) $(6-5i) - (5+6i) = 6 - 5i - 5 - 6i = 1 - 11i$

9. (3) $\dfrac{10i}{3+\sqrt{-1}} = \dfrac{10i}{3+i} \cdot \dfrac{3-i}{3-i} = \dfrac{10i(3-i)}{9+1} = 3i + 1$, or $1 + 3i$

10. (3) $i(2-\sqrt{-50}) = i(2-\sqrt{25(2)(-1)}) = i(2-5i\sqrt{2}) = 2i - 5i^2\sqrt{2} = 2i + 5\sqrt{2}$, or $5\sqrt{2} + 2i$

CHAPTER 5

(*Note:* There are no exercises for Section 1.)

Exercises 2 (page 124) A Answers

1. $3u - [2v - (2u - 5v) + 3v] = 3u - [2v - 2u + 5v + 3v] = 3u - 2v + 2u - 5v - 3v = 5u - 10v$

Therefore the left and right members are identical.

2. $(5v - w)(25v^2 + 5vw + w^2) = 125v^3 + 25v^2w + 5vw^2 - 25v^2w - 5vw^2 - w^3 = 125v^3 - w^3$

Therefore the left and right members are identical.

3. $\{2\}$ 4. $\{5\}$ 5. $\left\{-\dfrac{7}{2}, \dfrac{5}{2}\right\}$ 6. $\left\{\dfrac{23}{24}\right\}$ 7. $\left\{\dfrac{5}{2}, 6, 3\right\}$ 8. $\{0, -3\}$ 9. $\{-5, 6\}$

10. $\left\{-\dfrac{1}{7}, 3\right\}$ 11. $\{0, -5, 5\}$ 12. $\left\{0, \dfrac{1}{2}, -3\right\}$ 13. $\{-1, 7\}$ 14. $\left\{6, \dfrac{3}{2}\right\}$ 15. $\{3\}$

16. $\{3\}$ 17. $\left\{2, \dfrac{6}{5}\right\}$ 18. $\{1, 2\}$ 19. $\left\{\dfrac{1}{4}\right\}$ 20. $\{6\}$ 21. $\{1, 2, -3\}$

22. $\{1, -1, -2, 3\}$

Exercises 3 (page 129) A Answers

1. $\{1 \pm \sqrt{3}\}$ 2. $\left\{\dfrac{2 \pm \sqrt{3}}{2}\right\}$ 3. $\left\{\dfrac{2 \pm i}{3}\right\}$ 4. $\{3 \pm \sqrt{3}\}$ 5. $\{5 \pm \sqrt{2}\}$ 6. $\{2 \pm i\sqrt{2}\}$

7. $\left\{\dfrac{3 \pm \sqrt{5}}{2}\right\}$ 8. $\{-5 \pm i\}$ 9. $\{3.19, 0.31\}$ 10. $\{0.45i, -0.45i\}$

Exercises 4 (page 134) A Answers

1. Rational and unequal 2. Irrational and unequal 3. Rational and equal

4. Rational and unequal 5. Imaginary conjugates 6. $k = \pm\dfrac{3}{2}$ 7. $x^2 - x - 12 = 0$

8. $x^2 - 18 = 0$ 9. $2x^2 - 2x - 13 = 0$ 10. $9x^2 - 18x + 11 = 0$ 11. $x^3 - x^2 + 9x - 9 = 0$

12. $x^5 - 2x^4 + 23x^3 - 46x^2 - 50x + 100 = 0$

Exercises 5 (page 137) A Answers

1. $\{2, -2, 1, -1\}$ **2.** $\{8\}$ **3.** $\left\{1, 3, \frac{-1 \pm i\sqrt{3}}{2}, \frac{-3 \pm 3i\sqrt{3}}{2}\right\}$ **4.** $\left\{\frac{1}{27}, -8\right\}$

5. $\left\{4, -\frac{3}{2}\right\}$ **6.** $\{\pm 1.41\}$

Exercises 6 (page 140) A Answers

1. $k = \pm 6$ **2.** $-\frac{5}{2}$ **3.** 5 in. by 12 in. **4. a.** 12 in. by 10 in. **b.** 4 in. by 2 in. by 4 in.

5. a. 1 sec., 2 sec. **b.** 3 sec.

Exercises 7 (page 153) A Answers

1. a. $<$ **b.** $<$ **c.** $>$

2. $\{x \mid x < -1\}$ 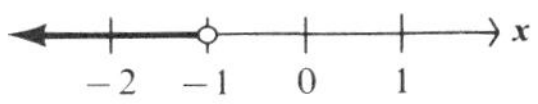

3. $\{2, -2\}$ 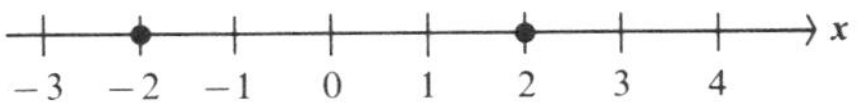

4. $\{x \mid x \geq 4\}$ 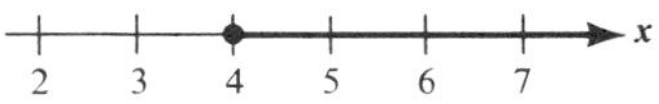

5. $\{4, -3\}$ 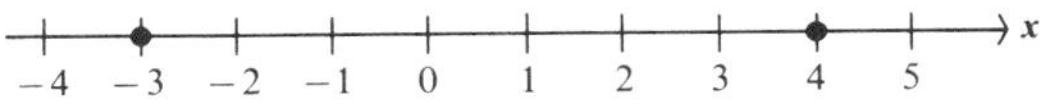

6. $\{x \mid 1 < x < 2\}$ 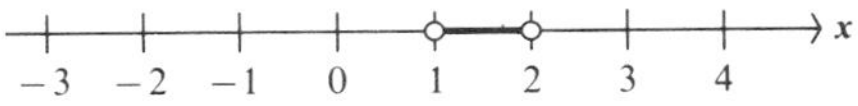

7. $\left\{\frac{4}{3}, 2\right\}$ 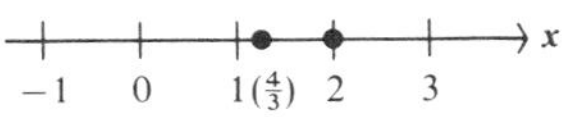

8. $\{x \mid x \leq 4\}$ 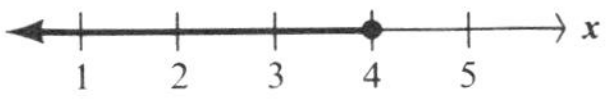

9. $\{x \mid -4 \leq x \leq -1\}$ 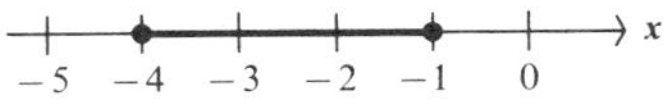

10. $\{x \mid -3 < x < 3\}$ 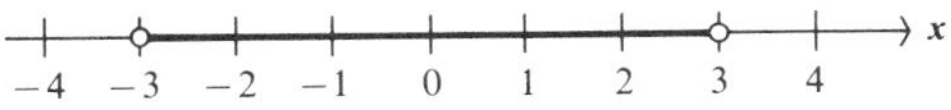

11. $(1, 3)$ 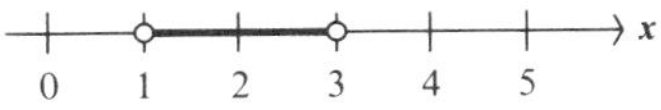

12. $\left(-\frac{3}{2}, -\frac{1}{2}\right)$ 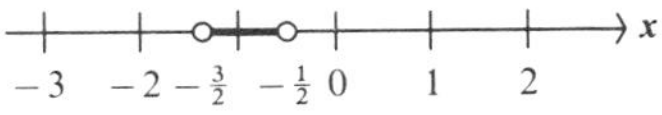

13. $(-\infty, 1) \cup (3, \infty)$ 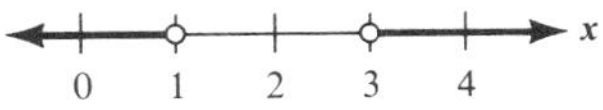

14. $[-5, 3]$ 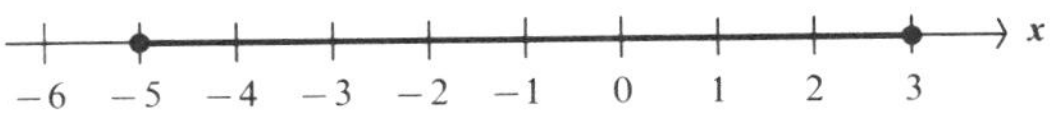

15. Given both a and b positive numbers and $a > b$, we multiply both sides of $a > b$ by a and then by b. $a > b$, so $a^2 > ab$; $a > b$, so $ab > b^2$. Therefore, $a^2 > ab > b^2$ or $a^2 > b^2$.

Review exercises (page 156) A Answers

1. Identity **2.** Conditional equation; $\left\{\frac{-7 \pm \sqrt{53}}{2}\right\}$

3. a. $[3, 1)$ 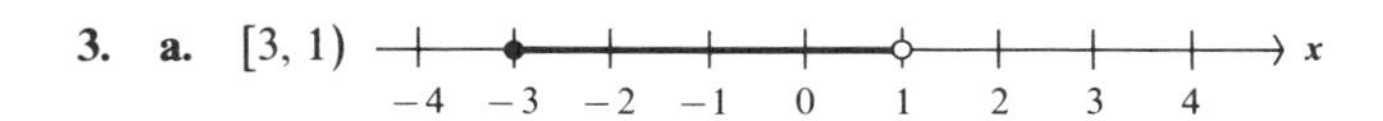

b. $[-1, 2] \cup [3, 5]$

4. $\{x \mid x \leq 1 \text{ or } x \geq 3\}$ **5.** $\left\{x \mid x > \frac{3}{2}\right\}$

6. $\left\{-\frac{2}{3}, \frac{4}{7}\right\}$ **7.** $\{x \mid x \geq 2\}$

8. $\left\{\frac{2}{5}, -\frac{1}{3}\right\}$ **9.** $\{0, 2, -7\}$ **10.** $\{3, 8\}$ **11.** $\left\{\frac{2 \pm 3i}{2}\right\}$ **12.** $\{2, -2, i\sqrt{3}, -i\sqrt{3}\}$

13. $\{4\}$ **14.** $x^3 - 27 = 0$ **15.** Rational and equal **16.** Base is 16, other leg is 12.

Diagnostic test (page 158)

(*Note:* Following each problem number in parentheses is the section number in which that kind of problem is discussed.)

1. (1) $3[x - x(x - 2) - 1] = 3[x - x^2 + 2x - 1] = -3x^2 + 9x - 3$
$3x(4 - x) - (3 + 3x) = 12x - 3x^2 - 3 - 3x = -3x^2 + 9x - 3$ Identity

2. (7) $|x - 3| \leq 1$
$-1 \leq x - 3 \leq 1$
$2 \leq x \leq 4$
$\{x \mid 2 \leq x \leq 4\}$

3. (7) $4x + 1 > 3x - 2$ and $3x - 2 \geq x + 6$
$x > -3$ $\quad 2x \geq 8$
$x \geq 4$
$\{x \mid x \geq 4\}$

4. (7) $3x - 2 = x + 2$ $\quad 3x - 2 = -x - 2$
$2x = 4$ $\quad 4x = 0$
$x = 2$ $\quad x = 0$
$\{2, 0\}$

5. (2) $10x^2 - x - 3 = 0$; $(5x - 3)(2x + 1) = 0$; $5x - 3 = 0$ or $2x + 1 = 0$
$x = \frac{3}{5}$ or $x = -\frac{1}{2}$; $\left\{\frac{3}{5}, -\frac{1}{2}\right\}$

6. (3) $4x^2 - 4x - 1 = 0$; $x = \frac{-(-4) \pm \sqrt{(-4)^2 - 4(4)(-1)}}{2(4)} = \frac{4 \pm \sqrt{32}}{8} = \frac{4 \pm 4\sqrt{2}}{8} = \frac{1 \pm \sqrt{2}}{2}$; $\left\{\frac{1 \pm \sqrt{2}}{2}\right\}$

7. (2) Multiplying through by the LCD, $(x - 1)(x + 1)$, we have $4(x + 1) - 1(x - 1) = 4 - 2x^2$
$4x + 4 - x + 1 = 4 - 2x^2$
$2x^2 + 3x + 1 = 0$
$(x + 1)(2x + 1) = 0$
$x + 1 = 0$ or $2x + 1 = 0$
$x = -1$ $\quad x = -\frac{1}{2}$
-1 does not check; $\left\{-\frac{1}{2}\right\}$

8. (2) $(\sqrt{2x - 1})^2 = (\sqrt{x - 1} + 1)^2$
$2x - 1 = x - 1 + 2\sqrt{x - 1} + 1$
$(x - 1)^2 = (2\sqrt{x - 1})^2$
$x^2 - 2x + 1 = 4x - 4$
$x^2 - 6x + 5 = 0$
$(x - 1)(x - 5) = 0$
$x - 1 = 0$ or $x - 5 = 0$
$x = 1$ $\quad x = 5$; $\{1, 5\}$

9. (5) $2x^{2/3} - 5x^{1/3} - 3 = 0$

$(2x^{1/3} + 1)(x^{1/3} - 3) = 0$

$x^{1/3} = -\frac{1}{2}$ or $x^{1/3} = 3$

$x = -\frac{1}{8}$ $\quad x = 27; \left\{-\frac{1}{8}, 27\right\}$

10. (4) $(x - 3)(x + 4)(x - 2i)(x + 2i) = 0$

$(x^2 + x - 12)(x^2 + 4) = 0$

$x^4 + x^3 - 8x^2 + 4x - 48 = 0$

11. (4) $b^2 - 4ac = (-7)^2 - 4(6)(3) = 49 - 72 = -23 < 0$

Therefore the roots are imaginary conjugates.

12. (6) Length × Width × Depth = Volume

$(x - 6)(x - 8)(4) = 252$

$(x^2 - 14x + 48) = 63$

$x^2 - 14x - 15 = 0$

$(x - 15)(x + 1) = 0$

$x - 15 = 0 \quad | \quad x + 1 = 0$

$x = 15 \quad | \quad x = -1$

-1 has no meaning.

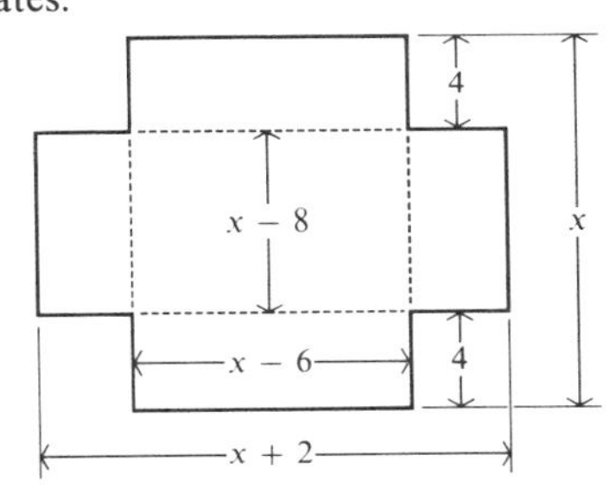

a. Dimensions of sheet of metal are 15 in. by 17 in.; **b.** Dimensions of box are 9 in. by 7 in. by 4 in.

CHAPTER 6

(*Note:* There are no exercises for Section 1.)

Exercises 2 (page 166) A Answers

1. a. Function **b.** Relation

2. a. Range: $\{-2, -1, 1, 3, 5\}$ **b.** Domain: $\{-5, -3, -1, 1, 2\}$

3.

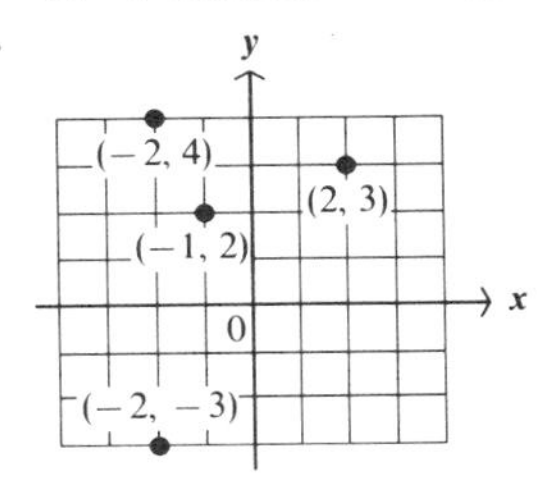

4. a, b, and d **5.** z

6. $y = \frac{-4x - 12}{3}$

7. a. -2 **b.** -17 **c.** 5

8. Domain: $x \geq -\frac{1}{2}$; Range: $f(x) \geq 0$

Exercises 3 (page 175) A Answers

1. $\frac{3}{7}$ **2.** $-\frac{3}{7}$ **3.** $\frac{13}{6}$ **4.** 0 **5.** 3 **6.** $-\frac{2}{5}$ **7.** $\frac{5}{4}$ **8.** 0

9.

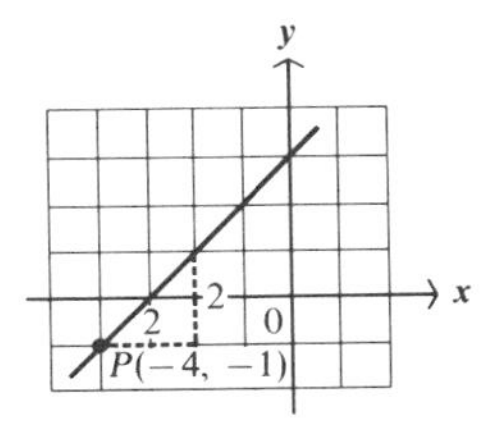

10.

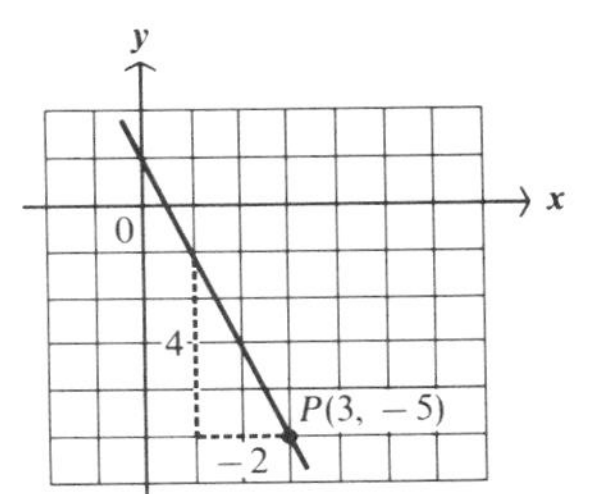

11.

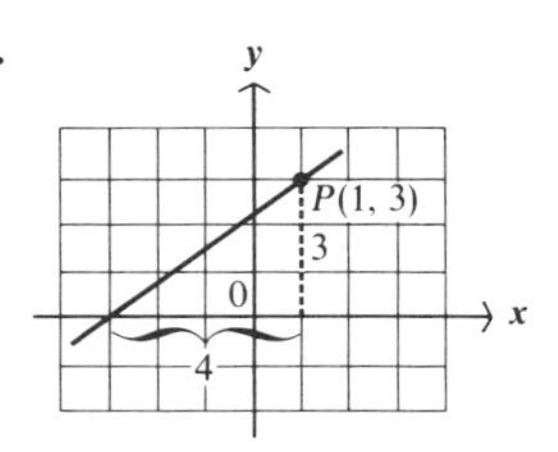

12.

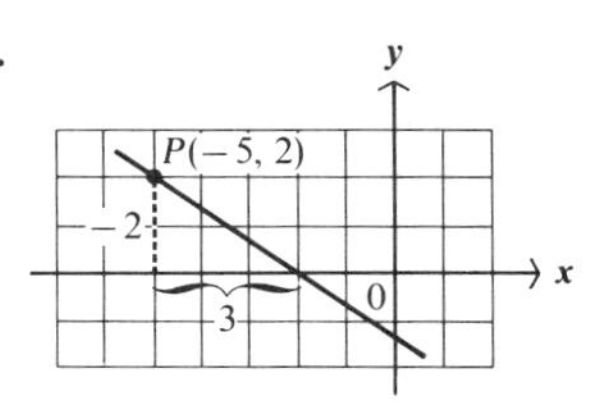

13. $2x - 5y - 17 = 0$ **14.** $2x + y + 6 = 0$ **15.** $3x - 4y + 12 = 0$ **16.** $x + 9y + 13 = 0$

17. $y - 5 = 0$

18.

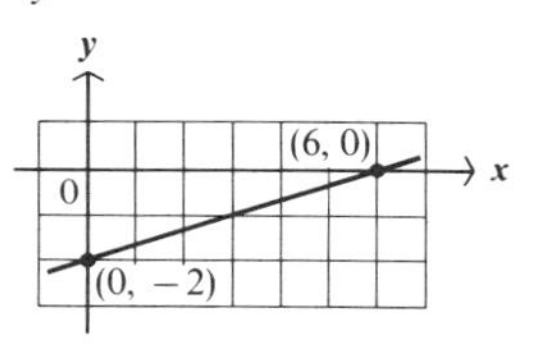

19.

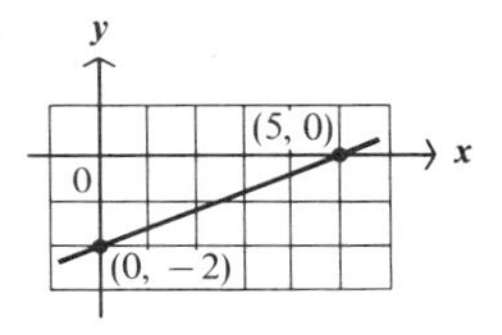

20.

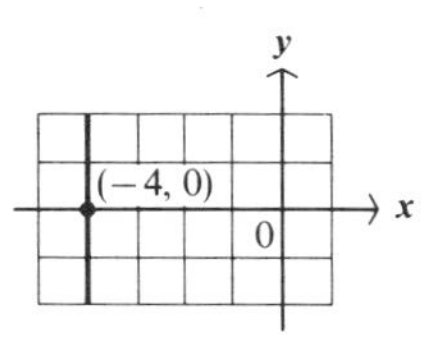

21.

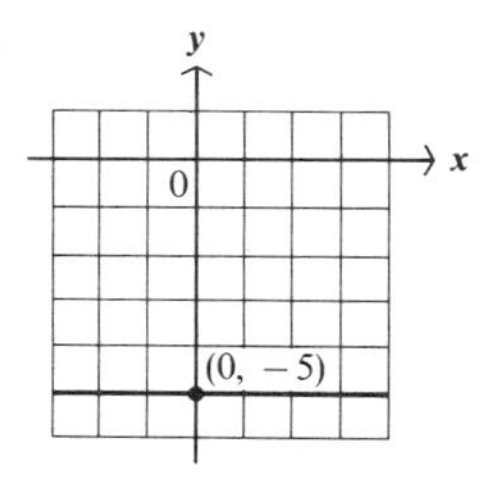

22.

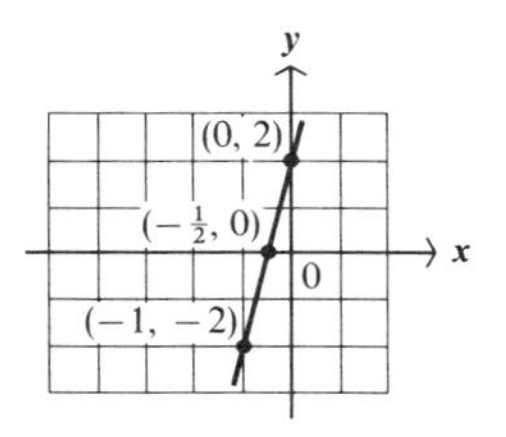

23.

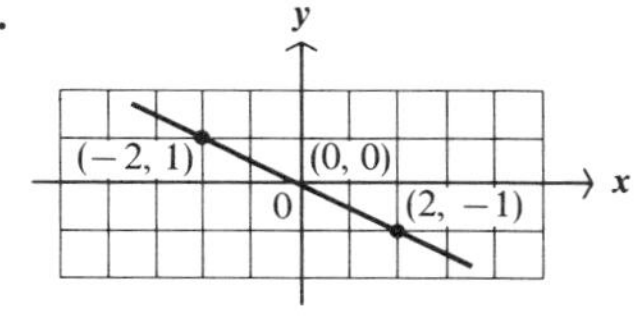

24. a. $m = -\frac{1}{4}$ **b.** $m = 4$ **c.** The slope of one is the negative reciprocal of the other.

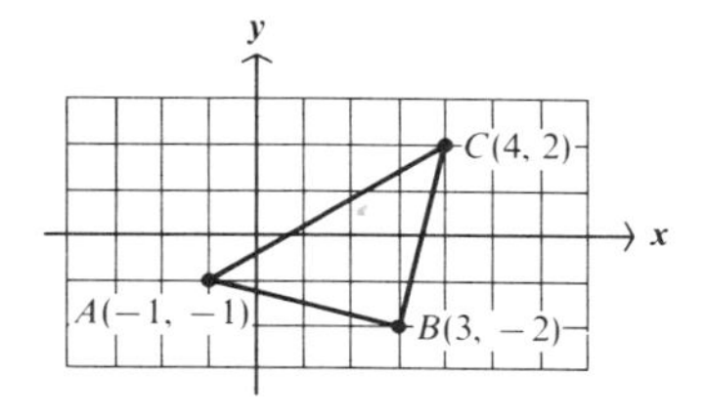

Exercises 4 (page 181) A Answers

1. a. $\sqrt{58}$ **b.** $\sqrt{130}$ **2.** 9.899 **3. a.** $(-1, 3)$ **b.** $(3, -2)$ **4.** 10
5. $(5, 1)$ **6.** 5

7. $AC = \sqrt{58}$; $AB = \sqrt{29}$; $BC = \sqrt{29}$

Thus $(AC)^2 = (AB)^2 + (BC)^2$

$(\sqrt{58})^2 = (\sqrt{29})^2 + (\sqrt{29})^2$

$58 = 29 + 29$

$58 = 58$

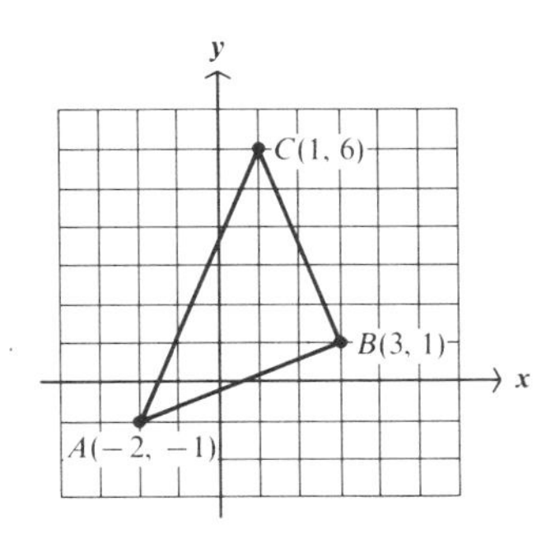

8. $OM = MB$ because M is the midpoint of OB; $OM = \sqrt{\left(\frac{a}{2} - 0\right)^2 + \left(\frac{b}{2} - 0\right)^2} = \frac{\sqrt{a^2 + b^2}}{2}$;

$AM = \sqrt{\left(\frac{a}{2} - a\right)^2 + \left(\frac{b}{2} - 0\right)^2} = \frac{\sqrt{a^2 + b^2}}{2}$; Therefore $OM = AM = MB$.

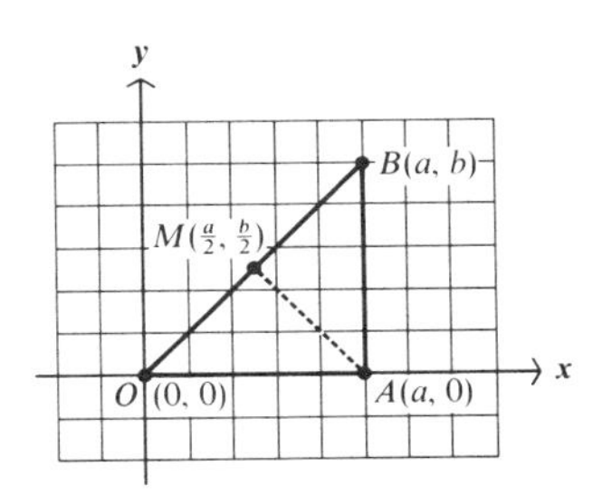

Exercises 5 (page 189) A Answers

1.

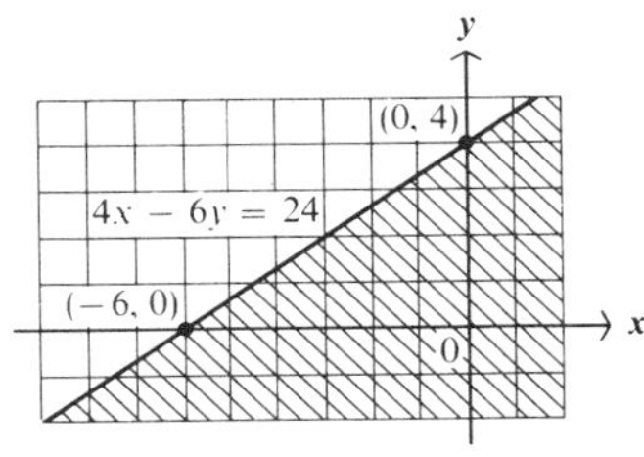

2.

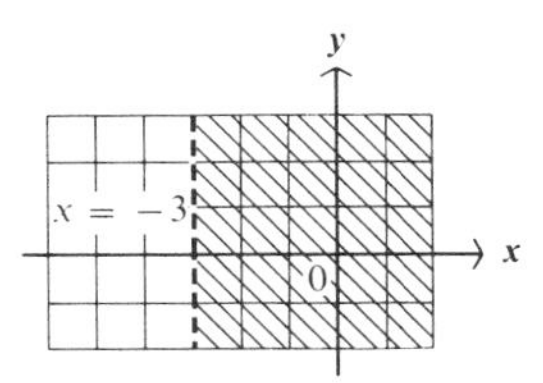

3.

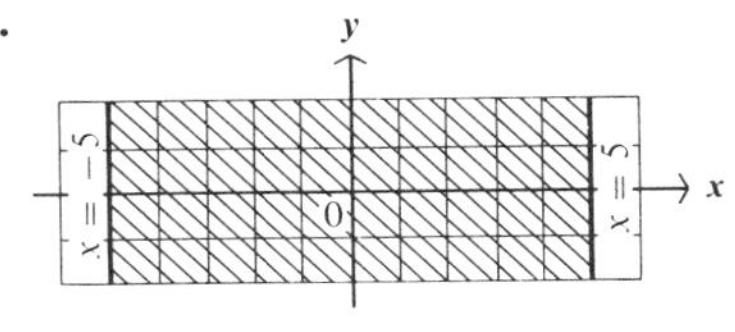

4.

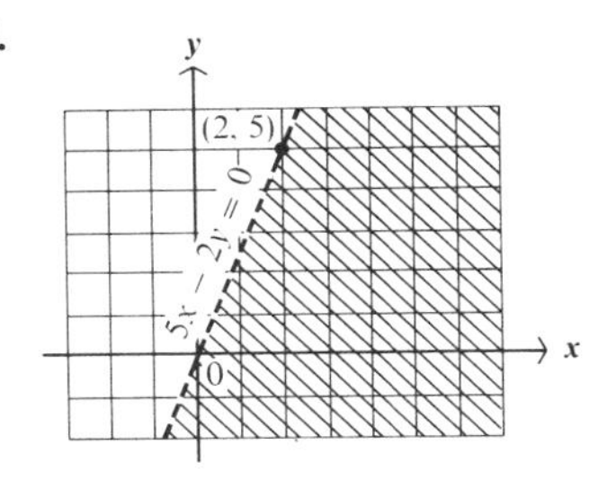

5.

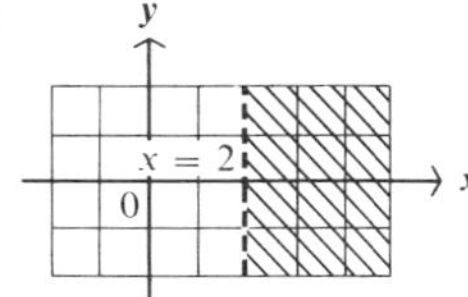

6.

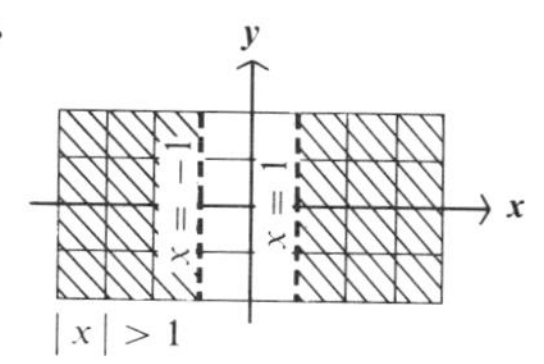

7.

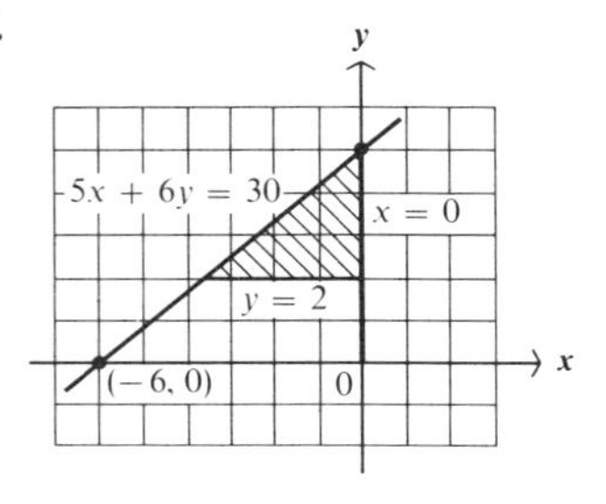

Review exercises (page 191) A Answers

1. 5 **2. a.** s **b.** t **3.** $\frac{3}{4}$

4.

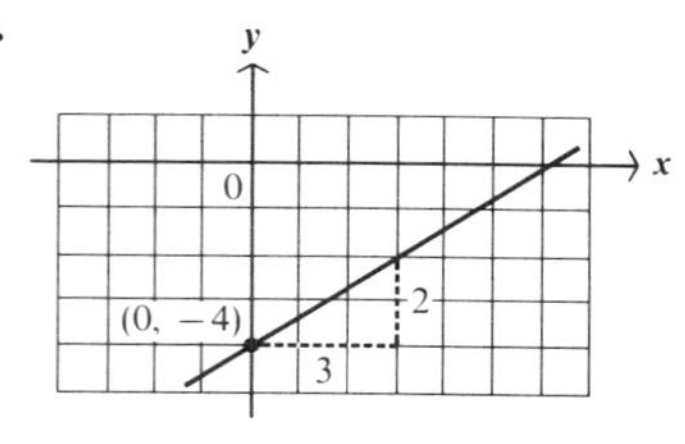

5.

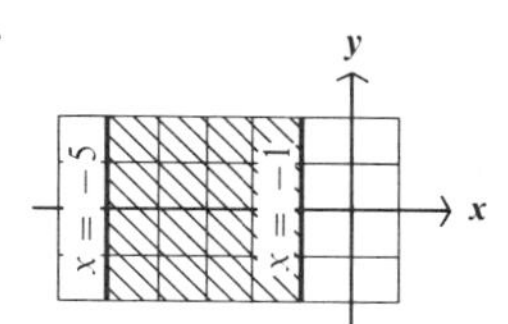

6.

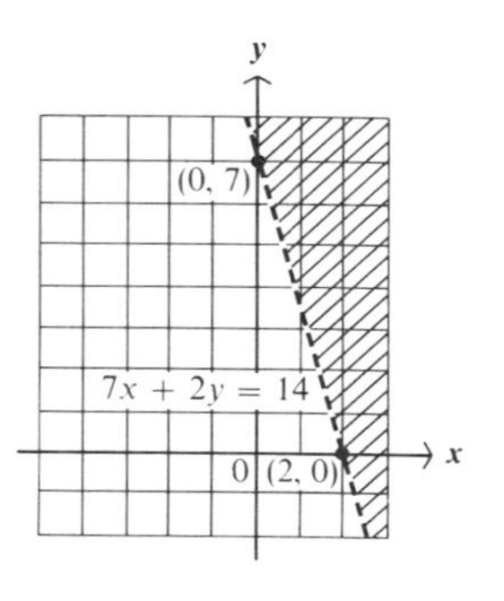

7.

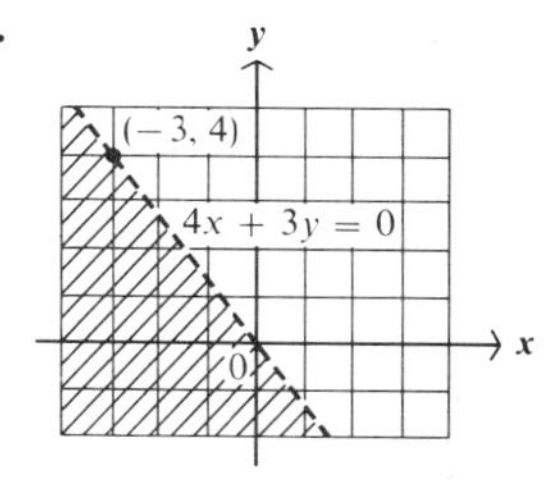

8. $$\frac{f(x+h)-f(x)}{h} = \frac{5(x+h)-3-(5x-3)}{h} = \frac{5x+5h-3-5x+3}{h} = \frac{5h}{h} = 5$$

9. **a.** 2
b. $2x - y = 0$
c. (4, 3)
d. $3x + y - 15 = 0$
e. Midpoint of both diagonals is (4, 3);
$MB = \sqrt{(8-4)^2 + (6-3)^2} = 5$ and
$MO = \sqrt{(4-0)^2 + (3-0)^2} = 5$; therefore
$MB = MO$ and M bisects OB;
$AM = \sqrt{(4-5)^2 + (3-0)^2} = \sqrt{10}$ and
$MC = \sqrt{(3-4)^2 + (6-3)^2} = \sqrt{10}$; therefore
$AM = MC$ and M bisects AC; thus the diagonals bisect each other.

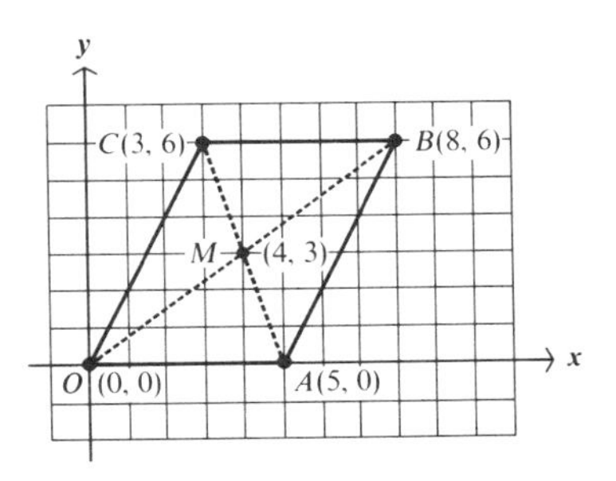

Diagnostic test (page 192)

(*Note:* Following each problem number in parentheses is the section number in which that kind of problem is discussed.)

1. (2) (b) and (d), because for each value of x there corresponds one and only one value of y.
2. (2) Implicit, because neither variable has been solved for in terms of the other variable and constants.
3. (2) x, because it is assigned independently of y.
4. (5) **a.**

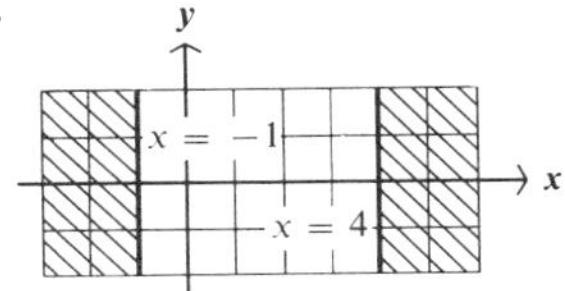

b.

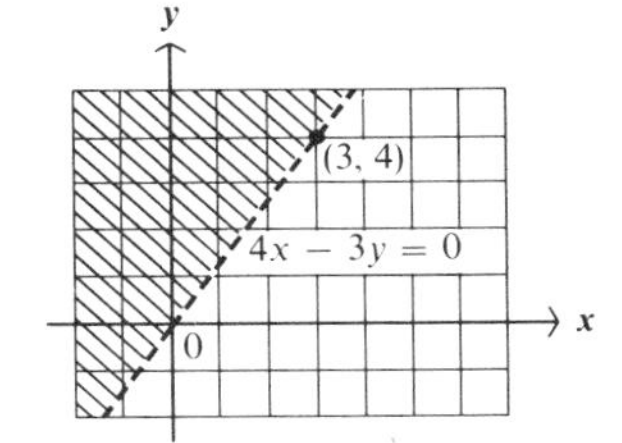

5. (4) **a.** $AB = \sqrt{[3-(-1)]^2 + [-2-(-4)]^2}$
$= \sqrt{20} = 2\sqrt{5}$

(3) **b.** AD has slope $\dfrac{2-(-4)}{-5-(-1)} = -\dfrac{3}{2}$;

$y - (-2) = \dfrac{-3}{2}(x - 3)$

$2y + 4 = -3x + 9$
$3x + 2y - 5 = 0$

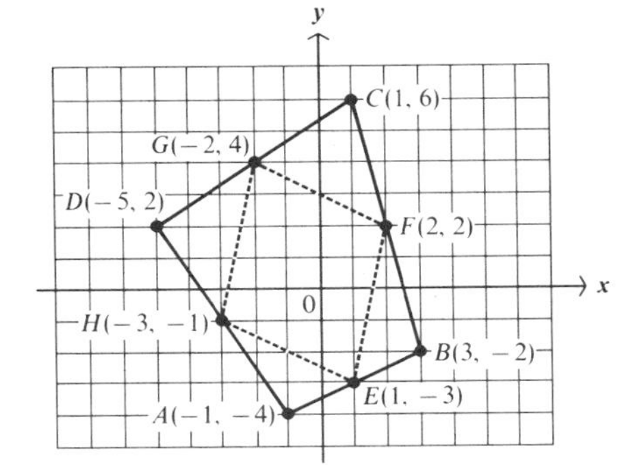

(4) **c.** Find the midpoints E, F, G, and H: $E\left(\dfrac{3-1}{2}, \dfrac{-4+(-2)}{2}\right) = (1, -3)$; $F\left(\dfrac{3+1}{2}, \dfrac{-2+6}{2}\right) = (2, 2)$; $G\left(\dfrac{1-5}{2}, \dfrac{6+2}{2}\right) = (-2, 4)$; $H\left(\dfrac{-1-5}{2}, \dfrac{2-4}{2}\right) = (-3, -1)$. Now prove that $EFGH$ is a parallelogram by showing that opposite sides have the same slope: slope of $EF = \dfrac{2-(-3)}{2-1} = \dfrac{5}{1}$; slope of $HG = \dfrac{4-(-1)}{-2-(-3)} = \dfrac{5}{1}$; slope of $HE = \dfrac{-3-(-1)}{1-(-3)} = -\dfrac{1}{2}$; slope of $GF = \dfrac{2-4}{2-(-2)} = -\dfrac{1}{2}$. Therefore $EFGH$ is a parallelogram because it is a four-sided figure with opposite sides parallel.

6. (2) Principal value of $\sqrt{5x - 1} \geq 0$; therefore $5x - 1 \geq 0$

$$5x \geq 1$$

$$\text{Domain:} \quad x \geq \frac{1}{5}$$

$$\text{Range:} \quad f(x) \geq 0$$

CHAPTER 7

Exercises 1 (page 198) A Answers

1.

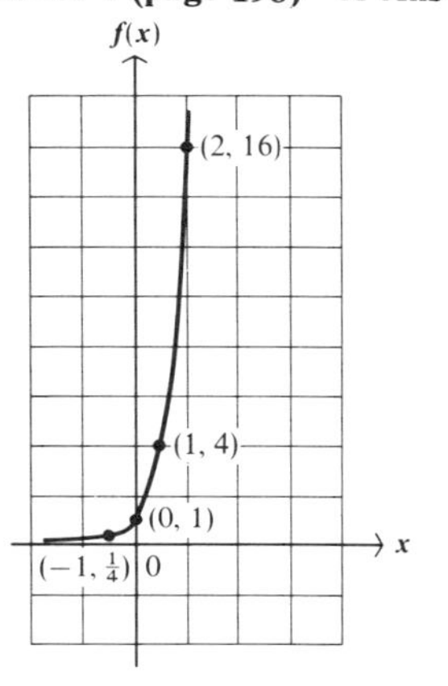

2.

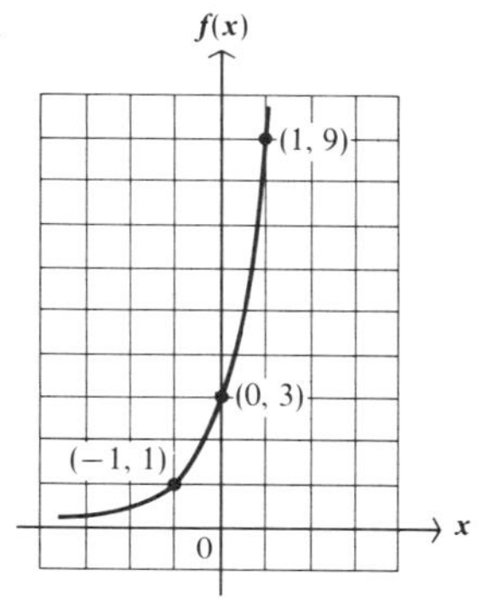

3.

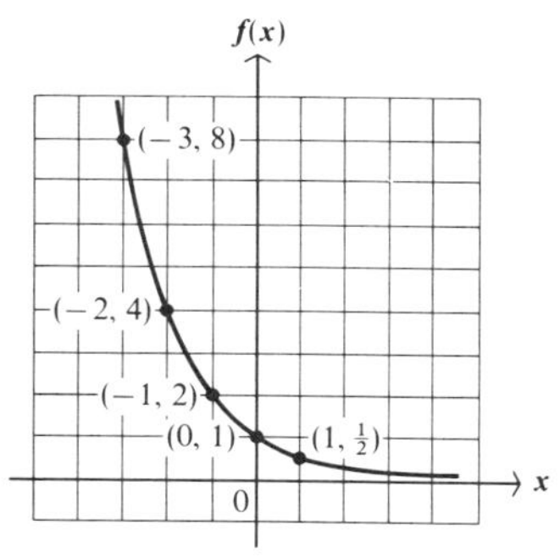

4.

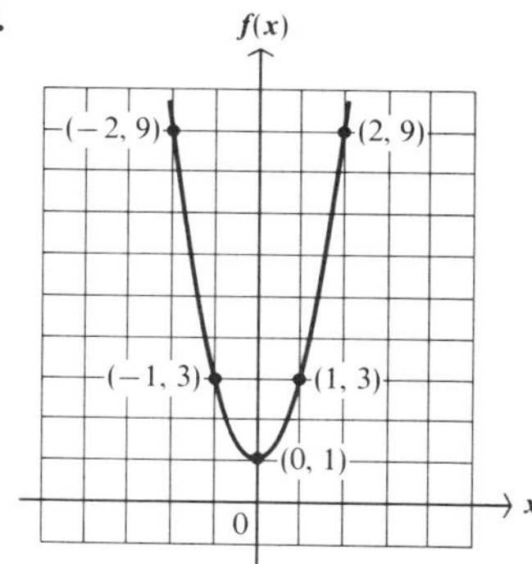

5.

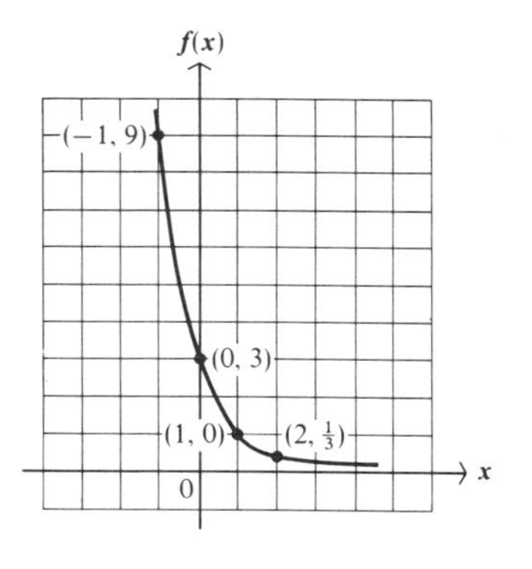

6.

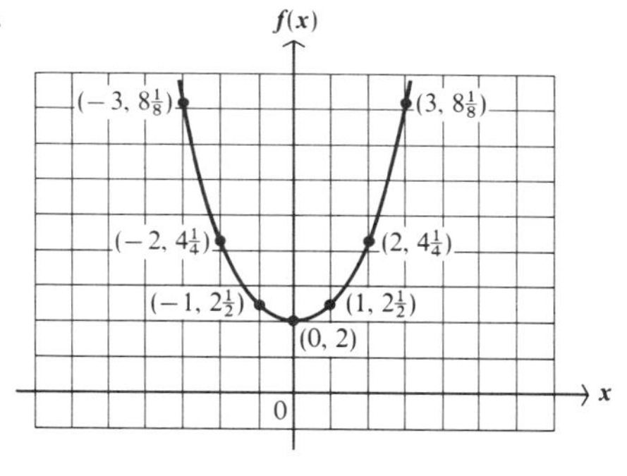

7.

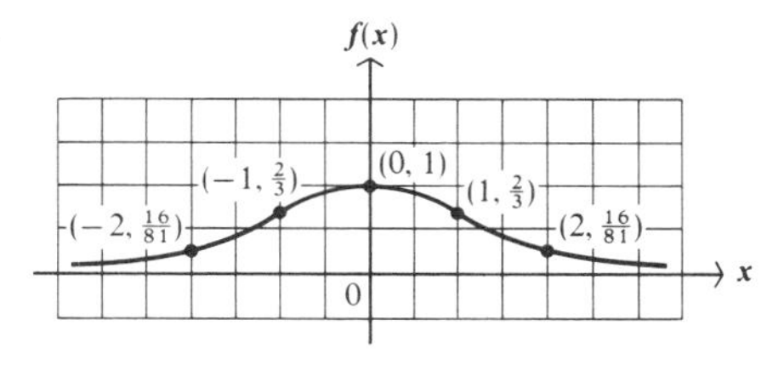

(*Note:* There are no exercises for Section 2.)

Exercises 3 (page 200) A Answers

1. $\log_3 9 = 2$ 2. $\log_{10} 0.001 = -3$ 3. $\log_{16} 4 = \frac{1}{2}$ 4. $\log_{27} \frac{1}{9} = -\frac{2}{3}$ 5. $\log_3 1 = 0$
6. $\log_3 3 = 1$ 7. $4^3 = 64$ 8. $16^{1/2} = 4$ 9. $10^{-2} = 0.01$ 10. $b^1 = b$ 11. 3
12. 0 13. −2 14. $N = 9$ 15. $b = 4$ 16. $x = 2$

Exercises 4 (page 204) A Answers

1. 1.146 2. 0.410 3. 1.699 4. 0.661 5. 1.778 6. 0.0301 7. $\log xy$
8. $\log \frac{x^2}{y^3}$ 9. $\log \frac{x^{(1/2)y}}{z}$ 10. $\log (x - y)$ 11. $y = 2^{3x} = 8^x$ 12. $y = x^3$
13. $y = x^3$ 14. $x = 11$

(*Note:* There are no exercises for Sections 5 and 6.)

Exercises 7 (page 209) A Answers

1. 2 2. 5 3. −2, or 8 − 10 4. −4, or 6 − 10 5. 0 6. −3, or 7 − 10
7. 4 8. 8 9. −3, or 7 − 10 10. 286 11. 24.5 12. 8.41 13. 0.745
14. 0.00547 15. 0.00207

Exercises 8 (page 212) A Answers

1. 1.7679 2. 2.5846 3. 7.7404 − 10 4. 8.9033 − 10 5. 2.3304 6. 7.010
7. 2.652 8. 505.0 9. 0.3565 10. 0.007363

Exercises 9 (page 214) A Answers

1. 90.12 2. 14.49 3. 2.101 4. 2.408 5. 6.995 6. 8.618 7. 9.035
8. 87.73 9. −14.94 10. 0.07555 11. 1.083 12. 18.58 13. 6.654
14. −2.782 15. 77.03

(*Note:* There are no exercises for Section 10.)

Exercises 11 (page 219) A Answers

1. 1.585 2. 1.481 3. −1.299 4. 10.66 5. 13.53 6. {1, 10} 7. 6
8. 3.424 9. 2.836
10.

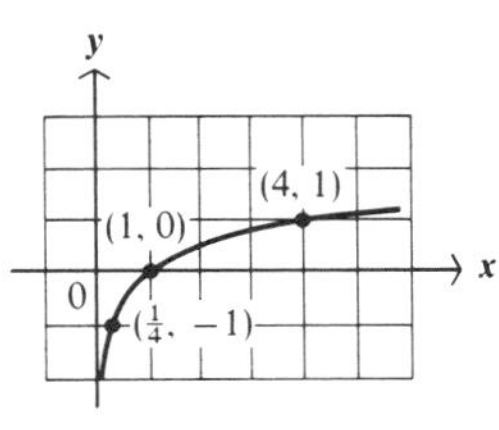

11. $x < -3.0801$ 12. \$335.68

Review exercises (page 221) A Answers

1. $\log_4 64 = 3$ **2.** $x = -2$ **3.** $N = 10^{-2} = \frac{1}{10^2} = 0.01$ **4.** $b = 4$ **5.** 3.010

6. 0.4771 **7.** 1.6811 **8.** $\log k^4$ **9.**

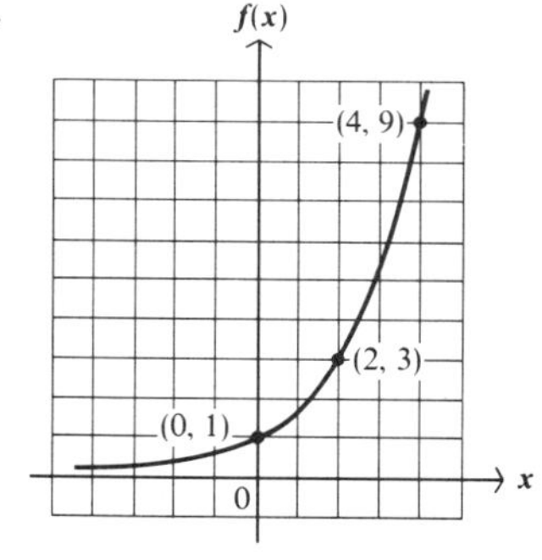

10. $x = \frac{5}{2}$ **11.** $x > -1$

12. 0.9456 **13.** 1.347

14. -0.3538

Diagnostic test (page 222)

(*Note:* Following each problem number in parentheses is the section number in which that kind of problem is discussed.)

1. (3) $25^{-1/2} = \frac{1}{5}$ **2.** (3) Rewriting in exponential form, we have $7^x = \sqrt{7} = 7^{1/2}$; Therefore, $x = \frac{1}{2}$.

3. (3) $b^{1/2} = 2$; $(b^{1/2})^2 = 2^2$; $b = 4$ **4.** (3) $3^{-4} = N$; $N = \frac{1}{81}$

5. (11)

$$\left(\frac{1}{6}\right)^x < 6$$

$$\log\left(\frac{1}{6}\right)^x < \log 6$$

$$x \log\left(\frac{1}{6}\right) < \log 6$$

$$x > \frac{\log 6}{\log \frac{1}{6}} > \frac{\log 6}{\log 1 - \log 6} > \frac{\log 6}{-\log 6} > -1$$

6. (4)

$$\log \frac{4x^2 - 1}{2x + 1} = \log \frac{x^2 - 1}{x - 1}$$

$$\log \frac{(2x - 1)\cancel{(2x + 1)}}{\cancel{2x + 1}} = \log \frac{\cancel{(x - 1)}(x + 1)}{\cancel{x - 1}}$$

$$\log (2x - 1) = \log (x + 1)$$

$$2x - 1 = x + 1$$

$$x = 2$$

7. (4) **a.** $\log \sqrt[5]{2} = \frac{1}{5} \log 2 = \frac{1}{5}(0.3010) = 0.602$ **b.** $\log 16 = \log 2^4 = 4 \log 2 = 4(0.3010) = 1.2040$

8. (4) $\log \sqrt{h^5} - 2 \log \sqrt[4]{h} + \log h^4 = \log h^{5/2} - \log (h^{1/4})^2 + \log h^4 = \log \frac{h^{5/2} \cdot h^4}{h^2} = \log h^{5/2 + 4 - 2}$

$= \log h^{9/2} = \frac{9}{2} \log h$

9. (1)

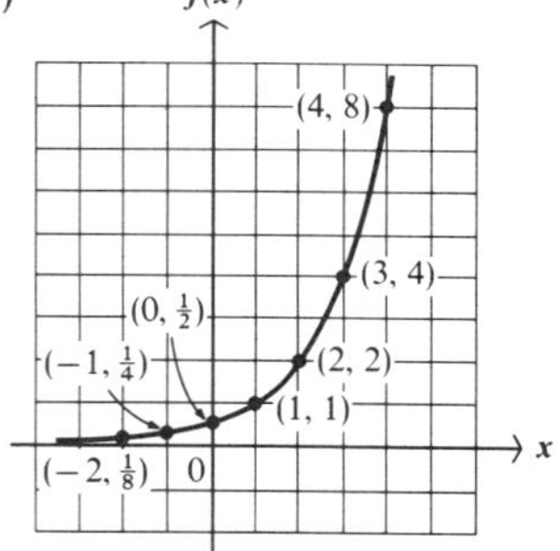

10. (9) Let $x = \frac{\log 2}{3.45} = \frac{0.3010}{3.45}$

$$\log x = \log 0.3010 - \log 3.45$$
$$= 9.4786 - 10 - 0.5378$$
$$= 8.9408 - 10$$
$$x = 0.08726$$

11. (9) Let $x = (69.4)^{2/5}$; $\log x = \log (69.4)^{2/5} = \dfrac{2}{5} \log (69.4) = \dfrac{2}{5}(1.8414) = 0.7366$; $x = 5.452$

12. (11) $(6.31)^x = 7.67$; $\log (6.31)^x = \log 7.67 = x \log 6.31$; $x = \dfrac{\log 7.67}{\log 6.31} = \dfrac{0.8848}{0.8000} = 1.106$

CHAPTER 8

Exercises 1 (page 227) A Answers

1.

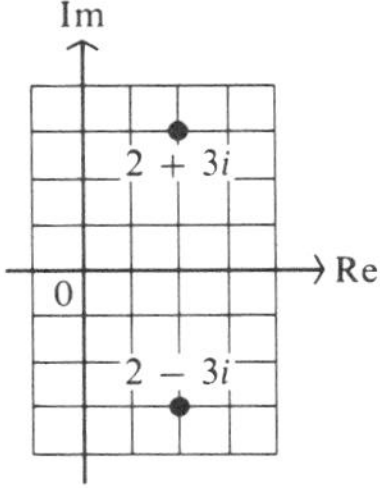

2.

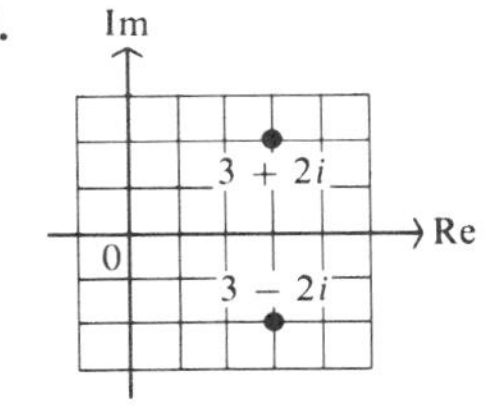

3.

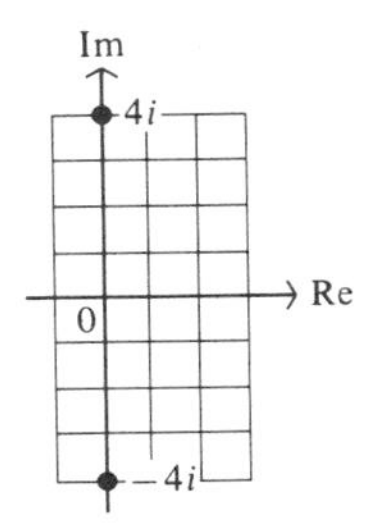

4.

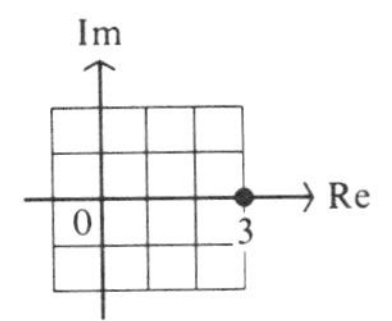

5.

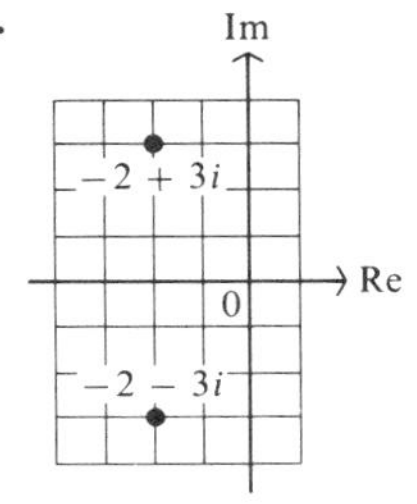

Exercises 2 (page 229) A Answers

1. $(1, \sqrt{3})$

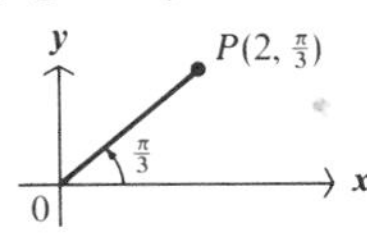

2. $(0, -8)$

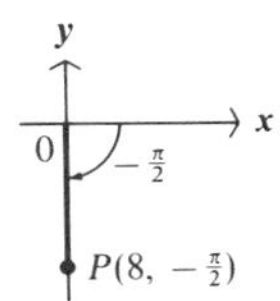

3. $(-1.736, 9.848)$

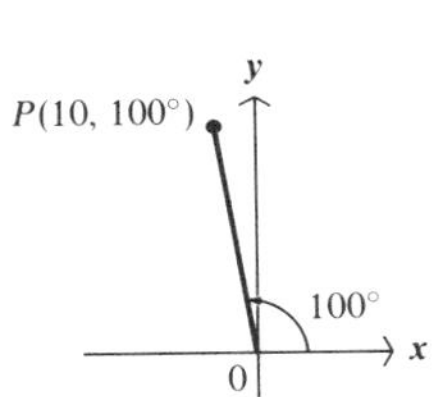

4. $(-2\sqrt{2}, 2\sqrt{2})$

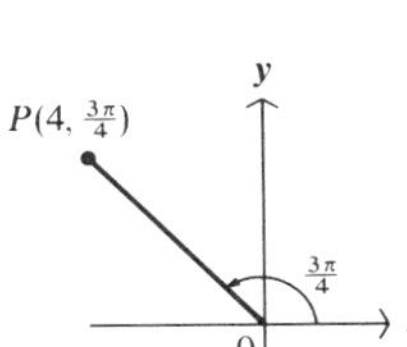

5. $\left(-\dfrac{5\sqrt{3}}{2}, -\dfrac{5}{2}\right)$

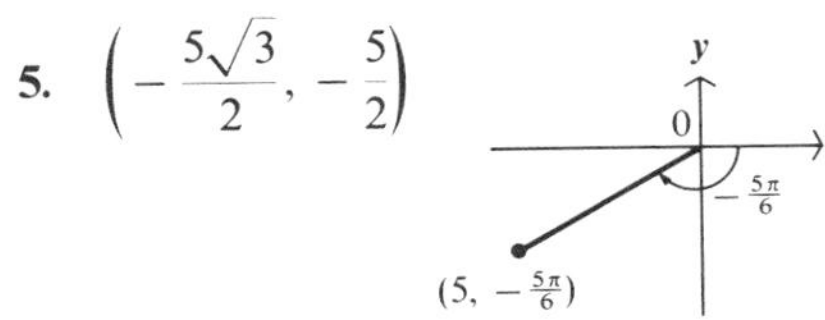

6. $\left(2, \dfrac{5\pi}{6}\right)$

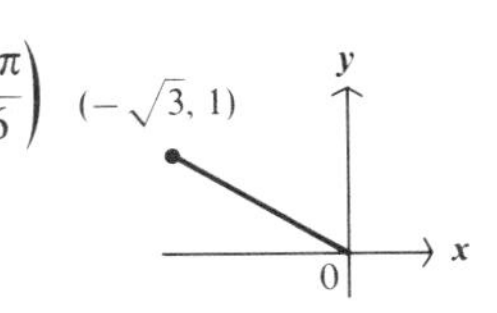

7. $(5, \pi)$

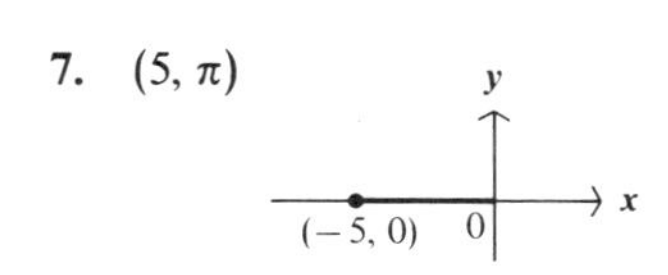

8. $\left(2, \dfrac{\pi}{2}\right)$

9. $(\sqrt{13}, 236^\circ\, 19')$

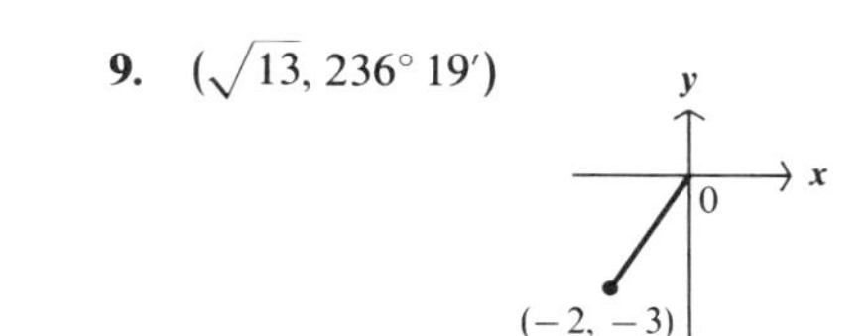

Exercises 3 (page 233) A Answers

1. Absolute value: 2; Amplitude: $\dfrac{2\pi}{3}$

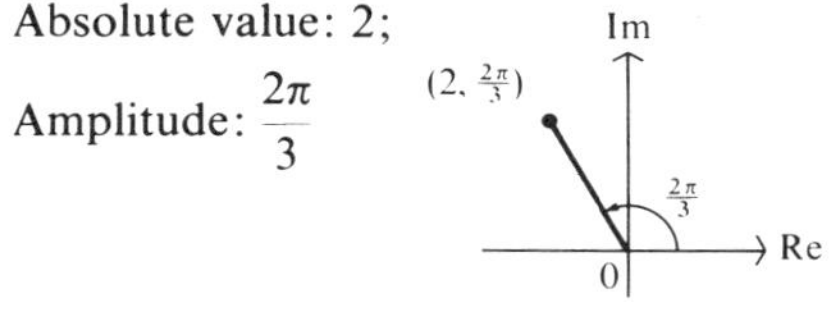

2. Absolute value: $\sqrt{2}$; Amplitude: $\dfrac{\pi}{4}$

3. Absolute value: 2; Amplitude: $\dfrac{3\pi}{4}$

4. Absolute value: 5; Amplitude: $-\dfrac{\pi}{2}$

5. Absolute value: 7; Amplitude: 0

6. $-2 + 0i$

7. $-1 + 0i$

8. $0 - 8i$

9. $0 - 2i$

10. $\dfrac{3}{2}\sqrt{2} - \dfrac{3}{2}i\sqrt{2}$

11. Polar form: $4\left(\cos\dfrac{\pi}{3} + i\sin\dfrac{\pi}{3}\right)$; Rectangular form: $2 + 2i\sqrt{3}$

12. Polar form: $4(\cos 2\pi + i\sin 2\pi)$; Rectangular form: $4 + 0i$

13. Polar form: $3\left[\cos\left(-\dfrac{2\pi}{3}\right) + i\sin\left(-\dfrac{2\pi}{3}\right)\right]$; Rectangular form: $-\dfrac{3}{2} - \dfrac{3\sqrt{3}}{2}i$

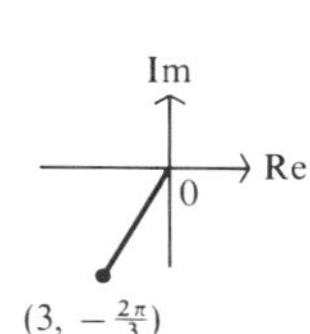

14. $|\cos\theta + i\sin\theta| = \sqrt{\cos^2\theta + \sin^2\theta} = \sqrt{1} = 1$

(*Note:* There are no exercises for Section 4.)

Exercises 5 (page 239) A Answers

1. $6(\cos 90^\circ + i\sin 90^\circ)$; $0 + 6i$
2. $4(\cos 240^\circ + i\sin 240^\circ)$; $-2 - 2i\sqrt{3}$
3. $10(\cos 140^\circ + i\sin 140^\circ)$; $-7.660 + 6.428i$
4. $\cos 150^\circ + i\sin 150^\circ$; $-\frac{\sqrt{3}}{2} + \frac{1}{2}i$
5. $2^{-5}[\cos(-60^\circ) + i\sin(-60^\circ)]$; $\frac{1}{64} - \frac{\sqrt{3}}{64}i$
6. $324(\cos 180^\circ + i\sin 180^\circ)$; $-324 + 0i$
7. $2(\cos 60^\circ + i\sin 60^\circ)$; $1 + i\sqrt{3}$
8. $4(\cos 60^\circ + i\sin 60^\circ)$; $2 + 2i\sqrt{3}$
9. $(\cos\theta + i\sin\theta)^4 = \cos 4\theta + i\sin 4\theta$

$\cos^4\theta + 4(\cos^3\theta\sin\theta)i + 6(\cos^2\theta\sin^2\theta)i^2 + 4(\cos\theta\sin^3\theta)i^3 + (\sin^4\theta)i^4 = \cos 4\theta + i\sin 4\theta$

$\cos^4\theta + 4(\cos^3\theta\sin\theta)i - 6\cos^2\theta\sin^2\theta - 4(\cos\theta\sin^3\theta)i + \sin^4\theta = \cos 4\theta + i\sin 4\theta$

$(\cos^4\theta - 6\cos^2\theta\sin^2\theta + \sin^4\theta) + (4\cos^3\sin\theta - 4\cos\theta\sin^3\theta)i = \cos 4\theta + i\sin 4\theta$

$[\cos^4\theta - 6\cos^2\theta(1 - \cos^2\theta) + (1 - \cos^2\theta)^2] + [4\sin\theta\cos\theta(\cos^2\theta - \sin^2\theta)]i = \cos 4\theta + i\sin 4\theta$

$[8\cos^4\theta - 8\cos^2\theta + 1] + [4\sin\theta\cos\theta - 8\sin^3\theta\cos\theta]i = \cos 4\theta + i\sin 4\theta$

Therefore $\cos 4\theta = 8\cos^4\theta - 8\cos^2\theta + 1$ and $\sin 4\theta = 4\sin\theta\cos\theta - 8\sin^3\theta\cos\theta$

Exercises 6 (page 243) A Answers

1.

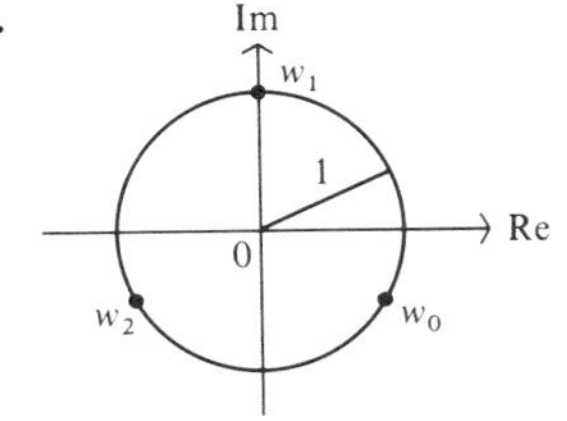

2.

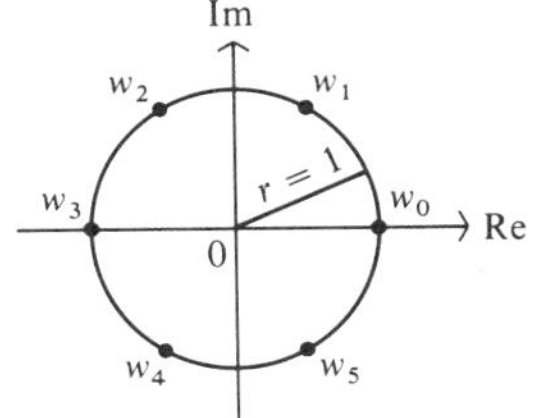

3.

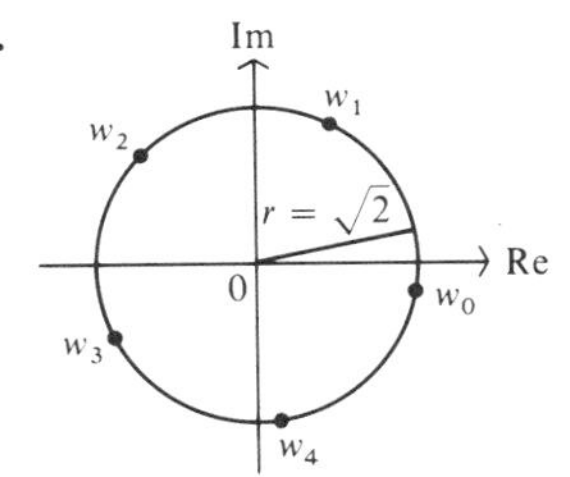

4.

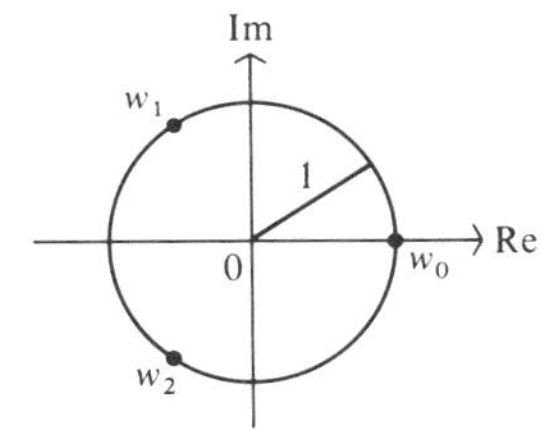

5.

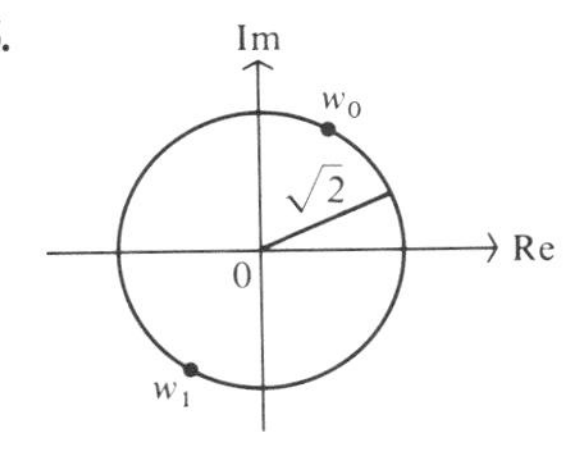

6.

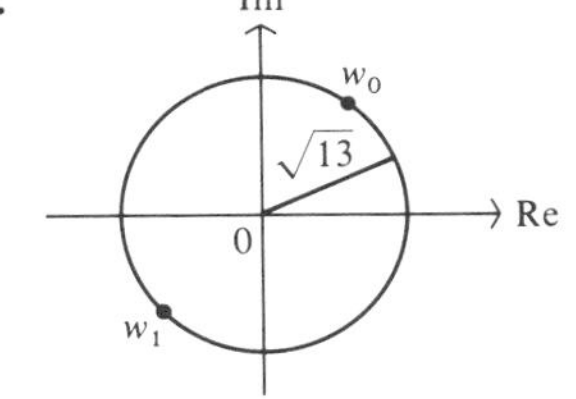

7. $\left\{-3, \frac{3}{2} + \frac{3}{2} i\sqrt{3}, \frac{3}{2} - \frac{3}{2} i\sqrt{3}\right\}$

8. $\left\{\frac{3}{2}\sqrt{2} + \frac{3}{2} i\sqrt{2}, -\frac{3}{2}\sqrt{2} + \frac{3}{2} i\sqrt{2}, -\frac{3}{2}\sqrt{2} - \frac{3}{2} i\sqrt{2}, \frac{3}{2}\sqrt{2} - \frac{3}{2} i\sqrt{2}\right\}$

Review exercises (page 245) A Answers

1. a.

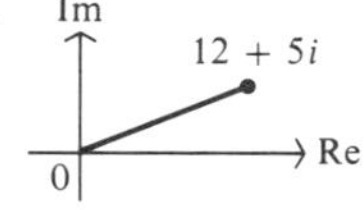

b. 13

c. 22° 37′

2. $3\sqrt{2}\,(\cos 315° + i \sin 315°)$
3. $w_0 = \cos 90° + i \sin 90°$;
   $w_1 = \cos 270° + i \sin 270°$

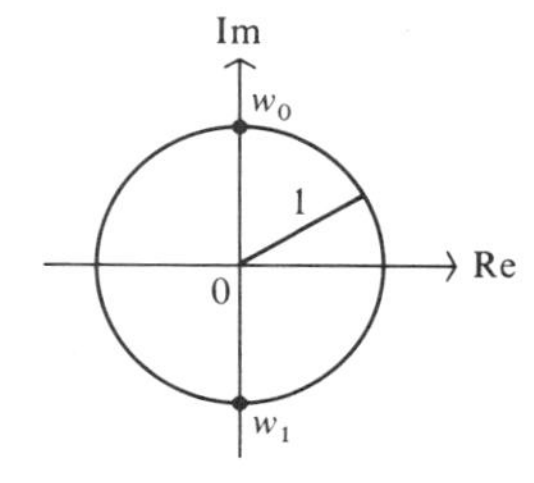

4. $15(\cos 180° + i \sin 180°) = -15 + 0i$
5. $2(\cos 360° + i \sin 360°) = 2 + 0i$
6. $32(\cos 60° + i \sin 60°) = 16 + 16i\sqrt{3}$
7. $2i$
8. $w_0 = 2(\cos 48° + i \sin 48°) = 1.3383 + 1.4863i$
   $w_1 = 2(\cos 120° + i \sin 120°) = -1.000 + 1.732i$
   $w_2 = 2(\cos 192° + i \sin 192°) = -1.9563 - 0.4158i$
   $w_3 = 2(\cos 264° + i \sin 264°) = -0.2091 - 1.9890i$
   $w_4 = 2(\cos 336° + i \sin 336°) = 1.8271 - 0.8135i$

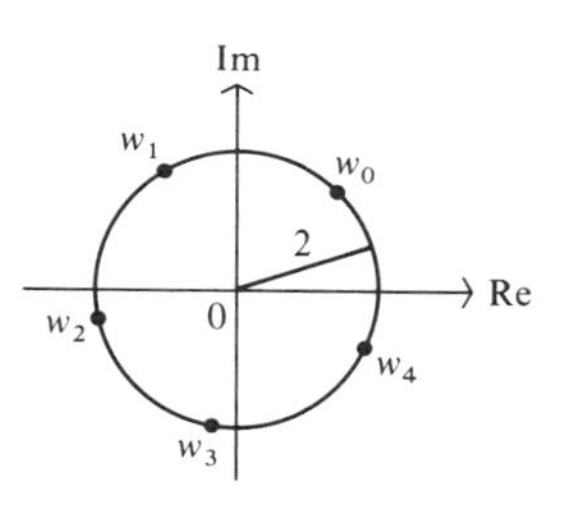

9. $x^4 + 4x^2 + 3 = 0$

$(x^2 + 1)(x^2 + 3) = 0$

$x^2 = -1$ or $x^2 = -3$

$x = \pm i \qquad x = \pm i\sqrt{3}$;

$\{0 - i, 0 + i, 0 - i\sqrt{3}, 0 + i\sqrt{3}\}$

Diagnostic test (page 246)

(*Note:* Following each problem number in parentheses is the section number in which that kind of problem is discussed.)

1. (3) $r = \sqrt{(-\sqrt{3})^2 + 1^2}$
   $= \sqrt{3 + 1}$
   $= 2$

$$\theta = \arc\tan\left(-\frac{1}{\sqrt{3}}\right)$$
$= 150°$

$-\sqrt{3} + i = 2(\cos 150° + i \sin 150°)$

Im
$-\sqrt{3} + i$
2
1
150°
Re
$-\sqrt{3}$
0

2. (3) $2(\text{cis } 120°) = 2(\cos 120° + i \sin 120°) = 2\left(-\frac{1}{2} + i\frac{\sqrt{3}}{2}\right) = -1 + i\sqrt{3}$

3. (3) $x = r\cos\theta = 2\cos 120° = 2\left(-\frac{1}{2}\right) = -1$; $y = r\sin\theta = 2\sin 120° = 2\left(\frac{\sqrt{3}}{2}\right) = \sqrt{3}$;
Rectangular coordinates: $(-1, \sqrt{3})$

4. (4) $[16(\cos 250° + i\sin 250°)] \div [4(\cos 70° + i \sin 70°)] = \frac{16}{4}[\cos(250° - 70°) + i\sin(250° - 70°)]$
$= 4(\cos 180° + i \sin 180°) = 4[-1 + i(0)] = -4 + 0i$

5. (4) $7(\cos 47° + i\sin 47°)\cdot 2(\cos 13° + i\sin 13°) = 14[\cos(47° + 13°) + i\sin(47° + 13°)]$
$= 14(\cos 60° + i\sin 60°) = 14\left(\frac{1}{2} + i\frac{\sqrt{3}}{2}\right) = 7 + 7i\sqrt{3}$

6. (5) $[2(\cos 135° + i\sin 135°)]^4 = 2^4[\cos 4(135°) + i\sin 4(135°)] = 16(\cos 540° + i\sin 540°]$
$= 16(\cos 180° + i\sin 180°) = -16 + 0i = -16$

7. (5) $1 + i\sqrt{3} = 2(\cos 60° + i\sin 60°)$

$$\frac{(1 + i\sqrt{3})^{10}}{2} = \frac{2^{10}(\cos 600° + i\sin 600°)}{2}$$
$$= 2^9(\cos 240° + i\sin 240°) = 512\left[-\frac{1}{2} + i\left(-\frac{\sqrt{3}}{2}\right)\right] = -256 - 256i\sqrt{3}$$

8. (5) $-1 + i = \sqrt{2}(\cos 135° + i\sin 135°)$

$$\frac{8}{(-1+i)^4} = 8(-1+i)^{-4} = 8[\sqrt{2}(\cos 135° + i\sin 135°)]^{-4}$$
$$= 8(2^{1/2})^{-4}[\cos(-4)135° + i\sin(-4)135°] = 8\left(\frac{1}{4}\right)[\cos(-540°) + i\sin(-540°)]$$
$$= 2(\cos 180° + i\sin 180°) = -2 + 0i = -2$$

9. (6) $i^{1/3} = [\cos(90° + k360°) + i\sin(90° + k360°)]^{1/3}$
$w_k = \cos(30° + k120°) + i\sin(30° + k120°)$
$w_0 = \cos 30° + i\sin 30°$
$w_1 = \cos 150° + i\sin 150°$
$w_2 = \cos 270° + i\sin 270°$

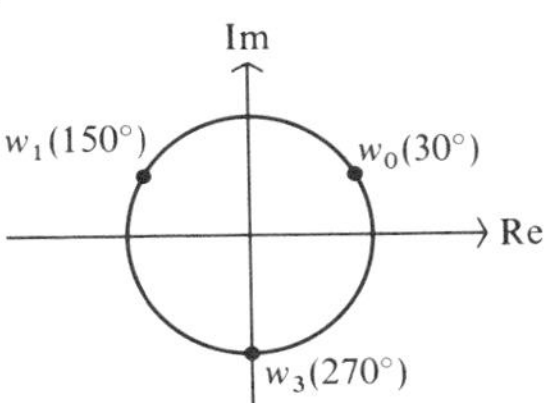

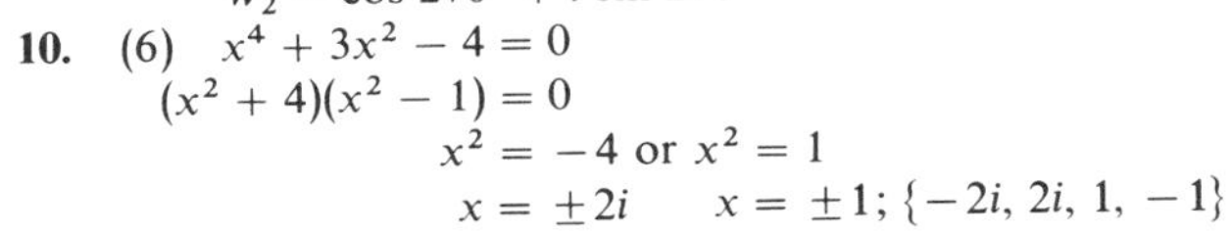

10. (6) $x^4 + 3x^2 - 4 = 0$
$(x^2 + 4)(x^2 - 1) = 0$
$x^2 = -4$ or $x^2 = 1$
$x = \pm 2i \qquad x = \pm 1$; $\{-2i, 2i, 1, -1\}$

CHAPTER 9

(*Note:* There are no exercises for Section 1.)

Exercises 2 (page 258) A Answers

1. Arctan $\frac{27}{14} \doteq 62°\,36'$

2. m (of AB) $= \frac{2}{3}$; m (of AC) $= -\frac{3}{2}$; $\left(-\frac{3}{2}\right)\left(\frac{2}{3}\right) = -1$; therefore A is the right angle.

3. Arctan $\frac{19}{11} \doteq 59°\,56'$

4. $5x + 3y + 7 = 0$ **5.** $m_1 = -\frac{1}{2}; m_2 = 2$

Exercises 3 (page 260) A Answers

1. y-axis **2.** x-axis; y-axis; origin **3.** x-axis **4.** x-axis **5.** None **6.** Origin

(*Note:* There are no exercises for Section 4.)

Exercises 5 (page 266) A Answers

1. $(x-2)^2 + (y-3)^2 = 4$ **2.** $x^2 + y^2 = 1$ **3.** $(x+1)^2 + (y-3)^2 = 20$

4. $(x-1)^2 + (y+1)^2 = 20$ **5.** $(x+4)^2 + (y-3)^2 = 9$ **6.** $x^2 + y^2 + 4x - 4y + 4 = 0$

7. $x^2 + y^2 - 6x - 14y + 49 = 0$ **8.** $x^2 + y^2 + 8x + 6y = 0$

9.
$$x^2 + y^2 + 6x - 16 = 0$$
$$(x^2 + 6x + \quad) + (y^2) = 16$$
$$(x^2 + 6x + 9) + (y^2) = 16 + 9$$
$$(x+3)^2 + (y-0)^2 = 25$$

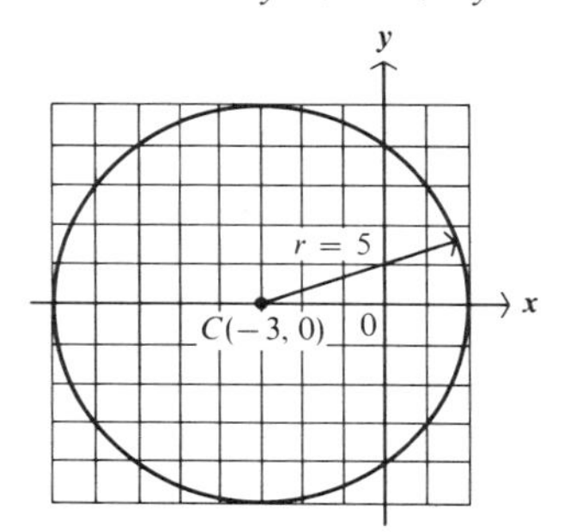

10.

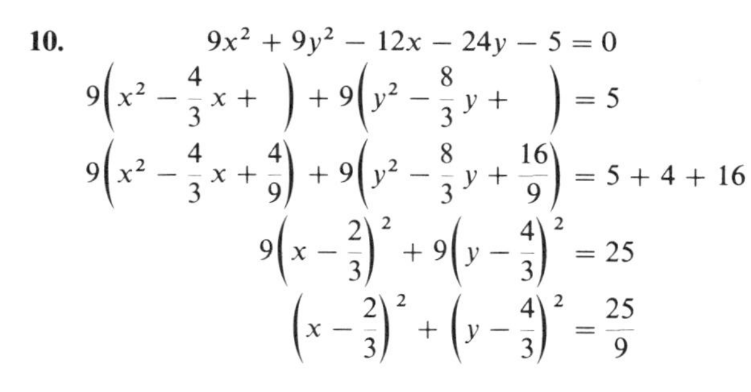
$$9x^2 + 9y^2 - 12x - 24y - 5 = 0$$
$$9\left(x^2 - \frac{4}{3}x + \quad\right) + 9\left(y^2 - \frac{8}{3}y + \quad\right) = 5$$
$$9\left(x^2 - \frac{4}{3}x + \frac{4}{9}\right) + 9\left(y^2 - \frac{8}{3}y + \frac{16}{9}\right) = 5 + 4 + 16$$
$$9\left(x - \frac{2}{3}\right)^2 + 9\left(y - \frac{4}{3}\right)^2 = 25$$
$$\left(x - \frac{2}{3}\right)^2 + \left(y - \frac{4}{3}\right)^2 = \frac{25}{9}$$

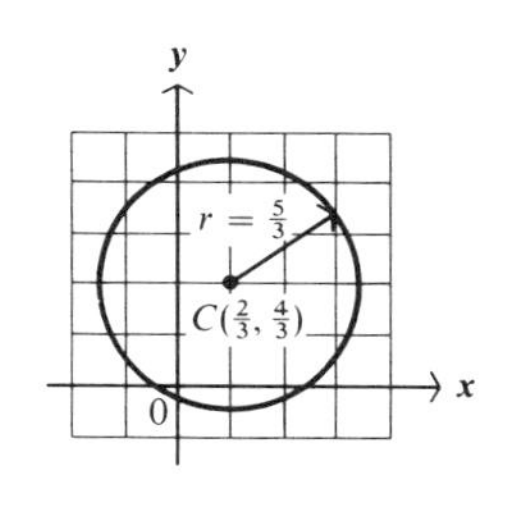

Exercises 6 (page 272) A Answers

1.

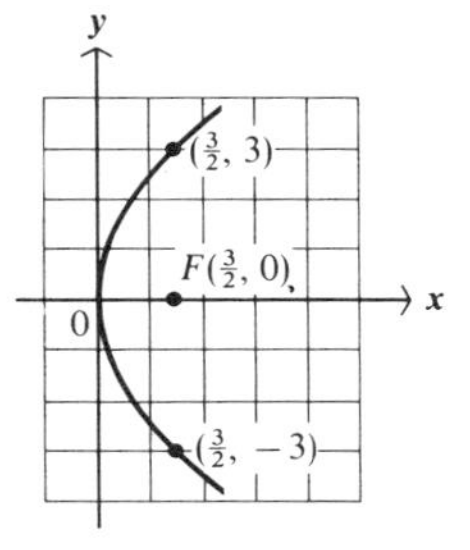

2.

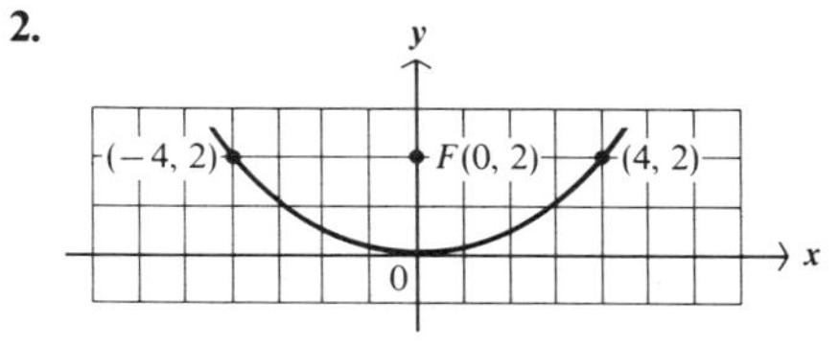

3.

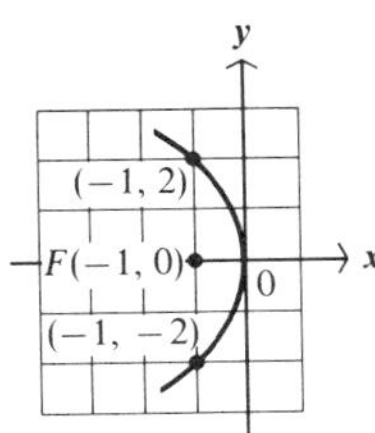

4.

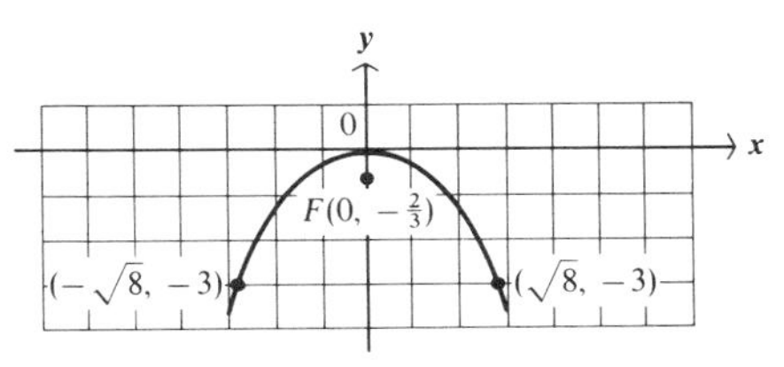

5. $y^2 = 8x$ **6.** $y^2 = 16x$ **7.** $3y^2 = 8x$ **8.** $x^2 = -8y$ **9.** $x = 4; x = -4$

10.

$\phi = 53°08'$

Exercises 7 (page 278) A Answers

1. Vertices: $(\pm 10, 0)$
Foci: $(\pm 5\sqrt{3}, 0)$
Semimajor axis: 10
Semiminor axis: 5

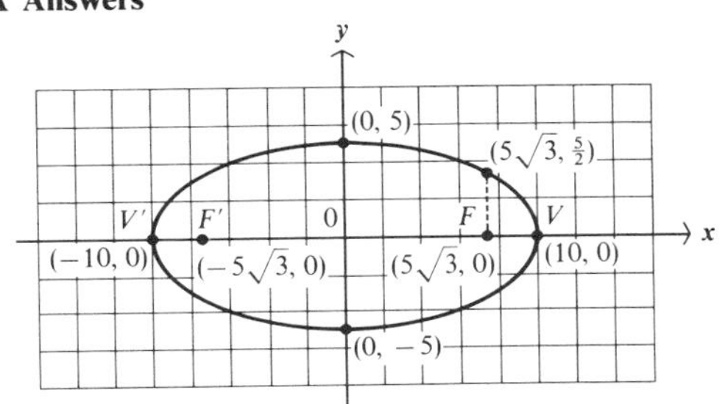

2. Vertices: $(0, \pm 3)$
Foci: $(0, \pm\sqrt{5})$
Semimajor axis: 3
Semiminor axis: 2

3. Vertices: $\left(\pm \frac{10}{9}, 0\right)$

Foci: $\left(\pm \frac{\sqrt{19}}{9}, 0\right)$

Semimajor axis: $\frac{10}{9}$

Semiminor axis: 1

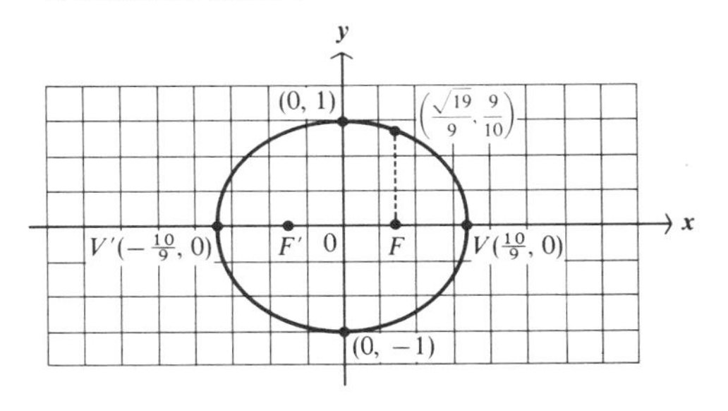

4. $\dfrac{x^2}{25} + \dfrac{y^2}{16} = 1$

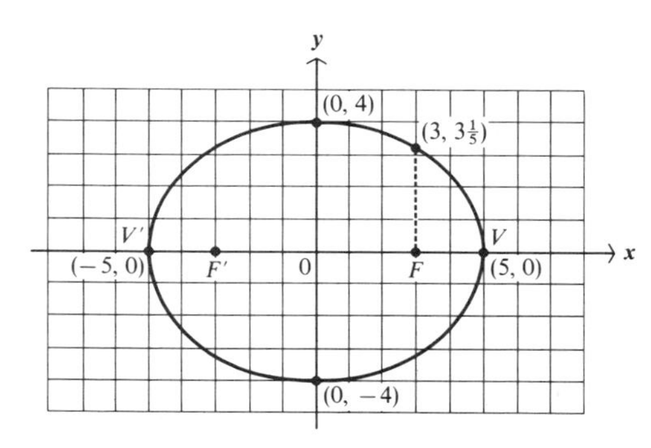

5. $\dfrac{x^2}{100} + \dfrac{y^2}{84} = 1$

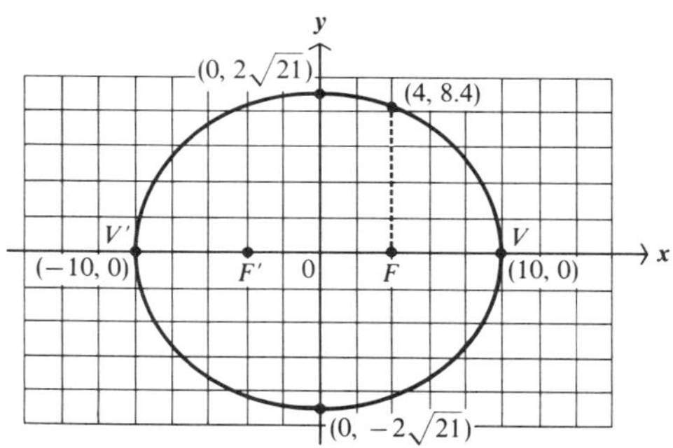

6. $\dfrac{x^2}{36} + \dfrac{y^2}{9} = 1$

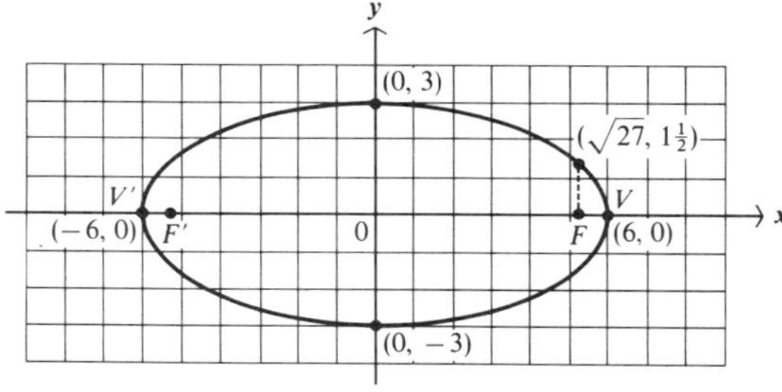

7. $\dfrac{x^2}{400} + \dfrac{y^2}{36} = 1$

8. $\dfrac{x^2}{9} + \dfrac{y^2}{5} = 1$

9.

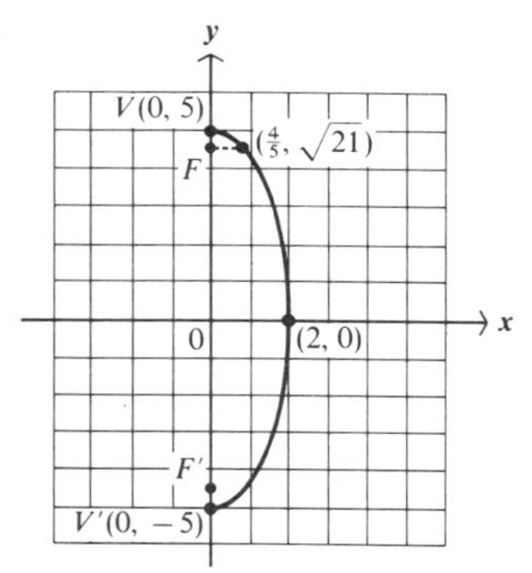

10. $G = 1.55 \times 10^8$ mi.; $L = 1.28 \times 10^8$ mi.

Exercises 8 (page 286) A Answers

1. $V(0, \pm 2\sqrt{2})$; $F(0, \pm 2\sqrt{7})$; $e = \frac{1}{2}\sqrt{14}$; L.R. $= 10\sqrt{2}$; $y = \pm\frac{1}{5}\sqrt{10}x$; $a = 2\sqrt{2} \doteq 2.8$; $b = 2\sqrt{5} \doteq 4.5$; $c = 2\sqrt{7} \doteq 5.3$

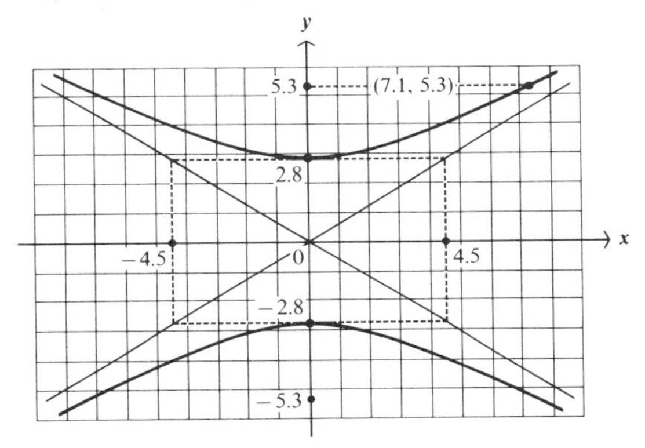

2. $V(\pm 4, 0)$; $F(\pm 4\sqrt{2}, 0)$; $e = \sqrt{2}$; L.R. $= 4$; $y = \pm x$; $a = 4$; $b = 4$; $c = 4\sqrt{2} \doteq 5.7$

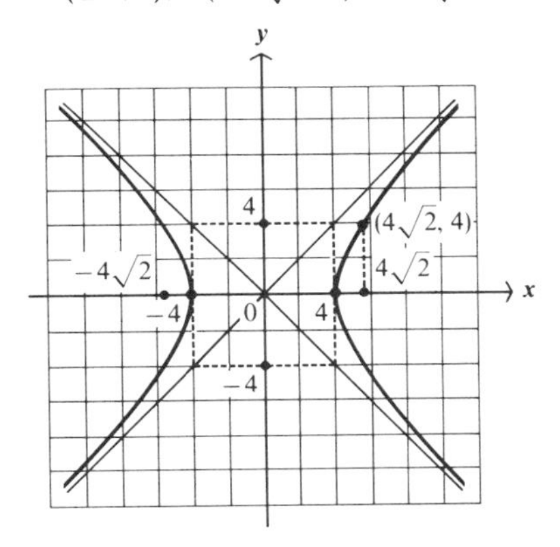

3. $\frac{y^2}{16} - \frac{x^2}{48} = 1$ **4.** $\frac{y^2}{48} - \frac{x^2}{16} = 1$ **5.** $\frac{x^2}{16} - \frac{y^2}{20} = 1$ **6.** $\frac{y^2}{16} - \frac{x^2}{36} = 1$

7. $\frac{x^2}{8} - \frac{y^2}{2} = 1$; $y = \pm\frac{1}{2}x$; $\sqrt{2} \doteq 1.4$; $2\sqrt{2} \doteq 2.8$; $\sqrt{10} \doteq 3.2$ **8.** $\frac{x^2}{5} - \frac{y^2}{4} = 1$

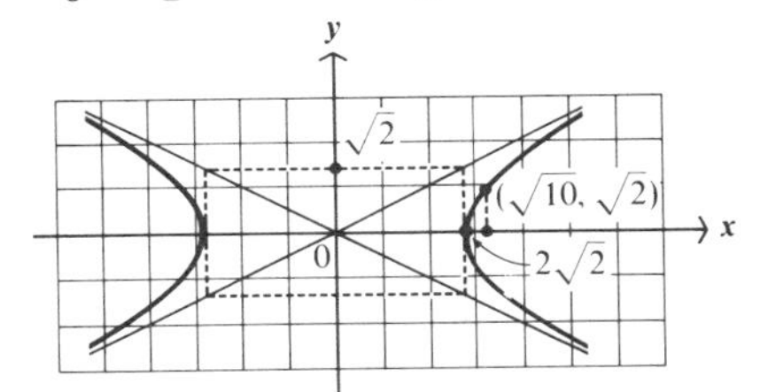

Exercises 9 (page 290) A Answers

1. $(-5, 2)$ **2.** $(-5, 8)$ **3.** $(x')^2 + (y')^2 - 16 = 0$; circle **4.** $4(x')^2 + (y')^2 - 8 = 0$; ellipse
5. $4(x')^2 - (y')^2 + 42 = 0$; hyperbola **6.** $(y')^2 = 4x'$

Exercises 10 (page 295) A Answers

1. $(y - 3)^2 = 16(x - 1)$ **2.** $\dfrac{(x-1)^2}{25} + \dfrac{(y-3)^2}{16} = 1$ **3.** $\dfrac{(x+2)^2}{16} - \dfrac{(y+3)^2}{9} = 1$

4. $\dfrac{(x-2)^2}{8} + \dfrac{(y-1)^2}{2} = 1$

$\dfrac{(x')^2}{8} + \dfrac{(y')^2}{2} = 1$

Ellipse

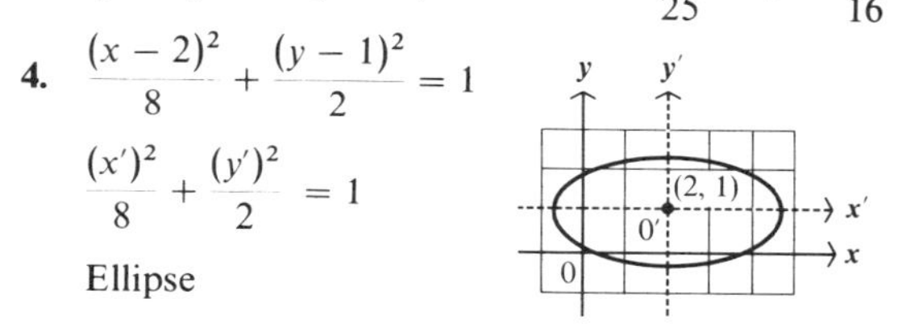

5. $(y + 1)^2 = -2(x - 2)$
$(y')^2 = -2x'$
Parabola

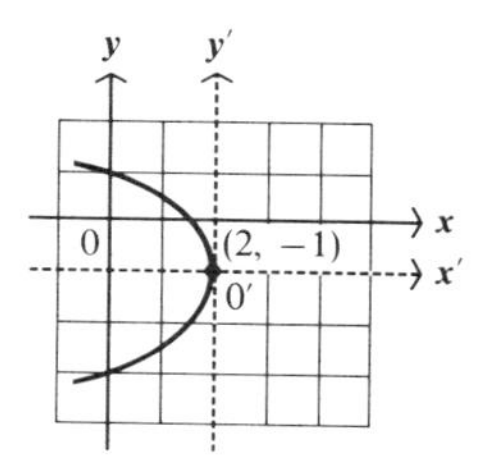

6. $\dfrac{(x-1)^2}{16} - \dfrac{(y-3)^2}{9} = 1$

$\dfrac{(x')^2}{16} - \dfrac{(y')^2}{9} = 1$

Hyperbola

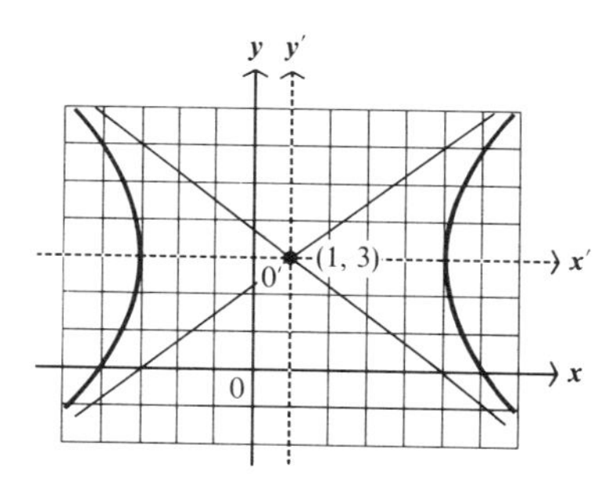

Review exercises (page 298) A Answers

1. a. $6x + 5y - 52 = 0$ **b.** $\left(4, -\dfrac{1}{2}\right)$ **c.** $\sqrt{122}$

d. m (of AC) $= -\dfrac{6}{5}$ and m (of AB) $= \dfrac{5}{6}$; $\left(-\dfrac{6}{5}\right)\left(\dfrac{5}{6}\right) = -1$, so A is a right angle. **2.** $(-2, -6)$

3. $(x - 3)^2 + (y + 4)^2 = 25$; $x^2 + y^2 - 6x + 8y = 0$ **4.** $(x + 4)^2 = -8(y - 2)$; $x^2 + 8x + 8y = 0$

5. $\dfrac{(x-2)^2}{4} + \dfrac{(y+3)^2}{9} = 1$; $9x^2 + 4y^2 - 36x + 24y + 36 = 0$

6. $\dfrac{(y-1)^2}{9} - \dfrac{(x-3)^2}{4} = 1$; $9x^2 - 4y^2 - 54x + 8y + 113 = 0$

7. $(x-2)^2 + (y-1)^2 = \dfrac{25}{2}$; circle: Center $(2, 1)$; Radius $\dfrac{5\sqrt{2}}{2}$; $(x')^2 + (y')^2 = \dfrac{25}{2}$

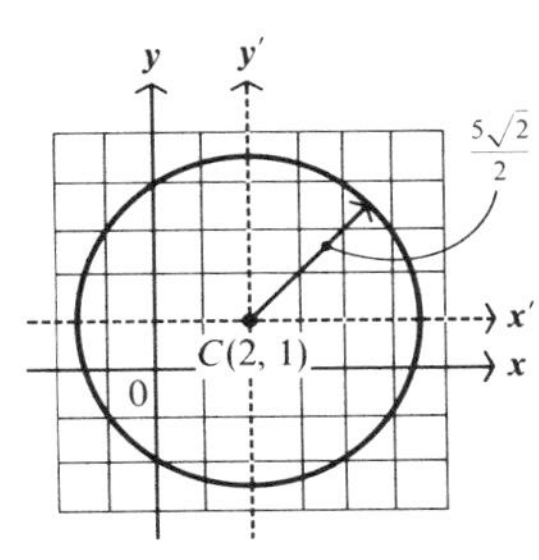

8. $\dfrac{(x+3)^2}{1} - \dfrac{(y-4)^2}{4} = 1$

Hyperbola with center at $(-3, 4)$ and vertical major axis; $\dfrac{(x')^2}{1} - \dfrac{(y')^2}{4} = 1$

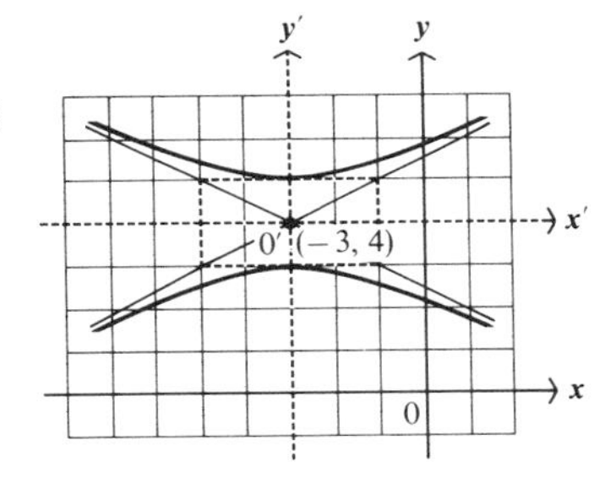

9. $(x+2)^2 = 8(y+5)$

Parabola with axis parallel to OY and vertex at $(-2, -5)$, opening upward; $(x')^2 = 8y'$

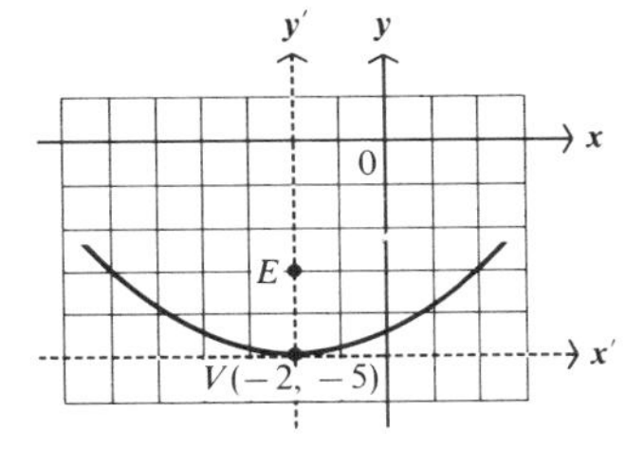

10. $\dfrac{(x+1)^2}{16} + \dfrac{(y-4)^2}{9} = 1$; Ellipse with center at $(-1, 4)$; $\dfrac{(x')^2}{16} + \dfrac{(y')^2}{9} = 1$

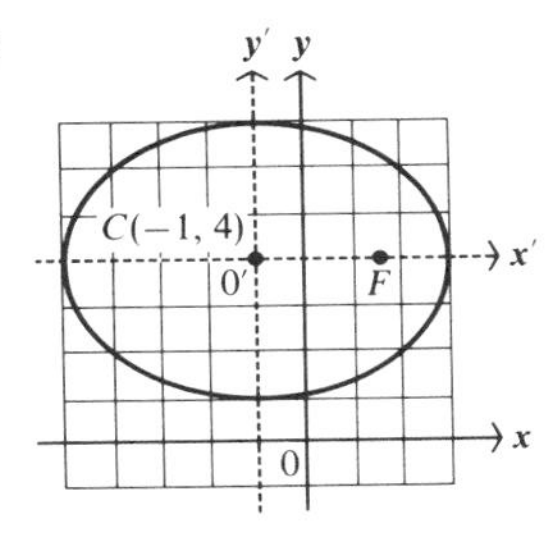

Diagnostic test (page 299)

(*Note:* Following each problem number in parentheses is the section number in which that kind of problem is discussed.)

1. **a.** (2) 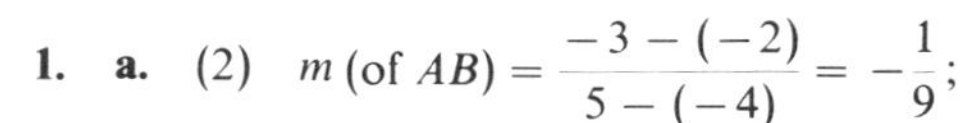
$m\,(\text{of } AB) = \dfrac{-3-(-2)}{5-(-4)} = -\dfrac{1}{9};$

Slope of line perpendicular to AB must have a slope of $\dfrac{9}{1}$;

$$\begin{aligned} y - y_1 &= m(x - x_1) \\ y - 2 &= 9(x - 1) \\ y - 2 &= 9x - 9 \\ 9x - y - 7 &= 0 \end{aligned}$$

b. (1) $AB = \sqrt{[5-(-4)]^2 + [-3-(-2)]^2} = \sqrt{81+1} = \sqrt{82}$

c. (2) 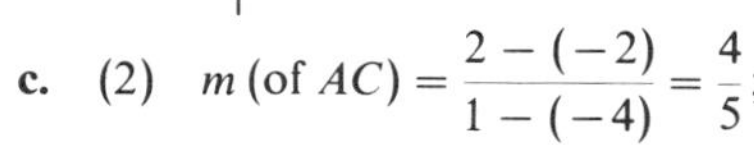
$m\,(\text{of } AC) = \dfrac{2-(-2)}{1-(-4)} = \dfrac{4}{5};$

$m\,(\text{of } BC) = \dfrac{2-(-3)}{1-5} = -\dfrac{5}{4};\ \left(\dfrac{4}{5}\right)\left(-\dfrac{5}{4}\right) = -1$, so AC and BC are perpendicular at C

d. (2) $m\,(\text{of } AB) = -\dfrac{1}{9};$

$m\,(\text{of } BC) = -\dfrac{5}{4}$; then $\tan B = \dfrac{-1/9 - (-5/4)}{1 + (-1/9)(-5/4)} = 1;\ B = 45°$

2. **a.** (5) 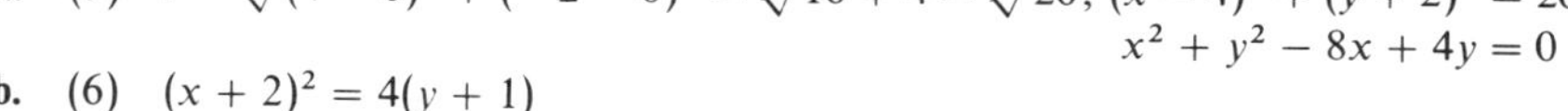
$r = \sqrt{(4-0)^2 + (-2-0)^2} = \sqrt{16+4} = \sqrt{20};\ (x-4)^2 + (y+2)^2 = 20$

$x^2 + y^2 - 8x + 4y = 0$

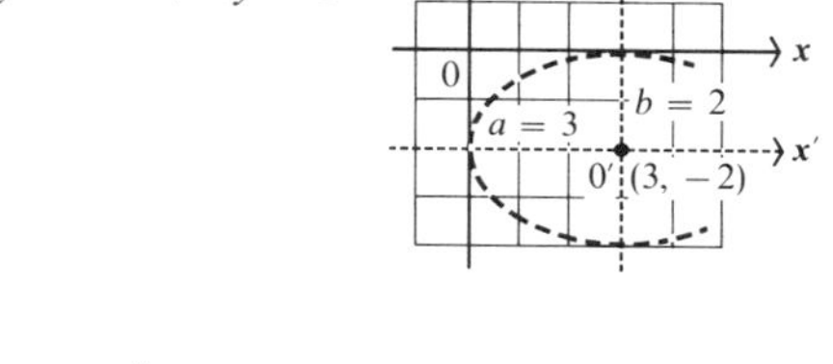

b. (6)
$$\begin{aligned} (x+2)^2 &= 4(y+1) \\ x^2 + 4x - 4 &= 4y + 4 \\ x^2 + 4x - 4y &= 0 \end{aligned}$$

c. (7) $\dfrac{(x-3)^2}{9} + \dfrac{(y+2)^2}{4} = 1;$

$4x^2 + 9y^2 - 24x + 36y + 36 = 0$

3. **a.** (9) 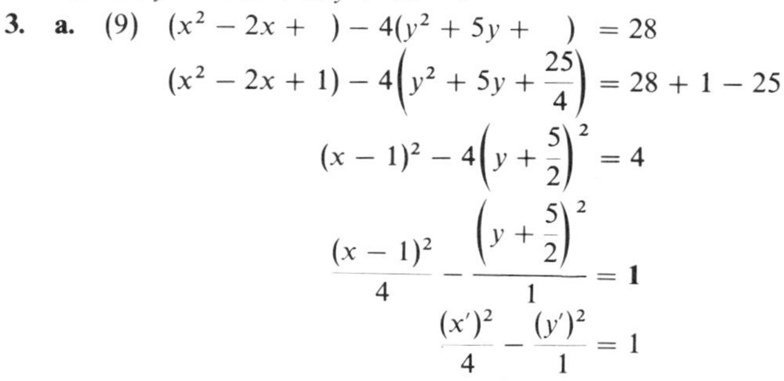
$$\begin{aligned} (x^2 - 2x + \quad) - 4(y^2 + 5y + \quad) &= 28 \\ (x^2 - 2x + 1) - 4\left(y^2 + 5y + \frac{25}{4}\right) &= 28 + 1 - 25 \\ (x-1)^2 - 4\left(y + \frac{5}{2}\right)^2 &= 4 \\ \frac{(x-1)^2}{4} - \frac{\left(y + \frac{5}{2}\right)^2}{1} &= \mathbf{1} \\ \frac{(x')^2}{4} - \frac{(y')^2}{1} &= 1 \end{aligned}$$

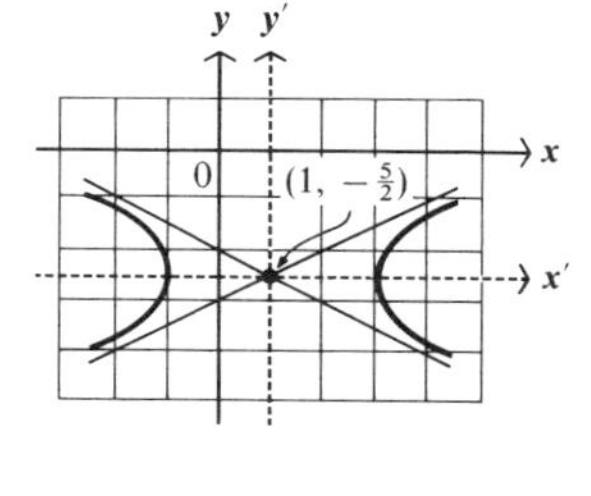

Hyperbola with center at $\left(1, -\dfrac{5}{2}\right)$ and major axis on line $y = -\dfrac{5}{2}$

b. (9)
$$\begin{aligned} y^2 - 6y + \quad &= 8x + 15 \\ y^2 - 6y + 9 &= 8x + 24 \\ (y-3)^2 &= 8(x+3) \\ (y')^2 &= 8x' \end{aligned}$$

Parabola with vertex at $(-3, 3)$, axis parallel to x-axis, and opening to the right.

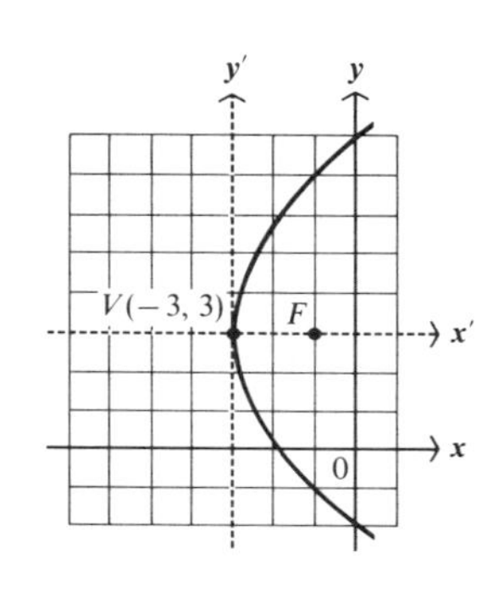

c. (9) $2(x^2 + 6x + \quad) + 2(y^2 - 4y + \quad) = -18$
$2(x^2 + 6x + 9) + 2(y^2 - 4y + 4) = -18 + 18 + 8$
$2(x + 3)^2 + 2(y - 2)^2 = 8$
$(x + 3)^2 + (y - 2)^2 = 4$
$(x')^2 + (y')^2 = 4$
Circle with center at $(-3, 2)$ and radius 2

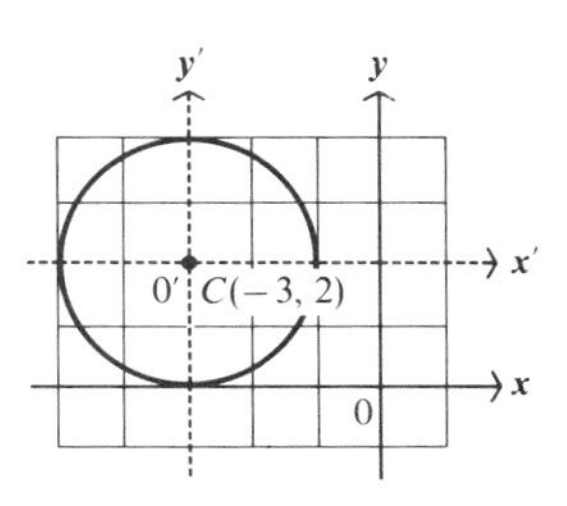

CHAPTER 10

(*Note:* There are no exercises for Section 1.)

Exercises 2 (page 305) **A Answers**

1. Consistent and independent system.

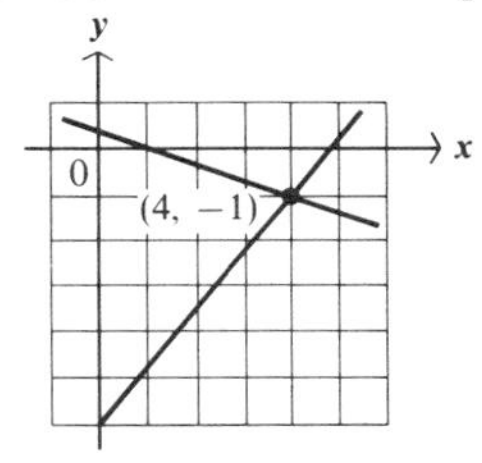

2. Consistent and independent system.

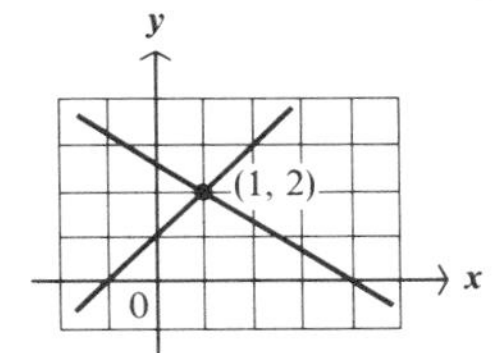

3. Consistent and independent system.

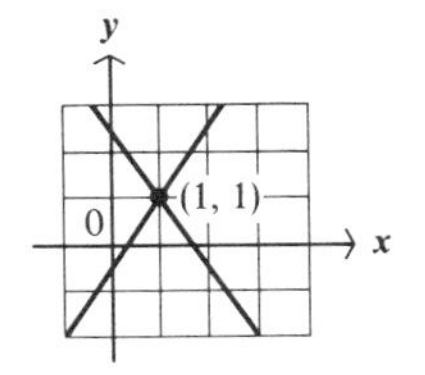

4. Consistent and independent system.

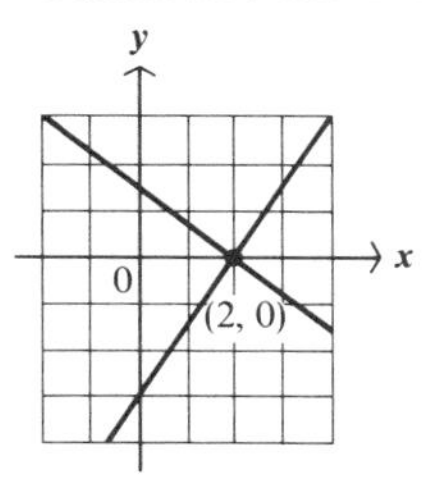

5. Consistent and dependent equations; infinitely many solutions.

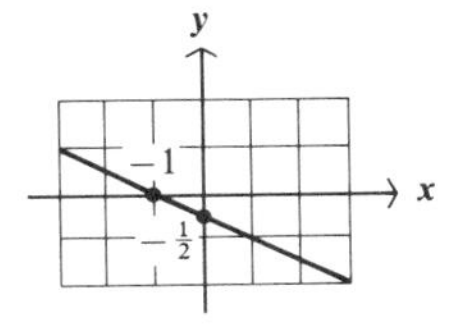

6. Inconsistent system; { }.

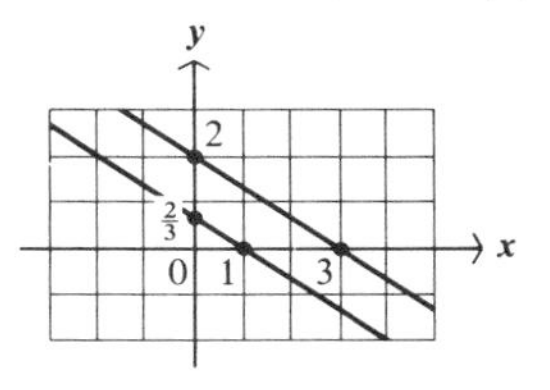

(*Note:* There are no exercises for Section 3.)

Exercises 4 (page 310) **A Answers**

1. $\{(-2, 1)\}$ **2.** $\{(4, -2)\}$ **3.** $\{(1, 2)\}$ **4.** $\{(13, 15)\}$ **5.** $\{(5, 0)\}$

6. Inconsistent system; parallel lines

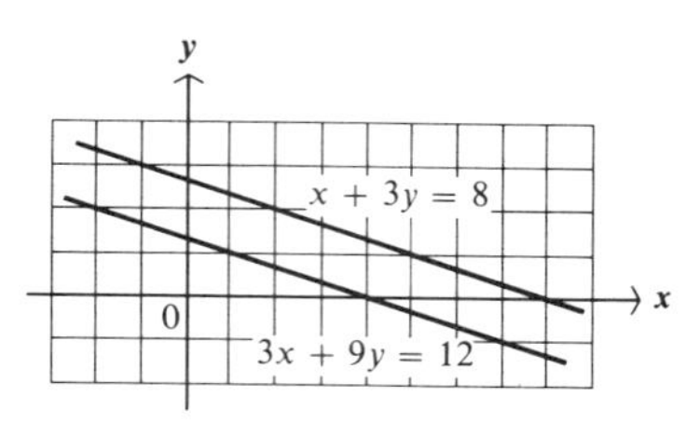

7. $\left\{\left(\frac{4}{3}, -\frac{5}{3}\right)\right\}$ 8. $\{(-1, 1)\}$ 9. $\{(2, 3)\}$

10. Dependent equations; infinitely many solutions.

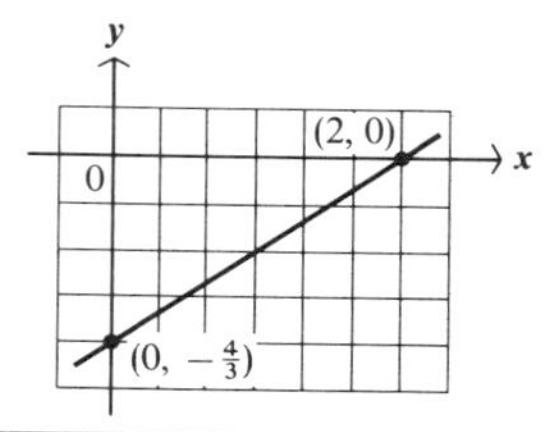

Exercises 5 (page 314) A Answers

1. $\{(2, 1, -3)\}$ 2. $\{(4, -2, 3)\}$ 3. $\{(-2, 5, -3)\}$ 4. $\left\{\left(-\frac{3}{2}, \frac{1}{2}, -\frac{1}{2}\right)\right\}$ 5. $\{(3, -4, 2)\}$

Exercises 6 (page 321) A Answers

(*Note:* See the answers to the exercises of Section 5.)

Exercises 7A (page 322) A Answers

1. 1 2. 13 3. 14 4. 26 5. -15 6. $4x - 15$

Exercises 7B (page 326) A Answers

1. -62 2. -78 3. 114 4. -130 5. 125 6. -55 7. 22.5

Exercises 8 (page 331) A Answers

1. $\{(-1, 3)\}$ 2. $\{(-1, 3)\}$ 3. $\{(10, -1)\}$ 4. $\left\{\left(\frac{29}{9}, \frac{5}{9}\right)\right\}$ 5. $\{\ \}$ 6. $\{(3, 4, -2)\}$
7. $\{(5, 2, -1)\}$ 8. $\{(2, -3, 4)\}$

Exercises 9 (page 334) A Answers

1. $\{(11k, 6k, 7k)\}$ 2. $\{(-11k, 5k, 7k)\}$ 3. $\{(-26k, 23k, k)\}$ 4. $\{(0, 0, 0)\}$

Exercises 10 (page 343) A Answers

1. $\{(1, -1), (4, 5)\}$ 2. $\left\{(3, 1), \left(\frac{9}{2}, \frac{7}{2}\right)\right\}$ 3. $\{(12 + 4\sqrt{6}, -9 - 4\sqrt{6}), (12 - 4\sqrt{6}, -9 + 4\sqrt{6})\}$
4. $\{(2, 1), (2, -1), (-2, 1), (-2, -1)\}$ 5. $\{(3, 2), (3, -2), (-3, 2), (-3, -2)\}$
6. $\{(11, 5), (11, -5), (-11, 5), (-11, -5)\}$ 7. $\left\{\left(\frac{2\sqrt{3}}{3}, 0\right), \left(-\frac{2\sqrt{3}}{3}, 0\right), (2, 4), (-2, -4)\right\}$
8. $\{65°, 25°\}$ 9. 7 rods by 18 rods or 9 rods by 14 rods 10. 4 hr. to fill, 6 hr. to drain

Review exercises (page 346) A Answers

1. 13 **2.** 0; row 4 is a multiple of row 3 **3.** $\{(6, -8)\}$ **4.** Inconsistent

5. $\{(-2, 5, -3)\}$ **6.** $\left\{(1, 1), \left(\frac{7}{11}, \frac{13}{11}\right)\right\}$ **7.** $\{(4, 3), (4, -3), (-4, 3), (-4, -3)\}$

8. $\{(4, 1), (-4, -1), (7, -2)(-7, 2)\}$ **9.** $\{(2, 3), (-2, -3), (i, 4i), (-i, -4i)\}$

10. $\{(1, 6), (-1, 6), (1, -4), (-1, -4)\}$ **11.** $\left\{\left(1, \frac{1}{2}\right), \left(-1, -\frac{1}{2}\right), \left(\frac{\sqrt{7}}{5}, \sqrt{7}\right), \left(-\frac{\sqrt{7}}{5}, -\sqrt{7}\right)\right\}$

12. 71 **13.** 540 m.p.h.; 60 m.p.h. **14. a.** $A = \frac{9}{4}$, $C = 1$, $D = 9$, $E = -2$

b. $\frac{9}{4}x^2 + y^2 + 9x - 2y + 1 = 0$

$9x^2 + 4y^2 + 36x - 8y + 4 = 0$

$\frac{(x+2)^2}{4} + \frac{(y-1)^2}{9} = 1$

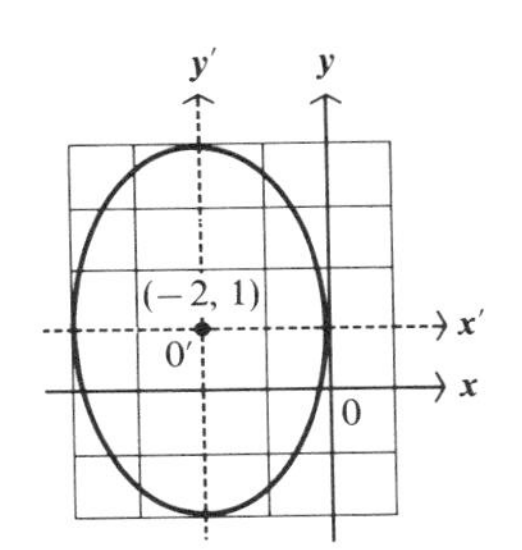

Diagnostic test (page 348)

(*Note:* Following each problem number in parentheses is the section number in which that kind of problem is discussed.)

1. (7) $\begin{vmatrix} 1 & -2 & 0 & 3 \\ -1 & 2 & 1 & 2 \\ 2 & 0 & 0 & 4 \\ -2 & -1 & 0 & 5 \end{vmatrix} = -1 \begin{vmatrix} 1 & -2 & 3 \\ 2 & 0 & 4 \\ -2 & -1 & 5 \end{vmatrix} = -1[-2(-7) - 4(-5)] = -1(+14 + 20) = -34$

2. (6) **a.** $\begin{bmatrix} 1 & -2 & -1 & 2 \\ 2 & 3 & 2 & 2 \\ 3 & 1 & -1 & -2 \end{bmatrix} \Rightarrow \begin{bmatrix} 1 & -2 & -1 & 2 \\ 0 & 7 & 4 & -2 \\ 0 & 7 & 2 & -8 \end{bmatrix} \Rightarrow \begin{bmatrix} 1 & -2 & -1 & 2 \\ 0 & 7 & 4 & -2 \\ 0 & 0 & -2 & -6 \end{bmatrix}$;

$x - 2y - z = 2$

$y + \frac{4}{7}z = -\frac{2}{7}$

$z = 3$

b. $y + \frac{4}{7}(3) = -\frac{2}{7}$

$y + \frac{12}{7} = -\frac{2}{7}$

$y = -2$

$x - 2(-2) - (3) = 2$

$x + 4 - 3 = 2$

$x = 1$; $\{(1, -2, 3)\}$

3. (1) Reduce each equation to slope-intercept form; $2x - 4y = 2$

$-4y = -2x + 2$

$y = \frac{1}{2}x - \frac{1}{2}$;

$-3x + 6y = 5$

$6y = 3x + 5$

$y = \frac{1}{2}x + \frac{5}{2}$;

the lines have equal slopes and unequal y-intercepts; their graphs are parallel lines

4. (10) $u^2 + v^2 = 20$

$u + v = 6$

$v = 6 - u$

$u^2 + (6 - u)^2 = 20$

$u^2 - 6u + 8 = 0$

$(u - 2)(u - 4) = 0$; $u = 2$ or $u = 4$.

When $u = 2$, $v = 6 - u = 4$; when $u = 4$, $v = 6 - u = 2$; $\{(2, 4), (4, 2)\}$

5. (10) 1] $17xy - 12x^2 = 56 \Rightarrow 17xy - 12x^2 = 56$
2] $3y^2 - 5xy = 28 \Rightarrow 6y^2 - 10xy = 56$

$$6y^2 - 10xy = 17xy - 12x^2$$
$$2y^2 - 9xy + 4x^2 = 0$$
$$(y - 4x)(2y - x) = 0$$
$$y = 4x \text{ or } y = \frac{1}{2}x$$

If $y = 4x$: $3y^2 - 5xy = 28$
$$3(4x)^2 - 5x(4x) = 28$$
$$48x^2 - 20x^2 = 28$$
$$x^2 = 1;\ x = \pm 1$$

If $y = \frac{1}{2}x$: $3y^2 - 5xy = 28$
$$3\left(\frac{x}{2}\right)^2 - 5x\left(\frac{x}{2}\right) = 28$$
$$\frac{3}{4}x^2 - \frac{5}{2}x^2 = 28$$
$$-\frac{7}{4}x^2 = 28$$
$$x = \pm 4i;\ \{(1, 4), (-1, -4), (4i, 2i), (-4i, -2i)\}$$

6. Let x = cost of tie and y = cost of pin;
$$x + y = 1.10$$
$$x - y = 1.00$$
$$2x = 2.10 \Rightarrow x = 1.05$$
The tie cost \$1.05 and the pin cost \$.05, or 5 cents.

CHAPTER 11

(*Note:* There are no exercises for Section 1.)

Exercises 2 (page 354) A Answers

1. $\frac{2}{x - 3} + \frac{1}{x + 4}$ **2.** $\frac{3}{x + 7} + \frac{1}{x - 2}$ **3.** $\frac{3}{x} - \frac{4}{x - 2} + \frac{3}{x + 1}$ **4.** $-\frac{2}{x} + \frac{1}{x - 2} - \frac{3}{x - 6}$
5. $x - 2 + \frac{1}{3x + 2} - \frac{2}{2x - 3}$

Exercises 3 (page 355) A Answers

1. $\frac{2}{x - 1} + \frac{1}{(x - 1)^2} + \frac{2}{(x - 1)^3} - \frac{1}{x}$ **2.** $\frac{1}{x^2} + \frac{1}{x} + \frac{1}{(x - 2)^2} - \frac{1}{x - 2}$
3. $\frac{5}{x - 2} + \frac{2}{(x - 2)^2} - \frac{3}{x + 2} + \frac{1}{(x + 2)^2}$ **4.** $\frac{2}{x} + \frac{5}{x^2} - \frac{1}{x^3} - \frac{2}{x + 1} - \frac{7}{(x + 1)^2} - \frac{11}{(x + 1)^3}$

Exercises 4 (page 357) A Answers

1. $\frac{5}{x + 1} - \frac{2x + 7}{x^2 + 3}$ **2.** $\frac{5}{y - 2} + \frac{4y - 2}{y^2 + 2y + 5}$ **3.** $\frac{2x}{(x^2 + 3x + 3)^2} - \frac{6}{x^2 + 3x + 3} + \frac{1}{x}$
4. $\frac{2}{x^2 + 2x + 10} + \frac{4x + 21}{(x^2 + 2x + 10)^2}$

Review exercises (page 358) A Answers

1. $\frac{4}{y + 5} - \frac{2}{y + 2}$ **2.** $\frac{7}{(z - 2)^2} + \frac{2}{z}$ **3.** $2 - \frac{3}{x} + \frac{3x - 2}{x^2 + 1}$ **4.** $\frac{1}{x} - \frac{3x + 2}{(2x^2 + x + 1)^2}$

Diagnostic test (page 358)

(*Note:* Following each problem number in parentheses is the section number in which that kind of problem is discussed.)

1. (2) $\dfrac{7x+4}{(3x-2)(x-5)} = \dfrac{A}{3x-2} + \dfrac{B}{x-5}$

$$7x+4 = A(x-5) + B(3x-2) = (A+3B)x + (-5A-2B) \qquad A+3B=7, \quad -5A-2B=4;$$

$A=-2$ and $B=3$; $\dfrac{7x+4}{3x^2-17x+10} = \dfrac{3}{x-5} - \dfrac{2}{3x-2}$

2. (3) $\dfrac{x^3+3x^2-3x-2}{x^2(x+1)^2} = \dfrac{A}{x} + \dfrac{B}{x^2} + \dfrac{C}{x+1} + \dfrac{D}{(x+1)^2}$

$$x^3+3x^2-3x-2 = Ax(x+1)^2 + B(x+1)^2 + Cx^2(x+1) + Dx^2 = (A+C)x^3 + (2A+B+C+D)x^2 + (A+2B)x + B$$

$$\begin{aligned} A \qquad + C \qquad &= 1 \\ 2A + B + C + D &= 3 \\ A + 2B \qquad &= -3 \\ B \qquad &= -2 \end{aligned}$$

$A=1$, $B=-2$, $C=0$, $D=3$; $\dfrac{x^3+3x^2-3x-2}{x^2(x+1)^2} = \dfrac{1}{x} - \dfrac{2}{x^2} + \dfrac{3}{(x+1)^2}$

3. (4) $\dfrac{2x^3+4x^2-x+1}{x^3+x} = 2 + \dfrac{4x^2-3x+1}{x^3+x}$

$$\frac{4x^2-3x+1}{x(x^2+1)} = \frac{A}{x} + \frac{Bx+C}{x^2+1}$$

$$4x^2-3x+1 = A(x^2+1) + (Bx+C)x = (A+B)x^2 + Cx + A$$

$$\begin{aligned} A + B &= 4 \\ C &= -3 \\ A &= 1 \end{aligned}$$

$A=1$, $B=3$, $C=-3$; $\dfrac{2x^3+4x^2-x+1}{x^3+x} = 2 + \dfrac{1}{x} + \dfrac{3x-3}{x^2+1}$

4. (4) $\dfrac{2x^4+13x^2-x+18}{x(x^2+3)^2} = \dfrac{A}{x} + \dfrac{Bx+C}{x^2+3} + \dfrac{Dx+E}{(x^2+3)^2}$

$$2x^4+13x^2-x+18 = A(x^2+3)^2 + (Bx+C)(x)(x^2+3) + (Dx+E)x = (A+B)x^4 + Cx^3 + (6A+3B+D)x^2 + (3C+E)x + (9A)$$

$$\begin{aligned} A + B &= 2 \\ C &= 0 \\ 6A + 3B + D &= 13 \\ 3C + E &= -1 \\ 9A &= 18 \end{aligned}$$

$A=2$, $B=0$, $C=0$, $D=1$, $E=-1$; $\dfrac{2x^4+13x^2-x+18}{x(x^2+3)^2} = \dfrac{2}{x} + \dfrac{x-1}{(x^2+3)^2}$

5. (4) $\dfrac{7x^3-4x^2+2x-1}{x^2(2x^2+1)} = \dfrac{A}{x} + \dfrac{B}{x^2} + \dfrac{Cx+D}{2x^2+1}$;

$$7x^3-4x^2+2x-1 = Ax(2x^2+1) + B(2x^2+1) + (Cx+D)x^2 = (2A+C)x^3 + (2B+D)x^2 + Ax + B;$$

$$\begin{aligned} 2A + C &= 7 \\ 2B + D &= -4 \\ A &= 2 \\ B &= -1 \end{aligned}$$

$A=2$, $B=-1$, $C=3$, $D=-2$; $\dfrac{7x^3-4x^2+2x-1}{x^2(2x^2+1)} = \dfrac{2}{x} - \dfrac{1}{x^2} + \dfrac{3x-2}{2x^2+1}$

CHAPTER 12

Exercises 1 (page 363) A Answers

1. 5,040 **2.** 336 **3.** 126 **4.** $r+1$ **5.** 20 **6.** $7! = 5{,}040$ **7.** $(n+1)!$
8. $(n+1)(n+2)$ **9.** $6(n+1)(n+2)$ **10.** $2(n+1)$

Exercises 2 (page 370) A Answers

1. $a^7 + 7a^6b + 21a^5b^2 + 35a^4b^3 + 35a^3b^4 + 21a^2b^5 + 7ab^6 + b^7$
2. $x^6 - 6x^5y + 15x^4y^2 - 20x^3y^3 + 15x^2y^4 - 6xy^5 + y^6$
3. $32x^5 + 40x^4y + 20x^3y^2 + 5x^2y^3 + \dfrac{5}{8}xy^4 + \dfrac{y^5}{32}$
4. $\dfrac{x^4}{81} - \dfrac{4x^3}{9y} + \dfrac{6x^2}{y} - \dfrac{36x}{y^3} + \dfrac{81}{y^4}$
5. $a^3 + 6a^2\sqrt[4]{a^2b} + 15a^2\sqrt{b} + 2a\sqrt[4]{a^2b^3} + 15ab + 6b\sqrt[4]{a^2b} + b\sqrt{b}$
6. $a^{-12} - 3\sqrt{2}a^{-10}b^2 + \dfrac{15}{2}a^{-8}b^4 - 5\sqrt{2}a^{-6}b^6 + \dfrac{15}{4}a^{-4}b^8 - \dfrac{3\sqrt{2}}{4}a^{-2}b^{10} + \dfrac{b^{12}}{8}$
7. $1{,}024x^{10} + 5{,}120x^9y + 11{,}520x^8y^2$
8. $\dfrac{a^{28}}{16{,}384} + \dfrac{7a^{26}b^3}{2{,}048} + \dfrac{91a^{24}b^6}{1{,}024}$
9. $a^4 - 8a(ab)^{1/2} + 28a^3b$
10. $\dfrac{1}{x^{12}} + \dfrac{12}{x^{11}y} + \dfrac{66}{x^{10}y}$

Exercises 3 (page 372) A Answers

1. $-6{,}435a^8x^7$
2. $210x^4y^6$
3. $38{,}808\dfrac{y^7}{z^{10}}$
4. $-15\dfrac{x^7}{y^3}$
5. $-\dfrac{2{,}176}{9}x^{-30}y^{-9}$

(*Note:* There are no simple answers for Exercises 4.)

Review exercises (page 378) A Answers

1. 720
2. 110
3. 21
4. 1
5. $8! = 40{,}320$
6. 40
7. $n(n-2)$
8. n
9. $x^4 - 12x^3y + 54x^2y^2 - 108xy^3 + 81y^4$
10. $\dfrac{x^5}{32} + \dfrac{5x^4}{8y} + \dfrac{5x^3}{y^2} + \dfrac{20x^2}{y^3} + \dfrac{40x}{y^4} + \dfrac{32}{y^5}$
11. $a^2 + 6a\sqrt[3]{a^2b} + 15a\sqrt[3]{ab^2} + 20ab + 15b\sqrt[3]{a^2b} + 6b\sqrt[3]{ab^2} + b^2$
12. $128x^{21} - \dfrac{7x^{18}y^2}{32} + \dfrac{21x^{15}y^4}{8} - \dfrac{35x^{12}y^6}{2} + 70x^9y^8 - 168x^6y^{10} + 224x^3y^{12} - 128y^2$
13. $x^{20} - 20x^{19}y + 190x^{18}y^2$
14. $4{,}620e^{-3}$
15. $\dfrac{1}{2} + \dfrac{1}{2^2} + \dfrac{1}{2^3} + \cdots + \dfrac{1}{2^n} = \dfrac{2^n - 1}{2^n}$

Step 1: When $n = 1$, $\dfrac{1}{2^1} = \dfrac{2^1 - 1}{2^1}$; $\dfrac{1}{2} = \dfrac{1}{2}$

Step 2: Assume $\dfrac{1}{2} + \dfrac{1}{2^2} + \dfrac{1}{2^3} + \cdots + \dfrac{1}{2^k} = \dfrac{2^k - 1}{2^k}$. Add $\dfrac{1}{2^{k+1}}$ to both sides:

$\dfrac{1}{2} + \dfrac{1}{2^2} + \dfrac{1}{2^3} + \cdots + \dfrac{1}{2^k} + \dfrac{1}{2^{k+1}} = \dfrac{2^k - 1}{2^k} + \dfrac{1}{2^{k+1}} = \boxed{\dfrac{2}{2}} \cdot \dfrac{2^k - 1}{2^k} + \dfrac{1}{2^{k+1}} = \dfrac{2^{k+1} - 2}{2^{k+1}} + \dfrac{1}{2^{k+1}}$

$= \dfrac{2^{k+1} - 2 + 1}{2^{k+1}}$. Thus $\dfrac{1}{2} + \dfrac{1}{2^2} + \dfrac{1}{2^3} + \cdots + \dfrac{1}{2^{k+1}} = \dfrac{2^{k+1} - 1}{2^{k+1}}$. Steps 1 and 2 have been proved; therefore the theorem is true for all positive values of n.

Diagnostic test (page 379)

(*Note:* Following each problem number in parentheses is the section number in which that kind of problem is discussed.)

1. **a.** (1) $5! = 1 \cdot 2 \cdot 3 \cdot 4 \cdot 5 = 120$ **b.** (1) $\dfrac{7!}{4!} = \dfrac{4!5 \cdot 6 \cdot 7}{4!} = 210$ **c.** (2) $\dbinom{6}{2} = \dfrac{6 \cdot 5}{1 \cdot 2} = 15$

   d. (2) $\dbinom{8}{0} = 1$ By definition

2. (1) **a.** $9!10 = 10! = 3{,}628{,}800$ **b.** $\dfrac{9!}{8!} = \dfrac{8!9}{8!} = 9$

   c. $\dfrac{n(n+1)[(n-1)!]}{n!} = \dfrac{n!}{n!} = 1$

3. (2) $a^{50} - 50a^{49}(2b) + 1{,}225a^{48}(2b)^2 = a^{50} - 100a^{49}b + 4{,}900a^{48}b^2$

4. (3) $\binom{12}{8}(\sqrt{a})^4(b)^8 = 495(\sqrt{a})^4b^8 = 495a^2b^8$

5. (4) *Step 1*: When $n = 1$, $[5(1) - 3] = \frac{1}{2}(1)[5 - 1]$; $2 = 2$

Step 2: Assume $2 + 5 + 12 + \cdots + (5k - 3) = \frac{1}{2}k(5k - 1)$. Add $[5(k + 1) - 3] = 5k + 2$ to both sides of the equation: $2 + 5 + 12 + \cdots + (5k - 3) + (5k + 2) = \frac{1}{2}k(5k - 1) + 5k + 2$

$= \frac{1}{2}[k(5k - 1) + 2(5k + 2)] = \frac{1}{2}(5k^2 + 9k + 4)$. Thus $2 + 5 + 12 + \cdots + (5k + 2) = \frac{1}{2}(k + 1)(5k + 4)$

$= \frac{1}{2}(k + 1)[5(k + 1) - 1]$.

This equation is precisely the original equation with $n = k + 1$. Therefore the theorem holds for all positive integral values of n.

6. (2) $(a - 2b)^5 = a^5 - 5a^4(2b) + 10a^3(2b)^2 - 10a^2(2b)^3 + 5a(2b)^4 - (2b)^5$
$= a^5 - 10a^4b + 40a^3b^2 - 80a^2b^3 + 80ab^4 - 32b^5$

7. (2) $(\sqrt{x} + 2)^6 = (\sqrt{x})^6 + 6(\sqrt{x})^5(2) + 15(\sqrt{x})^4(2)^2 + 20(\sqrt{x})^3(2)^3 + 15(\sqrt{x})^2(2)^4 + 6(\sqrt{x})2^5 + 2^6$
$= x^3 + 12x^2\sqrt{x} + 60x^2 + 160x\sqrt{x} + 240x + 192\sqrt{x} + 64$

8. (2) $\left(\frac{x^2}{2} - 2y^3\right)^7 = \left(\frac{x^2}{2}\right)^7 - 7\left(\frac{x^2}{2}\right)^6(2y^3) + 21\left(\frac{x^2}{2}\right)^5(2y^3)^2 - 35\left(\frac{x^2}{2}\right)^4(2y^3)^3$
$= \frac{x^{14}}{128} - \frac{7x^{12}y^3}{32} + \frac{21x^{10}y^6}{8} - \frac{35x^8y^9}{2}$

CHAPTER 13

(*Note:* There are no exercises for Section 1.)

Exercises 2 (page 386) A Answers

1. 2 **2.** 16 **3.** Remainder: 21 Quotient: $x^2 + 2x - 7$;
4. Quotient: $x^7 - x^6 + x^5 - x^4 + x^3 - 6x^2 + 9x - 4$; Remainder -2
5. Quotient: $2x^2 + 8x + 6$; Remainder: 11

Exercises 3 (page 388) A Answers

1. Yes **2.** Yes **3.** No **4.** No

(*Note:* There are no exercises for Section 4.)

Exercises 5 (page 391) A Answers

1. $\left\{-\frac{2}{3}, \frac{3}{5}, \frac{7}{2}\right\}$ **2.** $x^3 - 6x^2 + 5x + 12 = 0$ **3.** $x^4 - 1 = 0$
4. $4x^4 + 4x^3 - 59x^2 - 30x + 225 = 0$ **5.** $f(x) = (x + 3)(x - 1)(x - 7)$
6.

	1	−9	24	−20
		2	−14	20
2	1	−7	10	0
		2	−10	
2	1	−5	0	

$x = 5$ is the other root.

7. $x^4 + 2x^3 - 6x^2 + 2x + 1 = 0$

Exercises 6 (page 401) A Answers

1. 2 **2.** 4 **3.** $\{1, -1, i\sqrt{5}, -i\sqrt{5}\}$ **4.** $\{-2, -1, 0, 3\}$ **5.** $\left\{-4, -1, -\frac{1}{3}, \frac{1}{3}, 1\right\}$

6. $\{-4, -1, -3-i, -3+i\}$ **7.** $\left\{-1, 1, -\frac{5}{3}, \frac{5}{3}\right\}$ **8.** 2.82

Exercises 7 (page 404) A Answers

1.

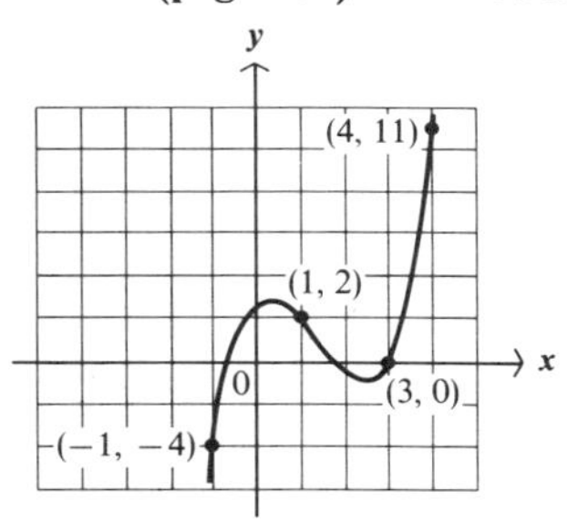

2.

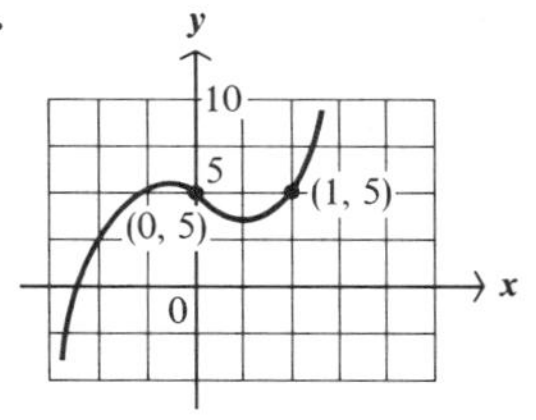

3.

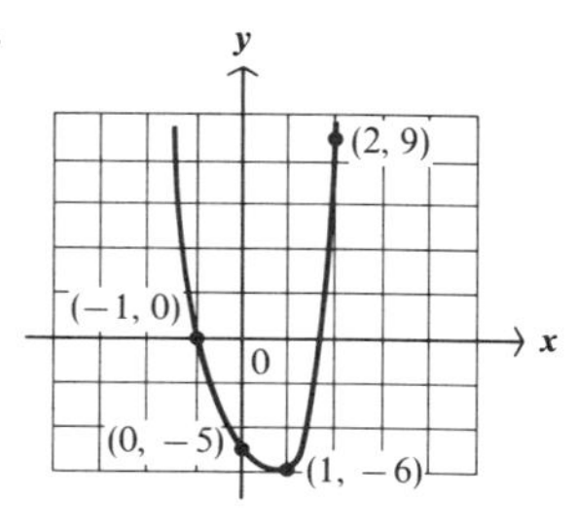

4.

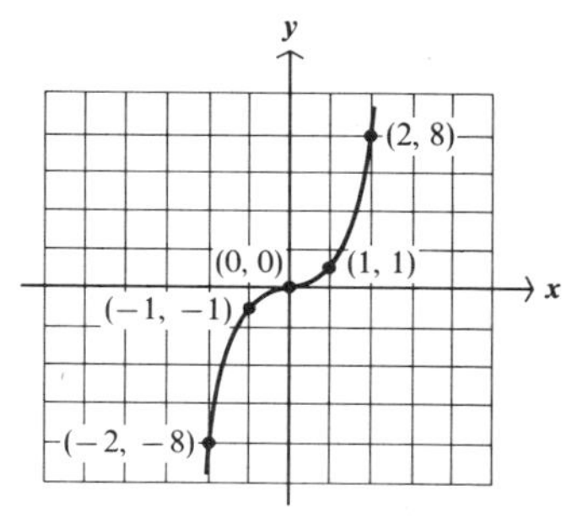

5.

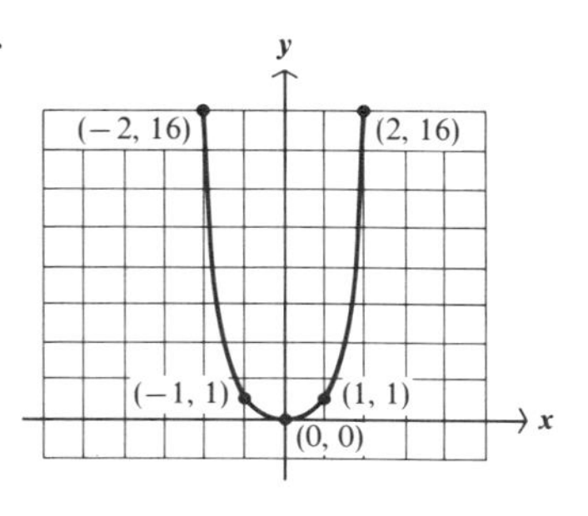

Review exercises (page 406) A Answers

1. 6 **2.** Quotient: $x^6 + x^5 - 3x^4 - 2x + 7$; Remainder: -15

3. Quotient: $x^5 + x^4 + x^3 + x^2 + x + 1$; Remainder: 2

4. Yes

Proof:

	1	6	0	2	14	12
		−6	0	0	−12	−12
−6	1	0	0	2	2	0

Remainder is 0; therefore $x + 6$ is a factor.

5. $x^5 - 3x^4 + 2x^3 + x^2 - 3x + 2 = 0$

6. $\left\{-1, \frac{1}{2}, 2, 3\right\}$

7. a. Two b. Two

8.

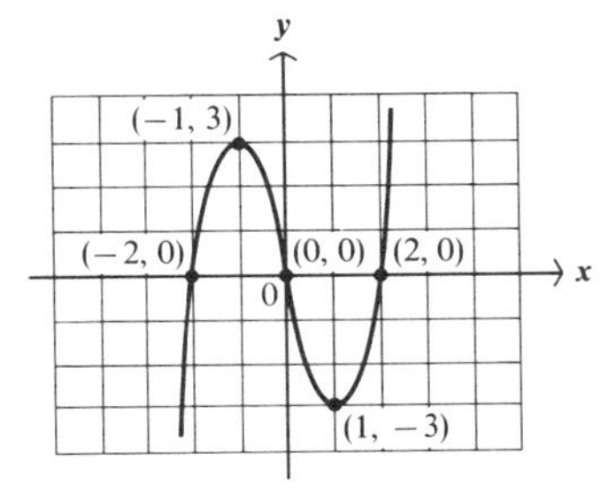

9. 3.42

Diagnostic test (page 407)

(*Note:* Following each problem number in parentheses is the section number in which that kind of problem is discussed.)

1. (2)

	1	0	−16	0	1	1	−10	−20
		4	16	0	0	4	20	40
4	1	4	0	0	1	5	10	20

$f(4) = 20$

2. (2)

	2	4	0	1	2	3	10
		−4	0	0	−2	0	−6
−2	2	0	0	1	0	3	4

Quotient: $2x^5 + x^2 + 3$; Remainder: 4

3. (3)

	1	0	−14	−9	4	6
		4	16	8	−4	0
4	1	4	2	−1	0	6

No, since $R \neq 0$.

4. (5) $a_0(x-1)(x+2)\left[x - \frac{1+5i}{2}\right]\left[x - \frac{1-5i}{2}\right] = 0$. Let $a_0 = 4$;

$$4(x^2 + x - 2)\left[\frac{2x-(1+5i)}{2}\right]\left[\frac{2x-(1-5i)}{2}\right] = 0$$

$$\not{4}(x^2 + x - 2)\left[\frac{(2x-1)^2 - (5i)^2}{\not{4}}\right] = 0$$

$$(x^2 + x - 2)(4x^2 - 4x + 6) = 0$$

$$2x^4 - 3x^2 + 7x - 6 = 0$$

5. (6)

	1	2	0	−8	−16
		2	8	16	16
2	1	4	8	8	0
		−2	−4	−8	
−2	1	2	4	0	

$f(x) = (x-2)(x+2)(x^2 + 2x + 4) = 0$;

$x^2 + 2x + 4 = 0$ has roots $x = \frac{-2 \pm \sqrt{4-16}}{2} = -1 \pm i\sqrt{3}$

$\{2, -2, -1 + i\sqrt{3}, -1 - i\sqrt{3}\}$

6. (6) a. Two, because the graph of the function crosses the x-axis in two places.

b. Two. A fourth-degree equation has four roots. If two roots are real, then there are two imaginary roots.

7. (7)

x	y
-3	-13
-2	3
-1	7
0	5
1	3
2	7

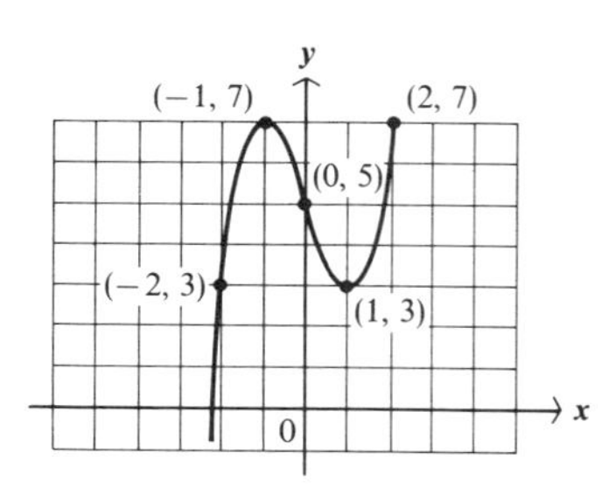

8. (7) Try $x = 1.5$:

$$\begin{array}{r|rrrr} & 3 & -3 & 1 & -7 \\ & & 4.5 & 2.25 & 3.375 \\ \hline 1.5 & 3 & 1.5 & 3.25 & -4.375 \end{array}$$

$f(1.5) = -4.375 < 0$

Try $x = 1.7$:

$$\begin{array}{r|rrrr} & 3 & -3 & 1 & -7 \\ & & 5.1 & 3.57 & 7.769 \\ \hline 1.7 & 3 & 2.1 & 4.57 & 0.769 \end{array}$$

$f(1.7) = 0.769 > 0$

Try $x = 1.65$:

$$\begin{array}{r|rrrr} & 3 & -3 & 1 & -7 \\ & & 4.95 & 3.2175 & 6.958875 \\ \hline 1.65 & 3 & 1.95 & 4.2175 & 0.041125 \end{array}$$

$f(1.65) = 0.041125 \doteq 0$. Further tests show that when expressed to two decimal places, $x = 1.65$.

CHAPTER 14

Exercises 1 (page 412) A Answers

1. 3, 5, 7, 9, 11; Sum = 35 **2.** 4, 6, 8, 10, 12; Sum = 40 **3.** $1^2, 3^2, 5^2, 7^2$; Sum = 84 **4.** $1, \frac{3}{2}, \frac{5}{3}$; Sum = $4\frac{1}{6}$ **5.** 0, 1, 4, 9; Sum = 14 **6.** $S_3 = \frac{5}{6}$ **7.** $S_4 = 4.1111$ **8.** $S_5 = 90$ **9.** $S_{11} = 11$ **10.** $S_4 = \frac{15}{16}$

Exercises 2 (page 414) A Answers

1. 40 **2.** $\frac{15}{2}$ **3.** $\frac{38}{3}$ **4.** $\frac{122}{105}$ **5.** 100π

Exercises 3 (page 417) A Answers

1. An A.P. **a.** 4 **b.** $a_n = 3 + (n-1)4$ **c.** 83 **d.** 253
2. An A.P. **a.** -2 **b.** $a_n = 6 + (n-1)(-2)$ **c.** -34 **d.** -44
3. An A.P. **a.** -13 **b.** $a_n = 67 + (n-1)(-13)$ **c.** -193 **d.** 22
4. An A.P. **a.** $-a - 6$ **b.** $a_n = 4a + 12 + (n-1)(-a-6)$ **c.** $-16a - 108$ **d.** $-11a - 198$ **5.** $a_{11} = 31a - 16e$; $a_{17} = 49a - 40e$ **6.** 2.385 **7.** $-16 + \sqrt{3}$
8. Either 2 terms or 19 terms. **9.** $n = 11$; $S_{11} = 22$ **10.** $a_1 = 53$; $S_{18} = 189$
11. $n = 6$; $a_6 = 15$ **12.** 231

Exercises 4 (page 423) **A Answers**

1. Not a G.P.
2. A G.P. with $r = 2$
3. A G.P. with $r = -2$
4. A G.P. with $r = 1 - x$
5. 270, 405, 607.5
6. $(1+x)^3, (1+x)^4, (1+x)^5$
7. -128
8. $a_{10} = \frac{1}{96}, S_{10} = \frac{341}{32}$
9. $n = 5, a_5 = 162$
10. $n = 5, S_n = 121$
11. $\frac{7^3}{16}, \frac{7^2}{8}, \frac{7}{4}, \frac{1}{2}$
12. 51,840
13. **a.** \$2.56 **b.** \$2,684,354.56 **c.** \$5,368,709.12
14. \$11,875.88

Exercises 5 (page 427) **A Answers**

1. Converges; 48
2. Converges; 256
3. Diverges
4. Converges; 25
5. Converges; 12.5
6. 6
7. 10
8. $\frac{3}{7}$
9. $\frac{1}{4}$
10. 54 in.
11. 110 ft.
12. $\frac{6}{11}$

Review exercises (page 429) **A Answers**

1. 55
2. 33
3. 40
4. $\frac{10}{7}$
5. $\frac{1}{3}$
6. **a.** 81 **b.** 961
7. **a.** $\frac{2}{2{,}187} \doteq 0.0009145$ **b.** About 26.9995
8. $\frac{149}{33}$
9. $r = 2$; 200% increase each day
10. \$8,755.64

Diagnostic test (page 430)

(*Note:* Following each problem number in parentheses is the section number in which that kind of problem is discussed.)

1. (2) $\sum_{n=0}^{3} (n+1)(n+2) = 2 + 6 + 12 + 20 = 40$
2. (2) $\sum_{n=0}^{3} (2)^n = 1 + 2 + 4 + 8 = 15$
3. (2) $\sum_{n=1}^{\infty} \left(\frac{1}{3}\right)^n = \frac{1/3}{1 - 1/3} = \frac{1}{2}$
4. (5) $2.121212\cdots = 2 + 0.12 + 0.12(0.01) + 0.12(0.01)^2 + \cdots$; $0.12 + 0.12(0.01) + 0.12(0.01)^2 + \cdots$ is a G.P. with $a_1 = 0.12$ and $r = 0.01$; $S_\infty = \frac{0.12}{1 - 0.01} = \frac{4}{33}$. Therefore $2.1212\cdots = 2 + \frac{4}{33} = \frac{70}{33}$.
5. (3) $a_{61} = 2 + (61 - 1)\frac{1}{2} = 32$
6. (4) 27, 9, 3, 1, . . . ; series is a G.P. with $a_1 = 27$ and $r = \frac{1}{3}$ **a.** $a_7 = 27\left(\frac{1}{3}\right)^6 = 3^3\left(\frac{1}{3^6}\right) = \frac{1}{3^3} = \frac{1}{27}$

   b. $S_6 = \dfrac{27\left[1 - \left(\frac{1}{3}\right)^6\right]}{1 - 1/3} = \dfrac{27\left(1 - \frac{1}{3^6}\right)}{\frac{2}{3}} = 40\frac{4}{9}$

7. (3) Let $a_1 = 3$ and $a_4 = 9$; $a_4 = a_1 + (4 - 1)d$

$$9 = 3 + (3)d$$
$$2 = d$$

In the original series, $a_4 = a_1 + (n - 1)d$, so $9 = a_1 + 3(2)$; $a_1 = 3$.
Therefore the original series is $-3, -1, 1, 3, \ldots$.

8. (4) $a_1 = \$12{,}000$ and $r = 1 + 20\% = 1.2$; $S_4 = 12{,}000\,\dfrac{[1 - (1.2)^4]}{1 - 1.2} = \$64{,}416$

9. (5) Series diverges, because $|r| = 1$.

10. (4) $a_1 = 1{,}296{,}000$ and $r = \dfrac{5}{6}$; $a_5 = 1{,}296{,}000\left(\dfrac{5}{6}\right)^4 = 625{,}000$

Index

Formulas

Pythagorean theorem (7)

$$a^2 + b^2 = c^2$$

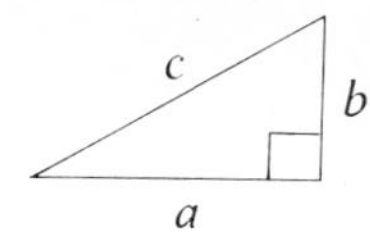

The roots of the quadratic equation (128)

$$ax^2 + bx + c = 0 \qquad \text{where } a, b, c \in R;\ a \neq 0$$

are

$$x = \frac{-b \pm \sqrt{b^2 - 4ac}}{2a}.$$

Binomial theorem (369)

$$(x + y)^n = \sum_{r=0}^{n} \binom{n}{r} x^{n-r} y^r$$

Polar coordinates (228)

$$x = r \cos \theta \qquad r = \sqrt{x^2 + y^2}$$

$$y = r \sin \theta \qquad \theta = \arctan \frac{y}{x}$$

Operations with complex numbers (245)

$$(r \text{ cis } \theta)(R \text{ cis } \phi) = rR \text{ cis } (\theta + \phi)$$

$$(r \text{ cis } \theta) \div (R \text{ cis } \phi) = \frac{r}{R} \text{ cis } (\theta - \phi)$$

$$(r \text{ cis } \theta)^n = r^n \text{ cis } n\theta$$

Roots of a complex number (241)

If n is a positive integer and $z \neq 0$, a complex number $z = r(\cos \theta + i \sin \theta)$, real or imaginary, has n and only n distinct nth roots, given by

$$w_k = \sqrt[n]{r}\left[\cos\left(\frac{\theta}{n} + \frac{k \cdot 360^\circ}{n}\right) + i \sin\left(\frac{\theta}{n} + \frac{k \cdot 360^\circ}{n}\right)\right],$$

where k takes the values $0, 1, 2, \ldots, n - 1$.